CRTB

USEFUL CONSTANTS

Electron charge	1.602×10^{-19} coulomb
Electron mass	9.108×10^{-31} kilogram
Proton mass	1.672×10^{-27} kilogram
One electronvolt	1.602×10^{-19} joule
Base of natural logs, ϵ	2.71828
ϵ^{-1} 0.3678	$\log_{10} \epsilon$ 0.4343
$\ln \pi$ 1.4473	$\log_{10} \pi$ 0.4971
$\ln x$ $2.3 \log_{10} x$	$\log_{10} x$ $0.4343 \ln x$

Introduction to
ELECTRIC CIRCUITS

Introduction to

John Wiley & Sons, Inc.
New York, London, Sydney, Toronto

Toppan Company, Ltd.
Tokyo, Japan

Wiley International Edition

ELECTRIC
CIRCUITS

H. Alex Romanowitz

Professor of Electrical Engineering

University of Kentucky

Wiley International Edition

Authorized reprint of the edition published by
John Wiley & Sons, Inc., New York and London.

Copyright © 1971, by John Wiley & Sons, Inc.

Wiley International Edition

Library of Congress Catalogue Card Number: 70–121913
ISBN 0 471 73260 5

Printed in Singapore by Toppan Printing Co. (S) Pte. Ltd.

Dedicated to my wife Mildred

PREFACE

The book should serve well for either a one-year course, three class hours per week in two semesters (or three quarters) in which the entire book is studied, or a semester course (or a two-quarter course) in which parts of Chapters 6, 7, 8, 12, 18, 20, 21, and most of 24 may be either omitted or touched lightly. If it is used for a one-quarter course. more omissions will be in order, such as Sections 10 to 13 in Chapter 4 and perhaps all of Chapters 20, 21, and 24.

Basic principles, theorems, circuit behavior, and problem-solving procedures are presented so that the average student can get a clear understanding by an adequate amount of private study. There are 200 worked-out numerical examples, 320 pictorial illustrations, 600 review questions, and 660 problems for class assignments. Items of historical interest concerning pioneers in science and electrical technology are introduced where appropriate. Answers to about half of the problems are in the back of the book.

Students using this book should have a working knowledge of simple algebraic equations by the time they start Chapter 3. Trigonometry does not come into use until Chapter 12. A brief look at sines, cosines, and tangents is presented. Calculus is avoided except in a few simple applications, where the algebraic forms needed for an understanding of principles and problem solving are furnished immediately. Solutions of the few equations in calculus, such as those needed to explain what happens when an RL or RC series circuit is suddenly connected to a constant-voltage dc source, are in the appendix.

vii

An instructor's manual containing solutions to all of the problems and answers to all of the review questions is available. Suggestions for making up quizzes and final examinations are given in the manual.

I am pleased to dedicate this volume to my loyal and competent wife, who has patiently borne the many and long periods of silence and, at times, inconvenience that were attendant to the writing of not only this manuscript but several others. She has typed every one of them in excellent fashion and has rendered much valuable assistance in their preparation.

H. Alex Romanowitz

Lexington, Kentucky
January 1971

CONTENTS

CHAPTER FOURTEEN
Phasors in Complex-Number and Polar Forms   339

CHAPTER FIFTEEN
Resistance, Reactance and Impedance   351

CHAPTER SIXTEEN
Parallel and Series-Parallel AC Circuits   365

CHAPTER SEVENTEEN
Power and Energy in AC Circuits   381

CHAPTER ONE

Introductory Concepts

In this chapter we shall determine just what electricity is, how it comes into being, and what is necessary to set it in motion and to cause it to move from one place to another. In order to "harness" it and utilize its capacity to producing light, heat, and power, the technicians, scientists, and engineers must understand terms like volts, amperes, ohms, and units of measurement of length, force, power, and energy. This chapter introduces many facts and concepts that we must know if we want to become capable technicians and aides to practicing engineers.

There are many review questions and problems at the ends of the chapters. Some of the problems for this chapter have nothing to do with electricity or electronics, but they will extend your general scientific knowledge and challenge you to think mathematically and to apply some practical reasoning.

1.1   What is Electricity?

Probably everyone has at one time or another, become curious about electricity and has asked himself, "Just what is this stuff, anyway?" Early philosophers called it a fluid. They could not detect it by sight, smell, weight, or by any method other than muscular reaction. One investigator, Professor M.

Musschenbroeck, reported in 1746, at the University of Leyden in Holland, that when he used a wire to draw sparks from a metal object that had been connected previously to his powerful "electric machine," his hand was struck so violently that all of his body "was affected as if it had been struck by lightning." He added, "In a word I thought it was all up with me." (We might even get a chuckle from the thought of how he must have looked and felt when he suddenly found himself sitting on the floor!) It is probably correct to say that one does not *feel electricity* when subjected to electric shock, but rather that he feels pain from muscular reaction.

If a dry glass rod is rubbed with a silk cloth it will acquire a **charge of electricity**. If the rod is suspended as in Fig. 1.1 and a second glass rod that has been rubbed with a silk cloth is brought near it, the first rod is found to be repelled. If, now, a hard rubber rod is first rubbed with a flannel cloth and then brought near the end of the suspended glass rod, the two will be found to attract each other.

Evidently the electrifications imparted to glass by rubbing it with silk and to hard rubber by rubbing it with flannel are of opposite kinds, since an electrified body (the first glass rod) is repelled by one and attracted by the other. Benjamin Franklin arbitrarily introduced the terms *positive* and *negative*, or + and −, to designate these two kinds of electrification. We then conclude that a *positively electrified body is one which acts with respect to other electrified bodies like a glass rod that has been rubbed with silk, and a negatively electrified body is one that acts like a piece of hard rubber that has been rubbed with flannel.* Rub a hard rubber comb on a woolen cloth or garment and note that you can pick up dry pits of paper with it.

Bodies charged with opposite kinds of electricity attract each other; bodies charged with the same kind of electricity repel each other.

How do we express the amount or quantity of electrification on an electrified body? Late in the last century, Lord Kelvin,[1] a famous English scientist, said, "I often say that when you can measure what you are speaking about and express it in numbers you know something about it. But when you cannot measure it, when you cannot express it in numbers, your knowledge is of a meagre and unsatisfactory kind." This is true in the study of technology of all kinds — electrical, mechanical, chemical, civil, metallurgical, and the others.

The question "what is electricity" should be answered with some *quantitative* information. Electrification involves the transfer of *negative charges* (ultimate, minute *quantities* of electricity) from one body to another. The small-

---

[1]Lord Kelvin's statement would be challenged today by political scientists, sociologists, and others who are not natural scientists or engineers. Much of what they must deal with cannot be expressed in numbers.

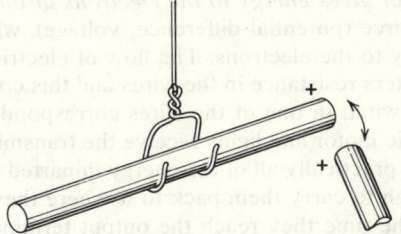

FIGURE 1.1   Two rods with positive charges. Like charges repel.

est bit of negative electrical quantity is the **electron.** It moves readily in metals; in fact, electrons in a metal are continually in motion, as will be learned later, and their movement in large numbers constitutes electric current.

The fundamental unit of quantity of electric charge is the coulomb.

The coulomb will be defined quantitatively in terms of the quantity of electricity required to produce a certain amount of action in a chemical solution. Without electricity, the world as we know it would not exist. Human beings and animals could not live, act, or think without the aid of electricity, since it has been established that our movements, speech, and senses of smell, touch, sight, and hearing involve electric impulses that are sent to the brain. Electric waves in the brain are detectable and analyzed with the help of a complicated electronic recording device called an *electroencehpalograph.*

The foregoing suggests that electricity serves as a carrier, or conveyor, by means of which energy may be transported from one point to another. Professor Musschenbroeck's machine charged the metal object with electrical energy (we shall find later that the energy was really in the space between the object and the earth), and the "discharge" of the object was the act of transporting this energy to his body which was in contact with the earth.

Since the earliest days of elementary textbook writing in the electrical field, some authors have used a circulating water system to compare the phenomenon of energy transfer *from source to load* (the water acting as the transfer medium) with energy transfer from a source (electric generator) to the load (motor or lights) by electricity. Water from a storage tank flows through a transmission pipe, through an open valve, to a water motor or a water wheel, which drives a machine such as a flour mill. The water flowing out of the water motor has spent its energy and falls into a basin. A pump drives the water from the basin back up into the storage tank, from which it can flow "around the circuit" again, delivering to the water motor the energy given to it by the pump. The water is not used up; only its energy is taken from it.

In the electric circuit, in which two transmission wires are required, *the electric generator gives energy to the electrons at the source end*. It develops electromotive force (potential difference, voltage), which is a form of energy, by giving energy to the electrons. The flow of electric charges, which we call current, encounters resistance in the wires and this corresponds to friction in a water pipe. A switch in one of the wires corresponds to a valve in the water pipe. The electric motor and lights receive the transmitted energy. The electric charges give up practically all of the energy imparted to them by the generator but retain enough to carry them back to it where they receive another supply of energy. By the time they reach the output terminal of the generator, they have been given their full amount of energy and are ready to make the circuit again. The potential difference between the two terminals of the electric motor corresponds to the difference in water pressure between the input port and output port of the water motor.

1.2   Electrons and Protons

All material substance is composed of tiny, invisible particles called atoms. The different kinds of material substance such as iron, copper, and gold, which are a few of the familiar chemical *elements* that are solid in their normal state. Mercury, which is liquid, and oxygen, nitrogen, and hydrogen which are gases, have atoms each of which contains a *nucleus* at its center. Atoms also contain one or more *electrons* that revolve around the nucleus like our earth and the other planets revolve around our sun.

Figure 1.2 shows Niels Bohr's (1885) concept of the make-up of hydrogen and helium atoms. The total amount of negative charge of the orbiting electrons must equal the total amount of positive charge in the nucleus.

The positive charge in the nucleus is a result of the presence of one or more *protons*, and they are present only in the nucleus of an atom. Research indicated that the weight of the nucleus of the helium atom was greater than the weight of two protons that must be in there, so after much experimentation and

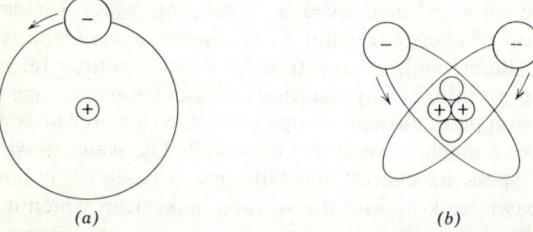

(a)   *(b)*

FIGURE 1.2   Atoms represented as Bohr models. (a) Hydrogen atom. One proton + in the nucleus, one electron — rotating around it. (b) Helium atom. Two protons and two neutrons in the nucleus, two electrons rotating around it.

calculation it was concluded that *two uncharged particles* — neutrons — each having the weight of a proton, must also be present in the helium nucleus.

In order that a particle may continue its rotation around a central mass and not fly away in a tangent direction, there must be a force of attraction between the particle and the central mass. It was concluded that the electron and proton must have charges of opposite kinds of electricity, since *opposite charges of electricity attract each other.*

The amount of negative electricity that we call *the electronic charge* is *1.602 × 10⁻¹⁹ coulomb* (C).[2] This number is 1.602 divided by 10,000,000,000, 000,000,000. That extremely small number is usually written as 1.602×10^{-19}. This method of writing numbers will be explained later. *We represent the charge of the electron by the symbol q_e and we say that*

$$q_e = -1.602 \times 10^{-19} \text{ C}$$

The amount of positive electricity of the proton is the same as the amount of negative electricity of the electron. Hence,

$$q_p = +1.602 \times 10^{-19} \text{ C}$$

As indicated before, the *neutron* is a particle without any net positive or negative charge.

The symbol for quantity of electricity, sometimes called "charge," is Q or q. It is expressed in coulombs.

An *electron* weighs very close to 9.109×10^{-28} gram (g) — a nickel weighs about 5 g — and a *proton* weighs very close to 1.672×10^{-24} g. The proton is about 1837 times heavier than the electron. The distance between the electron and the proton in a hydrogen atom is extremely large compared with the sizes of the particles. The radius of the electron has been calculated to be 2.82×10^{-13} cm, and the radius of the proton 1.5×10^{-16} cm. The average radius of a circle in which the lone electron rotates around the nucleus of the *hydrogen atom* is 5×10^{-9} cm and this is 17,700 times the radius of the electron, or more than 30 million times the radius of the proton around which it revolves. Sir James Jeans, a noted English astronomer, stated in one of his books that if one imagines that a baseball in a shop window in Dallas, Texas, represents the nucleus (proton) of a hydrogen atom, the electron might then be represented (comparing sizes and distance) by a huge balloon that would just fit into Yankee Stadium in New York City.

---

[2]See Section 1.8 for a discussion of this kind of numerical notation.

1.3 Atoms and Molecules

Figure 1.3 shows Bohr's concept of the structure of the copper atom. One should not think, as the sketch may indicate, that the electron paths are confined to one plane, such as the plane of the paper. The region occupied by the atom is three dimensional. The electrons of an atom rotate in regions called shells that are spherical and concentric with the center of the nucleus. Like the green shell (or hull) of a walnut, the shells occupied by the electrons have thickness in the direction of the radii drawn from the center.

Each shell is given a letter designation. The first, *K*, has 2 electrons, the second, *L*, has 8, the third, *M*, has 18, the fourth, *N*, has 32, etc. Uranium, with 92 electrons, is the heaviest atom found on our earth, although there are some manmade elements with as many as 102 or 103 electrons, but they are short lived and radioactive.

Electrons take up orbits in the shells according to their *kinetic energies—* $\frac{1}{2}mv^2$ joules (J)—in which *m* is the mass of the electron in *kilograms* and *v* is its velocity in *meters per second*. The shells, then, are regions representing certain *energy levels*. The shell farthest from the nucleus has the largest energy level assigned to it and, of course, the electrons in that shell have the highest energies of any in the atom.

The maximum number of electrons that each shell can contain is the same for all atoms. The first shell can contain only two, the second only eight, and so on.

Electrons in atoms may receive energy as a result of the material being heated or subjected to an electric field, or to high-energy radiation such as x-rays. When they receive this additional energy the electrons take up orbits in higher energy levels. Such an atom is then said to be *excited*. When the application of energy from the outside is discontinued, the energies of the electrons fall back to their original levels and the energy they had previously received is radiated out into the surrounding space. When all of the electrons of an atom are in shells as near to the nucleus as they can get, the atom is in its *lowest energy state*.

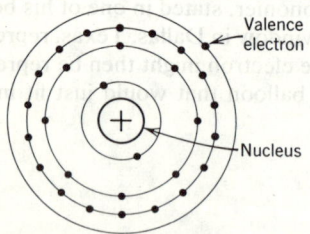

Valence electron

Nucleus

FIGURE 1.3 Bohr model of a copper atom. It has 3 filled shells with 2, 8, and 18 electrons rotating in them. These 28, and the valence electron, have orbits that are not confined to a single plane.

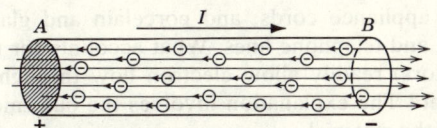

FIGURE 1.4 Section of wire carrying current *I*. Streamlines of the electric field extend through the cross-section surface *A* toward and through the cross-section surface *B*. Electrons are driven by the field force toward the positively charged surface. The force on each electron is $F = q\mathscr{E}$ in which q_e is the charge of the electron and $\mathscr{E}$ is the field strength.

An atom from which a valence electron has escaped, or from which it has been removed as a result of receiving added energy, is called an **ion**. Because the negative charge of the escaping electron has been removed, the ion is left positively charged and is called an *ionized atom* or a *positive ion*.

The maximum number of electrons a shell can accommodate is given by $2n^2$ where *n* is the number of the shell starting with the *K* shell (which is the only shell of the hydrogen and helium atoms) as number one.

Atoms combine with other atoms to form **molecules**. A hydrogen molecule consists of two atoms of hydrogen. The electrons in the outermost shell of an atom, the ones with the highest kinetic energies, are called *valence electrons*. They are capable of moving to other atoms, and when they do this we say an electric current flows. That is, a moving electric charge constitutes an electric current.

The image on a television picture tube is produced by electrons striking the *phosphors* on the inner surface of the tube. The *flow of electrons* from the electron-emitting surface in the neck of the tube to the front surface *is an electric current*. Because the phosphors are made highly positive with respect to the emitting surface, which is called the *cathode*, a strong *electric field* is produced and this field exerts forces on the electrons and drives them toward the phosphors on the front end of the tube. *The direction* (tube phosphor-to-cathode) *of the current thus produced is opposite to the direction of electron flow*. We owe this sometimes confusing concept of direction to Benjamin Franklin.[3] Electron flow in a section of the very narrow electron stream in the tube, and the current direction, are indicated in Fig. 1.4. Plane *A* is positive with respect to plane *B*.

1.4 Conductors, Insulators, and Semiconductors

It is common knowledge that conductors of electricity are usually metals; the most commonly used is copper. Some insulator materials are: rubber, asbestos cloth, and certain plastics, which we find in use as coverings of elec-

---

[3]Benjamin Franklin (1706–1790) is regarded as the first American physicist. He was also an inventor, a philosopher, and a famous statesman.

tric lamp and appliance cords, and porcelain and glass which are used on electric power and telephone lines. What accounts for the fact that some material (conductors) readily allow electron flow through them and others (insulators) do not? The explanation involves the outermost (valence) electrons of the atoms of the material.

In the case of copper, the single valence electron of an atom is not held very tightly to its "parent" atom's nucleus because it is farthest away from the nucleus. In fact, when another *atom* in the metal comes very close, this valence electron may be attracted to it and pass over to it, leaving the "parent" atom with only its 28 tightly bound electrons in its three completely filled shells. The parent atom will almost instantly capture the valence electron of a third atom that happens to come near. The valence electrons which can, and do, pass from one atom to another are called **free electrons**. If an *electric field* is established in a conductor, by making one end of it positive with respect to the other, this field will force the free electrons to move in a stream opposite to the direction of the field.

Good conductors have the largest number of *free electrons per unit volume*. This means that a cubic centimeter (or a cubic millimeter) taken *anywhere* inside a good conductor will have many times more free electrons than would exist in the same amount of volume inside a piece of rubber or other material that we call an insulator. A cubic centimeter of copper contains approximately 8.55×10^{22} free electrons; a cubic inch of copper contains about 1.4×10^{24} free electrons. Silver and gold are better electrical conductors than copper, by about 5 percent, but their cost prohibits their use except in applications in sensitive and expensive instruments. Aluminum is used as a conductor in high-voltage transmission lines mostly because of its lighter weight and lower cost. It is only about 60 percent as good as copper in conduction of electric current, since it has only about 60 percent as many free electrons per unit volume.

The internal structure of a highly conductive material is crystalline in nature. For example, the electrons in some metals are at the eight corners of a cube in space; sometimes there is a ninth one at the center of the cube. The outermost electrons are shared by *all* of the atoms of the material. This means they are free to wander from atom to atom. The reason silver is a better electrical conductor than copper is that its valence electron is in the fifth energy shell, even farther removed from the nucleus than the valence electron of copper, which is in the fourth shell. Less energy is required to remove the valence electron from silver than the amount required to remove the valence electron of copper from its atom.

Wires that must offer large resistance to the flow of electric current, and thus produce a lot of heat, are used in electric toasters and laundry irons. The metallic alloys of which these heater wires are made have only about 1 percent as many free electrons as copper. This means they have about 8×10^{20} free electrons per cubic centimeter. Such wires are classified as *resistance wires* but we do not call them poor conductors of electricity.

When electric current is supplied to an appliance or a light, it is necessary to run two parallel wires to the device. The reason two wires are needed instead of only one is explained in Section 1.6. To prevent the bare wires from touching each other and anything else, including the user of the appliance, they are covered with insulating material. The insulating material also holds the separate insulating wires together so that they may be handled as a single flexible unit.

In order to explain on a scientific basis why some materials are insulators, we must consider why free electrons do not pass from atom to atom in them as readily as they do in conducting materials. We learned that electrons in the various shells of an atom have corresponding amounts of energy. There are *energy bands* (each having an upper and lower limit on an energy scale) which are discrete ranges of energy. The free electrons in a conductor have energy values that are high enough to put them in the *conduction band* (see Fig. 1.5).

Whether a material behaves as a *conductor*, *semiconductor*, or *insulator* can be explained by the *permissible energy levels* of the atoms making up the material. The electrons of the copper atom fill the first three shells— 2 in *K*, 8 in *L*, 18 in *M*, and the valence electron (the 29th) is in the *N* band. Each shell contains electrons whose energies are in the energy band assigned to that shell.

There is a *forbidden region* or *forbidden energy gap* between each of the *allowed energy bands* from the first one (*K*) to the seventh one (*Q*). Furthermore, an energy gap (forbidden region) may exist between the energy of the outermost valence electrons and the energy needed for conduction (movement of electrons from one atom to another.) When some outside influence such as an electric field or addition of heat is applied, a valence electron may acquire sufficient energy to jump through the forbidden region and on into the conduction band of energy. Here it will have enough energy to free itself from any influence of its positive nucleus and become a *carrier of electricity*, ready to

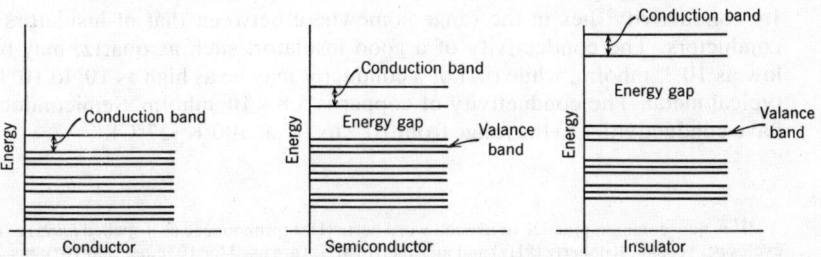

FIGURE 1.5 Relative magnitudes of energies in shells and gaps. The valence band is the uppermost energy band. In a conductor the valence band and conduction band overlap.

take the place of another electron that just left its own atom in the conductor in the same manner.

Energy can be given to electrons only in discrete amounts (*quanta*), according to the Bohr theory. Hence, there are certain energy values that an electron simply cannot acquire. As a rough example, if a man were climbing a vertical ladder that had rungs 1 ft apart, he could stand with both feet in a stable position only at 1-ft intervals above the base. He could not stand $1\frac{1}{2}$ ft or $2\frac{1}{4}$ ft above the base nor at any height involving a fraction of a foot. The *discrete quantum of energy* that can be acquired by an electron is given by hf, where h is Planck's constant (6.624×10^{-34} J-s) and f is the frequency of the energy in cycles/second or *hertz* (Hz).[4]

Figure 1.5 shows relative energy bands in conductors, semiconductors, and insulators. In conductors, there is no gap between the valence band and the conduction band. Researchers are convinced that they overlap.

In contrast to the action of conductors, insulators are poor carriers of electricity. The structure of solid insulators is such that almost all of the electrons remain bound to their parent atoms. The forbidden energy gap of these materials is very wide, so the conduction band remains almost empty. Few electrons can acquire enough energy to cross the forbidden region. Sulfur is a good insulator. Although it has 6 valence electrons in the M band, it has a very wide energy gap. It is very difficult to impart to these electrons sufficient energy for them to bridge the forbidden gap and enter the conduction band. This is why sulfur is an insulator.

Semiconductors, on the other hand, are neither good conductors nor good insulators, at normal room temperatures. More precisely, they are insulators at very low temperatures and reasonably good conductors at high temperatures. The elements *germanium* and *silicon* are the most important semiconductors used in electronics. Their use in crystal diodes almost excludes other semiconductors, although other materials are finding increasing use. However, these two elements appear to be the only ones suitable for the fabrication of transistors.

The element germanium has 4 electrons in the valence band of its atom. Its conductivity[5] lies in the range somewhere between that of insulators and conductors. The conductivity of a good insulator, such as quartz, may be as low as 10^{-16} mho/m, while that of a conductor may be as high as 10^7 to 10^8 for a typical metal. The conductivity of copper is 5.8×10^7 mho/m. Semiconductors have conductivities in the range from 10^{-3} to 10^5 at 300°K (27°C).

---

[4]It is now common practice to use the word hertz (Hz) (pronounced as if spelled *hurts*) to mean cycles per second. Kilohertz (kHz) and megahertz (MHz) are used for 10^3 c.p.s. and 10^6 c.p.s.

[5]Conductivity is a measure of the ability of a material to conduct electric current. It will be defined more precisely later.

1.5 Superconductors

In certain very pure metals the resistivity (resistance of a unit volume) suddenly falls to zero when they are refrigerated below a critical temperature. This state is called *superconductivity*. Lead becomes a superconductor at 7.2° Kelvin (256.8 centigrade degrees below 0°C), tin at 3.7°K, and tantalum at 4.4°K. Although this effect was discovered in 1911, it is not fully understood at this writing by research scientists who are working in this area. Results of their experiments have shown that currents of hundreds of amperes, which were induced into small circular paths of these refrigerated metals, continue for several years without any observable decrease in ampere value.

An important control feature associated with superconductivity is that the application of a magnetic field affects the temperature at which the metal becomes superconductive. At temperatures slightly above this critical temperature, the metal has a noticeable amount of resistance. The amount of the magnetic-field strength determines the resistance value. This characteristic makes small pieces of superconductive metal adaptable to digital-computer circuitry. The advantage is that practically negligible power will be required to run a computer, other than that required by the refrigeration equipment. Liquid helium, boiling at amospheric pressure, has beeh used as the refrigerant.

1.6 Electric Circuit

There are a great many examples that would serve to illustrate a simple electric circuit such as turning on a light either by operating a wall switch in our house, opening the door of our automobile, pressing the button or sliding the switch on a flashlight, or operating the switch of a radio or television receiver. Let's use the example of the dome light in the automobile (Fig. 1.6). The car battery is the source of energy in this case, but the room lights of a residence, and the radio and television receivers get their energy over a system

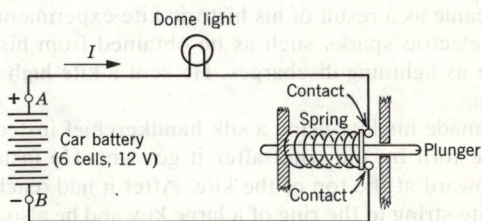

FIGURE 1.6 Simple electric circuit. Dome light in automobile is turned on by spring-loaded switch when door is opened. When door is closed, it pushes plunger to the left, breaking the circuit at the two contacts.

of circuits that goes all the way back to the electric generator of the power company.

The chemical action in the battery, which will be explained later, creates an *electrical potential difference* at its terminals so that one is + and the other − in *polarity*. For the time being, an easy way to understand the operation of an electrical circuit is to consider this potential difference as a *force* that pushes the electric charges along the wires and through the closed circuit, that is, from terminal A^+ to and through the dome light and back to terminal B^-. The *current*, designated by the arrow I, is *expressed in ampere units*.

In passing through the dome light the electric charges create heat and, of course, light. Both heat and light are forms of energy, and the electric charges convey energy from the battery to the light. As many charges as leave the battery must return to it because electric charges cannot be destroyed, nor are any lost in the process of completing their round trip from A to B and back to A again. Chemical energy of the battery is being used up in the process to create electrical energy which is carried to the light by the motion of electric charges.

Another name for potential difference is **electromotive force**, and both are expressed in **volts**. Later, it will be found that *the volt is a unit of energy, or of work*, and not actually a unit of force. Alas, it has been found that the charges that move in the closed circuit are the *electrons* rather than the positive charges, so although *we continue to let the current direction be as shown in Fig. 1.6*, actually the electrons go in the opposite direction. Why this ambiguity? It is blamed on Benjamin Franklin, who had no way of finding out that *it is the electrons that move in a conductor*. In fact, he had been dead nearly 100 years before the electron was discovered. He did announce, in 1749, however, that he suspected the existence of extremely small electric particles that move very easily. He said, "The electrical matter consists of particles that are extremely subtle, since it can penetrate common matter, even the densest metals, with such ease and freedom as not to receive any perceptible resistance."

Among the many other discoveries by Banjamin Franklin, two of which are bifocal spectacles and the Franklin coal stove for residence heating, one of his greatest came as a result of his historic kite experiment. In 1752 he demonstrated that electric sparks such as he obtained from his "electric machine" are the same as lightning discharges. He sent a kite high into the air during a thunderstorm.

Franklin made his kite using a silk handkerchief instead of thin paper so it would not be torn by the wind after it got wet. He fastened a pointed wire projecting upward at the top of the kite. After it had reached sufficient height, he tied the kite string to the ring of a large key and he also tied a dry silk ribbon to the ring at one opposite point. This he did under a shed so that he and the silk ribbon would not get wet. He kept the kite string firm by grasping the dry silk ribbon, so that any electric charge that might come down the wet kite string and appear on the key would not pass through his body to the earth. He then held the end of a wire in his hand (the other end of the wire touching the wet ground) and drew sparks from the key to the grounded wire (Fig. 1.7).

Franklin succeeded in charging a *Leyden jar* using the sparks from the key and, afterward, he used the charged jar to give a shock. He then succeeded in performing all the experiments with the charged jar that he had done with charges from his frictional electric machine. His experiment proved that lightning was caused by electric discharges from the atmosphere. This kite experiment is dangerous. Do not try it!

Although we shall learn quite a bit about *resistance* later, we may think about it as a sort of *electric friction* that opposes the passage of current (electrons in motion) through the lamp of Fig. 1.6. Friction creates heat, and the temperature of the high-resistance wire in the lamp is raised so high that *incandescence* results. The potential difference, V, required to send the proper current through the lamp is maintained by the energy conveyed from the battery to the lamp terminals. The first quantitative relationship we learn for the electric circuit is

$$\text{Current} = \text{potential difference} \div \text{resistance}$$
$$I = V/R \text{ (Ohm's law)} \tag{1.1}$$

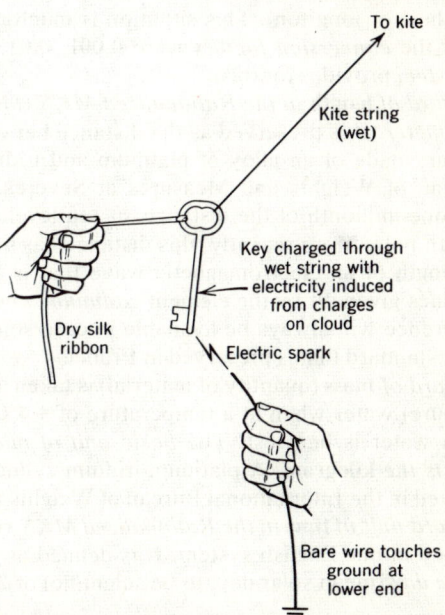

To kite

Kite string
(wet)

Key charged through
wet string with
electricity induced
from charges
on cloud

Dry silk
ribbon

Electric spark

Bare wire touches
ground at
lower end

FIGURE 1.7   Bengamin Franklin's kite experiment.

We shall use the letter E to represent the *potential difference at the terminals of a source* (the battery, in this case) and V to represent the potential difference at a *receiver* such as a lamp or an electric motor. The letter I represents current.

1.7   MKS Units, Rationalized

We are accustomed to measuring and expressing distances in inches, feet, yards, and miles, and weight in ounces, pounds, and tons. In order to convert a distance expressed in one unit to the same amount in another unit, we must divide or multiply by such numbers as 12, 3, 1760, or 5280. Circuit diagrams, often called *schematics*, are drawn with standard symbols that represent parts (elements) of the circuit. Fig. 1.8 shows many symbols, some of which will not have much meaning now. The direction of current is taken so that the flow of energy is from the source (at the left of the circuit diagram) toward the receiving unit, or system, on the right. The current direction is usually shown by an arrow pointing toward the right on the line (representing a conductor) connected to the top of the symbol for the source.

To convert weights we must use 16 (ounces per pound), 2000, or 2400 (pounds per short or long ton). This situation is much more troublesome than it would be if the *conversion factors* were 0.001, 0.01, 0.1, 10, 100 or 1000. The *metric system* provides for this.

The standard of **length** *in the Rationalized MKS (meter, kilogram, second) system is the meter.* It is preserved as the distance between two points marked on a metal bar, made of an alloy of platinum and iridium, kept in the International Bureau of Weights and Measures at Severes, France. Historically, it represents one-millionth of the distance, at sea level, from the earth's equator to the north pole. More recently, this distance has been expressed in terms of the wavelength of an electromagnetic wave that is associated with one of the spectral lines given off by the element *cadmium* when heated in an electric arc. This reference will always be available in case something happens to the original meter standard that is preserved in France.

The standard of **mass** (quantity of material) is taken as the mass of 1000 cm³ of distilled (pure) water when at a temperature of $+4°C$. This is the temperature at which water is heaviest. The *basic unit of mass in the Rationalized MKS system is the* **kilogram.** A platinum-iridium cylinder with a mass of 1 kg is also preserved in the International Bureau of Weights and Measures.

The standard unit of **time** *in the Rationalized MKS system is the* **second**, the same as is used in our English system. It is defined as 1/86,400 of the length of the *average day* (mean solar day, to be scientific) of 24 h. There are 86,400 s in 24 h.

The *unit of force in the Rationalized MKS system* is the **newton**, whose

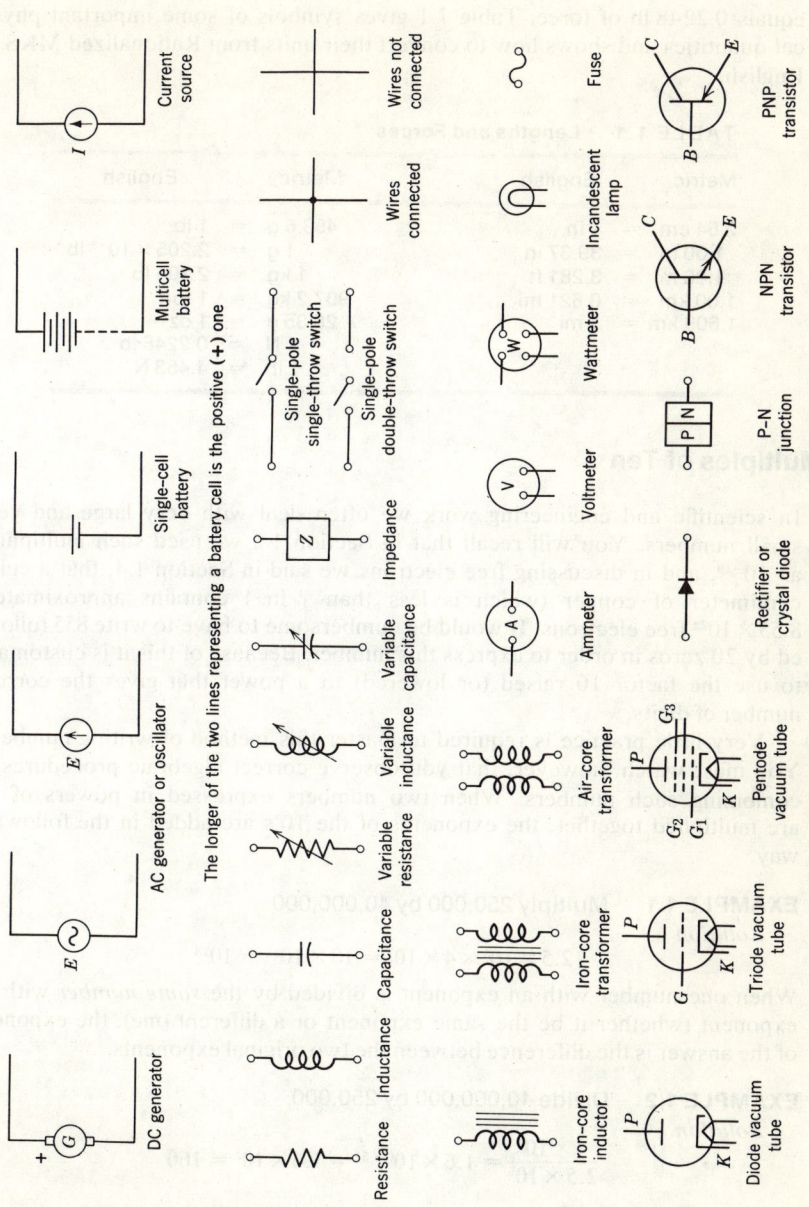

FIGURE 1.8 Symbols for circuit elements and devices.

15

symbol is **N**. The English pound of force is equal to 4.453 N. One newton equals 0.2248 lb of force. Table 1.1 gives symbols of some important physical quantities and shows how to convert their units from Rationalized MKS to English.

TABLE 1.1 **Lengths and Forces**

Metric		English	Metric		English
2.54 cm	=	1 in.	453.6 g	=	1 lb
1.00 m	=	39.37 in.	1 g	=	2.205×10^{-3} lb
1.00 m	=	3.281 ft	1 kg	=	2.205 lb
1.00 km	=	0.621 mi	907.2 kg	=	1 ton
1.609 km	=	1 mi	28.35 g	=	1 oz
			1 N	=	0.2248 lb
			1 lb	=	4.453 N

1.8 Multiples of Ten

In scientific and engineering work we often deal with very large and very small numbers. You will recall that in Section 1.2 we used such multipliers as 10^{-19}, and in discussing free electrons we said in Section 1.4, that a cubic centimeter of copper (which is less than $\frac{1}{16}$ in.³) contains approximately 8.55×10^{22} free electrons. It would be cumbersome to have to write 855 followed by 20 zeros in order to express this number. Because of this it is customary to use the factor 10 raised (or lowered) to a power that gives the correct number of digits.

Very little practice is required to master this method of writing numbers. You must watch, however, that you observe correct algebraic procedures in combining such numbers. When two numbers expressed in powers of 10 are multiplied together, the exponents of the 10's are added in the following way.

EXAMPLE 1.1 Multiply 250,000 by 40,000,000.
Solution
$$2.5 \times 10^5 \times 4 \times 10^7 = 10 \times 10^{12} = 10^{13}$$

When one number with an exponent is divided by the *same number* with an exponent (whether it be the same exponent or a different one), the exponent of the answer is the difference between the two original exponents.

EXAMPLE 1.2 Divide 40,000,000 by 250,000.
Solution.
$$\frac{4 \times 10^7}{2.5 \times 10^5} = 1.6 \times 10^{(7-5)} = 1.6 \times 10^2 = 160$$

EXAMPLE 1.3   Divide 25,400 by 0.0127.
  Solution.

$$\frac{2.54 \times 10^4}{1.27 \times 10^{-2}} = 2 \times 10^6$$

When we subtract or add exponents, they must be exponents of the same quantity.

EXAMPLE 1.4   Divide 10^5 by 5^2. It is incorrect to do it this way:

$$\frac{10^5}{5^2} = 2^3 = 8 \quad \text{(wrong)}.$$

Solution. Change 5^2 to 2.5×10^1:

$$\frac{10^5}{5^2} = \frac{10 \times 10^4}{2.5 \times 10^1} = 4 \times 10^3 = 4000 \quad \text{(right)}$$

Table 1.2 shows the names, symbols, and multiplier values of commonly used prefixes such as micro, milli, kilo, and mega. For example, a kilogram = 10^3 grams, a microampere is 10^{-6} A, and so on.

TABLE 1.2

Prefix	Symbol	Multiplier	Prefix	Symbol	Multiplier
atto	a	10^{-18}	deci	d	10^{-1}
femto	f	10^{-15}	decca	da	10
pico	p	10^{-12}	hecto	h	10^2
nano	n	10^{-9}	kilo	k	10^3
micro	μ	10^{-6}	mega	M	10^6
milli	m	10^{-3}	giga	G	10^9
centi	c	10^{-2}	tera	T	10^{12}

1.9 Units

One of the most serious handicaps you can encounter in your study of electrical technology is a lack of familiarity with the *names of the basic quantities and their units.* A knowledge of *their relations to one another* is just as important. The prefixes to their names: *kilo*gram; *milli*meter, etc. are listed in Table 1.2 for ready reference.

  Table 1.3 contains a list of frequently used units, their symbols, and conversion factors by means of which you can convert any amount of them in the MKS system to an equivalent amount in the English system. For example,

TABLE 1.3 **MKS and English Units**

Physical Quantity	Symbol	Defining Equation	mks Units	To Convert to English Units	English Units
Length	i, L	—	meter	$\times 3.281$	foot
			meter	$\times 39.37$	inch
Mass	m, M	—	kilogram	$\div 14.6^a$	slug
Time	t, T	—	second	1	second
Area	A	$l_1 l_2$	square meter	$\times (3.281)^2$	square foot
Volume	V	$l_1 l_2 l_3$	cubic meter	$\times (3.281)^3$	cubic feet
Velocity	v	l/t	meters per sec	$\times 3.281$	feet per sec
Acceleration	a	a	(meters per sec) per sec	$\times 3.281$	(feet per sec) per sec
Force	F, f	$m \times a$	newton	$\times 0.2248$	pounds
Mechanical work or energy	W	$f \times l$	joule	$\times 0.7376$	foot-pounds
Power	P	W/t	joules per sec	$\times 0.7376$	foot-pounds per sec
Power	P	$I^2 R$	watts	$\div 746$	horsepower
Electrical work or energy	W	$I^2 Rt$	watt-seconds	$\div (1{,}000 \times 3{,}600)$	kilowatt-hours
Temperature	T	—	degrees centigrade	$\times \frac{9}{5} + 32$	degrees Fahrenheit
Heat energy	Q_r	Heat required to raise unit mass of water 1 degree of temperature	kg calorie: 1 kg weight H_2O 1°C	$\div 0.252$	British thermal unit 1 lb weight H_2O, 1°F
Mechanical equivalent of heat	J_r	Ratio of mechanical energy to equivalent heat energy. Ratio of unit sizes	$\dfrac{\text{joules}}{\text{kg-calorie}} = 4{,}186$	—	$\dfrac{\text{ft-lb}}{\text{Btu}} = 778$
Electric charge	Q	—	coulomb	1	coulomb

^{a}This assumes that the mass which will be given an acceleration of 1 ft (30.48 cm) per sec per sec by 1 lb of force (445,000 dynes) is $m = f/g = 445{,}000/30.48 = 14{,}600$ grams or 14.6 kg. Therefore, divide kilograms of mass by 14.6 to get mass in slugs.
1 N of force = 101.92 g of force = 0.102 (approx) kg of force.

to convert a number of meters into feet, multiply by 3.281. To convert joules of mechanical energy into foot pounds, multiply by 0.7376. You need not try to memorize them, since frequent usage will help you to remember many of them.

The CGS (centimeter, gram, second) system is still in use by many scientists. You need to remember that there are 100 cm in a meter, 1000 m in a kilometer, 1000 g in a kilogram, and so forth.

The table shows us how to convert force from newtons to pounds, but it does not bring in grams of force which is a unit in the CGS system. One pound of force = 453.6 g of force, so after we get pounds from newtons:

$$F_{lb} = 0.2248 \times F_N$$

we simply multiply $(0.2248 \times F_N)$ by 453.6 to get grams:

$$F_g = 453.6 \times 0.2248 \times F_N$$
$$F_g = 101.97 \times F_N$$

The *dyne* is the *absolute unit of force* in the CGS system. It is very small and is generally taken equal to 1/980 g, that is,

$$1 \text{ g of force} = 980 \text{ dyn of force}$$

EXAMPLE 1.5 Verify that 1 hp is equivalent to 746 watts (W). Start with 33,000 ft-lb per min and show the units involved in each step in the transformation.

Solution.

$$33,000 \frac{\text{ft-lb}}{\text{min}} \times \frac{1}{60} \frac{\text{min}}{\text{s}} = 550 \frac{\text{ft-lb}}{\text{s}}$$

$$550 \frac{\text{ft-lb}}{\text{s}} \times \frac{1}{3.281} \frac{m}{\text{ft}} = 167.7 \frac{\text{lb-m}}{\text{s}}$$

$$167.7 \frac{\text{lb-m}}{\text{s}} \times \frac{1}{0.2248} \frac{\text{N}}{\text{lb}} = 746 \frac{\text{N-m}}{\text{s}}$$

$$\frac{\text{N-m}}{\text{s}} = \frac{\text{J}}{\text{s}} = \text{W}$$

The cancellation of names of units is an important part of this procedure. One can use this device to great advantage in performing conversions. Inasmuch as we know there are 60 s in a minute, we need the conversion factor in minutes per second which must be the reciprocal of 60, the seconds-per-minute factor. Likewise, we need newtons per pound so we are obliged to use the reciprocal of 0.2248 which is 4.453, the pounds equivalent of one newton (or pounds per newton) factor. Of course, we could have multiplied by 4.453 and still called this the newton-per-pound factor.

In the conversion process of this example there is no need to compute the

TABLE 1.4   Electric and Magnetic Units in the Rationalized MKS (Giorgi) System

Symbol	Name	Unit
ϵ_0	permittivity of free space	farads per meter
ϵ_r	relative permittivity	none
E, V	voltage, emf, potential difference	volts
$\mathscr{E}$	electric-field intensity (or strength)	volts per meter or newtons per coulomb
ρ_L	linear-charge density	coulomb per meter
ρ_S	surface-charge density	coulomb per square meter
ρ_v	volume-charge density	coulomb per cubic meter
ψ	electric flux	coulomb
D	electric-flux density	coulomb per square meter
I, i	current	ampere
R, r	resistance	ohm
ρ	resistivity	ohm per unit conductor[a]
G, g	conductance	mho
σ	conductivity	mho per meter
j	quadrature operator	none
C, c	capacitance	farad
H	magnetic-field intensity (or strength)	ampere (turns) per meter
M	mutual inductance	henry
θ, α	angle	radian or degree
Φ	magnetic flux	weber
$n\Phi$	magnetic-flux linkage	weber (turn)
B	magnetic-flux density	weber per square meter
mmf	magnetomotive force	ampere (turn)
S	susceptance	mho
μ_0	permeability of free space (vacuum)	henry per meter
μ_r	relative permeability	none
T	torque	newton-meter
T°	temperature	degree
L	inductance	henry
ω	angular velocity	radian per second
P	power	watt
W	energy	watt-second or joule
X, x	reactance	ohm
Z, z	impedance	ohm
Y, y	admittance	mho

$\epsilon_0 = 8.854 \times 10^{-12}$ farad per meter, also given as

$$\frac{1}{36\pi \times 10^9};$$

$\mu_0 = 4\pi \times 10^{-7}$ henry per meter

[a]"Resistivity is expressed in *ohm-meters*, which is understood to mean ohms resistance across opposite faces of a cubic meter of the material. Ohm-centimeters or ohm-millimeters may also be used, with corresponding meaning. In *wire measure* where length is in feet and cross-section area is in *circular mils*, the unit of resistivity is *ohm-circular-mil per foot*. This is the resistance of a round conductor 0.001 in. in diameter and 1 ft long. It is inconsistent, and confusing, to call this unit *ohms per circular-mil-foot*.

values of two intermediate steps. It can be written on a single line:

$$33,000 \frac{\text{ft-lb}}{\text{min}} \times \frac{1}{60} \frac{\text{min}}{\text{s}} \times \frac{1}{3.281} \frac{\text{m}}{\text{ft}} \times 4.453 \frac{\text{N}}{\text{lb}}$$

$$= 746 \frac{\text{N-m}}{\text{s}} = 746 \text{ W}$$

Table 1.4 shows the letter symbols, names, and units of most of the important quantities encountered in the study of electrical technology. You should not try now to memorize them. Your ability to recall them will develop as you progress in your course of study if you give them adequate attention and work diligently on the problems that will be assigned.

1.10   Review Questions

These are very important and should receive your serious and persistent attention.

1. How would you demonstrate the existence of an electric charge on an object, or show that a body is electrified?

2. How does the electric charge on a glass rod that has been rubbed with a silk cloth differ from the charge on a hard rubber comb that has been rubbed with a flannel cloth?

3. If you walk along a hotel corridor on a dry day and then attempt to press an elevator button with your finger, a spark is very likely to jump between the tip of your finger and the metal casing surrounding the button. Explain.

4. What is an electron? A proton? A neutron? How much electric charge does each have? Which has negative charge?

5. Describe the Bohr models of the hydrogen atom and the helium atom. Are the electron orbits in the helium atom, and in all other heavier atoms, in a single plane?

6. About how many times greater than the radius of a proton is the radius of the orbit in which the electron of a hydrogen atom rotates?

7. By what letters are the energy shells of an atom identified?

8. What determines the shell in which an electron finds itself, even though all electrons have the same amount of charge and the same[6] mass?

9. In an atom that has many electrons in orbits, where would we find those with the highest energies? Those with the lowest energies?

10. What characterizes an atom that is in its lowest energy state?

11. What, basically, is *electric current*?

12. A battery supplies current to a light. Do we say the direction of the current is *toward* the positive (+) terminal of the battery or *away* from it? How about the direction of the current at the negative terminal of the battery? Do the same number of *charges per second* that leave the battery return to it?

13. Which of the charged particles (electrons or protons or positive ions) flow in the light circuit of Question 12?

[6]According to the theory of relativity, the mass changes with the velocity, but this change need not affect the answer to this question.

14. What do we say exerts forces on electrons in a conductor and propels them along it?
15. In what respects do the valence electrons of an atom differ from the other electrons?
16. What are free electrons?
17. In what basic respect does the free electron content of a good conductor differ from that of a relatively poor conductor?
18. What are the forbidden regions, or forbidden energy gaps, in atoms?
19. Discuss, briefly, superconductors.
20. What is the unit of electric potential difference? Is it a force, a pressure, or an amount of energy? It is, really, only one of these.
21. How did the *choice of direction* of electric current flow as opposite to the direction of movement of free electrons in a circuit originate?
22. A fourth grade school class was required to write a story about Benjamin Franklin. A boy wrote as follows: "Benjamin Franklin was born in Boston and moved to Philadelphia. There he met a lady, married her, and discovered lightning." Explain the procedure and result of Franklin's historic kite experiment, which revealed to him the nature of electrical discharges in a storm cloud and on the earth's surface beneath the cloud.
23. Write Ohm's law in mathematical form.
24. What are the names of the basic units of length, mass, time, and force in the MKS system?
25. How many: (a) centimeters make an inch? (b) inches make a meter? (c) feet make a meter? (d) square centimeters make a square inch? (e) newtons of force make a pound of force?
26. Why is silver a better electrical conductor than copper?
27. What units do you come out with when you multiply (a) feet per second by meters per foot? (b) kilograms per cubic meter by pounds per kilogram and then divide the result by cubic feet per cubic meter?

1.11   Problems

You should learn to work as many of the problems in Group A in each chapter as are assigned by your instructor. Then you should work as many in Group B as you have time for (at least, those assigned by your instructor).

Group A

1. The population of China is said to be at least 600 million. Express this number using a power of 10 as a multiplier.
2. The latest value given for the mass of an electron is 0.000,000,000,000,000, 000,000,000,000,910.9 g. Express this using a power of ten as a multiplier.
3. Multiply 27,000,000 by 0.004 and divide the result by 3000 using powers of 10 for all three numbers.
4. Express 1/250 s in milliseconds and in microseconds.

5. Express 0.001 km in millimeters.
6. Express 6.45 cm² in square inches.
7. Convert 1 in.³ into cc (cubic centimeters).
8. Convert the approximate weight of a nickel (5 g) into milligrams, and then into kilograms.
9. Convert a velocity of 60 mi/hr into feet per second and then into meters per second.
10. Convert a velocity of 44 ft/s into (a) miles per hour and (b) kilometers per hour.
11. A 1-ft long piece of No. 12 AWG copper wire has a diameter of 0.081 in. How many free electrons does it have?

B Group

12. How many cubic centimeters are there in a cubic foot?
13. How many electrons would there be in a pound of electrons?
14. What is the density of the matter in a proton in (a) grams per cubic centimeter? (b) kilograms per cubic meter? (c) tons per cubic inch?
15. An ordinary 100 W electric light bulb takes 5/6 A (ampere) of current when at normal brightness. This means 5/6 C of electric charge passes through it every second. How many electrons per second must pass through the lamp?
16. When 1 N of force acts through a distance of 1 m, 1 J of work is done. We can also say that 1 J of energy is expended. How many dyne centimeters (ergs) of energy is this (1 N = 10⁵ dyn of force)?
17. When work is done at the rate of 1 hp, the rate is also 550 ft-lb/s. How many joules per second is this?
18. One horsepower is the equivalent of 746 W in electrical units. Express this in kilowatts, milliwatts, and megawatts.
19. One British thermal unit is an amount of energy equivalent to 778 ft-lb. Express this in kilogram-meters and in gram-centimeters.
20. Suppose you have a choice of either one orange 3 in. in diameter or three oranges 2 in. in diameter. Which would give you the most pulp if you neglect the thickness of the skins? How much more would this give you, expressed in cubic inch units? What percent of the smaller volume is this?
21. Suppose your choice in Problem 20 was the one 3-in. orange. Assuming the skins of all oranges have the same thickness, would your orange have less skin than that on the three other oranges? How much less, in percent, of the skin on the three 2-in. oranges?
22. Suppose you were a father and had three children with you when you were faced with the selection described in Problem 20. What would your choice be?
23. It is true that the volumes of two objects identical in shape and proportionate in all dimensions are in the same ratio as the cubes of like dimensions. For example, a sphere 4 in. in diameter has eight times the volume of a sphere 2 in. in diameter. Now consider two eggs that are identical in shape and proportionate in all dimensions. One is 10 percent longer and "thicker" than the other. By what percent does the volume of the larger egg exceed the volume of the smaller egg? Would it be more economical to buy a dozen of the larger ones for 55 cents than dozen of the smaller ones for 49 cents? *Suggestion.*

Express the percent increase in cost when the larger eggs are purchased, that is, 6 is what percent of 49?

24. Here is a problem in which the use of common logarithms is indicated. A horseman asked a blacksmith to put new shoes on his horse and first asked the price. The blacksmith said that he usually charges $10 for the job, but he would do it on the basis of 1 cent for the first nail, 2 cents for the second, 4 cents for the third, and so on. The cost per nail would double until the 32-nail job was finished. The customer *would have to pay only for the last (32nd) nail* if he had decided to pay on that basis. What would it have cost him.?

25. Here is another problem on which you should use logarithms. Assume you have a large sheet of plastic one-thousandth of an inch thick. You fold it double and continue doubling its thickness by folding. Assume you can fold it 25 times, doubling its thickness each time. If this were possible, how high would the stack be after the 25th fold?

CHAPTER TWO

Voltage, Current, and Electrical Energy Sources

We have learned that electric charge (quantity of electricity) can move from one point to another in a conductor and that when this happens we say there is an electric current. We use the term *charge* both because it is a shorter and simpler term than *quantity of electricity* and because a charged body has an excess of one kind (+ or −) over the other. Accordingly, we say that the charge of (or on) the electron is 1.602×10^{-19} C. We are now ready to furnish Lord Kelvin's answers to the questions: What is voltage? What is electric current? Where does the electrical energy required to light a lamp or run a motor come from?

2.1   A Simple Circuit

Figure 2.1 represents one of the simplest electric circuits of an automobile wiring system. This was selected not only for its simplicity but because it is perhaps one of the most widely used circuits that employs a battery as the energy source. It is common practice to have an *extra contact* on the dashboard switch that turns on the headlights of the car, so that *when the switch knob is turned counterclockwise*, the dome light of the car is turned on. This is called

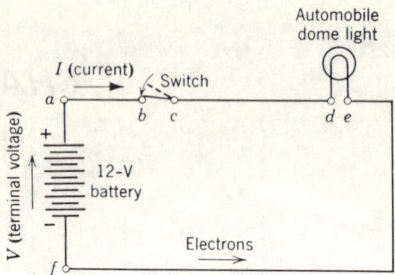

FIGURE 2.1   Simple series circuit.

a simple series circuit. It consists of three parts, or elements, in addition to the connecting wires: battery, switch, and dome light.

Let us assume that the light in the circuit of Fig. 2.1 has $\frac{1}{2}$ A of current flowing through it. At the instant it is turned on, the current through the switch (the current leaving the + terminal of the battery and the current entering the − terminal of the battery) is $\frac{1}{2}$ A. (The ampere will be defined in the next section.)

It is easy to see why the current in all of the elements is the same amount. Current (charges in motion) can come only from one source, the battery, and charges cannot leave the conducting wires and go anywhere else. If we compare a closed circuit of water pipes with a closed electrical circuit and say that a pump (in place of the battery) delivered $\frac{1}{2}$ gal/s to a water motor (in place of the light) through a valve (in place of the switch), each element would have water passing through it *at the rate of $\frac{1}{2}$ gal/s.* None leaks out of the closed-pipe circuit and none gets in. Similarly, the number of electrons per second that go through the light is the same number as that which goes through the switch and then on through the battery. What happens inside the battery will be discussed later in this chapter.

In a series circuit, there is only one amount of current and each element of the circuit has that amount flowing through it. It would be wrong to say that there are 1.5 A of current in the circuit of Fig. 2.1 (0.5 A for each of the three elements.) If we were to argue that this is true, someone would reply that each of the three connecting wires: *a* to *b*, *c* to *d*, and *e* to *f*, also has $\frac{1}{2}$ A in it, so why are there not 3 A of current flowing?

By current is meant flow of quantity per unit of time.

2.2   The Coulomb: Unit of Quantity of Electric Charge

How much quantity is a coulomb of charge? Figure 2.2 represents an *electro-plating bath* in which pure silver is deposited on a steel spoon. A solution of silver nitrate ($AgNO_3$) in water[1] is made an element in a series circuit in which a switch connects it to a current source. The molecules of the silver nitrate are broken up by the electric current into positive ions Ag^+ and negative ions NO_3^-. The positive ions pass to the spoon, where they are supplied with the negative charges of electrons and become neutral silver atoms and stick to the spoon. The NO_3^- ions go to the silver bar where they combine with silver atoms that leave the bar to form molecules of silver nitrate. These molecules then break up and new silver ions appear in the solution, each replacing one that became attached to the spoon as an atom of silver.

It has been determined that *when 0.001118 g of silver has been deposited* in this fashion *a coulomb of electric charge has passed through the solution and through the circuit.*

A coulomb is that quantity of electricity which will deposit 0.0001118 g of silver from a solution of silver nitrate.

This definition does not involve time. The quantity required to deposit this fixed amount of silver can be supplied rapidly by the source (stronger current flow) or very slowly (very weak current flow). At the instant the stated amount of deposit has been completed a coulomb of charge will have passed through the solution.

We shall use the capital C to represent coulombs.

[1] In industrial practice, silver cyanide and potassium cyanide are used instead of silver nitrate.

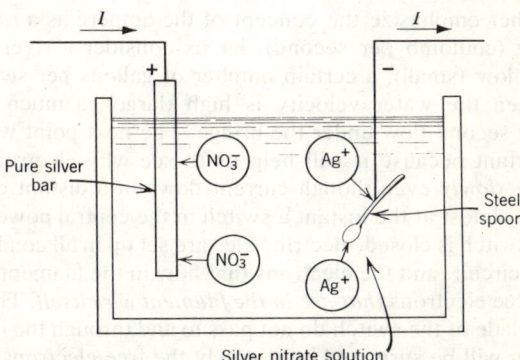

Pure silver bar

Steel spoon

Silver nitrate solution

FIGURE 2.2   Silverplating a spoon by transferring electric charge through a chemical solution.

2.3   The Ampere

When an electric charge passes through an element of a circuit, or past a point in the circuit, *at the rate of one coulomb per second* the current strength is called *one ampere*.

An ampere is the rate of flow of electric charge equal to one coulomb per second.

EXAMPLE 2.1   The current in an electrical appliance is 0.5 A. If this amount of current were kept constant for an hour in the electroplating bath of Fig. 2.2, how much silver would be deposited on the spoon?

Solution

$$0.5 \text{ A} = 0.5 \text{ C/s}$$
$$0.5 \times 3600 = 1800 \text{ C passed through}$$
$$1.8 \times 10^3 \times 1.118 \times 10^{-3} = 2.01 \text{ g}$$

Checking units

$$\frac{\text{Coulombs}}{\text{second}} \times \text{seconds} = \text{coulombs};$$

$$\text{coulombs} \times \frac{\text{grams}}{\text{coulombs}} = \text{grams}$$

The letter Q represents quantity of electricity and the letter **A** represents **amperes**.

$$Q = It \qquad \text{coulombs} = \text{amperes} \times \text{seconds} \tag{2.1}$$

$$I = \frac{Q}{t} \qquad \text{amperes} = \text{coulombs per second} \tag{2.2}$$

To further emphasize the concept of the ampere as a *rate of flow of electric charge* (coulomb per second), let us consider a river. When the water velocity is low (small), a certain number of gallons per second flow under a bridge; when the water velocity is high (large), a much larger number of gallons per second flow under the bridge. The next point we want to make is very important because it will help us to see why electrons in a metal wire *move quite slowly* even though current flows in a distant element, such as a street light, almost at the instant a switch in the central power station is closed. When the switch is closed, electric fields are set up in all conductors throughout the closed circuit, and the electrons that flow in the filament of the street lamp are those free electrons *that are in the filament wire itself*. The free electrons in the metal blade of the switch do not pass to and through the lamp at once. Read on, and you will be surprised how slowly the free electrons in the conductors do travel. In fact, they only *drift*.

The number of gallons of water that flow each second under a bridge that spans a rather narrow stream which has a large water velocity can be the same as the number that flow per second under another bridge that spans a very much wider but gently flowing river. That is, the gallons per second are the same, but the velocities of the water particles in the wide river are very much lower than the velocities in the narrow river.

We learned that a cubic inch of copper has 1.4×10^{24} free electrons. Let us calculate the number of free electrons in a 1-ft length of No. 12 AWG (American Wire Gage) copper wire such as is used to wire houses. We find the cross-section area of the wire to be 5.13×10^{-3} in². The volume of a 1-ft length is

$$5.13 \times 10^{-3} \times 12 = 61.56 \times 10^{-3} \text{ in.}^3/\text{ft}$$

The number of free electrons per foot length in this wire is given by

$$\frac{\text{inch}^3}{\text{foot}} \times \frac{\text{electrons}}{\text{inch}^3} = \text{electrons per foot}$$

$$61.56 \times 10^{-3} \times 1.4 \times 10^{24} = 86.184 \times 10^{21} \text{ electrons in 1 ft of the wire}$$

That's a lot of electrons, isn't it?

Assume that 1 A of current flows through an electric light bulb. This is 1 C of charge per second, or 6.24×10^{18} electrons per second.[2] We would like to know the velocity of the free electrons in the wire that carries them from their source to the lamp and back to their source again. Note the units as well as the numbers in the following solution.

Important

$$\frac{\text{electrons}}{\text{second}} \div \frac{\text{electrons}}{\text{foot}} = \frac{\text{electrons}}{\text{seconds}} \times \frac{\text{foot}}{\text{electrons}} = \frac{\text{feet}}{\text{second}}$$

$$\frac{6.24 \times 10^{18}}{0.86184 \times 10^{23}} = 7.24 \times 10^{-5} \text{ ft/s} = 0.000868 \text{ in./sec}$$

This is less than one thousandth of an inch per second! Billions of electrons therefore, merely drift in a conductor when current flows. No. 12 wire with rubber insulation can carry 20 A with safety. When this happens the free electrons must go 20 times as fast to deliver 20 C of quantity per second. Their velocity is then only

$$20 \times 0.000868 = 17.36 \text{ thousandths of an inch per second}$$

[2] 1.602×10^{-19} C per electron; $1/(1.602 \times 10^{-19}) = 6.24 \times 10^{18}$ electrons per coulomb.

2.4 The Volt

We used the concept of electric charges being forced by the electromotive force to move against resistance in a conductor and thus produce current. We said that the amount of current can be calculated by dividing the voltage (potential difference) by the resistance. The concept of potential difference (voltage) can be explained correctly by realizing that potential and potential difference are really *energy (or work) quantities* and not force or pressure quantities.

In the hydraulic system of Fig. 2.3 the pump *does work* in maintaining a pressure difference $(h_2 - h_1)$ between the inlet and outlet of the pump. Energy is given up by the water at the motor end, and, neglecting friction and other losses in the motor, the motor *does work*, that is, *delivers energy* in accordance with the loss in pressure undergone by the water $(h_3 - h_4)$. The loss of energy (work done against friction) in the forward pipe is represented by $(h_2 - h_3)$ and the loss of energy in the return pipe is represented by $(h_4 - h_1)$. Note that we say *represented by*, not equal to.

Looking at the electric circuit of Fig. 2.4 the generator does work in developing a potential difference at its terminals given by E, and the motor is supplied with energy at the potential difference $V_M = (V_3 - V_4)$. Note, again, that we use E to represent volts at the source and V to represent volts at the receiver or load. The *drop in potential* along the outgoing wire is $(E - V_3)$ and the drop in potential along the return wire is $V_4 = V_3 - V_M$.

It is important to understand that when water is forced to move from one point to another against some kind of resistance such as friction, or the opposing force of a blade of a water wheel, *work is done* and *energy is expended*. When electric current is forced through a conductor that has resistance, or

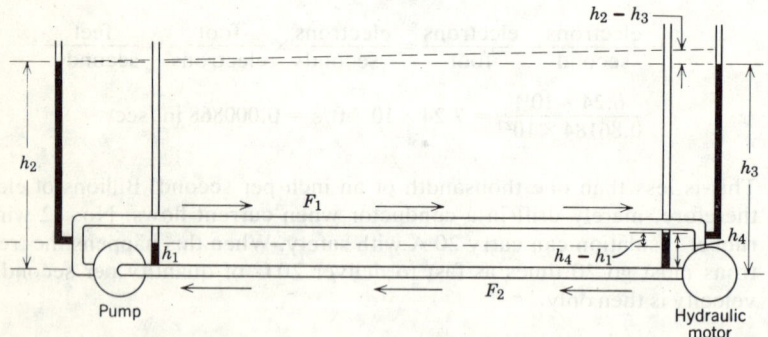

FIGURE 2.3 **Energy transfer from pump to hydraulic motor. Pressure at pump equals sum of three pressure drops: $(h_3 - h_4)$ through motor, $(h_2 - h_3)$ drop along supply pipe, and $(h_4 - h_1)$ drop along return pipe.**

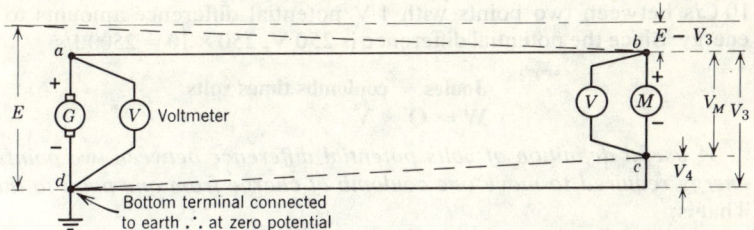

FIGURE 2.4   Energy transfer from generator to motor. Voltmeters measure potential difference (voltage) between terminals of each machine. Voltmeter reading on generator represented by E; voltmeter reading on motor: $V_M = V_3 - V_4$.

through a motor that *resists its passage*, work is done and energy is expended. Heat is developed in the resistance. Energy is transferred by the motor to an outside unit such as a machine.

Now, what about expressing this work, or energy, in volts? *The volt is an electrical unit of work or energy.* When an answer is given in *volts*, the symbol *V* is commonly used. *The fundamental unit* of work in the *MKS system* is the *joule*, whose symbol is *J*.

When the work done in moving 1 C of charge from one point to another in a circuit, or in a conductor, is 1 J, the potential difference between the two points is 1 V.

EXAMPLE 2.2   The electric motor in Fig. 2.4 must be supplied with energy at 240 V between its terminals. What must the voltage at the generator terminals be if a 5-V drop occurs in each of the two transmission wires?

Solution. Assume the negative terminal of the generator to be grounded. The negative terminal (*c*) of the motor has a potential of 5 V above ground (which is at zero potential) because there is a 5-V drop from the motor end of the return wire (point *c*) to the generator end (point *d*). There is also a 5-V drop from *a* to *b*.

$$\text{Potential of point } b = +5 + 240 = +245 \text{ V}$$
$$\text{Potential of point } a = +245 + 5 = +250 \text{ V}$$
$$\text{Voltage at generator terminals} = E = 250 \text{ V}$$

EXAMPLE 2.3   The generator in Fig. 2.4 sends 10 A of current through the circuit. How many joules of energy does it supply each second?

Solution

$$10 \text{ A} = 10 \text{ C/s}$$

10 C/s between two points with 1 V potential difference amounts to 10 J of energy. Since the potential difference is 250 V, $250 \times 10 = 2500$ J/s.

$$\text{Joules} = \text{coulombs times volts}$$
$$W = Q \times V \tag{2.3}$$

A useful definition of volts potential difference between two points is: the energy required to move one coulomb of charge from one point to the other. That is,

$$\text{Volts} = \text{energy per unit charge}$$

$$V = \frac{W}{Q} = \text{joules per coulomb} \tag{2.4}$$

If $W = 1$ J, and $Q = 1$ C, then $V = 1$ V. Employing this relationship, it is very useful to remember that:

Electrical energy expended, or work done, is given by potential difference times charge.

In order to calibrate voltmeters accurately, laboratory technicians use a *standard voltmeter* which has been calibrated previously against a *primary standard voltmeter*. Primary standard voltmeters (those in the U.S. Bureau of Standards, for example) are calibrated with the aid of a *potentiometer*. This instrument of extremely high accuracy has a *standard cell* in its circuit which provides a constant and unvarying electromotive force (EMF). The practical (and very useful) definition of the volt is given in terms of the EMF of the *Weston normal cell*, which is a special electrochemical "battery" that develops an EMF which remains constant. See Fig. 2.5. The EMF of this Weston normal cell is 1.018636 V, and this is determined by the metals and chemicals of which it is made. The fact that this cell is readily reproducible and holds its EMF over very long periods without changing makes it practical to define the volt as 1/1.018636 of its EMF. The international definition of the volt is as follows.

The volt is 1/1.018636 of the EMF of the Weston normal cell at 20°C.

After we define the ohm (Section 3.1) we shall be justified in defining the volt by making use of Ohm's law, thus:

The volt is the potential difference between the ends of a 1-Ω resistance when 1 A of current flows through it.

(The ohm is abbreviated by the Greek omega Ω.)

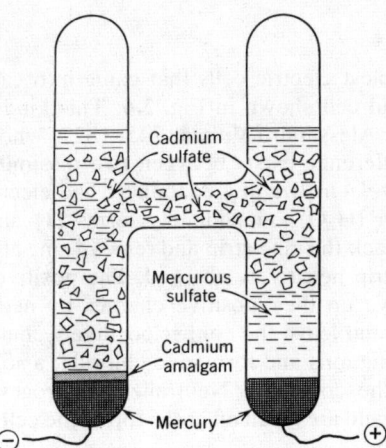

Cadmium sulfate

Mercurous sulfate

Cadmium amalgam

Mercury

⊖   ⊕

FIGURE 2.5   Construction of Weston Normal Cell. Dimensions when encased are approximately $3\frac{1}{2} \times 2\frac{1}{2} \times 1\frac{1}{4}$.

2.5   Sources of Electrical Energy

We now have the question of how electromotive force is produced. Where do the electrons that flow in a circuit come from? They come from what we shall call sources of energy, the most common of which are chemical cells and rotating machines called *electric generators*.

How does an electric cell differ from an electric battery? A battery is a group of two or more cells, but some people call one cell a battery. A small pocket flashlight may be operated by a single cell, but it is customary to call it a battery.

By chemical action, an electric cell produces *an excess* of electrons on an electrode called the cathode, making it negatively charged, and it simultaneously produces a deficiency of electrons on another electrode, called the anode. This is accompanied by the establishment of a potential difference such that the anode is positive with respect to the cathode. When connections are made from the two end points of an external circuit to the electrodes, current flows from the positive terminal (anode) of the cell through the external circuit and back to the negative terminal (cathode) of the cell. The cell thus delivers electrical energy to the external circuit. The cell may be called an *energy converter*, since it converts chemical energy into electrical energy.

2.6   Voltaic Cell

One of the simplest electric cells that came into early use is the copper-zinc hydrochloric acid cell shown in Fig. 2.6. This kind of cell is called a *voltaic cell* named after Alessandro Volta (1745–1827) who first showed that an electric potential difference exists between two dissimilar metals that are both in contact with an *electrolyte* in a container. The electrolyte — hydrochloric acid (HCl) and water (H_2O) — immediately forms H^+ and Cl^- ions. The negative chlorine ions attack the zinc strip and remove zinc atoms that become Zn^+ ions and leave the strip negatively charged. The positive hydrogen ions go to the copper strip, give up their positive charges by neutralizing electrons on the copper surface, and leave the copper positively charged. Some zinc ions combine with chlorine ions and form zinc chloride, a solid material that settles to the bottom of the container. Neutralized hydrogen ions become uncharged hydrogen atoms and are given off at the top of the cell as hydrogen gas.

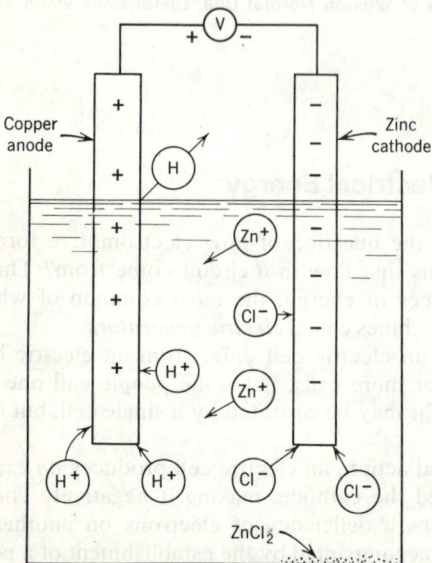

FIGURE 2.6   Simple voltaic cell produces electric potential difference (voltmeter reading) at its electrodes. The electrolyte is a mixture of hydrochloric acid and water. Zinc atoms take + charges with them from the zinc strip and leave it negative.

2.7   The Dry Cell

The most widely known and used electric cell is the so-called dry cell. Although the electrolyte of this cell is a moist paste containing ammonium chloride, zinc chloride, manganese dioxide, and graphite, there is very little liquid in the paste, so the name *dry cell* is not inappropriate. Figure 2.7 shows a cutaway view of one of these cells commonly used in a flashlight. The anode (positive terminal) is a small, flat, circular metal disc at the top center, and the cathode is the metal cylinder that encases the whole cell. Insulating material on the outside covers the cylindrical sides but the bottom is left exposed and serves as the negative terminal. When one of these cells is placed on top of another in a two-cell flashlight case, the bottom of the top cell rests on the center (+) terminal at the top of the bottom cell. Current flows out of the center (+) terminal of the top cell through the bulb of the flashlight, then back through the closed switch and on to the bottom of the bottom cell. The circuit is then completed when the current passes through both cells to the positive electrode where it started.

The chemical action in a dry cell is more complicated than that in a *wet cell*, like the copper-zinc hydrochloric acid type described earlier, but the production of a potential difference results from action of the electrolyte on the zinc. Electrons from the electrolyte are added to the zinc, and hydrogen ions that are supplied by the water in the paste go to the carbon electrode and take off electrons thus making that electrode positive.

As in the case of the copper-zinc wet cell, hydrogen atoms (neutralized hydrogen ions) gather around the positive electrode in the form of gas bubbles. These could surround the electrode with an *insulating layer of gas* and thus impede current flow. This is called *polarization*. To prevent this, manganese

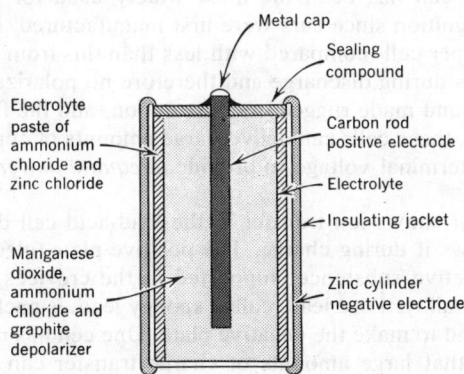

FIGURE 2.7   Internal view of a dry cell used in a flashlight.

dioxide added to an ammonium chloride and graphite paste is placed around the carbon electrode and serves as a *depolarizer*. This prevents the formation of hydrogen gas and, as a result, the cell can be sealed and thus made practical for many uses, since none of the material can spill out of the container. *The EMF generated in a dry cell is 1.5 V*. The copper-zinc cell generates only 1.1 V.

One may think that a carbon-zinc dry cell should last a very long time if left in its packing carton instead of being used. This is not the case because impurities inside the cell contribute to *local action* and, as a result, the *shelf life* is limited.

A manganese-alkaline cell, producing 1.5 V EMF also, is a newer type of dry cell. It lasts more than twice as long, under the same normal operating conditions, as the carbon-zinc cell and has a longer shelf life.

The mercury cell, which is used extensively in hearing aids, is superior to these other two but much more expensive. In the same sizes as the other dry cells, the mercury cell has a very much longer active life. Its positive electrode is zinc, and its negative electrode is graphite and mercuric oxide. The electrolyte is potassium hydroxide. Its EMF is 1.3 V.

The wet cells and dry cells just described, which are not rechargeable, are classified as *primary cells*. A rechargeable cell is one whose chemical action on discharge can be reversed by sending electric current backward through it. The automobile storage battery, made of three or six wet cells, is an example of a rechargeable battery.

2.8 Lead-Acid Battery

The lead-acid cell has been the most widely used for automobile starting, lighting, and ignition since cars were first manufactured. Its important advantages are 2 V per cell (compared with less than this from other kinds of cells), no gas bubbles during discharge and therefore no polarization, so that the cell can be sealed and made rugged in construction, and the fact that it can be recharged. Also, these cells can deliver large amounts of current while maintaining sufficient terminal voltage to provide *adequate power* to start a very cold engine.

Figure 2.8a shows ion transfer in the lead-acid cell during discharge and Fig. 2.8b shows it during charge. The positive plate (electrode) has lead peroxide as its active substance, supported in the crevices of a grid made of a lead-antimony alloy. Pure lead, called spongy lead, is packed into the crevices of a similar grid to make the negative plate. One cell has many plates—7 to 15 or more—so that large amounts of charge transfer can occur. The positive electrode may consist of eight plates solidly connected at an upper corner. In this case, seven negative plates sandwiched in between the positive plates, and

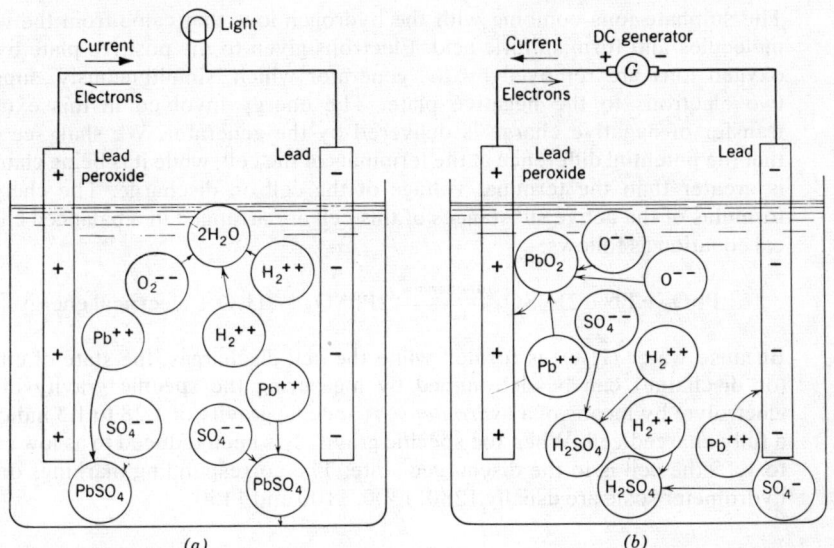

FIGURE 2.8 Chemical action in one cell of a lead, lead peroxide, sulfuric acid battery. (*a*) **Lead-acid cell discharging.** (*b*) **Lead-acid cell charging.**

kept apart by porous insulating sheets called separators. form the negative electrode.

Sulfuric acid *dissociates* into positive hydrogen ions H_2^{++} and negative sulfate ions SO_4^{--} when diluted with water. The sulfate ions. being very active chemically, combine with lead atoms in both plates and form lead sulfate ($PbSO_4$), a neutral molecule. When this happens. each plate receives two electron charges. The positive hydrogen ions that are released by the dissociation process find negative oxygen ions that were released from the breakup of lead peroxide at the positive plate and with these they form a water molecule. A net removal of two negative charges (electrons) from the lead peroxide plate and a deposit of two electrons on the negative plate results from the *internal action*. This continuous supply of electrons to the negative plate and removal of electrons from the positive plate constitute the flow of current *from minus to plus inside the cell*. Remember. the *circulating current* flows from plus to minus *outside* the cell.

While a cell is being charged. the current. supplied by an external source (dc generator). passes *through the electrolyte* from the positive plates to the negative plates. The current breaks up water molecules into H_2^{--} and O^{--} ions. This is called *hydrolysis*. The oxygen ions are attracted to the positive plate where they combine with lead from the lead sulfate that has been separated.

The sulphate ions combine with the hydrogen ions that came from the water molecules and form sulfuric acid. Electrons given to the positive plate by the oxygen ions are removed by the generator which, simultaneously, supplies two electrons to the negative plate. The energy involved in this external transfer of negative charge is delivered by the generator. We shall see later that the potential difference at the terminals of the cell, while it is being charged, is greater than the terminal voltage of the cell on discharge. The chemical formulas of the active substances of this cell are arranged in a balanced chemical equation as follows:

$$PbO_2 + Pb + 2H_2SO_4 \underset{\text{charging}}{\overset{\text{discharging}}{\rightleftarrows}} 2PbSO_4 + 2H_2O + \text{electrical energy}$$

Because water (H_2O) is formed while the cell discharges, the state of charge (or discharge) can be determined by measuring the specific gravity of the electrolyte by means of a *hydrometer*. A specific gravity of 1.28 to 1.3 indicates a fully charged cell. When the specific gravity has been reduced to as low as 1.1 to 1.15 the cell is in the discharged state. The corresponding markings on the hydrometer scale are usually 1280, 1300, 1100 and 1150.

2.9   The Nickel-Iron Cell

The *Edison cell* was developed in 1909[3] by Thomas Alva Edison (United States, 1847–1931). It uses plates of nickel oxide (+) and iron (−) in a solution of caustic potash (potassium hydroxide, KOH) and water. The positive plates are made of nickel oxide in perforated cylindrical capsules, and the negative plates are made of powdered iron held firmly in perforated flat rectangular containers.

These cells produce a voltage (EMF) of about 1.2 V. They weigh only about half as much as lead cells of equivalent energy storage. They have much larger internal resistance, however, which makes them unsuitable for automobile-starting service. They are also bulkier because 10 cells are required to produce 12 V instead of 6 lead cells. Other advantages of the Edison cell are (1) they are not damaged by being left in a fully discharged state for long periods of time, and (2) they are more rugged and will stand more mechanical and electrical abuse than the lead cell.

The state of charge of an Edison cell cannot be determined by the specific gravity of the electrolyte because this does not change. The equation of the chemical action is

[3]See George S. Bryan. *Edison, The Man and His Work*, Garden City, N.Y.: Garden City Publishing Co., page 211. This is a fascinating biography of a man who was America's greatest inventor; 1094 U.S. patents are in his name.

$$\underset{\substack{\text{Positive} \\ \text{plate}}}{Ni_2O_3} + \underset{\substack{21\% \\ \text{solution}}}{2KOH} + \underset{\substack{\text{Negative} \\ \text{plate}}}{Fe^+} \underset{\text{charging}}{\overset{\text{discharging}}{\rightleftharpoons}} \underset{\substack{\text{Positive} \\ \text{plate}}}{2NiO} + \underset{\substack{21\% \\ \text{solution}}}{2KOH} + \underset{\substack{\text{Negative} \\ \text{plate}}}{FeO} + \text{electrical}$$

energy

In the discharging process, each molecule of Ni_2O_3 gives up an ionized atom of oxygen (O^{2-}) which leaves the plate with a corresponding positive charge. This oxygen ion in solution takes the place of a similar one, formed by dissociation of the electrolyte, which moved to the negative (iron) plate and gave up its charge to form a molecule of iron oxide, FeO. The amount of KOH in water solution does not change.

2.10 The Nickel-Cadmium Cell

Although the *nickel-cadmium* storage battery had been known in Europe to be superior to the lead-sulfuric acid battery in several important respects, many years passed before the 1940's (and World War II) ushered the Ni-Cad battery into use in this country. One of its most popular and important features is its ability to last throughout the normal life of the vehicle in which it is used for starting service: automobile, truck, bus, etc. Periods up to twenty years of service can be expected from the general-purpose battery.

The average operating voltage of the nickel-cadmium cell is 1.2 V. The weight of the 100 amp-h, 6-V battery compares well with that of the lead-sulfuric acid battery—about 50 lb. The overall dimensions are about the same as those of the nickel-iron battery—$15\frac{1}{2} \times 16\frac{1}{2} \times 10\frac{1}{2}$ in. The 80 amp-h 12-V nickel-cadmium battery occupies less than half that volume.

To last so long the Ni-Cad cell would be expected to have unusually favorable operating characteristics. It can stand peak rates of discharge and charge up to 20 times the normal operating rate. It can be stored indefinitely in either a discharged or charged state, although the former state is more desirable.

Charged cells lose charge when stored, and the loss occurs more rapidly at higher storage temperatures. At room temperature a Ni-Cad cell will lose 40 percent of full-charge capacity in 3 months and 60 percent in 6 months; at 125°F it will lose 60 percent in one month, and 90 percent in 3 months. To return the cells to service it is only necessary to apply a normal charge. There is no permanent damage, such as sulfation, which occurs in a lead-sulfuric acid cell that stands for months in a discharged state.

Ni-Cad batteries can be 90 percent recharged in an hour, although the charging time may be adjusted for any value between 6 and 16 or more hours. The ampere-hour input required to recharge the battery is 120 percent of its rated capacity.

These batteries have very low internal resistance and therefore maintain their terminal voltage more nearly constant than do other kinds of batteries. This flat voltage-time characteristic curve is an asset. The range of operating temperatures is $-20°F$ to $+140°F$, and it may be extended to $-50°F$ by using a special electrolyte.

The chemical reactions of the Ni-Cad cell for charge and discharge may generally be written as follows:

$$2Ni(OH)_3 + Cd \xrightleftharpoons[\text{charge}]{\text{discharge}} 2Ni(OH)_2 + Cd(OH)_2 + \text{electrical energy}$$

When discharging, two negative hydroxyl ions $(2OH)^-$ combine with Cd and release two electrons $(2e^-)$. One of the electrons is used in forming a hydroxyl ion (OH^-) and changes a molecule of $Ni(OH)_3$ into $Ni(OH)_2$. The other electron is released to the load circuit.

The positive-plate active material is nickel hydroxide, $Ni(OH)_2$, and the negative-plate active material is cadmium hydroxide $Cd(OH)_2$. These are changed into $Ni(OH)_3$ and cadmium in the charging process. The electrolyte (used for conducting purposes only) is potassium hydroxide, KOH. Its specific gravity (as in the case of the nickel-iron cell) does not change with the state of charge or discharge of the cell.

Small rechargeable thin sintered-plate nickel-cadmium cell batteries are used in cordless electric appliances such as shavers, photography equipment (movie cameras, photoflash and photoflood lights), and in aerospace applications. A popular size is the 1.2 V, 4 A-h battery which has a 1-h discharge rate. It is a spiral-plate type with a recommended constant-current charge rate of 400 mA for 14 or 16 h.

2.11 Solar Battery

Certain materials will emit electrons when sunlight shines on them. Selenium and cesium do this in their pure state. Silicon does it also when treated properly to form an N-type material with a P-type material layer in contact with it.[4]

The construction process in the manufacture of a silicon photo-voltaic cell has three principal steps. First an ingot of silicon crystal is "grown" in a furnace where a specific amount of arsenic is introduced as an impurity into the "melt" of silicon. This makes it an N material. Arsenic is called a donor impurity because each of its atoms provides an extra electron which is free to move from atom to atom in the silicon and form *conduction current*. Such a situation exists in each of the many places in which an arsenic atom displaces a silicon atom. There is one more electron in the donor atom than is needed when the displacement occurs.

The second step in the construction process is the slicing of the cooled ingot

---

[4]Semiconductor materials that readily furnish electrons in solid-state electronic circuits are classified as N type and as *donors*. Others, which readily accept electrons, are classified as P type and as *acceptors*.

of N-type material into individual small slabs of a desired thickness. These become cells when, in the third step, they are diffused with boron from the gas of a boron compound.

Boron is an acceptor impurity. When one of its atoms displaces a silicon atom, there is a deficiency of one electron. Hence the material becomes P type and it will accept free electrons that come near it in the crystal structure.

The boron impregnation is removed from the bottom and from all sides of the cell, leaving only the top layer as P-type material. When sunlight shines on this thin layer and penetrates into the N material just below it, the free electrons receive energy and move across the P-N junction into the P material and thus constitute current. Figure 2.9a shows the cell construction and b shows the stacking method (called shingling) by which series connection is effected.

The operating voltage of one solar cell is about 0.39 V and the current may be between 30 and 40 mA. The maximum output with the sun directly overhead on a clear day is 8 or 9 mW/cm². The operating efficiency is about 10 percent. On satellites the power from the solar cells charges storage batteries (often they are nickel-cadmium), and storage batteries regulate the voltage. The lifetime of solar cells is estimated to be in thousands of years. They do not deteriorate when not in use.

Other uses of solar cells are: to provide power for transistor portable radios, to provide a constant charge for tiny nickel-cadmium batteries in spectacle-frame hearing aids, to provide power for clocks and other devices such as aperture control for movie cameras, microwave relay stations, flasher beacons, and control circuits in robot weather stations. They are also used as infrared radiation sensors and detectors of fire or flame.

2.12   Fuel Cell

The principle of the fuel cell was demonstrated over a century ago[5] by English scientist Sir William Grove. He fed hydrogen and oxygen into an arrangement of electrodes and an electrolyte whereby electrons were removed from an

[5]W. R. Grove, *Philosophical Magazine* III, **14**, 129 (1839).

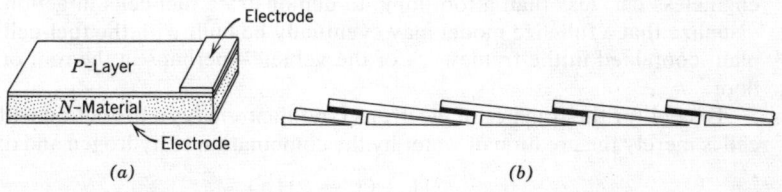

(a)   *(b)*

FIGURE 2.9   (a) Solar-cell construction. (b) Shingle-type stacking for convenient electrical connection. Wedges are used for support.

electrode connected to one end of a closed circuit, passed through the electrolyte, and deposited on the other electrode which was fastened to the end of the closed circuit. There was no organized effort to develop this idea until in relatively recent years, owing to the more readily available and cheaper sources of energy such as coal, oil, gasoline, water power, and storage batteries.

Fuel cells differ from storage batteries in that their electrode materials are not changed in chemical composition and, therefore, they do not require recharging. They possess advantages over other kinds of energy converters (engines, turbines, generators): they have no moving parts except certain control components, they operate at room temperature, produce no obnoxious fumes, use fuels that are not difficult to procure, are silent and more efficient than any other known type of energy converter. The electrochemical action in some cells yields water that may be used for drinking. This is an important reason for the use of fuel cells in space-travel applications, as are the high power-to-weight and power-to-volume ratios.

The fuel cell has an available practical efficiency of 70 percent as compared with the following: large power stations, 40 percent, gasoline and diesel engines, 25–35 percent; thermoelectric and solar sources, 15 percent.

In the conversions of heat energy of a fuel into mechanical or electrical energy, only a small fraction of the heat can be converted. Owing to the fact that fuel-cell operation does not involve the "heat cycle" — a basic principle in thermodynamics more often called the *thermal cycle* — the fuel-cell efficiency can be much higher than that of heat engines. About twice as much useful energy may be extracted from a pound of fuel in a fuel cell as when it is burned in the most efficient heat engine of today.

When a load is not connected (circuit is open), no energy is being consumed. The fuel cell delivers only direct current, which is a disadvantage where alternating current is needed. In such a case an *inverter* is required to convert the d-c output into a-c.

At present, fuel cells are not used to generate large amounts of power. Some systems require extensive installation of auxiliary equipment for their operation. The cost becomes prohibitive except for certain applications in military and industrial operations.

A group of automotive engineers are reported to have built a model of an engineless car, less than a foot long, to demonstrate fuel cells in action. They visualize that a full-size model may eventually be built with the fuel-cell power plant contained in the framework of the vehicle — perhaps in the roof or in the floor.

Except for the transfer of electrons (and their energy), the action within the cell is merely the creation of water by the combination of hydrogen and oxygen:

$$2H_2 + O_2 \rightarrow 2H_2O$$

A simplified sketch of a highly efficient fuel cell that uses a *solid-membrane*

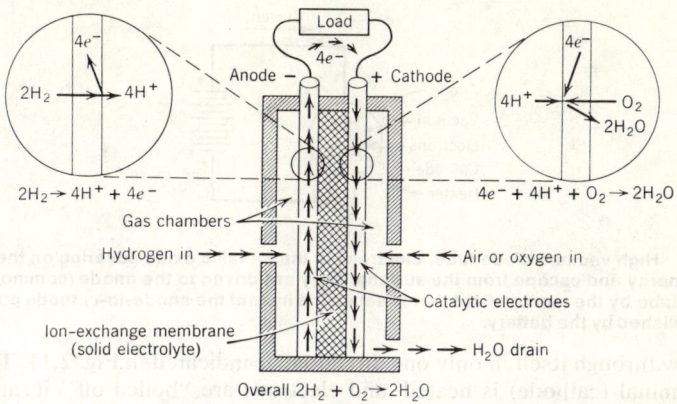

FIGURE 2.10   Principle of the Gemini fuel cell is based on the use of a solid-membrane electrolyte. It keeps separate the oxygen and hydrogen entering opposite sides of the cell ... but permits flow, or exchange, of hydrogen *ions*. This sets up a chain of chemical reactions that produces electricity *directly* ... without steam, combustion, machinery, noise, vibration, or toxic exhaust gases. A vital by-product is water. (Courtesy of General Electric Co., Schenectady, N.Y.) Fuel cells similar to the Gemini cell are used in the Apollo space flights.

electrolyte is shown in Fig. 2.10. It keeps separate the oxygen and hydrogen entering opposite sides of the cell, but permits exchange, or flow, of hydrogen ions.

Pairs of atoms of the hydrogen fuel give up their four electrons at the anode and the *four positive ions* thus formed pass through the membrane to the cathode. Meanwhile the *four electrons* pass through the external circuit to the cathode. This of course, means current flows through the load.

2.13   Current Flow in Electronic Devices

Experimenters in the 19th century studied the passage of electric current through evacuated glass tubes. Roentgen[6] discovered x-rays in 1895 while doing this. Edison[7] noticed, as early as 1883, that electrons given off by a hot filament, mounted in an evacuated glass bulb, pass over to an identical (but cold) filament mounted in the same space. He had connected a sensitive current-detecting instrument (galvanometer) outside between the cold filament wire and the positive terminal of the battery he used to heat the first filament. This discovery led to the development of the high-vacuum diode.

The *high-vacuum diode*, a two-electrode vacuum tube, allows current to

[6]Wilhelm Konrad Roentgen; Germany, 1845–1923.
[7]Thomas Alva Edison; United States, 1847–1931.

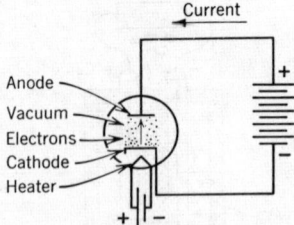

FIGURE 2.11 High vacuum diode tube. Electrons in the metallic oxide covering on the cathode are given heat energy and escape from the surface. They are driven to the anode (commonly called the *plate*) of the tube by the electric field that is set up the instant the anode-to-cathode potential difference is established by the battery.

flow through itself in only one direction, as indicated in Fig. 2.11. The negative terminal (cathode) is heated, and electrons are "boiled off" it, and enter the space very near to its surface. When the positive terminal (anode or plate) is connected to the positive terminal of a source, such as a battery, and the cathode is connected to the negative terminal of the source, electrons pass readily through the vacuum space from cathode to anode within the tube. The transfer of charge around the circuit constitutes current flow. The charge carriers in the vacuum are electrons.

A *gas triode* tube, the internal structure of which is illustrated in Fig. 2.12, may contain mercury vapor or an inert gas such as argon, neon, or helium. Electrons emitted from the cathode acquire high velocities on their way to the anode, and many of them strike gas atoms and knock electrons from them, leaving positive ions of the gas to find their way to the cathode of the tube. These "liberated" electrons may acquire enough energy (high enough velocity) to knock other electrons out of other gas atoms. In this manner many more

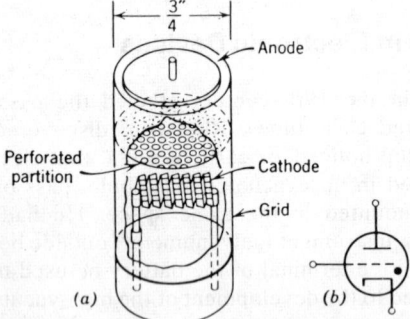

FIGURE 2.12 A gas triode tube, called a thyratron, has the structure shown (a) inside a glass or metal envelope. Mercury vapor or some other gas fills the tube. The grid is a metal cylinder with a horizontal perforated partition located between the cathode and the anode. The symbol used in circuit diagrams (schematic is shown in (b).

electrons per second reach the anode of the tube than leave the cathode per second. The positive ions collect electrons from the cathode at the instant they hit it and become neutral gas atoms again. The net result is a transfer of *negative charge* through the gas tube from cathode to anode, and thus a current flow is established through the tube from anode to cathode (+ to −). Fluorescent lights and so called neon signs (some of which do not contain neon gas, but helium gas or mercury vapor instead) conduct electricity in this way.

Electron motion accounts for current flow in semiconductors also. In a germanium or silicon diode (semiconductor *P–N* junction) electrons pass readily from the *N* type material to the *P* type. On the *N* side of the junction there is a high concentration of free electrons, while on the *P* side there is a great deficiency of electrons; this latter situation is described as a high concentration of *free holes*[8] which are vacancies in *valence bonds* between atoms of the crystal. Electrons cross the boundary region from the *N* material to the *P* material in diodes. In junction transistors there are two boundary regions as suggested by the symbol *PNP* for the transistor that has *N*-type material sandwiched between two layers of *P* type, or the symbols *NPN*, which has a layer of *P*-type material between two of the *N* type. Electrons cross these boundaries from the low-potential to the high-potential side.

2.14 Current Direction

In this book, as is customary in almost all electrical and electronics textbooks, *the positive direction of current* in a circuit, part of a circuit, or a device will be from + to − outside a *source of energy* and from − to + *inside the energy source*. We shall use *E* to represent this potential difference between terminals of a generator, battery, or other energy source and *V* to represent *potential drop*, often called *voltage drop*, between two points of an energy-receiving unit or circuit branch.

It is always true that current flows from the point of higher potential toward the point of lower potential in an external circuit. In a source, energy is utilized to build up a potential difference. *Instantaneous current in a source* flows from the point of lower potential to a point of higher potential because electrical energy is being made available for use at places external to the source. We must be reminded of the water-circuit analogy. A pump (source) forces water from the lower pressure point toward the higher pressure point, often in an elevated storage tank, so that the energy can be received by a "load" such as the extensive water supply system of a city through which the water flows from points of higher pressure to points of lower pressure.

[8]For an understandable discussion of the germanium atom, covalent bonding, and holes, see H. Alex Romanowitz and Russel E. Puckett. *Introduction to Electronics*, New York: John Wiley & Sons, Inc. 1968, p. 69 ff.

2.15   Review Questions

1. A simple electric circuit in an automobile consists of a battery as the source, a switch on the dash, a light in the roof of the car, and connecting wires. Why must there be only one amount of current in the circuit?

2. The ampere is the most commonly used unit of electric current. Is it a quantity of electricity, a quantity per second, or a velocity, perhaps in inches per second?

3. Distinguish between *rate of flow of charge* and *velocity of a charged particle* such as an electron. Give a practical unit of each.

4. Give a practical definition of the coulomb.

5. Define the ampere.

6. About how fast (feet per second) do the electrons travel parallel to the direction of the current in a wire that carries 1 A of current? Is feet per second 12 times inches per second or $\frac{1}{12}$ inches per second?

7. Considering that electrons drift at such slow velocities in a wire carrying current, how do we account for the fact that when a switch is closed in an electric generating plant the lights in the streets of a city light up almost instantly? (*Hint.* Are the free electrons in the metal of the lamp filaments ready to go?)

8. A pump feeds water through a pipe to a water wheel 100 ft away. The water that flows out of the wheel returns through another pipe to the pump. Does all the energy given to the water arrive at the pump? Why? What else might cause an energy loss in this system? Where is that lost energy *put into* the water? What could cause energy loss in wires that connect an electric generator to its load, which may be a system of lights or a motor some distance away?

9. Define the volt (a) using only words, (b) in a way that requires words and a certain numerical fraction.

10. State Ohm's law.

11. What three elements are required to construct a voltaic cell?

12. Assume you have one strip of each of three different kinds of metal and one acid solution that acts chemically on all three. How many different voltaic cells could you make with these four things?

13. How do the electrodes of a copper, zinc, hydrochloric acid cell acquire their electric charges? Which is the positive electrode?

14. What are the materials used in the construction of the most popular type of cell? Which electrode is positive?

15. What is the EMF of the (a) common dry cell? (b) lead-acid storage cell? (c) Edison nickel-iron cell? (d) nickel-cadmium cell? (e) mercury cell?

16. What are the materials of the + plates, − plates, and electrolyte of the conventional automobile battery?

17. Why does the specific gravity of the lead-acid battery decrease as the battery discharges?

18. How can one tell the conditions of charge, partial discharge, or practically total discharge of a lead-acid battery?

19. How does the Edison cell differ from the lead-acid cell? Name at least three differences. The Edison cell is also lighter in weight. It would occupy more space than the lead-acid cell if used in an automobile. Why?

20. Could you tell the condition of charge of an Edison cell by means of a hydro-meter? Why?

21. Name one big advantage of the nickel-cadmium battery when it is compared with other kinds of storage battery.

22. What are the ingredients of one type of silicon solar cell? What is the approximate operating voltage of a silicon solar cell?

23. Name some useful characteristics of a fuel cell.

24. What is the accepted direction of current flow in a vacuum diode tube when it is conducting current?

25. In a *P-N* junction diode, what is the direction of electron flow across the boundary region between them? How do the charge concentrations on the two sides of the boundary differ?

26. In a junction transistor in a circuit, one side of a boundary region is higher in potential than the other side. What is the direction of electron flow across the boundary? Does your answer agree with the statement that electrons *flow up a potential hill*?

2.16 Problems

Group A

1. In a simple series of circuit 50 C of electrical charge circulate at a constant rate every 5 s. How much current, in ampere units, flows in the circuit?

2. An electronic tube carries 1.25 mA (milliamperes) of current. How long does it take for a coulomb of charge to pass through the tube?

3. The smallest wire listed in the American Wire Gage table for copper conductors is No. 40 which is 3.1 mil (0.0031 in.) in diameter. Calculate the number of free electrons in 1 ft of this wire.

4. Two amperes flow through a silver plating bath for 1 h. How many coulombs pass through? How much silver is deposited on the cathode?

5. One thousand tons of water pass over a falls in 1 min. (a) Calculate the strength of the current just above the falls in gallons per second using 8.3 lb/gal as the density of the water. (b) How many gallons per second strike the rocks at the bottom of the falls? (c) If the same number of electrons per second flowed through a wire, what would the current be in amperes?

6. Two wires having a resistance of 0.1 Ω each supply electrical energy to a light bulb from a battery. The terminal voltage of the battery is 12.3 V, the resistance of the light is 4.72 Ω and the current is 2.5 A. Sketch the circuit. Use Ohm's law to calculate the potential difference between the ends of each of the supply wires (this is the voltage drop through each wire) and also the potential difference across the light bulb. What is the sum of the three voltage drops from the + terminal of the battery around the circuit to the − terminal?

7. Assume that aluminum is only 60 percent as good a conductor of electricity as copper. How many free electrons per cubic inch would you expect it to have?

8. It is desired to plate a silver knife with 8 g of pure silver using the electroplating bath shown in Fig. 2.2. A 100-W lamp carrying 5/6 A is connected in series with the bath. How long will it take?

9. An electric toaster has 10.91 Ω resistance when hot. How much current does it take on a 120-V line?

Group B

10. Assume that the dome light in Fig. 2.1 takes 1.5 A of current and that the copper wires connecting the light, switch, and battery are No. 20 AWG size (0.032 in. diameter). Calculate the electron drift velocity.

11. How many joules of energy are required to move 10 C between two points that have a potential difference of 100 V between them?

12. How much current flows through a load in a circuit if it receives energy at the rate of 50 J/s while the potential difference across the load is 120 V?

13. A gas tube used in industrial electronics has a heater coil requiring a potential difference of 5 V. Its resistance when hot is 0.4 Ω. How many joules of energy per second are supplied to the coil?

14. A hoist lifts a load weighing 1350 lb a distance of 20 ft, in 45 s. Calculate the rate of doing work in joules per second. One newton times one meter equals one joule.

15. The bulb in a flashlight carries 200 mA which is supplied by a 3-V battery. At what rate is chemical energy being converted into electrical energy in the battery?

16. A battery delivered 0.6 J of energy to a heater unit that operates at 12 V. How many electrons drifted through the heater during that time?

17. A chemical mixture must absorb 150 J of energy from a heating coil that operates at 6 V. How many coulombs of charge must be passed through the heater?

18. How much current must be passed through the heater of Problem 17 if the energy is to be supplied in 1 min?

19. A high-vacuum diode tube carries a plate current of 1.2 mA when the plate-to-cathode voltage is 150 V. How many electrons per second arrive at the plate?

20. Assume that each electron in Problem 19 starts from the cathode at rest (zero initial velocity). How many joules of energy does it have when it arrives at the plate?

CHAPTER THREE

Ohm's Law, Resistance

3.1 What is Resistance?

In Section 1.6, when we were studying the electric circuit, we mentioned the first law to be learned in studying the behavior of electric circuits, namely, *Ohm's law*. When I once asked my boss on a co-op job, "What determines how much current flows in an electric light?" The answer was "The amount of current will be as much as the resistance of the light will allow."

All electrical conductors have resistance. When heat or light is to be produced, conductors with high resistance are used in the manufacture of the heater and the light filament. When electrical energy is to be supplied to a heater or to a light, the conductors that carry current to the heating unit or the light should have *as little resistance as possible*. For example, wire that electricians use in wiring houses should not get warm when fully loaded with current that they supply to lights and appliances, so the wiring code specifies wire sizes and safety fuse (or circuit-breaker) capacities.

The opposition offered by a conductor to the flow of electric current is called its resistance.

It seems logical, that when electrical potential difference between the two end points of a conductor is constant the current that will flow in the conductor will be small if the resistance (of the conductor) to the current flow is large, and the current will be large if the resistance to its flow is small?

About 150 years ago the German scientist Georg Simon Ohm demonstrated by experiment that there is a constant proportionality between the potential difference applied to the ends of a conductor and the amount of current that flows in it. When he doubled the potential difference he found that the current was doubled, when he tripled the potential difference the current was tripled, and so on. He called the opposition offered by the conductor to the flow of current its *resistance*. He formulated his conclusions in what soon became known as *Ohm's law*:

$$I = \frac{E}{R}, \qquad \text{Amperes} = \frac{\text{Volts potential difference}}{\text{Ohms}} \qquad (3.1)$$

As indicated before, we shall use the letter E to represent potential difference *at the two terminals of a source* such as a battery or a generator, and the letter V to represent the potential difference (often called *voltage drop*) across a unit (called the load) that receives the energy. In using Ohm's law for the voltage-current relation *in a load*, or in part of a circuit that is receiving energy, we shall write $I = V/R$.

The ohm is the fundamental unit of resistance.

The international definition of the ohm specifies it as the resistance of a column of mercury at the temperature of melting ice (0°C), 14.4521 g in mass, of a constant cross-sectional area, and 106.300 cm in length.

Let us consider a single unit in which 1 A of current flows, and assume that in measuring the potential difference between the units terminals (voltage drop) we find it to be 1 V. The Ohm's law relation [Equation (3.1)] gives

$$R = \frac{E}{I}; \qquad R = \frac{1\ V}{1\ A} = 1\ \Omega$$

We may define the ohm as *the resistance of a conductor which will allow exactly 1 A to flow in the conductor when exactly 1 V EMF is applied between its terminals.*

3.2 Practical Uses of Resistance

We mentioned the use of resistance wire in heater units. Some electric heater units that are familiar to us are: the toaster that has resistance grids made of ribbonlike flat wire, the spiral grids of the heating elements on the kitchen range, the heating coils in coffee percolators, soldering irons, and heating pads. Incandescent light bulbs have filaments made of high-resistance wire.

Because the resistance of a conductor offers opposition to the flow of current, it is evident that if we should add a resistance unit to a closed circuit so that the circuit current would have to flow through the added resistance and also through all of the rest of the circuit, the current would be reduced in amount, that is, in ampere value.

EXAMPLE 3.1   The light bulb in Fig. 3.1 will receive too much current if it is connected to the 12-V battery because it was manufactured to operate on only 6 V. But no 6-V source is readily available. We can use the 12-V source if we connect a *current-limiting resistance R* in series with the 6-V bulb, provided R has the proper ohms value. If the ohms value of R is not large enough, the bulb will get too much current, and even though it would give off more light than it was made to provide, it would burn out too soon. If the ohms value of R is too large, the light bulb would not get its proper amount of current and it would not give off as much light as it should. In this case it would have longer than normal life, however.

Let us assume that the 6-V light bulb takes 0.2 A in normal use on 6 V. If a 12-V source is used, the extra 6 V must be taken up by the resistance R while R is carrying the 0.2 A needed by the bulb. Since there must be a 6-V drop across R, the resistance of R must be

$$R = 6/0.2 = 30 \, \Omega$$

This, of course, is the hot resistance of the light bulb.

Resistance units, such as R in Example 3.1, are used extensively to control current to a desired value. I decided to install an electric buzzer that would warn me that the emergency brake on my car was set, in the event I forgot to release it when I started the car. I could not find a 12-V buzzer so I had to purchase one that must operate on 6 V. It was a simple matter to measure the resistance of the coil of the buzzer and to procure a resistance unit like R in the above example. I connected the resistance unit in series with the buzzer, just as R is in series with the 6-V light bulb, and it works fine. We shall learn how to measure resistance in a later chapter.

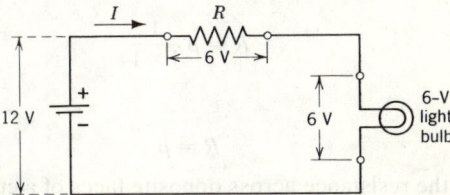

FIGURE 3.1   Six-volt light bulb gets correct amount of current from 12-V source when R has the same ohms value as the filament has when supplied with 6 V.

3.3   What Determines the Resistance of a Conductor?

Imagine that we have two coils of wire, one made of copper wire and the other of iron wire, but they are identical in all other respects. The total length of wire on each coil is the same, the cross sections of the wires are the same, and the temperatures are the same. The resistance of the copper-wire coil will be smaller than the resistance of the iron-wire coil because iron offers more resistance than does copper to the flow of current. The *resistivity* of iron is larger than that of copper. We shall learn about resistivity later in this chapter. *Specific resistance* is another name for resistivity.

The resistance of a metallic conductor depends on the kind of metal or alloy of which it is made.

The resistance of a conductor *increases* when its length is *increased, and in direct proportion*. One hundred feet of No. 12 copper wire has twice the resistance that 50 ft has.

The resistance of a conductor *decreases when its cross-section area is increased*. For example, two conductors are made of the same metal and have the same length. Conductor *A* has twice the cross section area of conductor *B*. The resistance of *A* is *one half that of B*.

We can combine the effects on resistance of changing the length, the cross-section area, and the material of a conductor and devise a formula that we can use to calculate its resistance. We shall use the Greek letter *rho*, ρ, *to represent resistivity*, or specific resistance.

The resistance of a given conductor, whether it be a straight wire, a coil or grid of wire, or a heavy cable or bus bar (a large bar of copper with rectangular cross section), may be computed by means of the equation

$$R = \frac{\rho L}{A} \text{ ohms} \tag{3.2}$$

in which *L* is the length, *A* is the cross-section area, and ρ is the resistivity.

Evidently the *units* of these three quantities must be compatible in order that *R* will come out in ohms. Let us first consider the meaning and units of ρ, the resistivity. If *L* is 1 cm, *A* is 1 cm square, the conductor will be a 1 cm cube, and

$$R = \rho \frac{1}{1 \times 1}$$

or

$$R = \rho$$

and ρ must be the resistance across opposite faces of a cubic centimeter of the material. This is a very small amount of resistance. For copper it is $1.724 \times 10^{-6} \ \Omega$ at 20°C. This is only $1/580,000 \ \Omega$.

3.4 Circular Mil

Although we, in the United States, may adopt the metric system in the next 10 or 15 years, we shall be obliged to use feet and inches in our calculations and measurements meanwhile. Therefore, the use of feet in expressing the length of a conductor has led to the use of a very convenient unit, the *circular mil*, to express *cross-sectional area*. See Fig. 3.2. This new unit gives us the convenience of not having to bother with $\pi/4$ in expressing the area of a circular cross section as $(\pi/4)d^2$, and also of not having to use large negative exponents in the process. For example, if we use L feet and then are obliged to use *square feet* for A, a popular size conductor such as No. 12 round wire would have a square-foot area of

$$\frac{\pi}{4} \times \frac{0.081^2}{144} = 35.80 \times 10^{-6}.$$

A circular mil is the area of a circle 0.001 in. (1 mil) in diameter.

In order to express the *cross-section area* of a circular conductor *in circular mils, we simply multiply the inch value of the diameter by 1000 and square the result.* Keep this in mind because we are going to use this information in the next example after we discuss the units the resistivity, ρ, must have when we express A in *circular mils* in Equation (3.2).

Because R must be in ohms, we can determine the units of ρ by solving for it in Equation 3.2, thus

$$\rho = \frac{RA}{L} = \frac{\text{ohms} \times \text{circular mils}}{\text{feet}}$$

If A is 1 circular mil (cmil) and L is 1 ft, we have a conductor 1 ft long and 1 cmil in area. In this case

$$\rho = R$$

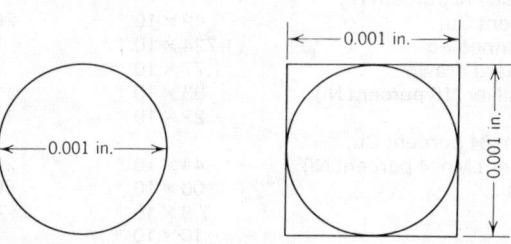

(a)   (b)

FIGURE 3.2 Circular mils and square mils. (a) A circle 1 mil in diameter. Its area in circular measure is 1 cmil. (b) A square 1 mil on a side. Its area in square measure is 1 mil². Its area in circular measure is $4/\pi$ cmil.

so that *the resistivity (ρ) is the resistance of a conductor 1 ft long and 1 cmil in cross-section area.* We say that the units of ρ in the special wire-measure system is ρ → *ohms circular mils per foot.* Remember that *the area of a round wire in circular mils is obtained simply by expressing its diameter in mils and squaring it.* That is, a wire 0.15 in. in diameter has a circular mil area of $150^2 = 22,500$ cmil. How much easier this is than to use square inches for area and have to compute $(\pi/4)(0.15)^2 = 0.01767$ in.². The value of ρ for annealed copper is 10.37 Ω cmil/ft at 20°C.

EXAMPLE 3.2 Compute the resistance at 20°C of a coil of annealed copper wire that contains 250 ft of wire 0.0201 in. in diameter. This is No. 24 in the American Wire Gage table.

Solution.

The cross-section area in circular mils is

$$A = (0.0201 \times 1000)^2 = 404.01 \text{ cmil}$$

$$R = \frac{\rho L}{A} = \frac{10.37 \times 250}{404.01} = 6.42 \ \Omega$$

Values of resistivity of different materials are given in Table 3.1 and the American Wire Gage data for solid annealed copper wire are given in Table 3.2.

TABLE 3.1 **Resistivities of Wire Materials**

Material	Resistivity, , ohm-m at 20°C	Resistivity, ρ, ohm-circular mils per foot at 20°C
Aluminum	2.83×10^{-8}	17
Brass	7×10^{-8}	42
Constantan (40 percent Ni, 60 percent Cu)	49×10^{-8}	295
Copper, annealed	1.724×10^{-8}	10.37
Copper, hard drawn	1.77×10^{-8}	10.65
German silver (18 percent Ni)	33×10^{-8}	14
Lead	22×10^{-8}	132
Manganin (84 percent Cu, 12 percent Mn, 4 percent Ni)	44×10^{-8}	265
Nichrome	100×10^{-8}	602
Nickel	7.8×10^{-8}	47
Platinum	10×10^{-8}	60
Silver	1.64×10^{-8}	9.9
Tungsten	5.5×10^{-8}	33

TABLE 3.2   **Standard Annealed Copper Wire, Solid**[a]
American Wire Gage (B. & S.). English Units

Gage Number	Diameter, mils	Cross section		Ohms per 1000 ft.		Ohms per Mile 25°C (=77°F)	Pounds per 1000 ft
		Circular Mils	Square Inches	25°C (=77°F)	65°C (=149°F)		
0000	460.0	212,000.0	0.166	0.0500	0.0577	0.264	641.0
000	410.0	168,000.0	0.132	0.0630	0.0727	0.333	508.0
00	365.0	133,000.0	0.105	0.0795	0.0917	0.420	403.0
0	325.0	106,000.0	0.0829	0.100	0.116	0.528	319.0
1	289.0	83,700.0	0.0657	0.126	0.146	0.665	253.0
2	258.0	66,400.0	0.0521	0.159	0.184	0.839	201.0
3	229.0	52,600.0	0.0413	0.201	0.232	1.061	159.0
4	204.0	41,700.0	0.0328	0.253	0.292	1.335	126.0
5	182.0	33,100.0	0.0260	0.319	0.369	1.685	100.0
6	162.0	26,300.0	0.0206	0.403	0.465	2.13	79.5
7	144.0	20,800.0	0.0164	0.508	0.586	2.68	63.0
8	128.0	16,500.0	0.0130	0.641	0.739	3.38	50.0
9	114.0	13,100.0	0.0103	0.808	0.932	4.27	39.6
10	102.0	10,400.0	0.00815	1.02	1.18	5.38	31.4
11	91.0	8,230.0	0.00647	1.28	1.48	6.75	24.9
12	81.0	6,530.0	0.00513	1.62	1.87	8.55	19.8
13	72.0	5,180.0	0.00407	2.04	2.36	10.77	15.7
14	64.0	4,110.0	0.00323	2.58	2.97	13.62	12.4
15	57.0	3,260.0	0.00256	3.25	3.75	17.16	9.86
16	51.0	2,580.0	0.00203	4.09	4.73	21.6	7.82
17	45.0	2,050.0	0.00161	5.16	5.96	27.2	6.20
18	40.0	1,620.0	0.00128	6.51	7.51	34.4	4.92
19	36.0	1,290.0	0.00101	8.21	9.48	43.3	3.90
20	32.0	1,020.0	0.000802	10.4	11.9	54.9	3.09
21	28.5	810.0	0.000636	13.1	15.1	69.1	2.45
22	25.3	642.0	0.000505	16.5	19.0	87.1	1.94
23	22.6	509.0	0.000400	20.8	24.0	109.8	1.54
24	20.1	404.0	0.000317	26.2	30.2	138.3	1.22
25	17.9	320.0	0.000252	33.0	38.1	174.1	0.970
26	15.9	254.0	0.000200	41.6	48.0	220.0	0.769
27	14.2	202.0	0.000158	52.5	60.6	277.0	0.61
28	12.6	160.0	0.000126	66.2	76.4	350.0	0.484
29	11.3	127.0	0.0000995	83.4	96.3	440.0	0.384
30	10.0	101.0	0.0000789	105.0	121.0	554.0	0.304
31	8.9	79.7	0.0000626	133.0	153.0	702.0	0.241
32	8.0	63.2	0.0000496	167.0	193.0	882.0	0.191
33	7.1	50.1	0.0000394	211.0	243.0	1,114.0	0.152
34	6.3	39.8	0.0000312	266.0	307.0	1,404.0	0.120
35	5.6	31.5	0.0000248	335.0	387.0	1,769.0	0.0954
36	5.0	25.0	0.0000196	423.0	488.0	2,230.0	0.0757

TABLE 3.2 (*cont.*)

Gage Number	Diameter mils	Cross section		Ohms per 1000 ft.		Ohms per Mile	Pounds per 1000 ft
		Circular Mils	Square Inches	25°C (=77°F)	65°C (=149°F)	25°C (=77°F)	
37	4.5	19.8	0.0000156	533.0		2,810.0	0.0600
38	4.0	15.7	0.0000123	673.0	776.0	3,550.0	0.0476
39	3.5	12.5	0.0000098	848.0	979.0	4,480.0	0.0377
40	3.1	9.9	0.0000078	1,070.0	1,230.0	5,650.0	0.0299

*From *Circ.* 31, U.S. Bureau of Standards

Note 1. The *fundamental resistivity* used in calculating the tables is the International Annealed Copper Standard, *viz.*, 0.15328 ohm (meter, gram) at 20°C. The *temperature coefficient* for this particular resistivity is $\alpha_{20} = 0.00393$, or $\alpha_0 = 0.00427$. The *density* is 8.89 grams per cubic centimeter.

Note 2. The values given in the table are only for annealed copper of the standard resistivity. The user of the table must apply the proper correction for copper of any other resistivity. Hard-drawn copper may be taken as about 2.7 percent higher resistivity than annealed copper.

Note 3. Pounds per mile may be obtained by multiplying the respective values above by 5.28.

3.5 Effect on Resistance of Changing Dimensions or Materials

Example 3.3   A copper conductor has 0.01 Ω resistance. What is the resistance of another copper conductor that has half the diameter and three times the length?

Solution

$$R_1 = \frac{\rho L_1}{d_1^2} = 0.01 \qquad R_2 = \frac{\rho 3 L_1}{d_1^2/4}$$

Dividing the second equation by the first, ρ and L_1 cancel out and we have

$$\frac{R_2}{0.01} = \frac{3}{d_1^2/4} \times \frac{d_1^2}{1} = 12$$

from which

$$R_2 = 0.12 \ \Omega$$

EXAMPLE 3.4   A copper conductor 100 ft long and 0.025 in. in diameter has 1.66 Ω resistance. What would its resistance be if it were made of aluminum?

$$1.66 = \frac{\rho_{cu} L}{A}, \qquad R_{al} = \frac{\rho_{al} L}{A}$$

$$\frac{R_{al}}{1.66} = \frac{\rho_{al}}{\rho_{cu}} = \frac{17}{10.37}$$

$$R_{al} = \frac{1.66 \times 17}{10.37} = 2.72 \ \Omega$$

EXAMPLE 3.5   Compute the resistance of a copper wire 100 ft long that has a square cross section 50 mil on each side.

Solution. First find the cross-section area in *circular mils*:

$$\text{Area in } square \ mils = 50^2 = 2500 \ \text{mil}^2$$

Now,

$$2500 = \frac{\pi}{4} d^2$$

in which d is the diameter of an equivalent round wire in mils and $\dot{d}^2$ is its cross-section area *in circular mils*:

$$d^2 = \frac{4}{\pi} \times 2500 = 3183 \ \text{cmil}$$

$$R = \frac{\rho L}{A} = \frac{10.37 \times 100}{3183} = 0.325 \ \Omega$$

Remember that the area of a circle *in square units* is given by $(\pi/4)d^2$. If d is in mils (1000 times inches) this area is in *square mils*. The area in *circular mils is only the d^2 part*. Can you see from this that *any area A has more circular mils than square mils?* A square mil covers more area than a circular mil as shown in Fig. 3.2. A *circle* 1 mil in diameter does not have as much area as a square mil. When you inscribe a circle *in a square 1 mil on a side* (which is a square mil), there is some area left over in all four corners. Yet, there is the correct amount of area inside the circle to make a *circular mil*. One square mil has the same amount of area as $4/\pi$ circular mils.

    To obtain the number of circular mils in *the cross section of* a solid round *wire, express the diameter in mils and then square it.*

    To obtain the diameter of the cross section of a solid round wire *that has a given number of circular mils, take the square root of the number of circular mils. The result will be the diameter of the wire in mils.*

    To obtain the number of circular mils in a noncircular area (a square, rectangle, triangle, or other shape), first get the area in square mils and then multiply it by $4/\pi$.

    To obtain the number of square mils in an area of a given number of circular mils, multiply that given number by $\pi/4$.

Problems

1. Compute the resistance of a spool of annealed copper wire that contains 700 ft of wire 0.05 in. in diameter.
2. What is the resistance of 300 ft of the wire specified in Problem 3.1?
3. A single-conductor copper cable ½ in. in diameter has a resistance of 1 Ω. How many miles long is it?
4. An aluminum conductor 200 ft long is to replace a copper conductor of the same length. The two must have the same resistance. The copper conductor is a No. 0 wire with 0.325 in. diameter. What must be the diameter of the aluminum conductor?
5. What length of 10 mil diameter nichrome wire is needed to make a heater coil for an electric coffee percolator if it is to have 40 Ω resistance?

The first column in Table 3.1 gives resistivities in *ohm meters* at 20°C. These values are used when dimensions of conductors are given in the *metric system*. Equation (3.2) requires that the conductor length L be in *meters* and the cross-section area A be in *square meters* *if resistivity is expressed in ohm-meters*. Dimensionally we then have

$$R \rightarrow \frac{\text{ohm-meters} \times \text{meters}}{(\text{meters})^2} = \text{ohms}$$

EXAMPLE 3.6   An annealed copper bar 3 m long has a cross section 5 mm by 2 cm. What is its resistance?

Solution

$$R = \frac{\rho L}{A} = \frac{1.724 \times 10^{-8} \times 3}{5 \times 10^{-3} \times 2 \times 10^{-2}}$$
$$R = 5.172 \times 10^{-4} \ \Omega = 517.2 \ \mu\Omega$$

EXAMPLE 3.7   How long must a round solid copper rod 5 mm in diameter be to have a resistance of 0.005 Ω?

Solution

$$L = \frac{RA}{\rho}$$
$$= \frac{5 \times 10^{-3} \times (0.005)^2 \pi/4}{1.724 \times 10^{-8}}$$
$$L = 5.69 \text{ m}$$

Note that resistivity is in *ohm-meters*, length in *meters*, and area in *square meters*.

3.6 Change of Resistance with Temperature

Coils of copper wire are used in innumerable applications in electrical and electronic engineering practice. If the resistance of a coil of wire is measured while it is at room temperature and measured again after it has been heated to a higher temperature (by heat absorption or by heat developed owing to the passage of current through it for an adequate length of time), its resistance will be found to have increased.

It can be demonstrated by experiment that points on a graph like that shown in Fig. 3.3, where ohms resistance of a piece of standard annealed copper wire at various temperatures between $-50°C$ and $200°C$ are plotted, lie on a practically straight line within that range. Assuming this to be true, the *increase* in resistance for each degree temperature rise is constant in that range. It can be seen, however, that the resistance is not directly proportional to the temperature because the straight line does not pass through the origin. That is, doubling the temperature does not double the resistance.

Note that the wire has a resistance of R_{20} ohms at 20°C. A temperature rise of 1°C results in an increase, ΔR, in the resistance. This increase (owing to a 1°C rise) divided by the resistance R_{20} at 20°C is called the *temperature coefficient of resistance* of the conducting material at 20°C. *Its units are ohms per centigrade degree per ohm.* It follows that if we had chosen our reference temperature as 0°C and used a ΔR for 1° temperature rise from there, we would have divided it by R_0 and obtained the temperature coefficient at the 0°C reference temperature.

The Greek letter alpha (α) is used as a symbol for temperature coefficient of

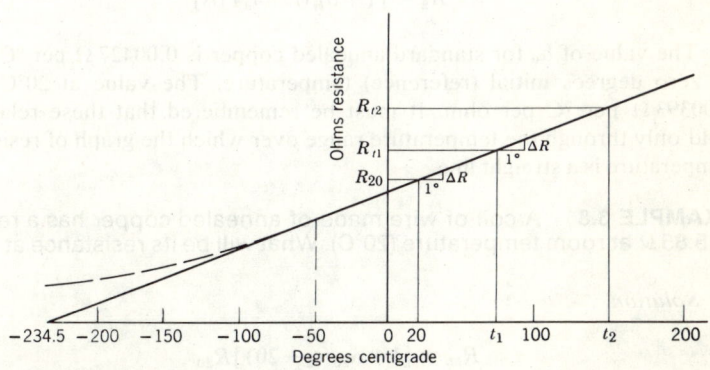

FIGURE 3.3   Resistance-temperature relation for standard annealed copper.

resistance, with a subscript denoting the reference temperature. In general then

$$\alpha_t = \frac{\Delta R}{R_t} \tag{3.4}$$

in which ΔR is the increase in resistance resulting from a 1°C increase in temperature above the reference temperature t.

Obviously, then, the coefficient values at t_1 and t_2 are

$$\alpha_{t_1} = \frac{\Delta R}{R_1}, \qquad \alpha_{t_2} = \frac{\Delta R}{R_2}.$$

In our equations we shall let $R_1 = R$ at t_1 and $R_2 = R$ at t_2. Because the graph is a straight line, it is obvious from similar right triangles that

$$\frac{\Delta R}{1} = \frac{R_2 - R_1}{t_2 - t_1} \tag{3.5}$$

Since $\alpha_{t_1} = \Delta R / R_1$

$$\alpha_{t_1} = \frac{R_2 - R_1}{(t_2 - t_1)R_1} \tag{3.6}$$

Now suppose we know the resistance of a coil of annealed copper wire to be R_1 ohms at t_1°C. We could calculate its resistance at another temperature, t_2, if we knew the value of α_{t_1}. Solving for R_2, we have

$$R_2 = [1 + \alpha_{t_1}(t_2 - t_1)]R_1 \tag{3.7}$$

The value of α_0 for standard annealed copper is 0.00427 Ω per °C per ohm at zero degrees initial (reference) temperature. The value at 20°C is $\alpha_{20} = 0.00393$ Ω per °C per ohm. It must be remembered that these relationships hold only through the temperature range over which the graph of resistance vs temperature is a straight line.

EXAMPLE 3.8   A coil of wire made of annealed copper has a resistance of 5.85 Ω at room temperature (20°C). What will be its resistance at 50°C?

Solution

$$R_{50} = [1 + \alpha_{20}(t_2 - 20)]R_{20}$$
$$R_{50} = [1 + 0.00393(50 - 20)]5.85$$
$$R_{50} = 6.54 \,\Omega$$

Obviously, if this coil had been made of iron wire, the temperature coefficient 0.005 would have been used. If it had been made of tungsten wire, α_{20} would have been taken as 0.0045. With different values of ΔR per degree centigrade rise, it follows that the *slopes* of the straight-line graphs for these metals, will be different from the slope of the line for copper in Fig. 3.8. In fact the straight line for iron would intersect the temperature scale at $-180°C$ and the one for tungsten at $-200°C$. These temperatures, as well as $-234.5°C$ for annealed copper, are sometimes called *inferred zero-resistance temperatures*. Nichrome wire has a very small temperature coefficient, which means that its straight-line graph of resistance versus temperature has a very small slope. It intersects the temperature scale at $-2480°C$. Values of inferred zero-resistance temperature are given in the second column of Table 3.3.

Let us ask ourselves whether Equation (3.7) will give the resistance of the wire if its temperature is decreased instead of increased. The answer is immediately forthcoming when we reflect that the graph is a straight line. This tells us that the decrease in resistance for every degree of temperature drop is the same as the increase for every degree temperature rise as long as we work with values on the straight-line portion of the graph.

To show this let's assume $6.54\ \Omega$ to be the resistance at 50°C and calculate

TABLE 3.3 **Temperature Coefficients of Resistance of Conductor Materials**

Material	Temperature Coefficient ohms per °C per ohm at 20°C	Inferred Zero-Resistance Temperature °C
Aluminum	0.0039	−236
Brass	0.002	−480
Carbon	0.0005	
Constantin (60% Cu, 40% Ni)	0.000008	—
Copper, annealed	0.00393	−234.5
Copper, hard-drawn	0.00382	−242
German silver	0.0004	−2,480
Iron	0.005	−180
Lead	0.0041	−224
Manganin (84% Cu, 12% Mn, 4% Ni)	0.000006	—
Mercury	0.00089	−1,100
Molybdenum	0.0034	−274
Nichrome	0.0004	−2,480
Nickel	0.006	−147
Platinum	0.003	−310
Silver (99.98% pure)	0.0038	−243
Steel, soft	0.0042	−218
Tin	0.0042	−218
Tungsten	0.0045	−200

the resistance of the wire at 20°C. We shall need to know the temperature coefficient, α_{50}, at 50°C, which is the temperature of the wire before it is changed. Later in this chapter we show that α_{t_1} is given by $\alpha_{t_1} = 1/(A + t_1)$ for annealed copper in which A is the inferred zero resistance temperature in degrees below zero. This is Equation 3.11. Using 50°C for t_1, we have

$$\alpha_{50} = 1/(234.5 + 50) = 0.00352$$

When the temperature of this wire is reduced from 50°C to 20°C, we let R_2 in Equation 3.6 represent the new value of resistance, t_2 the new temperature (20°C), t_1 the original temperature (50°C), and α_{t_1} the coefficient at the original temperature:

$$R_2 = [1 + 0.00352(20 - 50)]6.54$$
$$R_2 = (1 - 0.1056)6.54$$
$$R_2 = 5.85 \ \Omega$$

as expected.

3.7 Inferred Zero-Resistance Temperature

A practical and very useful equation for calculating resistance at a new temperature value is based on similar triangle relationships in Fig. 3.3. One is a right triangle with a base $234.5 + t_1$ units long and an altitude R_{t_1} units. The other has a base $234.5 + t_2$ units long and an altitude of R_{t_2} units. Obviously we may write

$$\frac{R_2}{R_1} = \frac{234.5 + t_2}{234.5 + t_1} \tag{3.8}$$

From this equation, one of the resistance values or one of the temperature values may be found if the other three are known. This equation holds for annealed copper only, but if we have a conductor made of a different kind of material, all we need to do is to use its inferred zero-resistance temperature in place of 234.5. Why do we not use the minus sign on these constants?

EXAMPLE 3.9   The coil of wire in the previous example had a resistance 5.85 Ω at 20°C. Calculate the resistance at 50°C by the method just described.

Solution

$$R_2 = \frac{234.5 + 50}{234.5 + 20} \times 5.85$$
$$R_2 = \frac{284.5}{254.5} \times 5.85 = 6.54 \ \Omega$$

Problems

6. A wire made of standard annealed copper has a resistance of 5.85 Ω at 20°C. Calculate its resistance at 0°C and at − 10°C.

7. A wire made of an alloy of metals has a temperature coefficient of resistance of 0.002 at 20°C. The resistance of a coil of this wire is 10 Ω at 20°C. What is its resistance to 75°C?

8. What will be the temperature of the wire in part *b* when its resistance is 10.5 Ω?

Suppose we let the letter A represent the number of degrees below zero centigrade, a straight-line graph, like that in Fig. 3.3 must be extended so that it will intersect the temperature axis. As an example, A would be 234.5 for standard annealed copper. We can now rewrite Equation (3.8) to fit any kind of conductor material for which we know the inferred zero-resistance temperature, thus

$$\frac{R_2}{R_1} = \frac{A + t_2}{A + t_1} \tag{3.9}$$

If A is known, the temperature coefficient can be found at any temperature. To achieve this we use the small right triangle at temperature t_1 in Fig. 3.3. This is similar to the large right triangle with base $234.5 + t_1$ units long. Therefore at t_1

$$\frac{\Delta R}{R_{t_1}} = \frac{1}{234.5 + t_1} \tag{3.10}$$

and since

$$\frac{\Delta R}{R_{t_1}} = \alpha_{t_1}$$

$$\alpha_{t_1} = \frac{1}{234.5 + t_1} \quad \text{for annealed copper} \tag{3.11}$$

In general, Equation 3.10 may be written

$$\frac{\Delta R}{R_{t_1}} = \frac{1}{A + t_1}$$

so that

$$\alpha_{t_1} = \frac{1}{A + t_1} \tag{3.12}$$

for any material whose inferred zero-resistance temperature is A degrees below zero (a *positive* number, here).

3.8 Precision Resistors

Small wire-wound resistors that must have extremely small resistance change with temperature may be made of manganin wire. This material has a temperature coefficient of only 6×10^{-6} Ω/°C/Ω at 20°C. The resistance of such a resistor at a different temperature cannot be calculated by the methods described for other kinds of conductors because its resistance-temperature curve is not a straight line. In fact, a piece of manganin wire has the same resistance at 30°C as it has at 15°C, and its resistance is only about 0.002 percent greater at the midpoint between these temperatures. When the temperature is raised above 30°C (86°F) the resistance *decreases* at a very small rate. At 70°C (158°F) it is still about 99.9 percent of its value at 20°C.

3.9 Nonlinear Resistances

A resistance in which current is always directly proportional to voltage obeys Ohm's law. It is called a *linear* resistance. Curve *a* in Fig. 3.4 is a volt-ampere curve of a linear resistance. It passes through the origin, it is a straight line, and therefore if the voltage applied to the resistance is doubled, trebled, or halved, the current will correspondingly be doubled, trebled, or halved.

Curve *b* is practically straight between the ordinates at 40 V and 120 V, but the current through the resistance it represents is not directly proportional to the applied voltage, even in this interval. Note that when the voltage is trebled the current is only doubled. The same voltage rise on the resistance represented by *a*, however, triples the current (from 0.167 A to 0.5 A). Curve *b* is the voltampere curve of a nonlinear resistance.

An incandescent lamp filament is a common variety of nonlinear resistance.

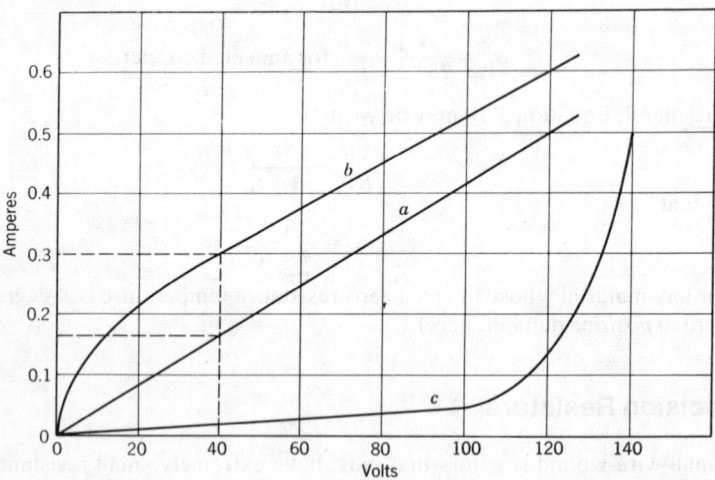

FIGURE 3.4   Volt-ampere characteristics of (*a*) a linear resistance, (*b*) a nonlinear resistance, (*c*) a Thyrite resistor.

When a dc voltage is suddenly applied, the initial current abruptly rises to about twelve times its normal operating value. This is because the resistance of a cold tungsten filament is only about one-twelfth the operating-temperature value.

Another practical nonlinear-resistance device is the *thyrite resistor*, used in lightning arresters. Thyrite is the manufacturer's name for a sintered mixture of carborundum and ceramic material. The current through the resistor is small and not objectionable at normal operating voltages, but when high-voltage surges are induced on a transmission line by lightning, or when there is a direct hit, the current increases to a very much higher value than it would in a linear-resistance unit. This quickly dispels, to earth, the energy received from the lightning. Curve *c* In Fig. 3.4 is the volt-ampere curve of a *thyrite* resistor. Many of such units are connected in series and built up into a stack several feet high in making a lightning arrestor for a transmission line. The three units represented in Fig. 3.4 have a normal operating voltage of 120 V.

3.10 Resistor Color Code

Many millions of small-size resistors made of a carbon composition are used in the electronics industry. They range from $\frac{1}{4}$ W to several watts in capacity and from a few hundredths of an ohm to many megohms resistance. Most of these have their individual resistance values identified by a standard color code. Bands of colors are printed near one end of the resistor as shown in Fig. 3.5*a*.

Three colored bands near one end of the resistor indicate the resistance value, and an added silver or gold band means the accuracy is ± 10 percent or ± 5 percent of the amount designated by the three bands. When there are only three bands, and the third one is neither silver nor gold, the tolerance is ± 20 percent.

Some resistors are coded by means of an overall body color that specifies the first digit, an end color that represents the second digit, and a dot color that indicates how many zeros follow the second digit. The colors and the numerals they represent are as follows:

Black	*Brown*	*Red*	*Orange*	*Yellow*	*Green*	*Blue*	*Violet*	*Gray*	*White*
0	1	2	3	4	5	6	7	8	9

In Fig. 3.5, part *a* illustrates the band type of code and part *b* illustrates the body-end-dot type of code. The former is much more widely used than the latter.

The first digit of the resistance value is designated by the color of the first band nearest the end (or by the body color). The second digit is given by the

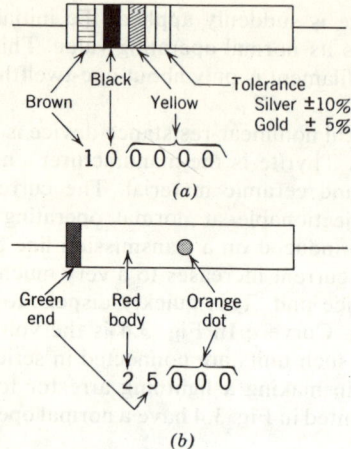

FIGURE 3.5   Color code designation of resistance values. (a) Band color coding. (b) Body-end-dot coding. This type of coding, (b), is going out of use.

middle color band, or by the end color, and the color of the third band, or of the dot, indicates the *number of zeros* that follow the first two digits. This completes the resistance value. If the *third band color* is gold, the two-digit number must be multiplied by 0.1, and if it is silver the multiplier is 0.01.

Examples of color codes on resistors:

(*a*) red, green, orange   25,000 Ω (20 percent tolerance)
(*b*) blue, black, brown, gold   600 Ω (5 percent tolerance)
(*c*) brown, black, green, silver   1 megohm (10 percent tolerance)
(*d*) brown, black, gold   1 Ω (20 percent tolerance)
(*e*) red, green, silver   0.25 Ω (20 percent tolerance)

3.11   Review Questions

1. Assume you want to send current through a coil of wire by connecting its ends to the terminals of a 12-V battery. What will determine how much current will flow in the coil?

2. One-half ampere flows through a certain electric light bulb when it is connected to a 115-V line. What is its hot resistance?

3. A large incandescent lamp takes 10 times as much current as the light bulb of Question 2 when it is connected to a 115-V line. What is its resistance?

4. Why must a 250-W electric light bulb have much less resistance than a 50-W bulb?

5. Is the filament wire in a 250-W light bulb larger or smaller in cross section than that in a 50-W bulb?

6. What would you do to make an 8-V relay coil work properly on a 12-V battery?
7. How does the resistance of a wire change when (a) its length is reduced 50 percent? (b) its diameter is reduced 50 percent? (c) its dimensions are unchanged but the material is changed to one with twice the resistivity?
8. Define resistivity.
9. Express mathematically the resistivity of a material in terms of the dimensions of a conductor made of it.
10. What is a circular mil? How many circular mils has a round bar that is 1 in. in diameter?
11. Is a square mil area larger or smaller than a circular mil area? What is the ratio of a square mil area to a circular mil area?
12. The temperature coefficient of resistance is expressed as ohms change per degree temperature change *per ohm at the initial temperature*. Why is it necessary to specify *per ohm at the initial temperature* instead of just stopping at the words *ohms change per degree temperature change*?
13. Assume that the resistance of a coil of wire made of a commonly used material such as copper, aluminum, or iron is measured at several temperatures between $-50°C$ and $100°C$. When these resistance values are plotted on a vertical axis with the temperatures on the horizontal axis, what is the nature of the line drawn through all of the points? Does it pass through the origin of the axes?
14. If the line of the graph described in Question 13 is extended (without changing its slope) to the temperature axis, what is the meaning of the reading at the intercept on that axis called?
15. What is known about two right triangles ABC and $A'B'C'$, which differ in size but have acute angles $A = A'$ and $B = B'$, that enables us to write Equation 3.8 and know it is correct?
16. Assume you know the resistance of an annealed copper wire coil to be $10 \, \Omega$ at $23°C$ and you desire to calculate what its resistance will be at $100°C$. Which of the equations in this chapter would be easiest to use?
17. Continuing with Question 16, if you were required to use Equation (3.7), how would you first find the value of α_{23}, which is the temperature coefficient of annealed copper at $23°C$?
18. Observe that manganin does not have an inferred zero-resistance temperature (Table 3.3). Why is this so?
19. Why do you suppose the inferred zero-resistance temperature of nichrome and German silver have such large values ($-2480°C$)?
20. What is a precision resistor?
21. What is a nonlinear resistor?
22. How does the nonlinear resistance characteristic of thyrite differ from that of an incandescent lamp?
23. What makes thyrite well adapted to use in a lightning arrester?
24. Why would a string of incandescent lamps be unsuitable for use as a lightning arrester? Note the current at and below normal operating voltage.
25. Refer to the wire table (Table 3.2). What is the change in cross section area when you skip two sizes in going from one size to another, such as from 0 to 0000, from No. 10 to No. 7, or from No. 27 to No. 24? This ratio holds through-

out the table for all practical purposes. A slight variation exists caused by practical manufacturing variations.

26. Describe the resistance ratings including tolerances of small carbon-composition resistors marked as follows: (a) orange, yellow, red, gold (b) brown, black, green, (c) green, black, gold (d) red, green, silver.

3.12   Problems

Group A

9. How much current will flow through a 15-Ω resistance when it is connected to a 120-V source?

10. How much resistance must a resistor have if its current should be 2 A when connected to a 100-V source?

11. An electromagnet coil is found to be carrying $\frac{1}{2}$ A while the volatge drop across it is 6 V. What is its resistance?

12. How much current must flow through a 75-Ω resistor to produce a 1.5 V drop across it? How much current will produce the same voltage drop across a 1500-Ω resistor?

13. How much voltage must you supply to a 40-Ω electric soldering iron so that it will receive its normal operating current of 3 A?

14. A resistor in a radio receiver must have 3 V drop across it when its current is 2 mA. How much resistance must it have?

15. A 500,000-Ω resistor in a television receiver has 5 V drop across it. How much current is it carrying?

16. A fuse in a section of an electronic circuit has a resistance of 0.015 Ω. How much is the voltage drop across it, in millivolts, when it carries 6 A?

17. How much current flows in a 2.5 kΩ resistor when it has a 125 mV drop across it?

18. A relay coil requires 10 mA for operation of the relay contacts. What voltage is needed across its terminals if it has a resistance of 1.2 kΩ?

19. How many megohms resistance has an electronic voltmeter that takes 10 μA while measuring 100 V?

20. How many circular mils are there in (a) a circle 0.5 in. in diameter, (b) the cross section of a wire 0.05 in. in diameter?

21. How many square mils are there in a rectangular area $\frac{1}{4}$ in. $\times \frac{3}{4}$ in.? How many circular mils are in that area?

22. Calculate the resistance of 150 ft of annealed copper round wire $\frac{1}{32}$ in. in diameter. What is the resistance of an aluminum wire of the same length and diameter?

23. How many feet of annealed copper wire $\frac{1}{32}$ in. in diameter are needed to make up 2 Ω of resistance?

24. Solve Problem 23 using nichrome wire.

25. How many feet of tungsten wire ($\rho = 33$ Ω-cmil/ft) are required to make a 1-Ω resistance coil of the wire if the wire is 11 mil in diameter?

26. Wire No. 1 is made of copper. Wire No. 2 is 3 times as long as No. 1, its diameter is $\frac{1}{3}$ that of No. 1, and it is made of nickel. Which wire has the larger resistance and what is the ratio of R_2 to R_1?

27. A strip of annealed copper 2 cm wide and 3 mm thick is 1 m long. How much resistance will it offer to current flow? How much resistance would it have if it were made of aluminum?

28. Assume that a cubic meter of annealed copper can be cut into 10,000 identical bars, each 1 cm² in cross section and 1 m long. What would be the resistance of each bar?

29. What is the amount of resistance between opposite faces of a cubic inch of aluminum? Work this problem two ways, first using 2.83×10^{-8} Ωm for the resistivity, then using 17 Ω-cmil/ft.

30. A copper winding in an electric motor has a resistance of 120 Ω at room temperature (20°C). The motor is put into service, and after the temperature has become stabilized at 70°C, the winding resistance is measured. Compute the measured value of the winding resistance.

31. An iron wire heater coil has 10 Ω resistance at 20°C. What is its temperature if, after becoming hot, its resistance has risen to 20 Ω?

32. A platinum wire used in a chemical test is heated from room temperature, 20°C to 1500°C. Use $\alpha_{20} = 0.003$ to compute how much its resistance increases if it is 2 Ω at room temperature. What is its resistance at 1500°C?

33. The cold resistance of the tungsten filament of an incandescent lamp is 16 Ω and the hot resistance at its operating voltage of 120 V is 240 Ω. How much *surge* current does the lamp take at the instant it is connected to the source and how much continues to flow after it has come up to operating conditions?

Group B

34. Prove that there are 10^6 cmil in the area of a 1-in. diameter circle. Divide the square-inch area of a 1-in. diameter circle by the square-inch area of a 1-mil diameter circle.

35. Convert the resistivity of copper from 10.37 Ω-cmil/ft to ohm-meters by calculation. To work this problem, remember that a circular mil foot of copper is 1 ft long and has 1 cmil cross section. Using dimensional conversion, you have

$$\frac{\text{ohms-circular mils}}{\text{foot}} \times \frac{\text{feet}}{\text{meter}} \times \frac{\text{square meters}}{\text{circular mil}} \to \text{ohm-meters}$$

The last factor is the reciprocal of the number of circular mils in a square meter. First get square inches per square meter and then change it to circular mils per square meter. The answer comes out 1.724×10^{-8}.

36. A power-line stranded cable consists of eight No. 14 AWG copper wires twisted together. A mile length of this cable requires 4 percent more than a mile of each strand. Calculate the resistance of a mile of this cable, considering that the resistance is one eighth of that of one strand because there are eight strands "in parallel."

37. If a solid copper conductor were to replace the cable of Problem 36 and offer the same resistance per mile, what would be its circular mil area and its diameter?

38. A copper-wire coil has $2.5\,\Omega$ resistance at 20°C. What will be its resistance after it is cooled to -20°C?

39. A piece of platinum wire the size of No. 40 AWG copper wire is to be heated to 350°C at which temperature its resistance must be $50\,\Omega$. What must be its length at 20°C?

40. To what temperature must a tungsten wire be raised to cause its resistance to increase 10 percent of its value at 20°C?

41. A copper telephone line wire has a resistance per mile of $34\,\Omega$ at -18°F which is -27.78°C. What is its resistance per mile on a hot summer day when the temperature is 113°F? Can you show that this is 45°C?

42. An electric furnace located 50 ft from a power switch requires 100 A at 120 V for operation. The voltage at the switch is constant at 124 V. What is the total resistance allowed in the two-wire line that connects the furnace to the switch? Calculate the circular mil size of the copper wire required. Select, from the wire table, a standard AWG number of wire you would use.

43. It is desired to determine which of the materials in Table 3.1 was used to make wire that was used in a certain coil. The resistance was measured at 20°C and found to be $8.5\,\Omega$. At 55°C it was found to be $9.1\,\Omega$. Can you determine from these data what the material was?

Power and Energy

4.1 Definitions

The word *power* is used by speakers and writers in a great many ways. There is brain power, power of persuasion, power of positive thinking, military power, the "power of the pen," and so forth. The meanings of those terms are different from the meanings of such electrical terms as *watt, kilowatt, megawatt, milliwatt*, and mechanical terms such as *horsepower, foot-pounds per minute*, and *joules per second*. The electrical and mechanical terms, all of which are units of power, can be defined precisely and measured accurately. It is impossible to measure accurately and define in a quantitative way the other kinds of power.

In order to define power properly we must first know about *energy* and *work*. We spoke of chemical energy in a storage battery and about a pump supplying energy to water and raising it to a higher level. When a pump does this we say it is doing work and the work it performs is the amount of energy it gives to the water. For example, if we say a pump raised the level of 1000 lb of water from the ground to an elevation of 50 ft, it is true that the pump did 50,000 ft-lb of work, and, in so doing, it gave 50,000 ft-lb of energy to the water. If all of that water is caused to turn a water motor which drives some other machine, such as a flour mill, the motor will receive the 50,000 ft-lb of energy. If, for convenience, we choose to neglect the opposing forces of friction in the motor in the flour mill, we can say that 50,000 ft-lb of work will be done in grinding the flour.

When we study the subject *mechanics* we learn that work is done when a force moves through a distance. To raise each pound of the 1000 lb of water in the above illustration required a force of 1 lb acting through a distance of 100 ft. This means that 100 ft-lb of work was done on each pound of water. All the water did was receive that amount of energy. At the higher level it was capable of doing 100 ft-lb of work, but before it was pumped up there it was not capable of doing any work on the ground.

Energy may be defined as the capacity to do work.

A gallon of gasoline has a lot of energy, that is, it has a lot of *capacity to do work*. Its energy may be utilized by the engine of a large truck and enable the truck to take its load up a hill a couple of miles long.

Energy and work are expressed in the same units. Common mechanical units are the *foot-pound*, *newton-meter*, *erg*, and *joule*.

The letter symbol for both energy and work is **W**.

4.2   Units of Work and Energy

DEFINITION. A joule is the amount of work done, or energy expended, when 1 N of force acts through a distance of 1 m. The joule is the basic unit of work and energy in the MKS system of units.

Correspondingly, *in the English system of units the foot-pound is the amount of work done, or energy expended, when 1 lb of force acts through a distance of 1 ft.*

The *erg* is the amount of work done, or energy expended, when *1 dyn of force acts through a distance of 1 cm*. Because 1 N equals 10^5 dyn, and 1 m equals 10^2 cm, *1 J equals 10^7 erg*.

Let's calculate the number of joules in a foot-pound:

$$\textbf{1 ft-lb} = 1 \text{ lb} \times 1 \text{ ft} = 4.45 \text{ N} \times 1/3.28 \text{ m} = \textbf{1.356 J}$$

The number of foot-pounds in a joule is just the reciprocal of this:

$$\textbf{1 J} = \textbf{0.7363 ft-lb}$$

EXAMPLE 4.1   A block of stone rests on a concrete street. A force of 10 N, acting parallel to the street, pushes the stone a distance of 3 m. How much work is done by the force?

Solution

$$10 \times 3 = 30 \text{ J}$$

EXAMPLE 4.2   A 100-lb weight is lifted from the ground to a height of 20 ft by a crane. How much energy was expended by the crane? Where did it go? Can use be made of this energy, and if so, how?

Solution
$$100 \times 20 = 2000 \text{ ft-lb of energy}$$

The weight has 2000 ft-lb of *potential energy*. Potential energy is sometimes called *energy of position*. The 100-lb weight can be caused to transfer its energy to a "pile" by letting the weight fall on the pile to drive it into the ground.

Power, **P**, *is work* (*or energy*) *per unit of time*, so the basic equation relating power and energy is

$$P = \frac{W}{t} \tag{4.1}$$

EXAMPLE 4.3   Assume the time required to do the 30 J of work of Example 4.1 is 10 s. How much power is required?

Solution
$$P = \frac{W}{t} = \frac{30}{10} = 3 \text{ J/s}$$

EXAMPLE 4.4   Forty seconds of time were required for the crane of Example 4.2 to lift 100 lb 20 ft. At what rate is the work being done?

Solution
$$P = \frac{W}{t} = \frac{2000}{40} = 50 \text{ ft-lb/s}$$

Problems

1. A force of 50 N is required to move an object, and the distance moved is 80 cm. How much work is done?
2. A 10 kg weight falls 5 m and lands on the ground. How much energy is expended? Give your answer in kilogram-meters and in joules. See the last line of Table 1.3.
3. How much power is required to do the work of Problem 1 in 5 s? In 10 s?
4. A lifting crane can do work at the rate of 150,000 ft-lb per minute. How long would it take to raise a 3-ton automobile a vertical distance of 15 ft?

4.3   Heat Energy

Vaporized gasoline, mixed with air, is exploded in the cylinders of an automobile engine to do work in driving pistons downward in their cylinders and thus turn the crank shaft. *Heat energy* is measured in units called *calories* and *British thermal units* (**Btu**).

DEFINITION. A calorie is the amount of heat energy needed to raise the temperature of 1 g of water 1°C. A kilogram calorie is 1000 cal.
DEFINITION. A British thermal unit is the amount of heat energy needed to raise the temperature of 1 lb of water 1°F.

EXAMPLE 4.5   Let's calculate the number of calories there are in 1 Btu. When 1 lb of water (453.6 g) is raised 1°F in temperature it is raised only 5/9°C:

Solution

$$453.6 \times \frac{5}{9} = 252 \text{ cal/Btu}$$

EXAMPLE 4.6   Now we'll calculate the number of joules of energy there are in 1 Btu. We shall need the number of foot-pounds per Btu. This was determined by James Prescott Joule, an English physicist, in a series of epoch-making experiments which extended from 1842 to 1870. He found that *778 ft-lb of energy must be expended to produce 1 Btu of heat.*

Solution. Lets look at the units in the conversion of Btu to joules. This kind of procedure is very helpful in converting a quantity from one kind of unit to another:

$$\text{Btu} \times \frac{\text{foot-pounds}}{\text{Btu}} \times \frac{\text{newtons}}{\text{pound}} \times \frac{\text{meters}}{\text{foot}} = \text{joules}$$

$$1 \times 778 \times 4.45 \times \frac{1}{3.28} = 1054.8 \text{ J/Btu}$$

$$\textbf{1 Btu} = \textbf{1054.8 J}$$
$$\textbf{1 cal} = \textbf{1054.8/252} = \textbf{5.185 J}$$

Checking the units,

$$\frac{\text{joules}}{\text{Btu}} \div \frac{\text{calories}}{\text{Btu}} = \text{joules per calorie}$$

EXAMPLE 4.7   How much heat is needed to bring 500 g of water at 20°C to the boiling point (100°C)?

Solution

$$500(100 - 20) = 40,000 \text{ cal}$$

To express this amount of heat in Btu we simply divide by the number of calories in 1 Btu:

$$40,000/252 = 158.73 \text{ Btu}$$

EXAMPLE 4.8   Friction in a device consumes 2109.6 J of energy per hour. How many Btu per hour are produced?

Solution

$$\frac{\text{joules}}{\text{hour}} \times \frac{\text{Btu}}{\text{joule}} = \frac{\text{Btu}}{\text{hour}}$$

Multiplying by Btu per joule is the same thing as dividing by joules per Btu:

$$2109.6 \div 1054.8 = 2 \text{ Btu/h}$$

4.4 Electrical Power

Power is defined as *the rate at which work is done, or energy expended.* This definition holds for mechanical power as well as for electrical power. When energy is expended in electric heating, the expenditure of energy is expressed in electrical units, namely *watt-seconds, watt-hours,* or *kilowatt-hours. The joule is the equivalent of the watt-second.*

The rate of doing work may be expressed in *joules per second.* This means that the *watt-second per second,* namely, the *watt,* is a unit of *power* in the electrical system. That requires a little extra thought for full understanding.

Let us summarize. Units of work or energy are: foot-pounds, joules, watt-seconds, watt-hours, and kilowatt-hours. Units of power are: foot-pounds per second, foot-pounds per minute, and joules per second = watts, kilowatts.

Units of heat energy are: Btu and calorie. There are 252 cal in one Btu and 778 ft-lb of energy, when converted into heat, produce 1 Btu. Sometimes the *great calorie* is used, and that is *1000 cal,* which is the amount of heat required to raise the temperature of *1 kg* of water *1°.* Another name for the great calorie, of newer usage, is the *kilocalorie.*

4.5 Efficiency

Energy-converting machines or devices cannot deliver all of the energy that is put into them. If they could, their efficiency would be 100 percent. Some of the energy put into a machine is lost either in overcoming friction or in some other way. The resistance of electrical conductors causes the production of heat when current flows. This is lost energy when it occurs in a current source such as a generator or a battery, or in a converter of energy such as an electric motor. Friction also accounts for energy loss in a motor.

Efficiency may be defined as the ratio of output energy to total input energy required to produce it. This must occur during the same time interval. The Greek letter eta (η) is commonly used as the letter symbol for efficiency:

$$\eta = \frac{W_{\text{out}}}{W_{\text{in}}} \tag{4.2}$$

It is usually more convenient to calculate efficiency as the ratio of output power to input power:

$$\eta = \frac{P_{out}}{P_{in}}$$

(4.3)

EXAMPLE 4.9 An electric motor drives a drum on which a hoisting cable is wound. While a load of 2000 lb is being lifted a distance of 25 ft, the input to the motor is 16.58 Wh (watthours). What is the efficiency?

Solution

$25 \times 2000 = 50,000$ ft-lb, output energy

$50,000 \times 1.356 = 67,800$ J, output

$16.58 \times 3600 = 59,688$ J, input

Efficiency, $\eta = \dfrac{59,688}{67,800} = 0.883 = 88.3$ percent

4.6   Electric Power Equations

In Chapter 2 we learned that work per unit charge when expressed in joules per coulomb is volts (Eq. 2.4):

$$V = \frac{W}{Q} \text{ volts} \qquad W = QV$$

and, since amperes = coulombs per second,

$$I = \frac{Q}{t} \quad \text{and} \quad t = \frac{Q}{I}$$

Using the basic power equation (4.1), we have

$$P = \frac{QV}{Q/I} = QV \times \frac{I}{Q} = V \times I$$

So that

$$P = VI, \quad \textbf{watts = volts times amperes}$$

(4.4)

in which I is the amperes of current in a resistance, V is the potential difference, in volts, across the resistance, and P is the power, in watts, expended in the resistance.

Using Ohm's law ($V = IR$), we substitute this for V and get

$$P = I^2 R$$

(4.5)

which is a second power equation usable when I amperes flow through R ohms of resistance. We may obtain a third equation for power from Equation (4.2) if we substitute $I = V/R$:

$$P = VI = V \times \frac{V}{R}$$

$$P = \frac{V^2}{R} \tag{4.6}$$

when V is the volts drop across a resistance R ohms in which current flows. Note that the amount of current need not be known.

EXAMPLE 4.10 When a current of 5 A flows through a certain resistance, a potential drop of 120 V exists across it. How much power is being delivered to the resistance?

Solution

$$P = 120 \times 5 = 600 \text{ W}$$

EXAMPLE 4.11 A 2 kΩ resistor is rated at 5 W power capacity. What is the maximum current it should be made to carry?

Solution

$$I^2 \times 2000 = 5$$
$$I = \sqrt{5/2000}$$
$$I = 0.05 \, A = 50 \text{ mA}$$

EXAMPLE 4.12 It is desired to deliver $\frac{1}{2}$ W of power to a 10 kΩ resistor by applying a voltage V across it. Calculate V.

Solution

$$P = \frac{V^2}{R} \qquad V = \sqrt{PR}$$
$$V = \sqrt{0.5 \times 10^4} = 70.7 \, V$$

EXAMPLE 4.13 It is desired to put 108,000 J of energy into a chemical mixture by means of an electric heater that will deliver it at a constant rate, that is, the power output of the heater will be constant. The heater takes 3 A at 120 V. Assume heat losses in the process are 20 percent. How long will it take to do the job?

Solution. Let W represent the energy output of the heater:

$$0.8 \, W = 108,000$$
$$W = 135,000 \, J = \text{watt seconds}$$
$$\text{Watts} = 120 \times 3 = 360 \text{ to heater}$$
$$\text{Time required} = 135,000/360 = 375 = 6.25 \text{ min}$$

As stated in Sec 5, efficiency is often determined simply from the power rates. This should be obvious when we consider that the ratio of output energy to input energy requires that the *time used in computing the energies be the same*. Using the energy rates:

$$\eta = \frac{P_{out} \times t}{P_{in} \times t} = \frac{P_{out}}{P_{in}} \tag{4.7}$$

EXAMPLE 4.14 An electric motor rated at 5 hp output delivers full load at 84.5 percent efficiency. How much power input is required?

Solution

$$P_{in} = \frac{P_{out}}{\eta} = \frac{5}{0.845} = 5.91 \text{ hp}$$

4.7 Horsepower

Automobile engines are rated in horsepower, which was originally conceived as a mechanical power unit. James Watt, a Scottish scientist who invented the steam engine, performed experiments in which he compared the output power of his engine with the power a horse could put out. He found that an "average" workhorse could do work at the rate of 33,000 ft-lb/min:

1 hp = 33,000 ft-lb/min = 550 ft-lb/s

Since the early days of electric power utilization, it has been necessary to classify electric motors in terms of horsepower output. This required the determination of the numerical relation between the mechanical power unit and the electrical power unit.

EXAMPLE 4.15 Let us calculate the *number of watts equivalent to 1 hp*.

Solution. Start with the mechanical unit — 33,000 ft-lb/min:

$$33,000 \frac{\text{(ft-lb)}}{\text{(min)}} \times \frac{1}{60} \frac{\text{(min)}}{\text{(s)}} = 550 \frac{\text{(ft-lb)}}{\text{(s)}}$$

$$550 \frac{\text{(ft-lb)}}{\text{(s)}} \times \frac{1}{3.281} \frac{\text{(m)}}{\text{(ft)}} = 167.7 \frac{\text{(lb-m)}}{\text{(s)}}$$

$$167.7 \frac{\text{(lb-m)}}{\text{(s)}} \times 4.45 \frac{\text{(N)}}{\text{(lb)}} = 746 \frac{\text{(N-m)}}{\text{(s)}}$$

$$\frac{\text{newton-meters}}{\text{second}} = \frac{\text{joules}}{\text{second}} = \text{watts}$$

$$\textbf{1 hp} = \textbf{746 W}$$

EXAMPLE 4.16   A crane is designed to lift 5 tons at a speed of 0.3 ft/s. The mechanical efficiency of the crane's pulleys and gearing is 75 percent. If the efficiency of the motor alone is 85 percent, how many kilowatts must be supplied to the motor?

Solution

$$5 \times 2000 \times 0.3 = 3000 \text{ ft-lb/s}$$

$$\frac{3000}{550} = 5.454 \text{ hp crane output}$$

$$\frac{5.454}{0.75} = 7.27 \text{ hp motor output}$$

$$\frac{7.27}{0.85} = 8.55 \text{ hp motor input}$$

$$\frac{8.55 \times 746}{1000} = 6.38 \text{ kW motor input}$$

Problems

5. A 10 hp motor working at full load requires 11.5 hp input. What is its full-load efficiency?

6. How many watts input does the motor of Problem 5 take?

7. A power resistor has 25 Ω resistance and carries 2 A of current. (a) What power does it take? (b) How much is the voltage drop across it? (c) Calculate the power in a manner different from that which you used in (a).

8. How many joules of energy does the resistance of Problem 7 dissipate in the form of heat in one hour?

9. A motor takes 2000 W of power and operates at 75 percent efficiency. What is its output in watts and in horsepower?

10. A 200 Ω resistance consumes energy at the rate of 2 W. What is the voltage drop across it? Can you calculate this using only one formula?

11. A motor does work at the rate of 275 ft-lb/s. What is its horsepower output?

12. What is the horsepower input to the motor of Problem 12 if its efficiency is only 50 percent?

4.8   Cost of Electrical Energy

Electrical energy used in a residence or in a commercial enterprise is metered in *kilowatt-hours* and charges are made, usually for each month of use, but in many cases for 2 months of use, on the basis of the number of kilowatt-hours used. A *watthour meter*, installed in the main lines of the customer's premises, registers the *total energy* (kilowatt-hours) *consumed* in the period. The meter reading at the end of the previous billing period is subtracted from the reading for the current billing period to obtain the amount used. Since energy is obtained by multiplying power by time, kilowatts times hours gives kilowatt-hours of electrical energy. Electric power companies charge for the energy they supply to a residence on the basis of a so-called sliding scale. For a residence in Lexington, Kentucky, that is not heated electrically (in which case the *rate* is lower), the cost to the customer is computed as follows:

> For the first 16 kWh or less $1
> For the next 34 kWh or less .... 5 ¢ per kWh
> For the next 50 kWh or less .... 3 ¢ per kWh
> For the next 100 kWh or less....... 2.2 ¢ per kWh
> For the next 200 kWh or less....... 2.0 ¢ per kWh
> For all in excess of 400 kWh....... 1.5 ¢ per kWh

EXAMPLE 4.17 A family in a residence heated with natural gas used 460 kWh of electrical energy in one month. How much was the electric bill when computed on the basis of the above scale?

Solution

kWh	Charge
16	$1.00
34 @ 5 ¢	1.70
50 @ 3 ¢	1.50
100 @ 2.2 ¢	2.20
200 @ 2 ¢	4.00
60 @ 1.5 ¢	.90
Cost for month	$11.30

The kilowatt-hour is the common unit of electrical energy. Since the watt-second is also an electrical energy unit,

$$1 \text{ kWh} = 1000 \text{ W} \times 3600 \text{ s} = 3.6 \times 10^6 \text{ W-s or J}$$

It is *incorrect* to interpret the kilowatt-hour as a kilowatt per hour. In fact, this would have no useful meaning. The kilowatt is already an *energy per unit time quantity*. Power per unit time can only mean that power is changing at a certain rate. Engineers and scientists are rarely required to deal with a situation in which power is changing at a specified rate.

4.9 Summary of Important Relations

For your convenience, a tabulation of the definitions and important equations developed in this chapter is given in Table 4.1. After memorizing the first two columns you should be able to write any of the expressions in the third column with little difficulty.

TABLE 4.1   **Electrical and Mechanical Relations**

Quantity	Mathematical Definition	Derived Relations
Voltage	$V = \dfrac{W}{Q}$	$V = IR$
	$\text{volts} = \dfrac{\text{joules}}{\text{coulomb}}$	$V = P/I$
		$V = \sqrt{PR}$
Current	$I = \dfrac{Q}{t}$	$I = V/R$
	$\text{amperes} = \dfrac{\text{coulombs}}{\text{seconds}}$	$I = P/V$
		$I = \sqrt{P/R}$
Resistance	$R = \dfrac{V}{I}$	$R = V^2/P$
	$\text{ohms} = \dfrac{\text{volts}}{\text{amperes}}$	$R = P/I^2$
Power	$P = \dfrac{W}{t}$	$P = VI$
		$P = I^2R$
	$\text{watts} = \dfrac{\text{joules}}{\text{seconds}}$	$P = V^2/R$
Energy (work)	$W = Pt$	$\text{joules} = \text{watts} \times \text{seconds}$
Horsepower	1 hp $= 33{,}000$ ft-lb/min	1 hp $= 746$ W
Kilowatt hours	kilowatts $\times$ hours	(watts $\times$ hours)/1000
Efficiency	$\dfrac{P_{\text{out}}}{P_{\text{in}}}$	$\dfrac{W_{\text{out}}}{W_{\text{in}}}$ in same time interval

4.10 Decibel

When the telephone came into use (Alexander Graham Bell invented it in 1875 with help from Elisha Gray), it soon became necessary to express loss of power in a telephone line in a way that would relate it to the amount of power put into the line. Power quantities in telephone work are quite small.

The decibel had its origin in the need for an accurate means of measuring and comparing powers required to produce sounds. The human ear responds to sound intensity in a logarithmic fashion. It readily recognizes the doubling of sound power, but it can hardly distinguish two sounds of the *same pitch* (middle C on the musical scale, for example) when their *powers* differ by as much as 20 percent.

The **bel** *is defined as the common logarithm of the ratio of two powers.* They do not have to be sound powers. Because the *bel* represents a relatively large power ratio, the **decibel**, which is **one-tenth of a bel**, is generally used.

The following equation is used to compute decibels, abbreviated **dB**:

$$\text{decibels} = 10 \log_{10} \frac{P_{\text{out}}}{P_{\text{in}}} \tag{4.8}$$

When the output power is larger than the input power, dB comes out positive, and we have a *gain*. When the input power is larger than the output power, dB comes out negative, and we have a *loss*. In this case we may use $\log_{10}(P_{\text{in}}/P_{\text{out}})$, get a positive value for dB which we must then interpret as a *loss*. The logarithm of a fraction is negative. See Fig. 4.1.

EXAMPLE 4.18   A test signal at the sending end of a telephone line applies 0.5 W to the input terminals. The output power is simultaneously measured and found to be 0.008 W. What is the *loss* in the line expressed in decibels?

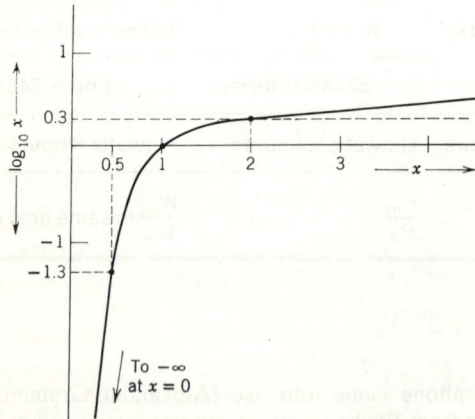

FIGURE 4.1   Variation of $\log_{10}x$ with x. $\log_{10} 1 = 0$, $\log_{10} 0 = $ minus infinity. Curve for natural log(ln) has the same shape, but vertical scale is different. ln 0.5 = $-$ 0.693, ln 2 = $+$ 0.693.

Solution

$$dB = 10 \log_{10} \frac{0.5}{0.008}$$

$$dB = 10 \log_{10} 62.5$$

$$dB = 10 \times 1.796 = 17.96 \, dB$$

The use of decibels to express loss is not confined to conditions of small amounts of power. Although rarely done, loss in decibels may be computed at kilowatt or even megawatt levels.

Gain may be expressed in decibels and, in fact, the gain that an electronic amplifier produces is customarily expressed in decibels.

EXAMPLE 4.19   An amplifier receives an input of 0.1 W and delivers 2 W at its output terminals. What is the gain in decibels?

Solution. In this case the output power is greater than the input power:

$$dB = 10 \log_{10} \frac{2}{0.1} = 10 \log_{10} 20$$

$$dB = 10 \times 1.301 = 13.01 \, dB \text{ gain}$$

Speaking of logarithms of fractional numbers, the curve shown in Fig. 4.1 is very informative and useful. You should examine it and use it as a reminder of the fact that the logarithms of fractions are negative. Furthermore, notice that *the logarithm of zero is minus infinity.*

This curve holds for natural (Naperian) logarithms, as well as for common logs, although the vertical scale would be different: $\ln 1 = 0$, $\ln 0 = -\infty$.

4.11   Decibel Reference Level

It is necessary to have a power value that will serve as a *base,* or *reference,* to which all other power values may be referred. This can be called the *zero-dB power* or the *zero-dB reference level.* Zero power could not be used for this because then either P_{out} or P_{in} in the dB equation would be zero and we would be confronted with the logarithm of zero, which is minus infinity, or the logarithm of infinity, which is infinity.

The *zero-dB-reference level* is generally taken to be 1 mW (0.001 W) although 6 mW is sometimes used. This may appear to be too small for practical use, but such is not the case. Power outputs of some microphones are as small as a few hundredths of a millionth of a watt. Their power outputs are of the order of -50 dB. In the other extreme, the power level of a 100,000 kW electric generator is $+110$ dB with respect to the 1 mW zero-decibel reference.

EXAMPLE 4.20   A microphone has an output of 0.01 μW. What is its output power level in decibels?

Solution

$$dB = 10 \log_{10} \frac{10^{-8}}{10^{-3}} = 10 \log_{10} 10^{-5}$$

$$dB = 10\,(-5) = -50\,dB$$

EXAMPLE 4.21   An electric generator has an output of 125,000 kW. What is the dB level of this power?

Solution

$$dB = 10 \log_{10} \frac{12.5 \times 10^7}{10^{-3}} = 10 \log_{10} 12.5 \times 10^{10}$$

$$dB = 10 \times 11.097 = 110.97$$

4.12   Decibel Equations Containing V and I

It is common practice to compute decibel gain or loss using voltage or current values instead of power values. However, power ratios determine the forms of the expressions. This means that the two voltages (V_1 and V_2) *must exist across equal resistances*, or the two current values *must exist in equal resistances*. This is practical. For example, the input voltage to an amplifier may be that taken from the terminals of a resistor R and the load on the amplifier may have the same resistance value as R. The powers would then be expressible as:

$$\text{input: } P_1 = \frac{V_1^2}{R}, \qquad \text{output: } P_2 = \frac{V_2^2}{R}$$

Then

$$\frac{P_2}{P_1} = \frac{V_2^2}{V_1^2}$$

and

$$dB = 10 \log_{10} \frac{V_2^2}{V_1^2}$$

$$dB = 20 \log_{10} \frac{V_2}{V_1}$$

Expressing power as I^2R leads to the following equation when the currents I_1 and I_2 exist in *equal resistances*:

$$dB = 20 \log_{10} \frac{I_2}{I_1} \tag{4.9}$$

EXAMPLE 4.22   An amplifier (repeater) in a telephone line takes a voltage of 0.01 V from the terminals of a 500-Ω resistor (to which it is perfectly matched) and delivers an output voltage of 10 V to a 500-Ω load resistance which is in the form of a continuation of the telephone line.

(a) Calculate the decibel gain, using 1 mW as the zero reference.
(b) Calculate the input and output powers of the amplifier.
(c) Verify your answer for part (a) by use of the power equation.

Solution

(a) $\text{Decibels} = 20 \log \dfrac{10}{0.01} = 20 \log 1000 = 20 \times 3 = 60$

(b) $P_{\text{in}} = V^2/R = 0.01^2/500 = 2 \times 10^{-7}\,\text{W}$
$P_{\text{out}} = 10^2/500 = 2 \times 10^{-1}\,\text{W}$

(c) $\text{Decibels} = 10 \log (2 \times 10^{-1})(2 \times 10^{-7})$
$\text{Decibels} = 10 \log 10^6 = 60$

4.13 Nepers

The term *attentuation* is used in telephone work and in some other applications to mean a *decrease* or a *loss* of power, voltage, or current. *Amplification* means an increase in these quantities.

If an output current, I_2, is delivered from an element or a network into which I_1 is sent, the current amplification, or gain, is defined as

$$n = \ln \frac{I_2}{I_1} \tag{4.10}$$

If I_2 is less than I_1, the natural logarithm is negative and there is a negative gain, which is an attentuation. Voltage values may be used instead of current values if such measurements are taken.

EXAMPLE 4.23   A network has an input of 50 mV and an output of 5 mV. Compute the attentuation in nepers (Np):

Solution

$$\text{Np} = \ln \frac{50 \times 10^{-3}}{5 \times 10^{-3}}$$

$$\text{Np} = \ln 10 = 2.3\,\text{Np}$$

Power gain in nepers is defined as $0.5 \ln (P_2/P_1)$, where P_2 is output power and P_1 is input power. This is obtained as follows:

$$\text{Since } P = I^2 R, \qquad I = \sqrt{\frac{P}{R}}$$

$$\text{Np} = \ln \frac{I_2}{I_1} = \ln \sqrt{\frac{P_2/R}{P_1/R}} = \ln \sqrt{\frac{P_2}{P_1}}$$

$$\text{Np} = 0.5 \ln \frac{P_2}{P_1} \tag{4.11}$$

1 Np = 8.686 dB

4.14   Review Questions

1. What is the amount of energy expended when 1 N of force acts through 1 M of distance?
2. What is the relation between the joule and the watt-second?
3. Name a unit of mechanical energy different from the joule, and two units of electrical energy different from the watt-second.
4. The horsepower is the equivalent of how many foot-pounds per minute? Per second?
5. Define the watt in terms of joules.
6. A 10 lb weight is lifted a vertical distance of 5 ft. How much work is done? How much potential energy was added to the weight?
7. Define British thermal unit.
8. Define (a) calorie, (b) kilogram calorie
9. How many calories are there in a Btu?
10. How many foot-pounds of energy are equivalent to 1 Btu? What special name has this quantity?
11. Define *power.*
12. Define *efficiency.*
13. A current of I amperes flows in a resistance of R ohms and a potential difference V exists across its terminals. Express the amount of power in three different ways.
14. At light load a certain motor needs an input of 746 W to deliver 0.5 hp. What is its efficiency at that load?
15. What is the logarithm of zero?
16. Why do electric power companies have a "sliding scale" of charge rates for supplying electricity to residence customers?
17. Define *bel* and *decibel.* How are decibels computed?
18. What is the power value at zero dB reference?
19. Define attenuation.
20. Define *neper.* How is the neper value computed?
21. Is the neper larger, or smaller, than the decibel?
22. How do you convert a loss of N nepers into decibel units?

4.15   Problems

13. A bag of groceries weighing 10 lb is carried from the ground up two flights of stairs and placed on a table. The elevation of the table top above the ground is 24 ft. How much has the potential energy of the contents of the bag been increased?
14. The bag of groceries of Problem 1 was lifted 1 ft above the table and held in that position until the person who lifted it got tired of holding it there. How much work was done (mechanically speaking) on the contents before it was

handed to someone else? If the second person put the bag of groceries back on the table, what was the net change in potential energy of the bag of groceries (net work done) from the instant it was lifted from the table?

15. How many joules of energy are given up by a 1-ton object that falls to the ground from an elevation of 10 ft?

16. A pump fills a 10,000 gal tank by pumping the water an average distance of 87 ft above the level of a reservoir. How many joules of energy were required, assuming the water to weigh $8\frac{1}{3}$ lb/gal?

17. The pump in Problem 16 filled the tank in half an hour. What was its rate of doing work expressed in horsepower?

18. The pump in Problem 16 has an efficiency of 70 percent. What was its horsepower input to the pump?

19. Calculate the number of joules in one calorie.

20. How much energy is required to heat 1 kg of water from freezing temperature to boiling temperature? Give the answer in calories and kilogram calories.

21. Express the answer of Problem 8 in joules.

22. An electric heater generates 45.36 kg-cal of energy in heating a mass of water. Heat losses are 20 percent of the energy input to the heater. How many joules of energy are supplied to the water? How many watt-hours does the heater consume?

23. An electric motor operates at 85 percent efficiency and delivers 5 hp for 4 h. Compute the motor input in kilowatt hours during that time.

24. A 5-W power resistor operates properly on 120 V. (a) What is its normal current capacity? (b) What is its resistance?

25. How much voltage should be applied to a 40-Ω, 10 W resistor to cause it to carry the maximum safe current?

26. What wattage rating should be specified in the selection of a 400-Ω resistor that must carry 25 mA?

27. How much resistance must a heater unit have in order to accept energy at the rate of 1 kW on a 120 V line?

28. An amplifier accepts 10 mW of power and delivers 5 W at its output. What is the decibel gain?

29. A communication network receives a 5-V signal at its 500-Ω input and delivers only 0.5 V at its 500-Ω output. Compute the decibel loss using voltage values. This method is applicable only when the voltages exist across equal resistances.

30. Compute the decibel loss in Problem 29, using power values.

31. Compute the neper loss in the network of Problem 29. Is the ratio of your calculated decibel loss to your neper loss correct?

32. A residence used electrical energy over a one-month period. The meter reader recorded 069,508 kWh as the reading at the end of the month. At the end of the previous month the reading was 068,954 kWh. Using the billing rate of Example 16, calculate the bill for the month's service.

33. An electric motor operates 8 hours a day, 23 days in a month, at an average output of 10 hp. Its efficiency is 87 percent. Calculate the cost of supplying this energy if the billing rate is constant at 1.5 ¢ per kWh.

Group B

34. How many Btu of heat energy are required to bring a quart of water from freezing temperature to boiling temperature?

35. A liter of water at 100°C cools to 20°C, the room temperature. How many calories of heat did it give off? How many Btu?

36. Friction in a manufacturing process wasted 70,020 ft-lb of energy. How much heat was produced in the operation?

37. An electric heater takes 15 A at 120 V. How long will it take to convert electrical energy into 1000 kg-cal of heat energy?

38. How much current is taken from a 240-V supply by a motor that is delivering mechanical energy at the rate of 7.5 hp and is 90 percent efficient?

39. An electric clock takes 2 W of power. Calculate the cost of operating it a year if the billing rate is 2¢ per kilowatt hour.

40. A small neon lamp in a wall switch at a stairway in a residence takes $\frac{1}{8}$ W. After working Problem 39, estimate closely the cost of continuous burning of this little neon lamp for a year.

41. An electric motor drives a pump that has 60 percent efficiency. The motor efficiency is 80 percent and it takes 5 A at 120 V from a supply line. What horsepower does the pump deliver?

42. A certain make of wire-wound resistor is available commercially in capacities of 1.5, 3.25, 5, and 11 W. What capacity would you choose if the resistor had to operate continuously on 15 V and have 50 Ω resistance?

43. How much gain, in decibels, does a preamplifier produce if it takes in 10^{-5} W and puts out 10^{-2} W?

44. A network has a loss of 3 dB when its input is 1 W. What is its output power?

45. An amplifier delivers twice the power it receives. What is its gain in decibels?

CHAPTER FIVE

Kirchhoff's Laws, Resistances in Circuits

5.1 Series Circuits

We have looked at some examples of simple circuits that contain a dc source and one or two resistance units. A circuit that contains two or more resistance units connected end to end, so that the same current flows through all units before it gets back to the source, is called a *series circuit*.

Figure 5.1 illustrates a series circuit that contains three resistance units with different ohms values. The current *I* (let's assume it to be 5 A) flows through all three units and returns to the source, *undiminished in amount*. If we were to open the circuit at point *a* and insert a current-measuring instrument (ammeter) the meter would read 5 A. If we were to do the same thing at points *b*, *c*, and *d* in succession, each reading would be 5 A. You are advised to read Section 1.1 again.

In a series circuit there can be only one, and the same, current value at every point.

The circuit of Fig. 5.1 can be represented by the circuit of Fig. 5.2 for the purpose of calculating the amount of current it carries. The resistance R_e is

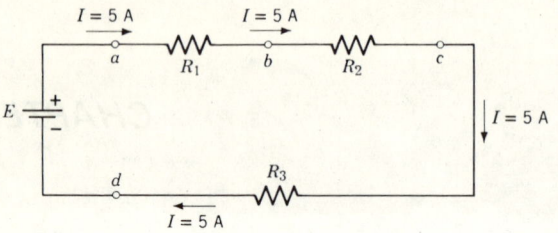

FIGURE 5.1   Series circuit.

called the *equivalent resistance* of all of those in Fig. 5.1, and we obtain its value by adding together the three resistances:

$$R_e = R_1 + R_2 + R_3 \qquad (5.1)$$

To calculate the current in the circuit we apply Ohm's law.

EXAMPLE 5.1   In the circuit of Fig. 5.1 the resistances are assumed to be $R_1 = 20\,\Omega$, $R_2 = 60\,\Omega$, $R_3 = 40\,\Omega$, and the applied potential difference 96 V. How much current does the source send through the circuit?

Solution

$$R_e = 20 + 60 + 40 = 120\,\Omega$$

$$I = \frac{E}{R_e} = \frac{96}{120} = 0.8\,\text{A}$$

The battery, quite obviously, would supply the same amount of current whether connected to a single resistance of 120 Ω or a set of separate resistances, all connected in series, that totaled 120 Ω.

The potential difference applied by the source to the set of series resistances is the sum of the potential drops across the individual resistance units. Let $R_1, R_2, R_3 \ldots R_n$ represent n resistance units in series. As explained, the equivalent resistance is obtained by adding all their values together.

Let E_T represent the total voltage applied to a series circuit containing n resistances in series. Then

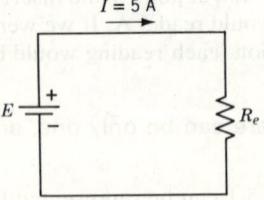

FIGURE 5.2   Circuit equivalent to that of Fig. 5.1. $R_e = R_1 + R_2 + R_3$.

$$E_T = IR_1 + IR_2 + IR_3 + \cdots + IR_n \tag{5.2}$$

And since the same current flows in all of the resistance units,

$$I = \frac{V_1}{R_1} = \frac{V_2}{R_2} = \frac{V_3}{R_3} = \cdots = \frac{V_n}{R_n} \tag{5.3}$$

in which the V's represent the voltage drops across the corresponding single resistance units. The total voltage drop is equal to the terminal voltage, E_{T_i} of the source.

Applying Eq. (5.2) to the circuit of Fig. (5.1)

$$E_T = 0.8 \times 20 + 0.8 \times 60 + 0.8 \times 40 = 96 \text{ V}$$
$$E_T = \quad 16 \quad + \quad 48 \quad + \quad 32 \quad = 96 \text{ V}$$

Applying Eq. (5.3),

$$I = \frac{16}{20} = \frac{48}{60} = \frac{32}{40} = 0.8 \text{ A}$$

5.2 Voltage Division in a Series Circuit

Figure 5.3 shows the voltage drops across the individual resistance units of another series circuit. Concentrating on two of the units, R_1 and R_2, we have

$$I = \frac{V_1}{R_1} = \frac{V_2}{R_2} \tag{5.4}$$

From this

$$\frac{R_1}{R_2} = \frac{V_1}{V_2} \tag{5.5}$$

This means that *when two or more resistances are in series, the ratio of any two of the resistance values, in ohms, is equal to the ratio of the corresponding voltage drops across them.*

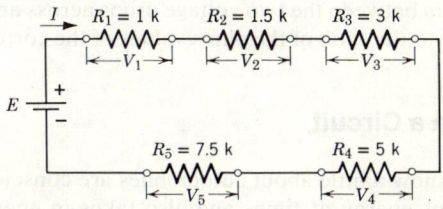

FIGURE 5.3 V_1 is known. All other voltages and the current can be calculated.

EXAMPLE 5.2 In the circuit of Fig. 5.3 the voltage across R_1 is 1 V. (a) How much voltage drop is there across R_2? (b) How much across R_5? (c) How much current is flowing? (d) What is the equivalent resistance? (e) What is the amount of the applied voltage V?

Solution

(a) $\dfrac{R_2}{R_1} = \dfrac{V_2}{V_1}$

$\dfrac{1500}{1000} = \dfrac{V_2}{1}$

$V_2 = 1.5$ V

(b) $\dfrac{V_5}{V_1} = \dfrac{7500}{1000}$ and since $V_1 = 1$ V,

$V_5 = 7.5$ V

(c) $I = \dfrac{1}{1000} = 0.001$ A $= 1$ mA

(d) $R_e = 10^3(1 + 1.5 + 3 + 5 + 7.5) = 18$ kΩ
(e) $V = IR_e$
$V = 1 \times 10^{-3} \times 18 \times 10^3 = 18$ V

Summarizing what we have learned about *series circuits*, let us remember that:

(a) The current in a series circuit exists in all units of the circuit at any time and it has only one value.
(b) The equivalent resistance of a group of resistance units connected in series is equal to the sum of their individual resistances.
(c) The voltage applied to a series circuit containing two or more resistance units is equal to the sum of all of the individual voltage drops across the resistance units.
(d) If the resistance of any unit is changed, but the applied voltage is not changed, the amount of current will change and so will the amount of voltage drop across each and every resistance unit.
(e) The *ratio* between the two voltage drops across any two resistance units is equal to the ratio of the ohms values of the corresponding resistances.

5.3 The Battery in a Circuit

All of us who know a little about automobiles are conscious of the fact that the battery puts out energy at times and also takes in energy at other times. In daytime driving, the electric generator of the car sends current into the battery at its + terminal and thus supplies energy to it. We say the battery is being charged. However, we shall first discuss the conditions in a battery circuit while it is discharging, and take up the charging situation afterward.

While a battery delivers current to a receiver, such as parking lights, the current encounters resistance inside each battery cell as it flows from plate to plate. In a circuit diagram we usually represent the total internal resistance of a battery as a single, "lumped" unit such as that labeled r in Fig. 5.4. In this circuit we assume that a lamp load having $0.57\ \Omega$ is connected across the output terminals of the battery. This is the equivalent resistance of several lamps in parallel. We shall calculate the discharge current.

It happens that each cell of this battery has an internal resistance of $0.005\ \Omega$. and therefore six cells in series have a total internal resistance of $6 \times 0.005 = 0.03\ \Omega$, which is represented by r in the circuit diagram.

The EMF of 12.6 V must send current through a total equivalent resistance of $R_e = 0.03 + 0.57 = 0.6\ \Omega$. Using Ohm's law the current is

$$I = \frac{12.6}{0.6} = 21\ A$$

Now what is the voltage V_T at the battery terminals during discharge? One way to get at this is to start with the total voltage (EMF) available and subtract from it the voltage drop inside the battery:

$$V_T = 12.6 - 21 \times 0.03$$
$$= 12.6 - 0.63 = 11.97\ V$$

Another way is simply to note that the *terminal voltage* is equal to the *voltage drop through the load* connected to the terminals:

$$V_T = 21 \times 0.57 = 11.97\ V$$

Study the circuit and convince yourself that these solutions are logical.

When a battery *is being charged* it is necessary to apply a *terminal voltage* larger than the EMF volts of the battery in order to drive current back through it. Before looking into this situation we should learn a basic law that deals with the sum of voltage drops and voltage rises around a closed circuit.

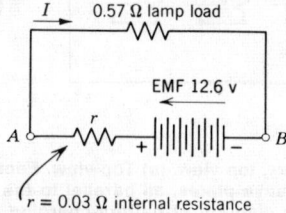

FIGURE 5.4   A storage battery with terminals *A* and *B* supplies current to a group of automobile lights. Their equivalent resistance is 0.57 Ω.

5.4 Kirchhoff's Voltage Law

Look at Fig. 5.3 and then at Fig. 5.2. You can see that if we start at some point in the circuit and add up all of the voltage drops around a closed path, this sum is equal to the total voltage applied, which we call a voltage rise.

Gustav Robert Kirchhoff (Germany, 1824–1887) discovered that *the sum of all of the voltage drops around a closed path in any electrical circuit must be equal to the sum of all of the voltage rises.* We consider the EMF of a battery a *voltage rise* if we pass in our imagination through the battery *from its negative to its positive terminal* as we "go around" the closed path in the process of analyzing the circuit.

EXAMPLE 5.3   Fig. 5.5 illustrates all of the elements and connections in a battery-charging circuit, and the equivalent circuit that is used in calculating the applied voltage required to send 8 A of charging current through the battery. *Each cell* has an EMF of 2.1 V and an internal resistance of 0.005 Ω. All six cells are connected in series, of course. We desire to know how much voltage must be applied to the battery terminals to cause 8 A of charging current to flow through it.

Solution. Starting with the applied voltage E and going clockwise around the equivalent circuit (Fig. 5.5c) we have 12.6 V as a voltage drop (which we must supply to overcome the EMF chemically generated, that is trying to

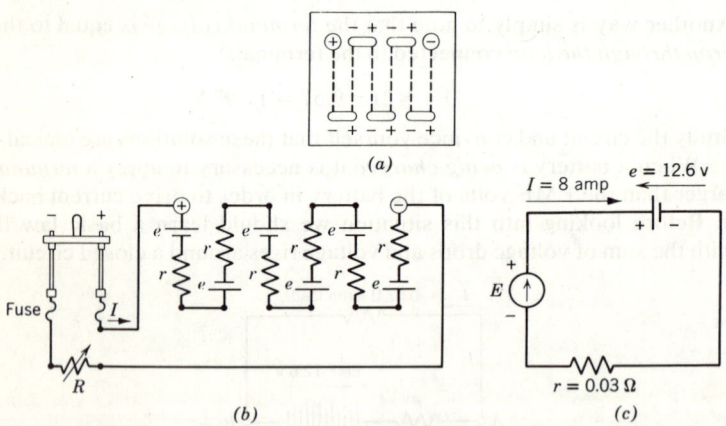

(a)

I = 8 amp e = 12.6 v

Fuse I

E

R

(b) r = 0.03 Ω

(c)

FIGURE 5.5 **Six-cell storage battery, top view. (a) Top view. Each cell has a set of positive plates sandwiched between a set of negative plates, all parallel to the dash lines between the + and − terminals of a cell. (b) Schematic diagram of battery-charging circuit. Internal emf of each cell represented by e, internal resistance by r. R is a current-regulating resistance. (c) Condensed schematic for Example 5.3.**

prevent current from entering) and I times internal r_t as a voltage drop:

$$E - 12.6 - 0.03 \times 8 = 0$$
$$E = 12.84 \text{ V}$$

This is enough voltage to overcome (or neutralize) the opposing EMF of the battery and to allow for a 0.24 V drop inside it.

EXAMPLE 5.4   How much charging current would 13.8 V send through the battery of Fig. 5.5?

Solution

$$13.8 - 12.6 - 0.03\,I = 0$$
$$I = 40 \text{ A}$$

In these two examples we note that the EMF (12.6 V) appeared in both equivalent circuits. It may be thought of as the "driving voltage" in the *discharge* case and the "bucking voltage" in the *charge* case.

It is obvious that *when no current* is passing from or to the battery, the terminal voltage of a battery is equal to the chemically generated EMF. If we open the circuit in either the discharging case (Fig. 5.4) or the charging case (Fig. 5.5) the current goes to zero and the *internal voltage drop* in the battery becomes zero. Hence, the potential difference across its terminals must be equal to the EMF.

5.5   Resistances Connected in Parallel

Resistances R_1, R_2, and R_3 in Fig. 5.6a are connected in parallel. Note that *junction a* is the upper terminal of all of them and *junction b* is the other terminal. If we choose to connect a fourth resistance, R_4, *in parallel* with these two, we should connect one end of it to junction a and the other to junction b.

Two or more circuit elements (in this case, resistances) are in parallel when all of them have the same two junction points.

Where two or more resistances are connected in parallel, it is often advantageous to calculate the value of *an equivalent resistance*, R_e, that could take their place in the circuit. By "taking their place" it is meant that R_e would carry current equal to the sum of the currents in the separate resistances acting together, and the voltage drop across R_e would be the same as the voltage drop across the group of parallel resistances at all times, whatever the current value would be.

Suppose a voltage V is applied to junctions a and b to which the three resistors in Figure 5.6a are connected. They would carry currents I_1, I_2, and I_3 given by

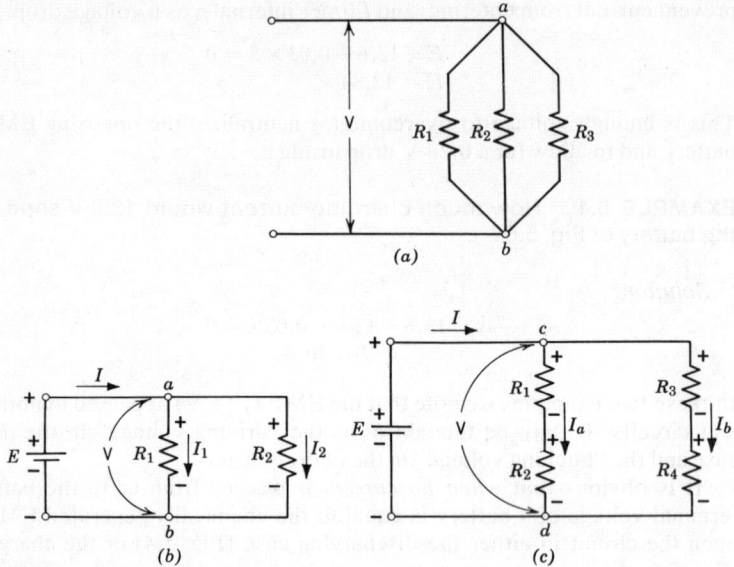

FIGURE 5.6 Parallel connections. (a) Three resistances in parallel. (b) Current supplied to two resistances in parallel. (c) Two series groups connected in parallel.

$$I_1 = \frac{V}{R_1} \quad I_2 = \frac{V}{R_2} \quad I_3 = \frac{V}{R_3}$$

The total current, I, entering junction a would be

$$I = \frac{V}{R_1} + \frac{V}{R_2} + \frac{V}{R_3} = V\left(\frac{1}{R_1} + \frac{1}{R_2} + \frac{1}{R_3}\right)$$

But the voltage divided by the total current gives the equivalent resistance of the group. That is, it gives the resistance of a single unit that can replace the three separate units taken as a group. That will allow the same total current to flow. We call this the equivalent resistance, that is, $R_e = V/I$:

$$R_e = \frac{1}{\frac{1}{R_1} + \frac{1}{R_2} + \frac{1}{R_3}} \tag{5.6}$$

If more than three resistances were in parallel, each would be represented by an additional $1/R$ term in the denominator of Equation 5.6.

EXAMPLE 5.5　　Assume four resistances are connected in parallel and their ohms values are: $R_1 = 10$, $R_2 = 20$, $R_3 = 25$, $R_4 = 40$. What would be the resistance of a single unit, called the equivalent resistance (R_e), that could substitute for them?

Solution

$$R_e = \frac{1}{\frac{1}{10} + \frac{1}{20} + \frac{1}{25} + \frac{1}{40}} = \frac{1}{0.1 + 0.05 + 0.04 + 0.025}$$

$$R_e = \frac{1}{0.215} = 4.65$$

Note that R_e is smaller than even the smallest of the four resistances it can replace. A little thought reveals why this must be so. Obviously, R_e must carry *more current* than any single one of the four. That is, when the same voltage V is applied, R_e must allow more current to flow through itself than even the smallest of the four allows. Therefore, it must have less resistance than even the smallest has.

Equation (5.6) can be rearranged into a form that may be easier to remember. The equivalent resistance, R_e, of any number (n) of resistances connected in parallel is related to them as follows:

$$\frac{1}{R_e} = \frac{1}{R_1} + \frac{1}{R_2} + \frac{1}{R_3} + \cdots + \frac{1}{R_n} \tag{5.7}$$

Problems

1. Three resistances with the following ohms values: $R_1 = 35$, $R_2 = 25$, $R_3 = 15$, are connected in series. (a) What is their total equivalent resistance? (b) How much resistance, R_4, must be connected in series with them to limit the current through them to 1 A when 100 V are impressed across the group after R_4 has been connected?

2. A resistance unit for which $R = 10\,\Omega$ must be connected to receive current from a 120-V source, but it must not be allowed to carry more than 10 A. (a) How much resistance must be connected in series with it? (b) After this is done, how much power must the added resistance unit be able to dissipate safely?

3. An old set of eight Christmas tree light bulbs have their sockets connected in series. On 120 V the current in the string is 1.5 A. In order to reduce the current in these lamps and make them last longer, the owner connected one more socket in series and thus had nine lights in the string. Assume that the resistance of each is the same as it was with only eight lamps burning and calculate the new value of current. (The resistance per lamp will actually be smaller when the current is smaller because the filament temperature will be lower.)

4. A storage battery has an EMF of 13 V and an internal resistance of 0.035 Ω. What is its terminal voltage as it supplies current to a load that has 1.265 Ω resistance?

5. In the battery circuit of Problem 5.4, how much power is delivered to the load and how much is lost in the battery?
6. Two resistors, $R_1 = 1000\,\Omega$, $R_2 = 3000\,\Omega$, are in parallel and have a 60-V drop across them. (a) What single resistance, R_3, could be used to replace them? (b) How much current would it carry?
7. Two points A and B must have $750\,\Omega$ of resistance connected between them. A $2000\,\Omega$ resistance is already connected. How could the resistance between them be brought down to $750\,\Omega$ without disturbing the $2000\,\Omega$ one? What value would you use?

5.6   Conductance

In Fig. 5.6b, I_1 is equal to V/R_1. We may also express I_1 in terms of the *conductance*, G_1, of the path from a to b, thus

$$I_1 = VG_1$$

where

$$G_1 = \frac{1}{R_1}\ \text{mho}$$

The conductance of a path is the reciprocal of its resistance.* The unit of conductance is the mho, which is ohm spelled backward.

The current in R$_2$ of Fig. 5.6b is given by

$$I_2 = V_2 G_2 \quad \text{where } G_2 = \frac{1}{R_2}$$

The total current I is

$$I = VG_1 + VG_2 = V(G_1 + G_2) = VG_T$$

and it may be obtained by multiplying the voltage V by the sum of the separate conductances. Note that *the potential difference across any one of a parallel group of resistances is the same as that across each one of the others.*

EXAMPLE 5.6   In Fig. 5.6b, $V = 100\,V$, $R_1 = 10\,\Omega$, $R_2 = 20\,\Omega$. Calculate the currents using conductances.

Solution

$$G_1 = \frac{1}{10} = 0.1\ \text{mho}, \ G_2 = \frac{1}{20} = 0.05\ \text{mho}, \ G_T = 0.15\ \text{mho}$$
$$I = VG_T = 100 \times 0.15 = 15\ A$$
$$I_1 = VG_1 = 100 \times 0.1 = 10\ A$$
$$I_2 = VG_2 = 100 \times 0.05 = 5\ A$$

---

*This is true when only direct current is present. We shall have a different expansion for conductance for the alternating current case in which elements other than pure resistance are present also.

The symbol for *mho*, which we shall not use here, is the *inverted Greek letter omega* (℧).

The total current may also be determined by dividing the applied voltage by the equivalent resistance of the parallel circuit. We may get a single expression for the total *conductance* in terms of the two resistances:

$$G_T = \frac{1}{R_1} + \frac{1}{R_2} = \frac{R_1 + R_2}{R_1 R_2}$$

R_e, the equivalent resistance of resistances that are in parallel is the reciprocal of their total conductance

$$R_2 = \frac{R_1 R_2}{R_1 + R_2} \tag{5.8}$$

From this we can see that *the equivalent resistance of two resistances* that are connected *in parallel is equal to their product divided by their sum. This is well worth memorizing.*

WARNING   When *three or more* resistances are in parallel, their equivalent resistance *is not* equal to their product divided by their sum. It is, however, equal to the reciprocal of their combined conductance.

Problems

8. What is the total conductance of a load circuit composed of resistors of 500 Ω, 1000 Ω, and 2000 Ω all connected in parallel?
9. What is the equivalent resistance of the load circuit in Problem 5.8?
10. In Figure 5.6c, $V = 200$ V and the ohms values are: $R_1 = 1$k, $R_2 = 2$k, $R_3 = 2$k, $R_4 = 4$k. Calculate (a) the conductance of each branch, (b) the total load current I, (c) the equivalent resistance of the whole load circuit.

When resistances connected in parallel are all equal, their equivalent resistance is equal to the resistance of one of them divided by the number that make up the parallel group. This may be proved as follows. Let $N = $ the number of *equal resistance* in parallel:

$$\frac{1}{R_e} = \frac{1}{R_1} + \frac{1}{R_2} + \frac{1}{R_3} + \cdots + \frac{1}{R_N} \tag{5.9}$$

Since the resistances R_1 to R_N are all equal, call each of them R.

$$\frac{1}{R_e} = \frac{1}{R} + \frac{1}{R} + \frac{1}{R} + \cdots + \frac{1}{R} = \frac{N}{R}$$

$$R_e = \frac{R}{N}$$

EXAMPLE 5.7   If six 30-Ω resistances are connected in parallel, what is their equivalent resistance?

Solution
$$R_e = 30/6 = 5 \; \Omega$$

The equivalent resistance of any number of resistances connected in parallel is *less than the value of the smallest one*. Common sense and a knowledge of what we have learned tell us this is true. Two or more resistances in parallel will carry more current, with a given applied voltage, than any single one of them carry, won't they? This means their equivalent resistance is less than that of any one of them, doesn't it?

5.7   Current Division at a Junction

It is common practice to use a simple procedure to determine how a current, such as I in Fig. 5.6b, divides between two paths at a junction point. In the example we found I to be 15 A, and we know that $R_1 = 10 \; \Omega$ and $R_2 = 15 \; \Omega$. Using these values we could have calculated I_1 in R_1 by simply multiplying the total current by the other resistance value and then dividing by the sum of the two resistances:

$$I_1 = I \frac{R_2}{R_1 + R_2} = 15 \frac{20}{10 + 20} = 10 \text{ A}$$

The following trick is well worth memorizing:

The amount of current in *one* of two parallel resistances is calculated by multiplying their total current by *the other* resistance and dividing by their sum.

Problem

11. A current of 150 mA flows into a junction of two resistances in parallel. One of them is 1.2 k and the other is 3.8 k. (a) How much of the current flows in the 1.2 k resistor? (b) Solve for the current in the 3.8 k resistor by the current division principle. Do your two answers add to give the total current exactly?

5.8   Series-Parallel Circuits

Examine Fig. 5.7. The upper part, into which I_1 is flowing, has a parallel section which has a 100-Ω resistance and a 200-Ω resistance in series with it. The lower part has 1200 Ω in parallel with 600 Ω, and as a group they are in series with a 500-Ω unit.

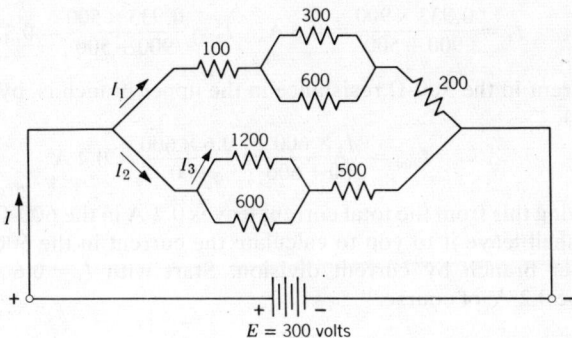

FIGURE 5.7 A series-parallel circuit. Resistances are in ohms.

The parallel pair in the top branch may be represented by R_{e1} their equivalent resistance:

$$R_{e1} = \frac{300 \times 600}{300 + 600} = 200\ \Omega$$

This, in series with 100 Ω and 200 Ω makes the total resistance of the upper branch $100 + 200 + 200 = 500\ \Omega$. The current I_1 in the upper branch is then

$$I_1 = \frac{300}{500} = 0.6\ \text{A}$$

The equivalent resistance of the parallel pair in the lower branch is

$$R_{e2} = \frac{600 \times 1200}{600 + 1200} = 400\ \Omega$$

The total resistance of the lower branch is $400 + 500 = 900\ \Omega$. The current I_2 in the lower branch is

$$I_2 = \frac{300}{900} = \frac{1}{3}\ \text{A}$$

The total current is the sum of the two which we may call 0.933 A. Using the conductances of the upper and lower branches:

$$G_1 = \frac{1}{500} = 0.002\ \text{mho}, \qquad G_2 = \frac{1}{900} = 0.00111\ \text{mho}$$

$$I = EG_T = 300 \times 0.00311 = 0.933\ \text{A}$$

Let us calculate the currents in the upper and lower branches using the current-division principle:

$$I_1 = \frac{0.933 \times 900}{900 + 500} = 0.6 \text{ A}; \quad I_2 = \frac{0.933 \times 500}{900 + 500} = 0.333 \text{ A}$$

The current in the 300-Ω resistance in the upper branch is, by current division (Sec 5.7),

$$I_{300} = \frac{I_1 \times 600}{300 + 600} = \frac{0.6 \times 600}{900} = 0.4 \text{ A}$$

Subtracting this from the total current leaves 0.2 A in the 600-Ω resistance.

We shall leave it to you to calculate the current in the 600-Ω resistance of the upper branch by current division. Start with $I_1 = 0.6$ A. Your answer should be 0.2 A, of course.

Problem

12. The current I_2 in the lower branch of Fig. 5.7 is $\frac{1}{3}$ A. Calculate the current in (a) the 1200-Ω resistance, (b) the 600-Ω resistance, both by current division. Do your results add to give the total current I_2 in the lower branch?

As a final example, let us calculate currents, voltages, and power in a simple series-parallel circuit.

EXAMPLE 5.8   Calculate all currents, voltages, and powers in the circuit of Fig. 5.8.

Solution. The equivalent resistance of the *parallel pair* is

$$R_e = \frac{R_2 R_3}{R_2 + R_3} = \frac{50 \times 25}{75} = \frac{50}{3} = 16\frac{2}{3}\ \Omega$$

Total resistance of the circuit is

$$R_T = R_e + R_1 = 16\frac{2}{3} + 13\frac{1}{3} = 30\ \Omega$$

Total current I is

$$I = E/R_T$$
$$I = 90/30 = 3 \text{ A}$$

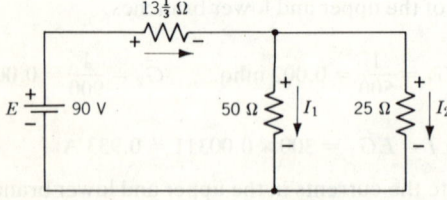

FIGURE 5.8   For Example 5.8.

Using the current-division principle,

$$I_1 = 3 \times \frac{25}{75} = 1 \text{ A} \qquad I_2 = \frac{3 \times 50}{75} = 2 \text{ A}$$

Power in $13\frac{1}{3}\,\Omega$ is $3^2 \times 13\frac{1}{3} = 120$ W

Power in $50\,\Omega$ is $1^2 \times 50 = 50$ W

Power in $25\,\Omega$ is $2^2 \times 25 = 100$ W

Total power $= 120 + 50 + 100 = 270$ W

Power delivered by the source $= EI = 90 \times 3 = 270$ W

Voltage drop across the $13\frac{1}{3}\,\Omega$ series resistance is

$$V = IR = 3 \times 13\frac{1}{3} = 40 \text{ V}$$

This leaves $90 - 40 = 50$ V for the parallel resistances. Let's get this in another way. The voltage drop across the $50\text{-}\Omega$ resistance is $50I_1 = 50 \times 1 = 50$ V. The voltage drop across the $25\text{-}\Omega$ resistance is $25I_2 = 25 \times 2 = 50$ V.

Verifying Kirchhoff's voltage law, let's choose a closed path up through the battery, to the right through $13\frac{1}{3}\,\Omega$ and down through the $50\,\Omega$ back to the bottom of the battery:

$$E - I \times 13\frac{1}{3} - I_1 \times 50 = 0$$
$$90 - 3 \times 13\frac{1}{3} - 1 \times 50 = 0$$
$$90 = 40 + 50$$

We may also verify this law by going around the complete path from the junction of the three resistors at the top, down through $25\,\Omega$ to the bottom junction, then up through $50\,\Omega$ to the starting point. We encounter a *voltage drop* when we go down through $25\,\Omega$ and a *voltage rise* when we go up (against the current arrow) through the $50\,\Omega$ to our starting point. We shall consider $E = 0$ on the left side of our equation because there is no *source* in this loop:

$$0 = -1 \times 50 + 2 \times 25$$
$$50 = 50$$

5.9 Kirchhoff's Current Law

In Example 5.6, Fig. 5.8, we obtained the values of three currents that were unknown at the beginning of the solution. I, the current entering the top junction, was calculated to be 3 A and the two currents leaving this junction were $I_1 = 1$ A, $I_2 = 2$ A. These results verify **Kirchhoff's current law** which states:

At any junction point in a circuit, the sum of all currents entering the junction equals the sum of all currents leaving the junction

This law is easy to understand, and very useful in analyzing complicated circuits as well as simple ones.

EXAMPLE 5.7   The circuit of Fig. 5.9 contains two known source voltages, three known resistances, and three unknown currents. Set up equations based on Kirchhoff's laws and solve them for the current values.

Solution. The first thing we do is label the currents at a junction. We draw a short arrow pointing toward junction *a* representing the current in R_1 and label it I_1. We do the same thing to represent the currents in the other elements connected to that junction: thus we designate them as I_2 and I_3. Having *assumed* these directions we place polarity marks. There is no rule that tells which way to point any of the current arrows. Evidently we choose to show I_1 entering the junction and I_2 and I_3 leaving it. *If we have guessed wrong* in one or more of these indicated directions, *the numerical answer* for the current *will come out negative.* As a matter of fact, this will be the case for one of the currents here.

Applying the voltage law to the closed path at the left, we start at the bottom junction and proceed clockwise. E_1, which is 105 V, is a source voltage and a voltage rise, so we use a positive sign. The other voltages in this path are I_1R_1 and I_3R_3; both are voltage drops.

$$105 - 3000\,I_1 - 6000\,I_3 = 0 \tag{1}$$

Starting at the lower right corner and proceeding clockwise around the closed path at the right, we find $E_2 = 500$ V to be a source, I_2R_2 a voltage drop, and I_3R_3 a *voltage rise*. We shall find it very helpful to place polarity marks on every resistance symbol showing the current passing through it *from plus to minus*. Then we may quickly and accurately determine if the IR voltage is a

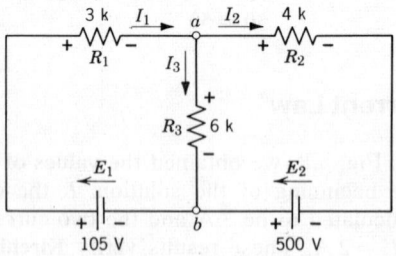

FIGURE 5.9   For Example 5.9.

drop or a rise. We must use the *plus sign* every time we go *through a resistance from minus to plus* as we mentally pass around a closed mesh:

$$500 - 4000\,I_2 + 6000\,I_3 = 0 \tag{2}$$

Now, let's rewrite these two equations after dividing every term by 1000:

$$3I_1 + 6I_3 = 0.105 \tag{1'}$$
$$4I_2 - 6I_3 = 0.500 \tag{2'}$$

Because there are three unknowns and only two equations, we must write another independent equation, making use of the current law. At junction *a* at the top,

$$I_1 - I_2 - I_3 = 0 \tag{3}$$

We may readily evaluate I_3 by solving for I_1 in Equation (1') and I_2 in Equation (2'), and placing the results in Equation (3).

$$\text{From (1')}\quad I_1 = \frac{0.105 - 6I_3}{3} = 0.035 - 2I_3 \tag{4}$$

$$\text{From (2')}\quad I_2 = \frac{0.500 + 6I_3}{4} = 0.125 + 1.5I_3 \tag{5}$$

Substituting into (3)

$$0.035 - 2I_3 - 0.125 - 1.5I_3 - I_3 = 0$$

From this,

$$I_3 = -0.020\ \text{A}$$

Because I_3 comes out *negative* we say we guessed the wrong direction for it when we drew the current arrow that denotes its direction in the circuit diagram. We should have pointed its arrow upward. I_3 is actually 0.020 A, or 20 mA, but it actually flows upward into junction *a*.

It is easy to evaluate I_1 and I_2. From (4), $I_1 = 0.075$ A and from (5), $I_2 = 0.095$ A.

The voltages across the resistances are:

$$V_{R_1} = I_1 R_1 = 0.075 \times 3000 = 225\ \text{V drop}$$
$$V_{R_2} = I_2 R_2 = 0.095 \times 4000 = 380\ \text{V drop}$$
$$V_{R_3} = I_3 R_3 = -0.020 \times 6000 = -120\ \text{V drop}$$

V_{R_3} is a negative drop. This means the polarity marks on R_3 should be reversed, and the arrow for I_3 should be reversed.

The powers in the resistors are

$$P_1(\text{in } R_1) = V_1 I_1 = 225 \times 0.075 = 16.875 \text{ W}$$
$$P_2(\text{in } R_2) = V_2 I_2 = 380 \times 0.095 = 36.100 \text{ W}$$
$$P_3(\text{in } R_3) = V_3 I_3 = 120 \times 0.020 = \underline{2.400 \text{ W}}$$

$$\text{Total power} \qquad = 55.375 \text{ W}$$

IMPORTANT R_3 receives energy and therefore is classified with R_1 and R_2 as a receiver and not as a source, even though the voltage across it had to be treated as a *rise*, simply because in writing the voltage equation for *clockwise* travel around the path we were obliged to move *against the I_3 arrow*. Note that we moved *with* the arrow in writing Equation (1) so the term $6000 I_3$ represented a *voltage drop* and was given a positive sign.

Don't let this apparent complication throw you. You will have time to get it straightened out when you do some exercises using the Kirchhoff laws. How about doing Problem 13 now?

Problem

13. Solve for all of the currents in the circuit of Fig. 5.10.

5.10 Slide-Wire Resistance (Voltage Divider)

A useful device that finds application in various forms in physics and electronics laboratories is represented, schematically, in Fig. 5.11. Coils of resistance wire are wound uniformly on an insulating tube and a sliding contact is mounted on a conducting bar fastened about half an inch above the tube. With terminal C open, that is, not connected to anything, the voltage V_{CB} can be changed in small increments from zero to V_{AB}. The *resistance per inch of wire between A and B is constant throughout its length* and, therefore, the resistance between points A and C is the same fraction of the total resistance (R_{AC}) as the distance from A to C is a fraction of the distance from A to B. Expressing the distances as lengths of wire between the points, we have

$$\frac{R_{AC}}{R_{AB}} = \frac{L_{AC}}{L_{AB}}$$

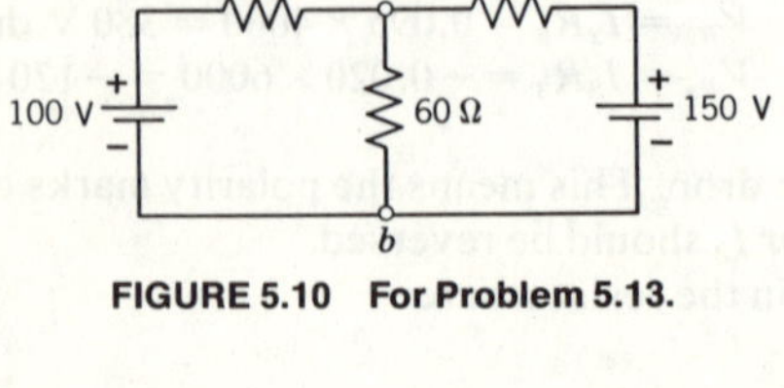

FIGURE 5.10 For Problem 5.13.

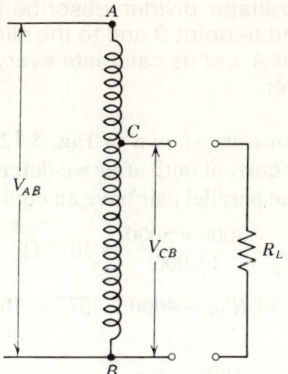

FIGURE 5.11 Slide-wire resistance (voltage divider) for providing a variable voltage V_{CB}. R_L is a load resistance that may be connected.

Since $V_{AC} = IR_{AC}$ and $V_{AB} = IR_{AB}$ when nothing is connected to the slider, we have

$$\frac{V_{AC}}{V_{AB}} = \frac{R_{AC}}{R_{AB}} = \frac{L_{AC}}{L_{AB}} \tag{5.10}$$

EXAMPLE 5.10   The total resistance of the wire in Fig. 5.11 is 12,000 Ω. The slider is set one third the way down from *A* to *B*. (a) What are the resistances of the parts of the device that are above and below point *C*? (b) What is the voltage V_{CB} when $V_{AB} = 100$ V? (c) How much current flows from *A* to *C* and from *C* to *B*?

Solution. Refer to Fig. 5.11:

(a) $\dfrac{R_{AC}}{12{,}000} = \dfrac{1}{3}$; $R_{AC} = 4{,}000$ Ω

(b) $\dfrac{V_{CB}}{100} = \dfrac{2}{3}$; $V_{CB} = 66.67$ V

(c) $I_{AC} = \dfrac{100}{12{,}000} = \dfrac{1}{120} = I_{CB}$

Remember that these relations hold *only when nothing is connected between the slider and either end of the device.* If a load, represented by the resistance R_L, is to be supplied with current and is connected between *C* and *B*, the current in the upper section (*A*, *C*) will no longer be equal to that in the lower section. It will be greater. It will be the sum of the current in the lower section and the current in R_L, which we usually call I_L. In this case we have a series-parallel circuit.

EXAMPLE 5.11   The voltage divider described in Example 5.10 has a load of 5000 Ω connected to point *B* and to the slider *C* when it is one-third the way down from point *A*. Let us calculate everything we can about currents, voltages, and power.

Solution. The conditions are shown in Fig. 5.12. We cannot calculate the input current or any other current until after we determine the equivalent resistance between *A* and *B*. The parallel pair have an equivalent resistance of:

$$R_{CB} = \frac{8000 \times 5000}{13,000} = 3077 \ \Omega$$

$$R_e = R_{AB} = 4000 + 3077 = 7077 \ \Omega$$

The total current is

$$I_T = \frac{100}{7077} = 0.01413 = 14.13 \text{ mA}$$

$$I_L = 14.13 \times \frac{8000}{13000} = 8.69 \text{ mA}$$

$$I_{CB} = 14.13 \times \frac{5000}{13000} = 5.44 \text{ mA}$$

These total 14.13 mA as they should.

$$V_{CB} = 8.69 \times 10^{-3} \times 5000 = 43.45 \text{ V}$$

This is the same as $5.44 \times 10^{-3} \times 8000$, within $\frac{1}{5}$ of 1 percent.
Exact equality can be achieved by using common fractions in place of decimals.

$$V_{AC} = 100 - 43.45 = 56.55 \text{ V}$$
Check: $4000 \times 14.13 \times 10^{-3} = 56.52$ V within $\frac{1}{10}$ of 1 percent.

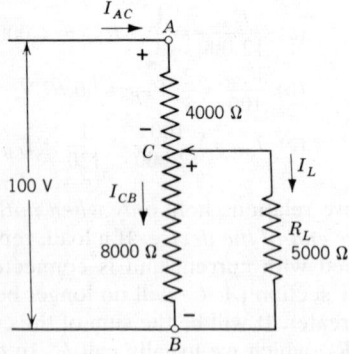

FIGURE 5.12   For Example 5.11 and Problem 14.

Power to the load is

$$P_L = E_L L_L = 43.45 \times 8.69 \times 10^{-3} = 0.377 \text{ W}$$
$$\text{Check: } P_L = I_L^2 R_L = 8.69^2 \times 10^{-6} \times 5000 = 0.377 \text{ W}$$

Power that produces heat in sections of the slidewire:

$$P_{AC} = V_{AC} I_{AC} = 56.55 \times 14.13 \times 10^{-3} = 0.799 \text{ W}$$
$$P_{CB} = V_{CB} I_{CB} = 43.45 \times 5.44 \times 10^{-3} = 0.236 \text{ W}$$

Total power in the system:

$$P_T = 0.377 + 0.799 + 0.236 = 1.412 \text{ W}$$
$$\text{Check: } P_T = 100 \times 14.13 \times 10^{-3} = 1.413 \text{ W}$$

Slide-wire resistances like the one described are often used to apply a reduced voltage to a resistance such as R_L in the above example. The wire of which the slide-wire resistance is made can carry only a definite maximum amount of current safely. A larger amount causes it to overheat, and there is danger of burning out the section between A and C. In this example, the slide-wire should have a capacity of not less than 15 mA and perhaps the ability to carry 25 mA would be much safer. If, in making adjustments in this example, the slider were moved somewhat nearer to terminal A, the current in AC would go above 15 mA. Find out about this by working Problem 14 that follows.

Problem

14. Assume that you are working with the system in Fig. 5.12 with R_L connected and V_{AB}, R_{AB}, and R_L values as given in Examples 5.8 and 5.9. In order to obtain the desired amount of current in R_L it is necessary to set the slider C only one-tenth of the distance down from A to B. Calculate the current in the top section — AC — and in the load.

CAUTION   When using a slide-wire resistance as a voltage divider, or as a variable resistance, or in any other application, *be sure its wire can safely carry the maximum current it may be required to carry.*

5.11   Review Questions

1. When two or more resistance units are connected in series, why must there be only one and the same value of current in each and every unit?
2. How would you determine the equivalent resistance of a series circuit that contains a number of resistance units not all having the same ohms value?
3. A certain series circuit has N resistance units, each having R ohms resistance. What is the value of the single resistance that can be used as their equivalent?

4. $R_1 = 80\,\Omega$ is in series with $R_2 = 20\,\Omega$. A potential difference of 120 V is applied to the series group. Determine mentally the voltage drop across R_1.

5. A string of 5 resistance units are connected in series and carry current. What is the ratio of V_3, the drop across R_3, to V_5, the drop across R_5?

6. In a series circuit containing R_1 and R_2, their resistance values are equal and, of course, they carry the same current. If the voltage applied to the series pair is not changed, how much resistance *must be added* to one of them (or connected in series) to reduce the current to half its first value?

7. Assume 10 V potential difference is applied to a series pair $R_1 = 5\,\Omega$, $R_2 = 15\,\Omega$. Occasionally a beginning student will calculate the current in R_1 as $10 \div 5 = 2$ A. What is wrong with this procedure?

8. Inside a storage battery, where does the internal resistance exist?

9. A special storage battery has 12 cells. The internal resistance in each cell is $0.001\,\Omega$. What is the total internal resistance of the battery?

10. When a storage battery delivers 10 A to a load, its terminal voltage is 12 V. If the total internal resistance is $0.06\,\Omega$, what is the EMF of the battery?

11. A storage battery has an EMF of 12.6 V. What can you say about the amount of charging current that will go into the battery if the charging device, properly connected, has a terminal voltage of 12.6 V?

12. State Kirchhoff's voltage law.

13. State Kirchhoff's current law.

14. Two parallel-connected resistances of $10\,\Omega$ each are supplied with current at a potential difference of 40 V. How much total current is supplied to them from the source? What is their equivalent resistance?

15. Define conductance of a circuit that contains only resistance units. In what unit is conductance expressed?

16. What is the conductance of each of the resistances of Question 14, and what is the equivalent conductance of the group? Does the voltage applied multiplied by the equivalent conductance give you the same answer for total current as your answer for total current in Question 14?

17. What is a short way to calculate the equivalent resistance of two resistances that are connected in parallel?

18. Five amperes of current flow in a wire. The current divides at a junction point, some of it flowing through $R_1 = 10\,\Omega$ and the remainder flowing through $R_2 = 40\,\Omega$. How much current flows through R_1 and how much through R_2?

19. A resistance R_2 is connected to the middle point of a resistance R_1. The other end of R_2 is connected to one end of R_1. Sketch this situation. Then sketch a third resistance R_3 and connect its ends to the two ends of R_1. Is it correct to consider R_1 and R_3 in parallel when calculating the equivalent resistance of all three? To replace R_1 and R_3 with a single resistance, R_e, equal to their product divided by their sum, where would we connect R_2? In the sketch you have drawn, label $R_1 = 60\,\Omega$, $R_3 = 30\,\Omega$, R_2 is in parallel with $30\,\Omega$ of the $60\,\Omega$ of R_1, and $(R_1 R_3)/(R_1 + R_3) = 20\,\Omega$. Can you see that it is impossible to connect R_2 to this $20\text{-}\Omega$ replacement resistance and have it in parallel with $30\,\Omega$, as it is in the original circuit? Evidently R_1 and R_3 must not be treated as if they are in parallel when calculating the equivalent resistance of all three. R_2 and half of R_1 are in parallel and they may be replaced by the single equivalent resistance obtained from $(0.5\,R_1 R_2)/(0.5 R_1 + R_2)$.

20. Four wires, A, B, C, and D, meet at a junction. A carries 2 A and B carries 4 A, both flowing into the junction. D carries 12 A away from the junction. How much current is carried by C and what is its direction with respect to the junction?

21. In analyzing a complicated circuit it was found that the value of the current I_5 in a certain branch came out to be -4 A. What is the significance of the minus sign?

5.12 Problems

Group A

15. A relay is designed to operate when its coil is supplied with 300 mA, and it must operate on a 120-V supply line. How much additional resistance must be used if the coil resistance is 12 Ω? How must it be connected with respect to the relay coil?

16. $R_1 = 100\,\Omega$ joins points A and B, $R_2 = 250\,\Omega$ joins points B and C, $R_3 = 650\,\Omega$ joins points C and D. How much voltage must be applied to points A and D to send 100 mA through the circuit?

17. When 200 V potential difference is applied to the set of resistors described in Problem 16, how much current flows? What is the potential difference between B and C? Obtain the answer by two different methods.

18. In the circuit of Fig. 5.13, $I_3 = 0.5$ A. What amounts of current flow through R_1 and R_2?

19. How much resistance does R_e have if it is the equivalent of the three resistors of Problem 18? How much current will flow in R_e if it is connnected in place of the three resistors?

20. What is the conductance of the parallel group in Fig. 5.13?

21. $R_1 = 30\,\Omega$, $R_2 = 50\,\Omega$ and they are connected in series. What is the conductance of this series branch of a circuit? How much current will the branch take if 120 V are impressed across it?

22. A current of 100 mA enters the junction of two resistors connected in parallel. $R_1 = 2500\,\Omega$ and $R_2 = 4000\,\Omega$. How much current flows through each resistor?

23. Three resistance units are connected in parallel. Their ohms and wattage are: $R_1 = 1000$, 20 W; $R_2 = 2000$, 10 W, $R_3 = 2500$, 10 W. What is the largest voltage that can be applied to this group without overheating one of the units?

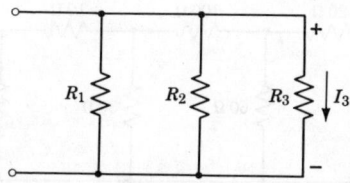

FIGURE 5.13   For Problem 18. $R_1 = 1000\,\Omega$, $R_2 = 400\,\Omega$, $R_3 = 500\,\Omega$.

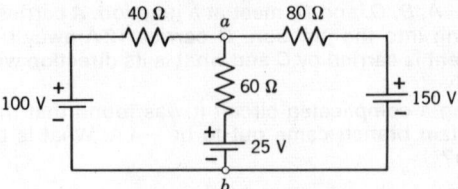

FIGURE 5.14   For Problem 27.

24. An immersion heater that dissipates energy at the rate of 100 W is designed to operate on 120 V. Calculate the series resistance necessary to cause it to operate normally from a 208-V source.·

25. What should be the wattage rating of the added resistance in Problem 24?

26. A storage battery has an internal resistance of 0.04 Ω and an *open-circuit* terminal voltage of 12.3 V. (a) What will be the voltage across a 2-Ω resistance connected to it? (b) How much voltage drop occurs inside the battery? (c) Considering the energy delivered to the external resistance (called the load) as useful, how much energy is wasted, per hour, inside the battery? (d) Calculate the efficiency of conversion)

27. Calculate the currents in the branches of the circuit of Fig. 5.14.

Group B

28. The picture tube of a television receiver receives 5×10^{-4} A of current from its high-voltage power supply which has an internal resistance of 2.5 MΩ. When the power supply is unloaded it delivers 20 kV. What amount of voltage is applied to the tube when in operation?

29. The brightness control on the television set (Problem 28) is turned so that the tube current increases 25 percent. What, then, is the voltage applied to the picture tube?

30. If an accidental short circuit occurred at the output terminals of the high-voltage transformer that supplied current to the picture tube of the television receiver of Problem 28, how much current would flow in the high voltage winding of the transformer?

31. Assume 120 V applied to terminals A, B in Fig. 5.15. Calculate the current in the 50-Ω resistance.

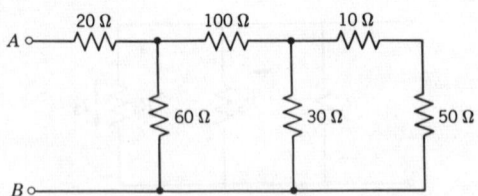

FIGURE 5.15   For Problems 31 and 32.

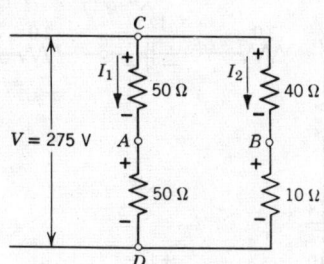

FIGURE 5.16   For Problem 36.

32. When the circuit of Fig. 5.15 was used in a different application, with a different source connected to the input terminals, it was found that 60 V existed across the terminals of the 30-Ω resistor. What, then, was the source voltage at A, B?

33. How many times faster would heat be developed in the transformer winding (Problem 30) in the case of a short circuit than would develop under normal operating conditions (Problem 28)?

34. If the resistors of Problem 23 are connected in series, what is the largest voltage that can be applied to the group without overheating one of them?

35. How much resistance must be connected in parallel with a 3000-Ω unit to produce an equivalent resistance of 2000 Ω?

36. Assume $V = 275$ V in Fig. 5.16. Compute the potential difference between points A and B and specify which is higher in potential.

37. Calculate the currents in the load resistances, and also the line currents in Fig. 5.17.

38. The stator windings of a three-phase motor are connected as represented in Fig. 5.18. The resistance measured between any two pairs of terminals, T, is 5 Ω. What is the resistance R of each winding?

39. Compute the current and the Ir drops around the circuit of Fig. 5.19. Plot the potentials (with respect to $V = 0$ at o) of the labeled points a to j, using a voltage scale on the vertical and a distance scale on the horizontal. Let about $\frac{1}{2}$ in. represent the distance between adjacent points. Show point o at $V = 0$ at both ends of the horizontal scale.

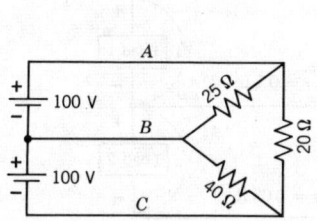

FIGURE 5.17   For Problem 37.

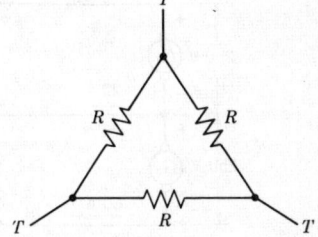

FIGURE 5.18   For Problem 38.

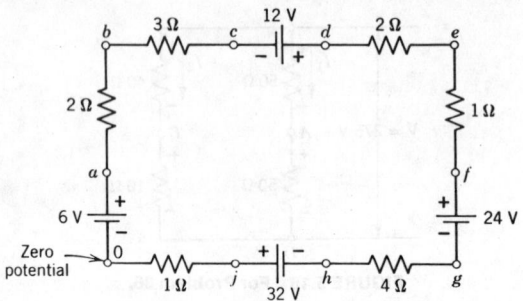

FIGURE 5.19   For Problem 39.

40. Two generators with 250 V at their terminals supply load currents, as shown in Fig. 5.20, over cables that have resistances as indicated. (a) Calculate the voltages at the loads. (b) What percent of the power delivered by the generators is lost in the transmission? (c) What is the efficiency of transmission?

41. During the operation of the system shown in Fig. 5.20, the current in Load 1 was not changed, but that in Load 2 was reduced to 20 A. (a) Calculate the terminal voltages at the loads under these conditions. (b) Compare the voltage at Load 2 with the terminal voltage of G_2. (c) Must the sum of the load voltages in this system always be less than the sum of the two generator terminal voltages?

42. Sketch the circuit of Problem 36 (Fig. 5.16) and connect an ammeter between points *A* and *B*. Assuming the meter resistance is zero, calculate the current through the meter and state its direction.

43. In the circuit of Fig. 5.21, the connecting leads *AB, CD, DE, EF, FG, KJ*, and *JH* have zero resistance. Calculate the currents in all of the resistors and also in the leads *DE* and *EF*.

44. Solve for the current, voltage, and power in the 25-Ω resistor of Fig. 5.22.

45. A 1000-Ω, 10-W resistor in an electronic circuit developed a defect. As a temporary measure it was replaced by using 1000-Ω, 5-W resistors. What minimum number could be used and how were they connected? What was the total wattage capacity?

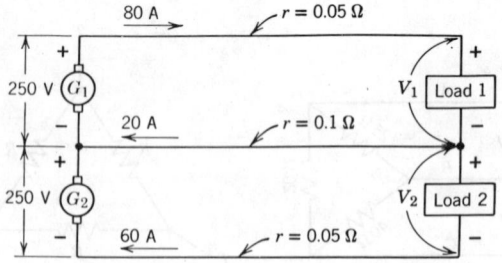

FIGURE 5.20   For Problem 41.

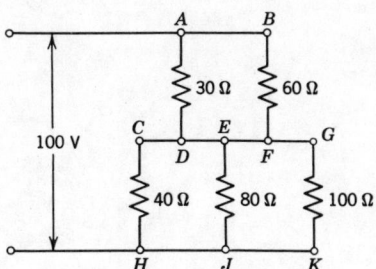

FIGURE 5.21   For Problem 43.

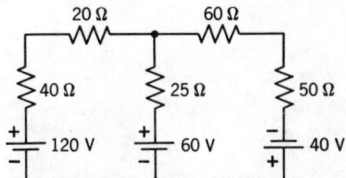

FIGURE 5.22   For Problem 44.

46. Two resistors, $R_1 = 60$ k, $R_2 = 120$ k are in series across 270 V. (a) What is the voltage across R_2? (b) A voltmeter having a resistance of 5 MΩ is used to measure the voltage across R_2. What will the meter reading be? (c) If the reading of voltage across R_2 is made with a voltmeter that has 250 k resistance, what would its reading be? (d) Calculate the percent error in each measurement, using the answer to part (a) as the correct voltage value.

47. Determine the voltage E in Fig. 5.23.

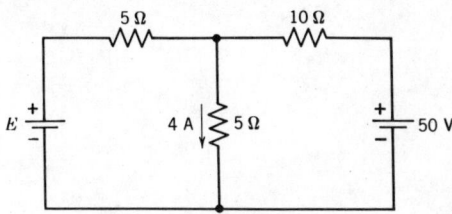

FIGURE 5.23   For Problem 47.

CHAPTER SIX

Resistance Networks

We learned in Chapter 5 that Kirchhoff's laws enable us to write current and voltage equations from which we can determine currents and voltages in circuits that do not permit complete simplification by means of substituting an equivalent resistance for two or more resistance units. Review Question 19 at the end of Chapter 5 brings out this limitation. In this chapter we shall learn how to apply the mesh method and the nodal method of writing equations, both of which will result in a minimum of unknown terms. The result is a reduction in the amount of algebra necessary to get the answers.

We shall learn, also, how to set up a simple equivalent circuit to replace a complicated original circuit, with the result that a load current, or a load voltage, is easily obtained.

6.1 Mesh Analysis

Mesh-current notation is used in a powerful method of circuit analysis. It involves representing a current that is assumed to circulate around a closed loop by a curved arrow, and labeling the arrow with its identifying current symbol I with a subscript. We are using the circuit for Problem 13 in Chapter 5 to illustrate the method.

117

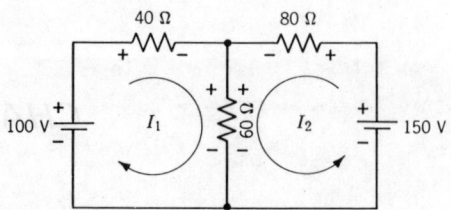

FIGURE 6.1   Circuit analyzed by mesh-current method.

In Fig. 6.1, the curved arrow I_1 represents the current that flows out of the 100-V source, through the 40-Ω resistance, *and it represents part of the current* flowing downward in the 60-Ω resistance. We say I_1 is the *mesh current*, or loop current, in the closed path made up of the 100-V battery and the 40-Ω and 60-Ω resistances. Mesh current I_2 has a corresponding meaning for the closed path containing the 150-V source, the 80-Ω resistance and the other part of the total current in the 60-Ω resistance.

It is usually advantageous to point the curved arrows, so that the current they represent comes out of the positive terminal of the source at which you start writing the equation involving that source. There are times when the arrow may go through another *source* from + to −, meaning that the current is assumed to be going through this source in the wrong direction. If this source is represented as an EMF, that value of volts will then be treated as a *drop* and will be given the same sign as IR drops in the equation. If this source has an internal resistance, r, then the current times r, that is, IR, will be treated as any other IR drop. Obviously, the current in the 60-Ω resistance in Fig. 6.1 is the sum of I_1 and I_2. You will find it very helpful to put polarity marks on the resistors on the side near the curved arrow to indicate a *fall* in potential when going with the arrow, and a rise in potential when going against it. In Fig. 6.1, I_1 goes through a fall in potential (+ to −) through the 40-Ω and 60-Ω resistors. I_2 goes through falls in potential in the 80-Ω and 60-Ω resistors.

Looking ahead at Figure 6.2, I_1 goes through a fall (top to bottom) through

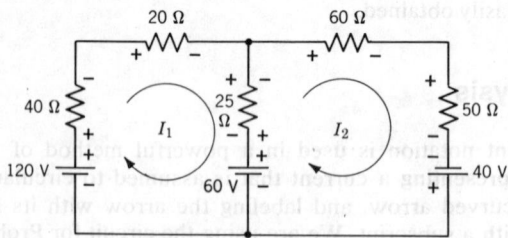

FIGURE 6.2   For Example 6.2.

25 Ω, and I_2 *contributes a rise in potential* when the Kirchhoff voltage equation is written for mesh 1.

EXAMPLE 6.1   Using mesh analysis, find the currents in the network of Fig. 6.1.

Solution. We shall now write two independent equations, since we have only two unknowns:

$$100 - 40I_1 - 60(I_1 + I_2) = 0$$
$$150 - 60(I_1 + I_2) - 80I_2 = 0$$

Collecting terms,

$$100 = 100I_1 + 60I_2 \qquad (1)$$
$$150 = 60I_1 + 140I_2 \qquad (2)$$

Multiply (1) by 0.6,

$$60 = 60I_1 + 36I_2 \qquad (3)$$

Subtracting (3) from (2),

$$90 = 104I_2$$
$$I_2 = \frac{90}{104} = 0.865 \text{ A}$$

Substituting into (1),

$$I_1 = \frac{100 - 60 \times 0.865}{100}$$
$$I_1 = 0.481 \text{ A}$$

The current in the 60-Ω resistance:

$$I_1 + I_2 = 1.346 \text{ A}$$

EXAMPLE 6.2   We shall now apply the mesh-current method of analysis to the circuit of Fig. 6.2. Note that the 60-V middle source has mesh-current arrows through it in both directions. Recall that when we pass *through a source from minus to plus* we record a *rise* in potential and when we pass *through a source from plus to minus*, we record a *drop* in potential.

Solution. Around the mesh on the left, we have

$$120 - 60 = 40I_1 + 20I_1 + 25(I_1 - I_2)$$

And on the right

$$60 + 40 = 25(I_2 - I_1) + 60I_2 + 50I_2$$

Transposing,

$$60I_1 + 25I_1 - 25I_2 = 60$$
$$-25I_1 + 25I_2 + 110I_2 = 100$$
$$85I_1 - 25I_2 = 60 \qquad (1)$$
$$-25I_1 + 135I_2 = 100 \qquad (2)$$

From (1)

$$I_1 = \frac{25 I_2 + 60}{85} = 0.294 I_2 + 0.706 \qquad (3)$$

Substituting into (2),

$$-25(0.294 I_2 + 0.706) + 135 I_2 = 100$$

Multiplying and collecting terms,

$$127.65 I_2 = 117.65$$
$$I_2 = 0.922 \text{ A}$$

From (3), after dividing through by 5,

$$I_1 = \frac{12 + 5 I_2}{17} = \frac{12 + 5(0.922)}{17} = 0.977 \text{ A}$$

The current in the 25-Ω resistance flows downward, as indicated by the fact that I_1 is larger than I_2:

$$I_1 - I_2 = 0.977 - 0.922 = 0.055 \text{ A}$$

EXAMPLE 6.3   Find the currents in all elements of the bridge network in Fig. 6.3.

Solution. Note that the mesh for I_1 includes the source and three resistances. Three independent voltage equations will be needed.

For mesh 1, going around in the direction of the I_1 arrow,

$$7 I_1 - 2 I_2 - 4 I_3 - 10 = 0 \qquad (1)$$

Note that we go opposite the directions of the I_2 and I_3 arrows:

For mesh 2,

$$2 I_1 - 10 I_2 - 6 I_3 = 0 \qquad (2)$$

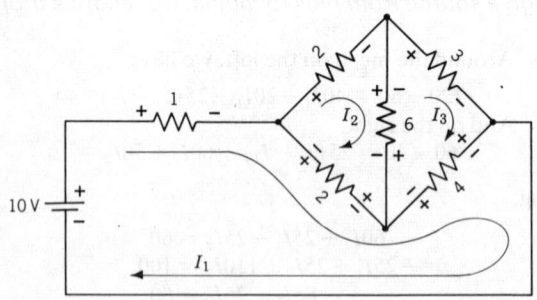

FIGURE 6.3   For Example 6.3. Resistances are in ohms.

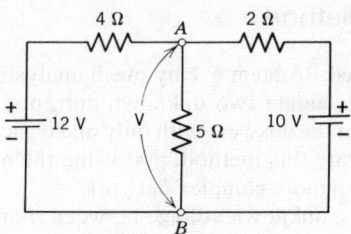

FIGURE 6.4   For Problem 1.

For mesh 3,

$$4I_1 + 6I_2 - 13I_3 = 0 \tag{3}$$

Divide through (2) by 2,

$$I_1 - 5I_2 - 3I_3 = 0 \tag{1'}$$

After multiplying (1)' by -4, combine it with (3) and get

$$26I_2 - I_3 = 0 \tag{4}$$

After multiplying (1)' by -7, combine it with (1) and get

$$33I_2 + 17I_3 = 10 \tag{5}$$

Solving (4) and (5) simultaneously gives

$$I_2 = 0.021 \text{ A}, \quad I_3 = 0.546 \text{ A}$$

Putting these into (1)' gives $I_1 = 1.743$ A, the battery current.

Problems

1. By mesh analysis, determine the amount of current in the 5-Ω resistance in Fig. 6.4.
2. Solve, by mesh analysis, for the line currents in the network of Fig. 6.5.

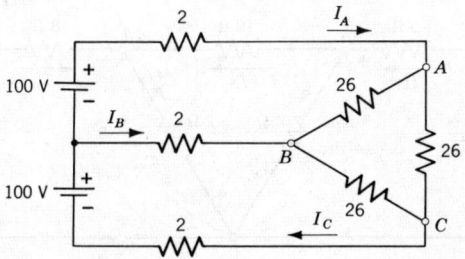

FIGURE 6.5   For Problem 2. Resistances are in ohms.

6.2   Nodal-Voltage Method

If you have worked Problem 6.1 by mesh analysis you were obliged to solve two equations to evaluate two unknown currents. The nodal-voltage method enables us to obtain the answers with only one basic equation.

We shall illustrate this method, first using the network of Fig. 6.4 and then we shall apply it to a more complex network.

Let's denote the unknown voltage between A and B by V and assume point A to be positive with respect to point B. Can you see that the current in the 4-Ω resistor is then given by $(12 - V)$ divided by 4? And the current in the 2-Ω resistor is given by $(10 - V)$ divided by 2. The current from A to B is easily denoted by $V/5$. Now we can write *one current equation* with only one unknown. The sum of the currents at junction A is zero:

$$\frac{12 - V}{4} + \frac{10 - V}{2} - \frac{V}{5} = 0$$
$$3 - 0.25V + 5 - 0.5V - 0.2V = 0$$
$$V = 8/0.95 = 8.42V$$

In the 5-Ω resistance, $I_5 = 8.42/5 = 1.684$ A.

Any *junction* of two or more branches of a network is called a *node*. It is customary to select one as a *reference node*, and to express the potential difference between each of the other nodes and the reference node in terms of an unknown voltage (symbolized as V_x, V_y, or other symbols). Sometimes the voltage is given. We shall now look at an example in which two unknown voltages V_x and V_y will be evaluated. Then they will be used in determining unknown currents.

EXAMPLE 6.4   Apply nodal analysis to the network in Fig. 6.6.

Solution. At node *a*,

$$I_1 = I_2 + I_3 .$$

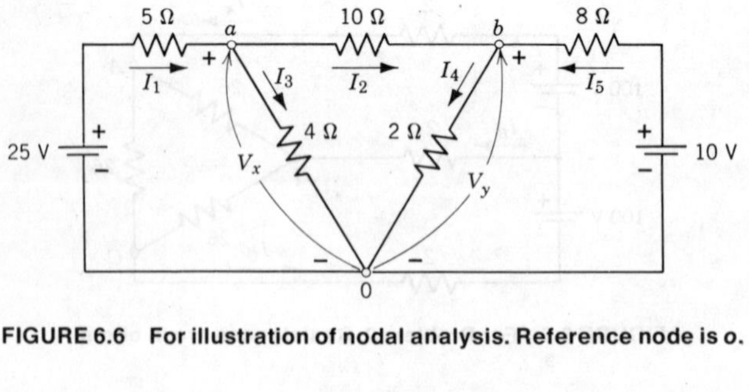

FIGURE 6.6   For illustration of nodal analysis. Reference node is o.

Observe that

$$I_3 = \frac{V_x}{4}, \quad I_1 = \frac{25 - V_x}{5}, \quad I_2 = \frac{V_x - V_y}{10}$$

This assumes $V_x > V_y$.
Substituting,

$$\frac{25 - V_x}{5} = \frac{V_x - V_y}{10} + \frac{V_x}{4} \tag{1}$$

At node b,

$$I_5 + I_2 = I_4$$

and

$$I_4 = \frac{V_y}{2}, \quad I_5 = \frac{10 - V_y}{8}, \quad I_2 = \frac{V_x - V_y}{10}$$

Substituting,

$$\frac{10 - V_y}{8} + \frac{V_x - V_y}{10} = \frac{V_y}{2} \tag{2}$$

Multiplying (1) by 20 and rearranging,

$$11V_x - 2V_y = 100 \tag{3}$$

Multiplying (2) by 40 and rearranging,

$$-4V_x + 29V_y = 50 \tag{4}$$

Solving (3) and (4) gives

$$V_x = 9.64 \text{ V} \quad V_y = 3 \cdot 05 \text{ V}$$

Using these voltages,

$$I_1 = (25 - 9.64)/5 = 3.07 \text{ A}$$
$$I_2 = (9.64 - 3.05)/10 = 0.659 \text{ A}$$
$$I_3 = 9.64/4 = 2.41 \text{ A}$$
$$I_4 = 3.05/2 = 1.525 \text{ A}$$
$$I_5 = (10 - 3.05)/8 = 0.869 \text{ A}$$

Checking:

$$I_1 = 0.659 + 2.41 = 3.069 \text{ A}$$
$$I_4 = 0.659 + 0.869 = 1.528 \text{ A}$$

We needed only two independent equations to determine V_x and V_y. Had we used the mesh-analysis method, three independent equations would have been required. After obtaining values of mesh currents, it would have been necessary to solve for I_3 and I_4 from them. You can usually reduce the amount of work and time required to analyze a network by using the nodal method instead of the mesh method.

Problems

3. Solve by the nodal voltage method for the voltage drop across the 300-Ω resistor in Fig. 6.7.

4. (a) Add a 30-V battery in series with the 300-Ω resistor in Fig. 6.7 so that its positive terminal is connected to node *a* and its negative terminal is connected to the top of the resistor. Solve for the current in R_L. (b) Verify your result by means of mesh method.

6.3   Determinants

It is worthwhile to use determinants in solving a group of simple simultaneous equations. Even if you use them only when you have just two equations and two unknowns, the method saves time and lessens chances of error.

First we shall look into the method, using very simple equations, and then apply it to the determination of two, and then three, unknowns.

EXAMPLE 6.5   Solve the following equations for *x* and *y* by determinants:

$$2x + 3y = 26$$
$$5x - 4y = -4$$

Solution. We get the values of *x* and *y* by dividing a particular quantity for each by a determinant, represented by *D*, as a *denominator*. *D* is set up using the coefficients of *x* and *y*:

$$D = \begin{vmatrix} 2 & 3 \\ 5 & -4 \end{vmatrix}$$

The numerical value of *D* is obtained by multiplying the number at the *top left* by that at the *bottom right* and from this subtracting the product of the lower left number and the *upper right* number. The arrows indicate how the products are taken, but they are not written after you learn what to do. The value of *D* is then

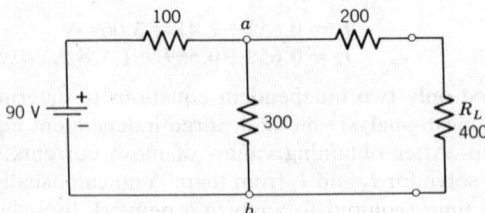

FIGURE 6.7   For Problem 3. Resistances are in ohms.

$$D = 2(-4) - (5)(3) = -8 - 15 = -23$$

A separate determinant is written for each unknown. To get x, use the numbers at the right of the equation instead of the coefficients of x, thus

$$x = \frac{\begin{vmatrix} 26 & 3 \\ -4 & -4 \end{vmatrix}}{\begin{vmatrix} 2 & 3 \\ 5 & -4 \end{vmatrix}} = \frac{-104 - (-12)}{-8 - 15} = \frac{-92}{-23} = 4$$

And the numerator for computing y is set up using 26 and -4 in place of the coefficients of y:

$$y = \frac{\begin{vmatrix} 2 & 26 \\ 5 & -4 \end{vmatrix}}{\begin{vmatrix} 2 & 3 \\ 5 & -4 \end{vmatrix}} = \frac{-8 - 130}{-8 - 15} = \frac{-138}{-23} = 6$$

Now let's solve for I_1 and I_2 in the following equations:

$$100I_1 + 60I_2 = 100$$
$$60I_1 + 140I_2 = 150$$

First divide by 10:

$$10I_1 + 6I_2 = 10$$
$$6I_1 + 14I_2 = 15$$

$$I_1 = \frac{\begin{vmatrix} 10 & 6 \\ 15 & 14 \end{vmatrix}}{\begin{vmatrix} 10 & 6 \\ 6 & 14 \end{vmatrix}} = \frac{140 - 90}{140 - 36} = \frac{50}{104} = 0.481 \text{ A}$$

$$I_2 = \frac{\begin{vmatrix} 10 & 10 \\ 6 & 15 \end{vmatrix}}{104} = \frac{150 - 60}{104} = \frac{90}{104} = 0.865 \text{ A}$$

Problem

5. Solve for I_1 and I_2 by determinants:

$$2.5I_1 + 3.2I_2 = 17.84$$
$$7.1I_1 - 3.9I_2 = 2.61$$

Slide rule computation is suggested.
If you want to take the time, solve the equation by another method and compare the amounts of computation required.

The use of determinants to solve first degree simultaneous equations with three or more unknowns saves time and, usually, a substantial amount of work.

In the case of three equations, there is no need to multiply two equations by factors that will lead to the elimination of one unknown, or to solve for one unknown in terms of the other two and substitute a two-term quantity in the two remaining equations. Instead, we deal with coefficients only, as in the case of second-order determinants.

Before we examine procedures and few simple rules for solving third- and higher-order determinants, let's set up a determinant form for three equations containing three unknowns and go through the solution:

$$2I_1 - 3I_2 + 2I_3 = 7$$
$$4I_1 + 5I_2 + I_3 = 45$$
$$-I_1 + 4I_2 - 3I_3 = -7$$

The denominator determinant that will be used to solve for the three currents represented in these equations by I_1, I_2, and I_3 is

$$D = \begin{vmatrix} 2 & -3 & 2 \\ 4 & 5 & 1 \\ -1 & 4 & -3 \end{vmatrix}$$

A process called expanding into minors, along the first row, gives the following form:

$$D = 2\begin{vmatrix} 5 & 1 \\ 4 & -3 \end{vmatrix} - (-3)\begin{vmatrix} 4 & 1 \\ -1 & -3 \end{vmatrix} + 2\begin{vmatrix} 4 & 5 \\ -1 & 4 \end{vmatrix}$$

Note that alternate + and − signs are used *in front of* the top-row numbers as these become coefficients of the minors.

From what we know about second-order determinant solutions,

$$D = 2(-15 - 4) + 3(-12 + 1) + 2(16 + 5) = -29$$

Now we undertake the task of writing the determinant to be put into the numerator of the expression that will give us the value of I_1. We use the same procedure we followed in the second-order case. We write the determinant for the numerator as we wrote one for D, *except* we replace the coefficients of I_1 in D with the constants to the right of the equal signs. This give us

$$I_1 = \frac{1}{D}\begin{vmatrix} 7 & -3 & 2 \\ 45 & 5 & 1 \\ -7 & 4 & -3 \end{vmatrix}$$

$$= \frac{1}{-29}\left[7\begin{vmatrix} 5 & 1 \\ 4 & -3 \end{vmatrix} - (-3)\begin{vmatrix} 45 & 1 \\ -7 & -3 \end{vmatrix} + 2\begin{vmatrix} 45 & 5 \\ -7 & 4 \end{vmatrix}\right]$$

$$I_1 = \frac{1}{-29}[7(-15 - 4) + 3(-135 + 7) + 2(180 + 35)] = \frac{-87}{-29} = 3\ A$$

Expressing I_2 in determinant form, we replace the coefficients of I_2 in D with the constants to the right of the equal signs:

$$I_2 = \frac{1}{D} \begin{vmatrix} 2 & 7 & 2 \\ 4 & 45 & 1 \\ -1 & -7 & -3 \end{vmatrix}$$

$$= \frac{1}{-29}\left[2\begin{vmatrix} 45 & 1 \\ -7 & -3 \end{vmatrix} - 7\begin{vmatrix} 4 & 1 \\ -1 & -3 \end{vmatrix} + 2\begin{vmatrix} 4 & 45 \\ -1 & -7 \end{vmatrix} \right]$$

$$I_2 = \frac{1}{-29}[2(-135+7) - 7(-12+1) + 2(-28+45)] = \frac{-145}{-29} = 5 \text{ A}$$

Expressing I_3 in determinant form, we replace the coefficients of I_3 in D with the constants to the right of the equal signs:

$$I_3 = \frac{1}{D} \begin{vmatrix} 2 & -3 & 7 \\ 4 & 5 & 45 \\ -1 & 4 & -7 \end{vmatrix}$$

$$= \frac{1}{-29}\left[2\begin{vmatrix} 5 & 45 \\ 4 & -7 \end{vmatrix} - (-3)\begin{vmatrix} 4 & 45 \\ -1 & -7 \end{vmatrix} + 7\begin{vmatrix} 4 & 5 \\ -1 & 4 \end{vmatrix} \right]$$

$$I_3 = \frac{1}{-29}[2(-35-180) + 3(-28+45) + 7(16+5)] = \frac{-232}{-29} = 8 \text{ A}$$

We have seen by this example how a third-order determinant is converted into a series of second-order determinants by expanding across the top row into three minors.

The element of a row, which is written as if it were a coefficient of a minor, is called a *cofactor*. For example, in the expansion of the last determinant D, the cofactors are $2, -3$, and 7. The signs that must be used in front of the cofactors may be determined from $(-1)^{i+k}$ where i represents the number of the column. Using this equation, an array of signs which may be used as a guide may be set up thus

$$
\begin{array}{l}
(i = 1) \quad + \quad - \quad + \quad - \quad \cdots \\
(i = 2) \quad - \quad + \quad - \quad + \quad \cdots \\
(i = 3) \quad + \quad - \quad + \quad - \quad \cdots \\
(i = 4) \quad - \quad + \quad - \quad + \quad \cdots \\
\qquad\qquad \cdot \qquad \cdot \qquad \cdot \qquad \cdot
\end{array}
$$

$$(k = 1) \ (k = 2) \ (k = 3) \ (k = 4)$$

As a check, take $i = 2$, $k = 3$. $(-1)^{2+3} = -1$ which indicates that the minus sign located in the second row and the third column is correct.

We have shown the basic procedure for solving third-order determinants, but we must add that expansion into minors may be done downward *along a column* also. In doing this (as well as in expanding along a row) we may find it helpful to imagine a line drawn through the column and another drawn through the whole row containing the number when it is used as the coefficient of a minor.

Let us expand the denominator determinant D again, but use the middle column rather than the top row:

$$D = -(-3)\begin{vmatrix} 4 & 1 \\ -1 & -3 \end{vmatrix} + 5\begin{vmatrix} 2 & 2 \\ -1 & -3 \end{vmatrix} - 4\begin{vmatrix} 2 & 2 \\ 4 & 1 \end{vmatrix}$$
$$D = 3(-12+1) + 5(-6+2) - 4(2-8) = -29$$

Whenever a determinant contains one or more zeros as elements it is desirable to expand along the row or column that contains the most zeros. This eliminates the minors that have the zeros as cofactors. Zero appears as an element when one of the variables is missing in an equation. The first equation in Problem 6 will have zero at the place where the coefficient of I_3 is to be placed.

Let us expand a determinant containing zeros, first along a row and then downward along a column,

$$D = \begin{vmatrix} 20 & 0 & -15 \\ -10 & 0 & 5 \\ 25 & 15 & 10 \end{vmatrix}$$

Along the first row, remembering the sign sequence is $+, -, +,$

$$D = 20\begin{vmatrix} 0 & 5 \\ 15 & 10 \end{vmatrix} + (-15)\begin{vmatrix} -10 & 0 \\ 25 & 15 \end{vmatrix} = 20(-75) - 15(-150) = 750$$

Along the middle column, remembering the sign sequence is $-, +, -,$

$$D = -15\begin{vmatrix} 20 & -15 \\ -10 & 5 \end{vmatrix} = -15(100 - 150) = 750$$

We should add that whenever it is possible to divide a number which is a common factor into the *elements of a whole row or a whole column*, it should be done because this makes the elements smaller for computation. The devisor becomes a factor in the answer. For example, we may divide every row of the determinant by 5, and thus obtain 125 as a factor:

$$D = (5)(5)(5)\begin{vmatrix} 4 & 0 & -3 \\ -2 & 0 & 1 \\ 5 & 3 & 2 \end{vmatrix} = 125\left(-3\begin{vmatrix} 4 & -3 \\ -2 & 1 \end{vmatrix}\right)$$
$$= -375(4-6) = 750$$

Problem

6. Solve for I_1, I_2, and I_3 using determinants:

$$12I_1 + 8I_2 = 28$$
$$10I_1 - 5I_2 + 15I_3 = 45$$
$$9I_1 + 6I_2 - 3I_3 = 12$$
$$\text{Answer: } I_1 = 1\text{A}, I_2 = 2\text{A}, I_3 = 3\text{ A}$$

The student may wish to study determinants more completely in a book on algebra. There he will find Cramer's rule, which expresses the value of any one of the unknowns in a general equation involving the denominator determinant D and the sum of a series of minors and their cofactors. Fourth-order determinants may be reduced to a sum of four third-order ones by the same type of procedure that is used to reduce third-order determinants to second-order ones.

6.4 Transformations: Delta to Wye or Wye to Delta

Examine carefully the circuit of Fig. 6.8 and imagine that we have not drawn circles 1, 2, and 3, or the dash lines connecting them. If we are asked to determine the amount of current, I, supplied by the battery, using only the methods we now know, we would either write three loop equations and have three simultaneous equations to solve or choose two nodal points* and set up two simultaneous equations containing two unknown voltage symbols.

There is a way to *transform the three resistance units* joining points b, c, and d (R_G, R_3, R_4 are said to be *connected in delta*) *into three new resistance*

---

*Problem 45 at the end of this chapter specifies the application of this method of solution to a bridge network.

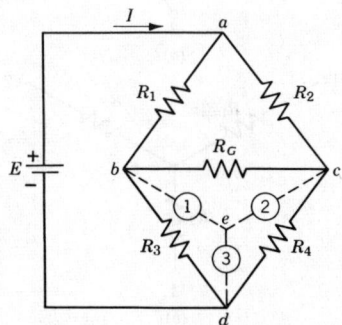

FIGURE 6.8   Network to illustrate delta-to-wye transformation.

units, 1, 2, and 3 joined together at *e* and with their other ends connected to *b*, *c*, and *d*. These three new ones are said to be connected in *wye*. This *replacement procedure* is called a Δ-*Y transformation*.

Now, what's the point in doing this? The point is that it simplifies the analysis and lets us obtain the value of the input current, I, without using simultaneous equations. If R_G, R_3, and R_4 are no longer there, R_1 will be in series with ①, and R_2 will be in series with ②. Then we can add R_1 and the resistance of ① together and get a single resistance unit which will be in parallel with another single one that will be the sum of R_2 and the resistance of ②. The rest is easy.

First we must obtain *transformation formulas* by which we can calculate the three replacement resistances. We shall work with both networks in Fig. 6.9 and begin by hunting for values of R_1, R_2, and R_3 that we can use as replacements for R_A, R_B, and R_C. When we get their individual amounts of resistance we must connect the units to terminals 1, 2, 3 as shown.

The resistance between terminals 1 and 3 *before* the replacement must be exactly the same as the resistance between the same two terminals *after* the replacement. Well, what is this resistance *before* the replacement? It is the resistance presented by R_A in parallel with $R_B + R_C$, isn't it? This is

$$\frac{R_A(R_B + R_C)}{R_A + R_B + R_C}$$

After the replacement with R_1, R_2, and R_3, what will be the resistance between *the same two terminals*? It will be $R_1 + R_3$, so we can equate these and have

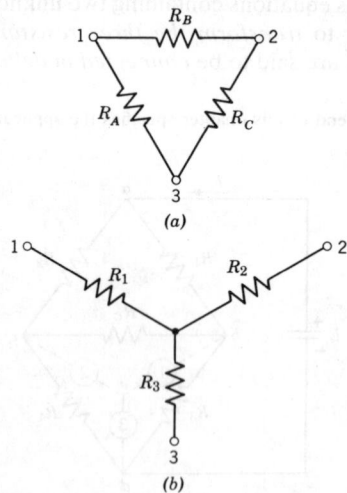

(a)

(b)

FIGURE 6.9   Networks for Δ-Y and Y-Δ transformations. (a) Delta. (b) Wye.

one of three useful equations we shall need:

$$\frac{R_A(R_B + R_C)}{R_A + R_B + R_C} = R_1 + R_3 \tag{6.1}$$

In the same manner, we equate the two expressions for the resistance between terminals 1 and 2:

$$\frac{R_B(R_A + R_C)}{R_A + R_B + R_C} = R_1 + R_2 \tag{6.2}$$

and between terminals 2 and 3.

$$\frac{R_C(R_A + R_B)}{R_A + R_B + R_C} = R_2 + R_3 \tag{6.3}$$

From these we can get formulas for R_1, R_2, and R_3, the Y-connected replacement resistance units. Notice that the denominators are all equal.

The first thing we'll do is subtract (6.3) from (6.1):

$$\frac{R_A R_B + R_A R_C - R_A R_C - R_B R_C}{R_A + R_B + R_C} = R_1 - R_2$$

From these,

$$\frac{R_A R_B - R_B R_C}{R_A + R_B + R_C} = R_1 - R_2 \tag{6.4}$$

Adding this to Equation (6.2) gives

$$2R_1 = \frac{2R_A R_B}{R_A + R_B + R_C}$$

which becomes

$$R_1 = \frac{R_A R_B}{R_A + R_B + R_C} \tag{6.5}$$

By the same kind of procedure we get

$$R_2 = \frac{R_B R_C}{R_A + R_B + R_C} \tag{6.6}$$

$$R_3 = \frac{R_A R_C}{R_A + R_b + R_C} \tag{6.7}$$

Note the symmetry of these equations. Use your imagination and place the Y network on top of the Δ network (or draw them together on a piece of paper) and connect R_1, R_2, and R_3 to their proper terminals. R_1 now occupies space between R_A and R_B. It is easy to remember that R_A times R_B is in the numerator of the formula for R_1. Likewise, R_B times R_C is in the numerator of the formula for R_2, and R_C times R_A is in the numerator of the formula for R_3. Their denominators are all alike: $R_A + R_B + R_C$!

Now, let's say R_1, R_2, R_3 in *wye* are given, and we want to obtain formulas for R_A, R_B, R_C, the equivalent *delta* group. This can be done by a straight-forward algebraic manipulation, which we shall not worry you with now. The results are:

$$R_A = \frac{R_1 R_2 + R_2 R_3 + R_3 R_1}{R_2} \tag{6.8}$$

$$R_B = \frac{R_1 R_2 + R_2 R_3 + R_3 R_1}{R_3} \tag{6.9}$$

$$R_C = \frac{R_1 R_2 + R_2 R_3 + R_3 R_1}{R_1} \tag{6.10}$$

Note the symmetry of these equations and the physical *location* of each resistance represented in a denominator with respect to the location of the delta resistance being evaluated. Normally we would not be expected to memorize these transformation equations, but if someone wanted us to it would not be difficult. If we were obliged to use them very often we could commit the simple system of remembering them to memory without much effort. In fact, some of my students have done it voluntarily.

EXAMPLE 6.6   The battery voltage E in Fig. 6.8 is 26 V and the ohms resistance values are $R_1 = 4.5$, $R_2 = 2$, $R_3 = 3$, $R_4 = 2$, $R_G = 5$. (a) Determine the resistances of the arms of the Y network that may be connected to junctions b, c, and d. (b) Calculate the battery current. (c) Calculate the current in R_G.

Solution. The resistance of the Y arm connected to junction b is obtained by means of Equation (6.5):

$$R_{1Y} = \frac{R_3 R_G}{R_3 + R_G + R_4}$$

$$R_{1Y} = \frac{3 \times 5}{3 + 5 + 2} = 1.5\,\Omega$$

Using Equations (6.6) and (6.7),

$$R_{2Y} = \frac{R_G R_4}{R_3 + R_G + R_4}; \quad R_{2Y} = \frac{5 \times 2}{3 + 5 + 2} = 1\,\Omega$$

$$R_{3Y} = \frac{R_3 R_4}{R_3 + R_G + R_4}; \quad R_{3Y} = \frac{3 \times 2}{3 + 5 + 2} = 0.6\,\Omega$$

Using the Y arms in place of the delta arms, the resistance of the path *abe* is $R_1 + R_{1Y}$, and the resistance of the path *ace* is $R_2 + R_{2Y}$:

$$R_{abe} = 4.5 + 1.5 = 6 \, \Omega$$
$$R_{ace} = 2 + 1 = 3 \, \Omega$$

This parallel pair has an equivalent resistance $R_{ae} = (3 \times 6)/(3 + 6) = 2 \cdot \Omega$, and this is in series with R_{3Y}, the arm from e to d. Therefore the equivalent resistance of the whole circuit is $R_{eq} = 2 + 0.6 = 2.6 \, \Omega$.

The battery current is

$$I = \frac{26}{2.6} = 10 \text{ A}$$

The current in R_1 and R_{1Y} is given by the current-division principle:

$$I_1 = I \frac{R_2 + R_{2Y}}{R_2 + R_{2Y} + R_1 + R_{1Y}}$$

$$I_1 = 10 \frac{2 + 1}{2 + 1 + 4.5 + 1.5}$$

$$I_1 = 3\tfrac{1}{3} \text{ A}$$

The current in R_2 and R_{2Y} is

$$I_2 = 10 - 3\tfrac{1}{3} = 6\tfrac{2}{3} \text{ A}$$

The potential drop from junction a to junction b is

$$E_{a,b} = I_1 R_1 = 3\tfrac{1}{3} \times 4.5 = 15 \text{ V}$$

And from junction a to junction c, it is

$$E_{a,c} = I_2 R_2 = 6\tfrac{2}{3} \times 2 = 13\tfrac{1}{3} \text{ V}$$

These two voltage drops tell us that junction c is $1\tfrac{2}{3}$ V higher in potential than junction b. The current in R_G must then flow toward the left (from c to b) and its value is

$$I_{RG} = 1\tfrac{2}{3} \div 5 = \tfrac{1}{3} \text{ A}$$

Problems

7. Draw the original bridge circuit (Fig. 6.8) and label all elements with resistance, current, and voltage values already found. Continue the analysis of the example just given and determine the currents in R_3 and R_4. This can be done mentally.

8. Determine the resistances which may be used to replace R_1, R_2, and R_G of the group connected to junctions a, b, and c of Fig. 6.8.

9. (a) Calculate the currents in R_3, R_4, and G of Fig. 6.8, using the replacement values found in Problem 8. (b) Calculate the currents in R_1 and R_2.

6.5   Ideal and Practical Sources of Voltage and Current

In our study of circuits we have used batteries with *constant terminal voltages* as sources. We did take internal resistance into account, but in most of our discussion of circuit analysis we have regarded the battery terminal voltage as constant and unaffected by changes in current value.

An ideal voltage source is one whose terminal voltage is independent of the current through it. No matter how much current an ideal voltage source is called upon to deliver, a milliampere or a million amperes, the terminal voltage does not change. Of course, there is no such voltage source in our practical world because it is impossible to have a source of voltage with zero internal resistance. Some sources do, however, maintain practically constant voltage at their terminals when small amounts of current are drawn from them.

Suppose an automobile battery has an *open-circuit* voltage of 12.5 V at its terminals. While starting the cold engine it delivers 150 A and its terminal voltage drops to 11 V. The internal resistance is responsible for the 1.5-V drop. A schematic representation of this source would be that shown in Fig. 6.10 in which e_{se} is the open-circuit voltage (12.5 V in this case) and r_{se} is the internal resistance (0.01 Ω in this case). R_L represents a resistive load of any value.

Using voltage division, the terminal voltage, which is also the load voltage, is given by

$$e_T = e_L = \frac{R_L}{r_{se} + R_L} e_{se}$$

and the load current is

$$i_L = \frac{e_{se}}{r_{se} + R_L} \tag{6.12}$$

An ideal current source is another impossibility in practice because, by definition, it can deliver a constant amount of current no matter how large or small a load resistance may be connected to its terminals. The current versus plate-voltage curve of a pentode vacuum tube is a practically flat horizontal straight line when the grid-to-cathode voltage is held constant at any one of many practical values. Exceptionally high values of load resistance, however,

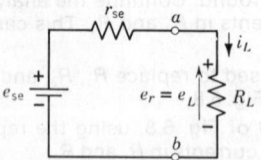

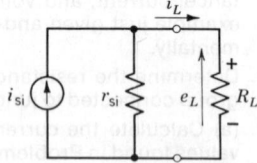

FIGURE 6.10 Practical "constant-voltage" source with load resistor connected to its terminals a, b.

FIGURE 6.11 Practical constant-current source with load resistor connected to its terminals.

will cause the plate current to decrease appreciably below normal values. The plate-to-cathode voltage of the tube becomes very small when this condition prevails.

The schematic diagram of a practical current source (Fig. 6.11) has an ideal current source and an internal resistance r_{si} in parallel. The ideal source current, i_{si}, is constant, but the load current must depend on the value of R_L, of course. The voltage e_L must also depend on R_L and therefore e_L is not constant.

We can see that

$$i_L = i_{si} \frac{r_{si}}{r_{si} + R_L} \tag{6.13}$$

and

$$e_L = i_{si} \frac{r_{si} R_L}{r_{si} + R_L} \tag{6.14}$$

6.6   Equivalent Practical Current and Voltage Sources

Two sources are *equivalent* if they produce identical current in the same *load* resistance, when it is connected to their terminals. This implies that they produce the same voltage value across that resistance. This also means that if we have a container with a pair of terminals protruding from it, we cannot tell whether it is a current source or a voltage source by measuring current in or voltage across a resistive load.

Since the load currents of equivalent circuits are identical, let us equate the two expressions we found for load current [Equations (6.12), (6.13)].

$$i_L = \frac{e_{se}}{r_{se} + R_L} = \frac{r_{si} i_{si}}{r_{si} + R_L} \tag{6.15}$$

To make these expressions the same for *any value* of R_L, the following relations must be true:

$$e_{se} = r_{si} i_{si}$$
$$r_{se} = r_{si}$$

The second equation indicates that, for equivalence, we need not distinguish between the internal resistance of the voltage source (r_{se}) and the internal resistance of the current source (r_{si}). We may represent each by r_s.

Let us consider a given practical current source and its equivalent voltage source. These are shown in Fig. 6.12. Note that the source resistances are equal (2 Ω), and that $e_s = r_s i_s = 2 \times 5 = 10$ V. To check their equivalence, let

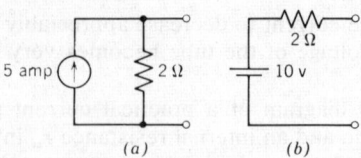

FIGURE 6.12   (a) A practical current source. (b) The equivalent practical voltage source.

$3\,\Omega$ be connected to each pair of terminals. The load current in part a is

$$i_L = 5 \times \frac{2}{2+3} = 2\,\text{A}$$

and in part b

$$i_L = \frac{10}{2+3} = 2\,\text{A}$$

Each source delivers 12 W of power to its 3-Ω load, but the ideal current source puts out 30 W, 18 of which are lost in the internal resistance, and the ideal voltage source puts out 20 W, only 8 of which are lost internally. This shows the voltage and current sources *are equivalent only insofar as the load is concerned.*

EXAMPLE 6.7   Change the two ideal voltage sources of Fig. 6.13a which may be considered as practical voltage sources if the series resistances are regarded as internal resistances, into equivalent practical current sources. Solve for the current in the 8-Ω resistor.

Solution. The equivalent current source for the 9-V source has a current value of $e_s \div r_{se} = 9 \div 3 = 3\,\text{A}$. For the 12-V source, the equivalent current source has $12 \div 6 = 2\,\text{A}$. We can replace the 3-Ω and the 6-Ω resistors with a 2-Ω resistor, which will be in parallel with the 8-Ω resistor.

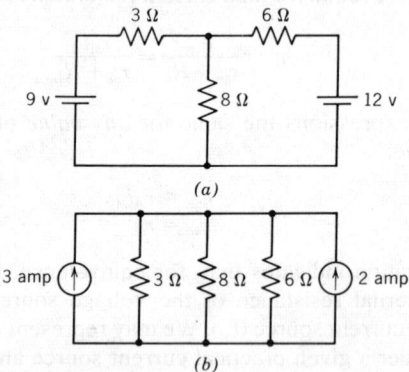

FIGURE 6.13   (a) Circuit with constant-voltage sources. (b) Equivalent to circuit in (a); constant-current sources.

The current in the 8-Ω resistor is then

$$I_8 = 5 \times \frac{2}{2+8} = 1 \text{ A}$$

The student should check this result by analyzing Fig. 6-13a with another method.

EXAMPLE 6.8   Let's determine the current I_1 in Fig. 6.14 to show how to handle a network that has both kinds of source. Kirchhoff's current law enables us to label the branch currents as shown, using the voltage law as we go counter-clockwise around the inside mesh,

Solution

$$20 - 4I_1 - 10I_2 = 0$$
$$4I_1 + 10I_2 = 20 \tag{1}$$

going around the path containing E_2 and the 10-Ω branch,

$$30 - 6(I_2 - 3 - I_1) - 10I_2 = 0$$
$$6(I_2 - 3 - I_1) + 10I_2 = 30 \tag{2}$$

Simplifying (2),

$$-6I_1 + 16I_2 = 48 \tag{3}$$

From (1) and (3),

$$I_1 = 1.29 \text{ A}$$

The minus sign indicates that we drew the current arrow in the wrong direction. I_1 actually flows downward.

Problems

10. (a) Convert the constant-current generator and its associated shunt-resistance in Fig. 6.14 into an equivalent practical voltage source.
   (b) Solve for the current in the 4-Ω resistor, using the mesh-current method.
11. Solve for the current in the 10-Ω resistor in Fig. 6.15 by applying Kirchhoff's voltage law. Currents are labeled for your convenience.

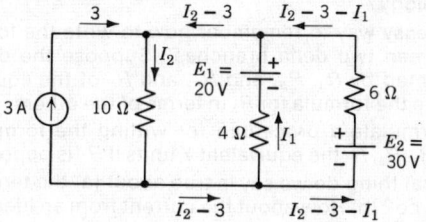

FIGURE 6.14   For example 6.8.

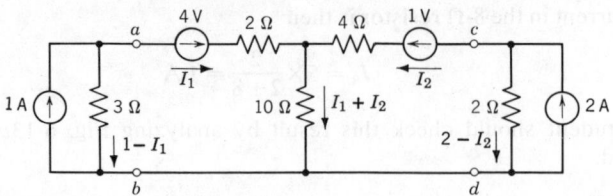

FIGURE 6.15   For Problem 11.

12. Solve for the current in the 10-Ω resistor of Fig. 6.15 after changing the current sources into voltage sources.

Constant-voltage generator representation is employed in the analysis of triode vacuum-tube circuits. See Problem 23 at the end of this chapter. Constant-current generator representation is used in the analysis of electronics circuits employing pentode tubes or transistors. See Problem 24 at the end of this chapter.

6.7   Review Questions

1. How are mesh currents designated in a circuit containing two or more meshes?
2. If you get a negative answer when you solve for the amount of current in an element of a circuit, what is the significance of the negative sign?
3. In going around a closed loop in the direction of the I_2 arrow, suppose one of resistances (R_3) carries I_2 toward the right and I_3 toward the left. How do you write the voltage drop term for this element in the voltage equation?
4. When the nodal-voltage method is used to analyze a network, do we write voltage equations or current equations?
5. Do you use determinants to solve two equations with two unknowns? If so, and you prefer this method, why not look up how to handle three equations with three unknowns? The method is straightforward and saves time.
6. Name an important advantage of transforming a delta network to a wye network or vice versa. Which of these is a T network and which a π network in that terminology?
7. What is an easy way to remember how to write the formula for a wye branch that is between two delta branches? Suppose the delta branch resistances are represented by R_A, R_B, and R_C, and R_2 of the equivalent Y is between R_B and R_C. Write the formula for R_2 in terms of the others.
8. Can you formulate a procedure for writing the formula for R_A, a delta unit, in terms of R_1, R_2, R_3 the equivalent Y units if R_A is opposite R_2?
9. What unusual thing do we say is true about (a) the terminal voltage of an ideal voltage source? (b) How about the current from an ideal current source?
10. How does a practical voltage source differ from an ideal voltage source? How is it represented in a circuit diagram?

11. How does a practical current source differ from an ideal current source? How is it represented in a circuit diagram?

6.8 Problems

Group A

13. Use mesh analysis and solve for the currents in all three resistance units in Fig. 6.16.

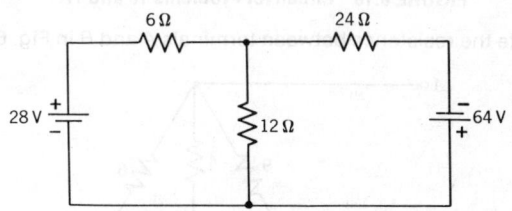

FIGURE 6.16   For Problem 13.

14. Solve for the current in the 8-Ω resistor of Fig. 6-17 by the mesh-current method.

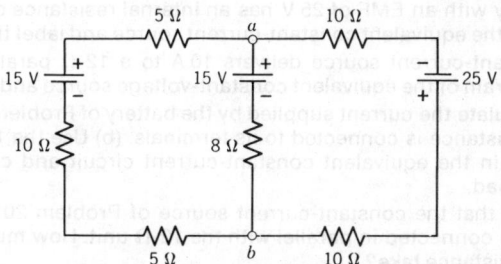

FIGURE 6.17   For Problem 14.

15. Apply the nodal voltage method to determine the current in the 12-Ω resistance of Fig. 6.16 (Problem 13) letting V represent the voltage rise across that resistance.

16. In the circuit of Fig. 6.18, transform the delta group connected to junctions a, b, d, e to an equivalent Y group, and draw the transformed network. (a) Solve for the current in the source. (b) How much current flows in the 2-Ω resistor?

17. The 5-, 2-, and 4-Ω resistances connected in Y form to junctions a, c, and d in Fig. 6.18 can be transformed into a delta-connected set that will take their place. Do this, redraw the circuit, and calculate the current delivered by the battery.

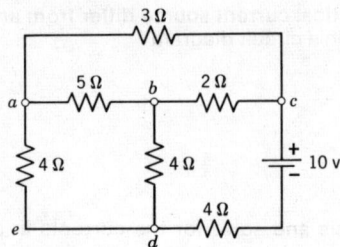

FIGURE 6.18   Circuit for Problems 16 and 17.

18. Calculate the resistance between terminals A and B in Fig. 6.19.

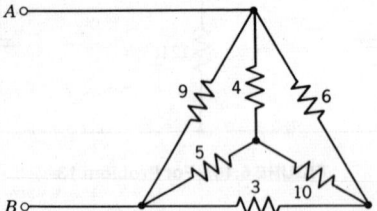

FIGURE 6.19   Resistances are in ohms. For Problem 18.

19. A battery with an EMF of 25 V has an internal resistance of 1 Ω. Draw the diagram of the equivalent constant-current source and label the parts.

20. A constant-current source delivers 10 A to a 12-Ω parallel resistance. Draw the diagram of the equivalent constant-voltage source and label the parts.

21. (a) Calculate the current supplied by the battery of Problem 19 when an 11.5-Ω load resistance is connected to its terminals. (b) Use the 11.5-Ω unit as a load resistor in the equivalent constant-current circuit and calculate the current in that load.

22. Assume that the constant-current source of Problem 20 has a 36-Ω load resistance connected in parallel with the 12-Ω unit. How much current does the 36-Ω resistance take?

23. A vacuum tube amplifier employs a triode (3-electrode) tube that has an amplification factor (μ) of 20 and an internal plate resistance (R_p) of 10 kΩ.

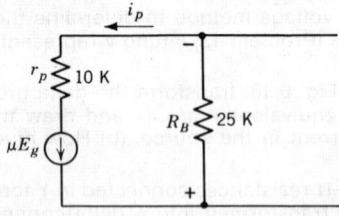

FIGURE 6.20   Thévenin equivalent circuit for a triode-tube amplifier. For Problem 23.

Its plate current flows through an external (load) resistance (R_B) of 25 kΩ. The input voltage E_g between the grid and cathode of the tube is 1 V. The constant-voltage equivalent circuit is shown in Fig. 6.20. Assume E_g has an instantaneous maximum value of 1.414 V. (a) Calculate the maximum instantaneous voltage that appears across the plate-load resistor R_B. (b) What is the ratio of the voltage across R_B to the voltage E_g? (c) Assuming E_g was measured across an input resistance equal to R_B, calculate the decibel gain of the amplifier.

24. The constant-current network in Fig. 6.21 is the equivalent of the emitter-collector circuit of a common-emitter transistor amplifier. Known quantities are $h_{fe} = 32.3$ mhos, $h_{oe} = 2 \times 10^{-5}$ mho conductance, $I_1 = 100 \, \mu$A, $R_2 = 1$ KΩ. What is the current in the load resistance?

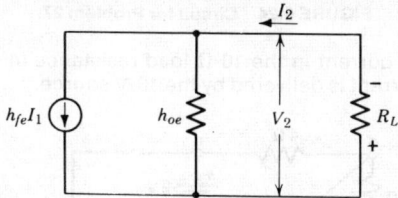

FIGURE 6.21   Norton equivalent network for the emitter-collector circuit of a common-emitter amplifier. For Problem 24.

Group B

25. Solve for the voltage V in Fig. 6.22 using Kirchhoff's laws.

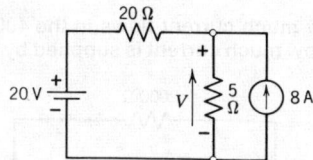

FIGURE 6.22   Circuit for Problem 25.

26. Convert the constant-current generator into a constant-voltage generator in Fig. 6.23 and determine the current in R_L.

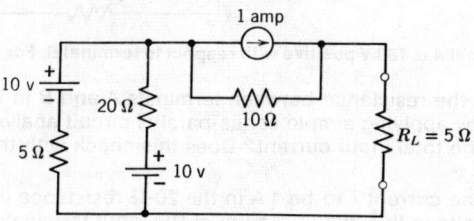

FIGURE 6.23   Circuit for Problem 26.

27. The 5-Ω electromagnet in Fig. 6.24 must carry 200 mA for proper operation. What must be the resistance of *R*?

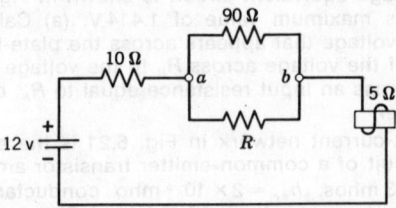

FIGURE 6.24 Circuit for Problem 27.

28. Solve for the current in the 10-Ω load resistance in Fig. 6.25. Also determine how much current is delivered by the 10-V source.

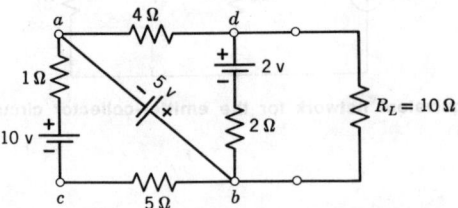

FIGURE 6.25 Circuit for Problem 28.

29. (a) Determine how much current flows in the 4000-Ω resistor in Fig. 6.26, and its direction. (b) How much current is supplied by the source?

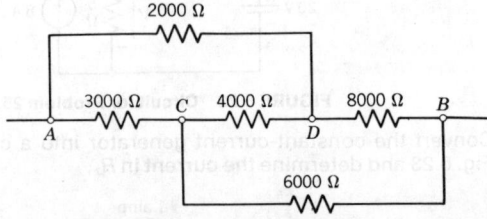

FIGURE 6.26 Terminal *A* is 180 V positive with respect to terminal *B*. For Problems 29 and 30.

30. Calculate the resistance between terminals *A* and *B* in Fig. 6.26. This cannot be done by applying simple series-parallel circuit analysis. Why? What is the value of the total input current? Does this check with the answer to Problem 29?

31. Assume the current *I* to be 1 A in the 20-Ω resistance in Fig. 6.27 and calculate the voltage this would require at the input terminals. Then, by direct proportion, calculate the *actual value* of *I* in the given circuit.

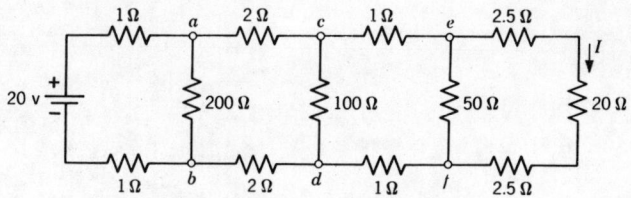

FIGURE 6.27   For Problem 31.

32. Use nodal analysis to determine the load voltage, V, of the transmission system of Fig. 6.28. Then solve for the generator currents.

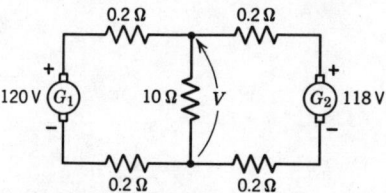

FIGURE 6.28   Circuit for Problem 32.

CHAPTER SEVEN

Network Theorems

After learning to apply Kirchhoff's laws to simple circuits in Chapter 5, we studied mesh-and nodal analysis in which we used the laws in Chapter 6. We also learned the advantages of delta-wye transformation and something about equivalent practical voltage and current sources.

In this chapter we examine some principles and special theorems that we can use to simplify complex circuits and thereby make it easier to analyze them and reduce the amount of work required. Although we shall restrict our applications to circuits containing resistance only, the same procedures will also enable us to analyze ac circuits that contain impedances, which are elements that do not exist in dc circuits to which constant voltages are applied.

7.1   Linearity and Response of Networks

Most of the resistance elements encountered in networks we have analyzed up to now had constant ohms values. These are *linear* in nature because the current they will carry is *directly proportional* to the voltage applied to them. That is, doubling the voltage applied to the terminals of a resistance whose ohms value is constant will double the current it carries, tripling the voltage will triple the current. Furthermore, a resistance is classified as a *passive* ele-

145

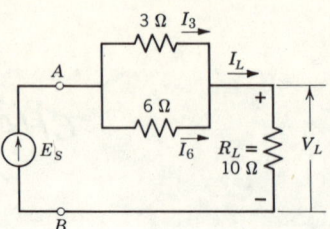

FIGURE 7.1   A passive linear network.

ment because it cannot act as an energy source. A battery is an *active* element because it is a source of energy.

A linear network *responds* in direct proportion to the magnitude of the stimulus applied to it. The *response* of a network (or even of an element of a network) is a term used to designate the nature of the way the network, or the element, reacts to an *input*. The input, which may be a kind of applied voltage or current, is usually called the *forcing function*.

To illustrate, let us consider the circuit of Fig. 7.1. The network connected to the right of junctions A, B is passive because it does not contain an energy source. It is linear because its *response* (in this case the load current in R_L) is *directly proportional* to the magnitude (amount) of the *forcing voltage E_s*. If E_s is 12 V, I_L is 1 A and V_L is 10 V; if E_s is 24 V, I_L is 2 A and V_L is 20 V. It is left to the student to show that I_3 will be doubled when E_s is doubled and that the same kind of response will occur with reference to I_6.

A *linear circuit* is made up entirely of linear elements and ideal sources. We have seen that an ideal voltage source has zero internal resistance. Theoretically it can supply an unlimited amount of energy without a decrease in its terminal voltage, so it does not exactly represent any physical device. An ideal current source delivers current that is completely independent of the voltage across it. In fact, the voltage across it is that across an element (or a network) in parallel with it.

7.2   Superposition Principle

We already know how to solve for the current in any element of a linear circuit with more than one source, such as finding the amount of current I in the 6-Ω resistor in Fig. 7.2. We have used Kirchhoff's laws to write the required equations.

We can obtain the correct answer for the current without solving simultaneous equations by applying the *principle of superposition*. If we solve first for the current that E_1 alone would produce if the source E_2 were replaced by a zero-resistance conductor (its internal resistance is zero), and then for the

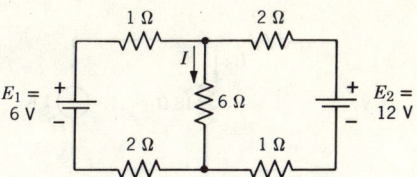

FIGURE 7.2 A linear circuit containing two voltage forcing functions used to illustrate the super-position principle.

current that E_2 alone would produce while E_1 is replaced by a zero-resistance conductor, the sum of the separate currents will give the current that flows with both sources acting simultaneously. We shall now illustrate this method.

EXAMPLE 7.1   How much current flows in the 6-Ω resistor in Fig. 7.2?

Solution. Replacing E_2 with a zero-resistance conductor, the current supplied by E_1 is

$$I_1 = \frac{6}{3 + \frac{3 \times 6}{9}} = \frac{6}{5}\,\text{A}$$

The current in the 6-Ω resistor is

$$I_6 = \frac{6}{5} \times \frac{3}{9} = \frac{2}{5}\,\text{A}$$

Replacing E_1 with a zero-resistance conductor, the current supplied by E_2 is

$$I_2 = \frac{12}{3 + \frac{3 \times 6}{9}} = \frac{12}{5}\,\text{A}$$

The current in the 6-Ω resistor is

$$I_6 = \frac{12}{5} \times \frac{3}{9} = \frac{4}{5}\,\text{A}$$

The sum of the currents is

$$I = \frac{2}{5} + \frac{4}{5} = 1.2\,\text{A}$$

To set up the equations for the solution *when both sources are acting*, let us call the 12-V source current I_{12} and the 6-V source current $I - I_{12}$:

$$6 = 6I + 3(I - I_{12})$$
$$12 = 6I + 3I_{12}$$

Solution of these equations may be accomplished by simply adding them to eliminate I_{12}. This gives

$$15I = 18, \qquad I = 1.2\,\text{A}$$

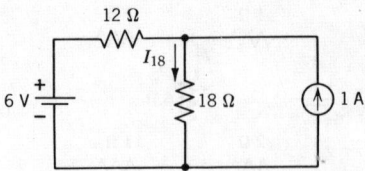

FIGURE 7.3 A circuit used to illustrate the superposition theorem.

EXAMPLE 7.2 The circuit of Fig. 7.3 will be used for illustration *when a constant-current source is present.*

Solution. When the current source is removed,

$$I_{18} = \frac{6}{30} = 0.2 \text{ A}$$

When the voltage source is replaced by a zero-resistance conductor, we get, by current division,

$$I_{18} = 1 \frac{12}{12 + 18} = 0.4 \text{ A}$$

The actual current in the 18-Ω resistor is thus $0.2 + 0.4 = 0.6$ A.

When the two sources are acting together, we may assume for analysis purposes that the voltage source is replaced by a $\frac{1}{2}$-A current source with a 12-Ω resistor *in parallel* ($6 \div 12 = 0.5$ A). The circuit is shown in Fig. 7.4. The total current, which is 1.5 A, will divide as follows:

$$I_{18} = 1.5 \times \frac{12}{12 + 18} = 0.6 \text{ A}$$

Note that a current of 1.5 A flows through zero-resistance paths to the junction of a 12-Ω and an 18-Ω resistor. It will divide into two parts whose values are to each other inversely as the resistances of the respective paths.

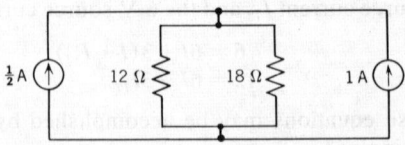

FIGURE 7.4 Circuit equivalent to that of Fig. 7.3.

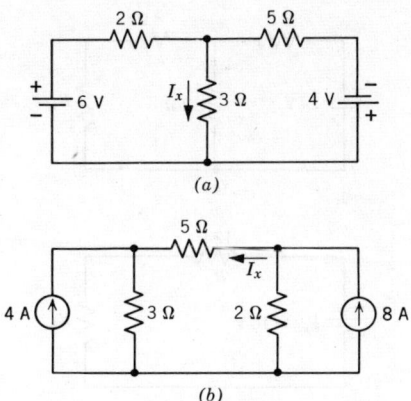

(a)

(b)

FIGURE 7.5   For Problems 1 and 2.

STATEMENT OF THE SUPERPOSITION THEOREM: In a linear resistive network containing two or more voltage sources, the current through any element (resistance or source) may be determined by adding together algebraically the currents produced by each source acting alone, when all other voltage sources are replaced by their internal resistances. If a voltage source has no internal resistance, the terminals to which it was connected are joined together. If there are current sources present they are removed and the network terminals to which they were connected are left open.

Problems

1. Find the current I_x in the circuit shown in Fig. 7.5a, using the superposition principle. (Ans: 22/31 A downward.)
2. Find the current I_x in 5-Ω resistor in Fig. 7.5b, using the superposition principle. Check by means of another solution method. (Ans: 0.4 A toward the left.)

7.3   Thévenin's Theorem

Suppose we wish to determine the current, voltage, or power in only one element, R_L, of a circuit containing several resistors, such as that in Fig. 7.6a Application of a theorem devised in 1883 by M. L. Thévenin, a French engineer, makes possible the conversion of this circuit into the simple series circuit of Fig. 7.6b. This new circuit will have the same current and power values in R_L

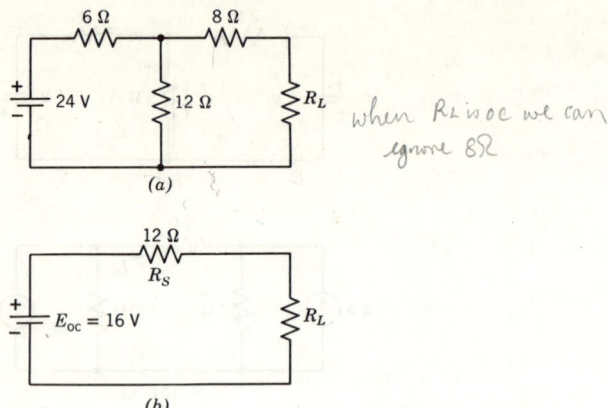

when R_L is oc we can ignore 8Ω

FIGURE 7.6 **Illustrating Thévenin's theorem.** *(a)* **Original circuit.** *(b)* **Thévenin equivalent of the circuit in** *(a)*.

as will the original circuit and, of course, the same voltage across R_L. The source voltage and the series resistance values were obtained as follows.

To get the source voltage in the *Thévenin equivalent circuit* we simply imagine R_L disconnected in Fig. 7.6*a*. The open-circuit voltage *at the two open terminals* is then

$$V_{oc} = 24 \times \frac{12}{6+12} = 16 \text{ V} \quad Voltage \ divider$$

To get the value of the *equivalent series resistance R_s*, we consider the 24-V (zero-resistance) source in Fig. 7.6*a* to be replaced by a zero-resistance conductor and "look into" the circuit at the open terminals from which R_L has been removed. We see $8\,\Omega$ in series with a parallel pair made up of $6\,\Omega$ and $12\,\Omega$. This parallel pair is then replaced by $4\,\Omega$ (product ÷ sum) which would then be in series with $8\,\Omega$, making $12\,\Omega$ the value of the equivalent series resistance R_s.

Inasmuch as R_L is a passive unit, it can have any value, and its current, voltage, and power will be

$$I_L = \frac{16}{12+R_L}, \quad V_L = \frac{R_L}{R_L+12}16, \quad P_L = \left(\frac{16}{12+R_L}\right)^2 R_L$$

Application of Thévenin's theorem has thus enabled us to get load resistance information without first obtaining currents or voltages of other elements of the circuit. It is important to state here that we could have considered any other resistance of Fig. 7.6*a as a load resistor* instead of R_L, and obtained current, voltage, and power values for it alone. In such a case it would be necessary to know the ohms resistance of R_L, which would be a permanent element in the network, so that current, voltage, and power values could be obtained for the other element chosen.

EXAMPLE 7.3   Assume $R_L = 20\,\Omega$ in Fig. 7.6. Continue with the calculations already made and obtain values of load current, voltage, and power. Verify the load current by another method of solution.

Solution

$$I_L = \frac{16}{12+20} = 0.5\ \text{A}$$

$$V_L = \frac{20}{20+12} \times 16 = 10\ \text{V}$$

$$P_L = 0.5^2 \times 20 = 5\ \text{W}$$

The total resistance connected to the source in Fig. 7.6a is

$$R_T = 6 + \frac{12 \times 28}{12+28} = 14.4\,\Omega$$

The total current is

$$I_T = \frac{24}{14.4} = 1\frac{2}{3}\ \text{A}$$

The load current is

$$I_L = 1\frac{2}{3} \times \frac{12}{40} = 0.5\ \text{A}$$

We'll have to admit that no advantage is gained by using Thévenin's theorem to solve a simple problem like this one, but you'll find it a time and work saver when we encounter more complicated circuits. It is extremely useful in converting circuits containing vacuum tubes and transistors into equivalent circuits for purposes of analysis and design.

Problem

3. If $R_L = 16\,\Omega$ in Fig. 7.6a, calculate the current in the 8-Ω resistance using Thévenin's theorem. Imagine *this 16 Ω resistor* to be the load resistor which is to be disconnected so that you may determine E_{oc} and R_s for the required equivalent circuit.

STATEMENT OF THEVENIN'S THEOREM: The current in any passive circuit element (which may be called R_L) in a network is the same as would be obtained if R_L were supplied with a source voltage E_{oc} in series with an equivalent resistance R_s, E_{oc} being the open-circuit voltage at the terminals from which R_L has been removed and R_s being the resistance that would be measured at these terminals after all sources have been removed and each has been replaced by its internal resistance.

It will be found later, when we study ac theory, that Thévenin's theorem and other network theorems hold if the word *resistance* is changed to *impedance* in the statements of the theorems. Impedance is expressed in ohms. In addition to representing the effect of resistance in a circuit, it embraces additional ohmic effects, induced voltages, and electric charges that vary in amount.

7.4 Norton's Theorem

Let us examine Fig. 7.6*b* which may be used as a substitute for Fig. 7.6*a* if we are interested only in the circuit response (load voltage, current, and power) at the terminals of R_L. We have learned that a constant-current source may be substituted for a constant-voltage source, so the circuit of Fig. 7.7 will supply R_L with the same electrical quantities.

To obtain this constant-current circuit configuration, we need not first employ Thévenin's theorem, which would give us Fig. 7.6*b*. We can apply *Norton's theorem* at once, in the following manner.

In Fig. 7.6*a*, replace R_L with a *zero-resistance conductor*. This is the same thing as short circuiting R_L. Calculate the current value in this conductor. This is sometimes called the *short-circuit current* because *Norton's theorem* is sometimes called the *short circuit theorem*. Applying what we know about parallel circuits and current division, we have

$$I_{sc} = \frac{24}{6 + \dfrac{8 \times 12}{8 + 12}} \times \frac{12}{8 + 12} = 1\tfrac{1}{3} \text{ A}$$

This is taken as the *current put into the Norton equivalent circuit* by a *constant-current* generator.

The next step is to determine the ohm value of a parallel resistance R_p to use with the constant-current generator. This is found in the same way the series resistance is determined for the Thévenin equivalent circuit. We remove the generator voltage source and replace it with a zero-resistance conductor. Note that, in applying both of these theorems, we do not remove from the circuit the internal resistance of a generator if it has one. We just imagine its EMF is zero and leave its internal resistance, if it has any, to represent the generator. So, R_p equals 12 Ω, which was the value of the series resistance of

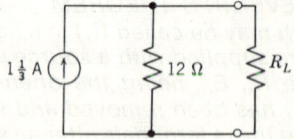

$1\tfrac{1}{3}$ A   12 Ω R_L

FIGURE 7.7 For illustration of Norton's theorem. This is the Norton equivalent of the network of Fig. 7.6*b*.

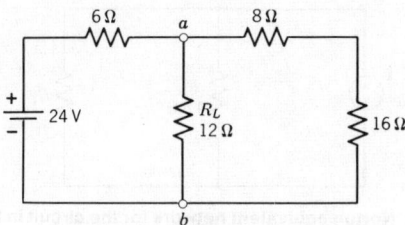

FIGURE 7.8   For Example 7.4.

the Thévenin equivalent circuit. The circuit of Fig. 7.7, therefore, is the Norton circuit equivalent to Fig. 7.6*a*, insofar as the current and voltage of the load resistance, R_L, are concerned.

Let's make R_L equal to 16 Ω, as in Problem 3, and calculate the current it will take, using Norton's theorem:

$$I_L = 1\frac{1}{3}\left(\frac{12}{12+16}\right) = \left(\frac{4}{3}\right)\left(\frac{12}{28}\right) = \frac{4}{7}\text{ A}$$

*STATEMENT OF NORTON'S THEOREM: The current in any **passive circuit element** (which may be called R_L) in a network is the same as would flow in it if it were connected in parallel with R_p and the parallel pair were supplied with a constant current I_{sc}. R_p is the resistance measured "looking back" into the original circuit after R_L has been disconnected and all the sources have been replaced by their internal resistances; I_{sc} is the current which will flow in a short circuit at the terminals of R_L in the original circuit.*

EXAMPLE 7.4   Redraw the circuit of Fig. 7.6*a*, replacing R_L with a 16-Ω resistance and letting R_L be the 12-Ω resistance in the middle branch (see Fig. 7.8). Use Norton's theorem to determine the current in this R_L. This will require removing it from the circuit, getting the short-circuit current at the terminals from which it has been removed. When junctions *a* and *b* are connected by a zero-resistance conductor, the load on the battery is only 6 Ω. It is also necessary to calculate the resistance "looking back" into the circuit at the open terminals *a* and *b* after replacing the 24-V source with zero resistance.

Solution. The circuit is shown in Fig. 7.9:

$$I_{sc} = \frac{24}{6} = 4\text{ A}$$

$$R_p = \frac{6 \times 24}{30} = 4.8\ \Omega$$

$$I_{12} = \frac{4 \times 4.8}{4.8 + 12}$$

$$I_{12} = 1.142\text{ A}$$

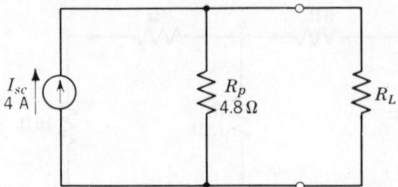

FIGURE 7.9   Norton equivalent network for the circuit in Fig. 7.8.

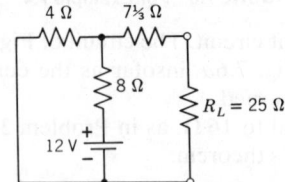

FIGURE 7.10   For Problem 4.

Problem

4. Use Norton's theorem to develop an equivalent circuit to supply R_L in Fig. 7.10. Calculate the current and power delivered to R_L. Be careful, this is tricky. (Ans: $I_L = 0.114$ A, $P_L = 0.325$ W.)

7.5   Maximum Power Transfer Theorem

In communications engineering it is usually desirable to deliver *maximum power* to a load. Electronic amplifiers in radio and television receivers are designed to do this, and it does not matter that an equal amount of power is lost in the process. Obviously, it is important to make maximum use of power available for driving a loudspeaker. One can readily agree, however, that such a situation would not be tolerable in power engineering because the consumer should be provided with as large a fraction as possible of the generated power, and the losses in the generator and transmission system should be kept to a minimum.

We shall find that *a resistance load, served through a resistance network, will receive maximum power when the load resistance value is the same as the resistance "seen by the load as it looks back into the network."* This is also called the *output resistance* of the network, or the resistance presented to the output terminals by the network. It is the resistance we called R_s in explaining Thevenin's theorem.

The load resistance R_L in Fig. 7.11 will receive maximum power if R_L has

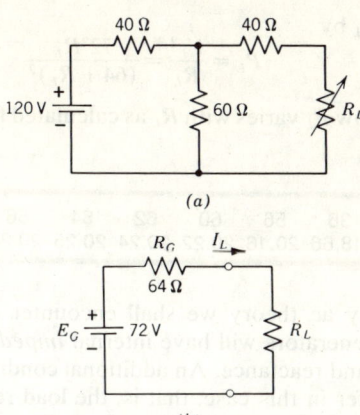

(a)

(b)

FIGURE 7.11 (a) **Illustrating the maximum-power-transfer theorem.** (b) **A Thévenin equivalent circuit for part a. For maximum power in R it must have 64 Ω resistance.**

only one particular ohms value. This circuit, or any other passive, linear, bilateral network, may be represented (insofar as load current, voltage, and power are concerned) by a Thévenin equivalent circuit. The student may verify that the equivalent series resistance for this circuit is 64 Ω and the open-circuit output voltage is 72 V.

From what has been said, a variable resistance used for R_L would have to be set at 64 Ω so that it would receive maximum power. This can be proved by using simple calculus.[1] We shall be content with calculating the power transferred to the load R_L when it is set at several values above and below 64 ohms and show that, if a curve were plotted with power in R_L on the vertical axis and with ohms resistance of R_L on the horizontal axis, the peak of the curve would be at $R_L = 64$ ohms.

The voltage across R_L will always be given by

$$V_L = \frac{R_L}{64 + R_L} \times 72 \text{ V}$$

[1]In Fig. 7.11b $I_L = E_G/(R_G + R_L)$. The power, P_L, to the load is $I_L{}^2 R_L$, $P_L = E_G{}^2 R_L/(R_G + R_L)^2$. Assume that R_L is varied until P_L is a maximum. For this condition, $dP_L/dR_L = 0$.

$$\frac{dP_L}{dR_L} = \frac{E_G{}^2(R_G + R_L)^2 - 2E_G{}^2 R_L(R_G + R_L)}{(R_G + R_L)^4} = 0$$

From this,

$$E_G{}^2(R_G + R_L)^2 = 2E_G{}^2 R_L(R_G + R_L)$$
$$R_G + R_L = 2R_L$$
$$R_G = R_L$$

and the power in R_L by

$$P_L = \frac{V_L{}^2}{R_L} = \frac{72^2 R_L}{(64 + R_L)^2} \qquad (7.1)$$

Table 7.1 shows how P_L varies with R_L as calculated from Equation (7.1).

TABLE 7.1

R_L (ohms)	16	36	56	60	62	64	66	68	72	92	112
P_L (watts)	12.96	18.66	20.16	20.22	20.24	20.25	20.24	20.23	20.12	19.59	18.74

When we study ac theory we shall encounter *reactances* in addition to resistances, and generators will have internal *impedance* which is a combination of resistance and reactance. An additional condition must be met for maximum power transfer in this case, that is, the load reactance must be equal to the generator reactance but opposite in sign.

Note that, although the value used as the resistance of R_L for maximum power transfer is the resistance looking back into the circuit, with the voltage source replaced by a zero-resistance conductor, the power in the original circuit is not equally divided between the load resistor and the network between it and the source. This is understandable because Thévenin's theorem only guarantees to evaluate the current in the actual load. An element in series with the load (the 2-Ω resistor in Fig. 7.12, for example) has no effect on the *open-circuit* voltage at the load. However, it does have an effect on the value of the Thévenin circuit series resistance computed as the "output resistance" presented to the load terminals by the original circuit. The point is this: the load receives maximum power, so that the generator can send it through the connecting network, but Thévenin's theorem does not guarantee that this will be half of the generated power.

Problems

5. It is desired to deliver the maximum possible power to R_L in the network of Fig. 7.12. (a) What should be the value of R_L? (b) Calculate the power in R_L and the output power of the generator. (c) Calculate the power lost in the three wye-connected resistances. Can you do this in two ways?

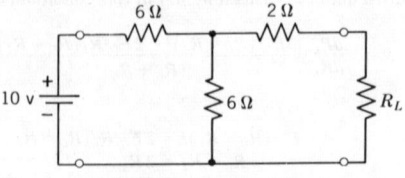

FIGURE 7.12   For Problem 5.

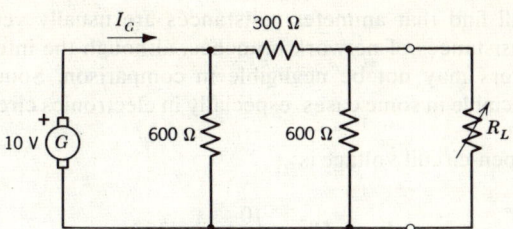

FIGURE 7.13   For Problems 6 and 7.

6. What should be the value of R_L in Fig. 7.13 so that it will receive maximum power? How much current is delivered by the generator?
7. Convert the delta network in Fig. 7.13 to an equivalent wye and then calculate the value of R_L for maximum power. Compare the generator current in this case with the generator current in Problem 6.

7.6   Reciprocity Theorem

In a *linear, passive network* an interesting and important relation exists between a source voltage in one branch and the current indicated by a zero-resistance ammeter in some other branch. The reciprocity theorem states, in effect, that if you interchange the source voltage and the ammeter, the amount of current through the ammeter will be the same, no matter how complicated the network. Indeed the same principle tells the radio engineer that the directional characteristics of a receiving antenna are the same as the directional characteristics of the same antenna when it is used for transmitting. This is a highly useful relation.

EXAMPLE 7.5   To demonstrate the reciprocity theorem let's calculate the current in the ammeter in Fig. 7.14a. Then we'll interchange the battery voltage (leaving its internal resistance where it was, if present) and the ammeter and calculate the meter current.

Solution. If the source has an internal resistance, r_s, or the ammeter has an internal resistance, r_m, these values must remain in the network branches. This is easily accomplished when we are solving a problem on paper. But if we have a physical network on which we are to take measurements in a laboratory, we must provide a way to have the total branch resistances the same as they were before we interchanged the source and the meter. The branch into which the meter is to be placed will have had the source resistance removed and the meter resistance added. The branch into which the source is to be placed will have had the meter resistance removed and the source resistance added.

We shall find that ammeter resistances are usually very small compared with the resistances of network branches, although the internal resistances of milliammeters may not be negligible in comparison. Source resistances are quite appreciable in some cases, especially in electronics circuits.

In (a) the open-circuit voltage is

$$E_{oc} = 50 \frac{10}{5+5+10} = 25 \text{ V}$$

Internal resistance, looking back at a, b, is

$$R_s = 4 + \frac{10(5+5)}{10+5+5} = 9 \, \Omega$$

$$I = \frac{25}{9+4} = 1\tfrac{12}{13} \text{ A}$$

We now make the exchange of source voltage and ammeter (Fig. 7.14c):

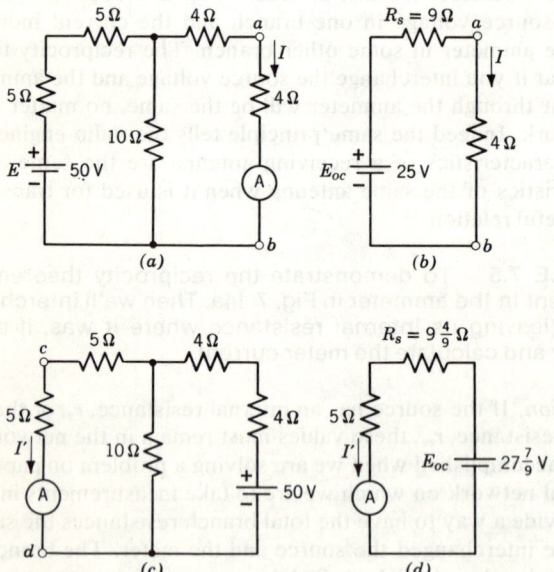

FIGURE 7.14   For illustration of the Reciprocity theorem. (a) Original network. (b) Thévenin equivalent of (a). (c) Source and ammeter interchanged. (d) Equivalent of (c).

$$E_{oc} = 50 \frac{10}{4+4+10} = 27\tfrac{7}{9} \text{ V}$$

$$R_s = 5 + \frac{8 \times 10}{8+10} = 9\tfrac{4}{9}\Omega$$

$$I' = \frac{27\tfrac{7}{9}}{14\tfrac{4}{9}} = 1\tfrac{12}{13} \text{ A}$$

The reciprocity theorem is useful occasionally in solving network problems. We can obtain one or two more independent equations by applying it and working with the new network it provides. We shall make use of this theorem to obtain Equation (7.9).

Problem

8. In Fig. 7.1, assume $E_s = 48$ V and an ammeter reads the current in the 3-Ω resistance. Show that the reciprocity theorem holds.

7.7 Pads and Attenuators

In our discussion of decibels in Section 4.10, we learned the meaning of attenuation as applied to the loss of power in a telephone line or other kind of transmission system. We also inferred that the input resistance of a line, or a network, should be the same as the output resistance of a source that supplies energy to it; also that the resistance of a load should, for good transmission conditions, equal the resistance looking back into the line that supplies it. We are ready to discuss the properties of a three-terminal network, called a *pad*, that is useful in *matching a source to a load*.

The three resistance units that make up a pad that is used for matching and to insert attenuation in a communications line may have the form of a wye or a delta network. Three elements connected in wye form what is called a *T* network and delta-connected units form what is called a *π* network by communications engineers. You can see that if the arms of a *Y* are lowered they can be made to form a *T* and if any two units of a Δ configuration are separated at a junction until they are parallel they may be said to resemble, roughly, the letter *π*. We shall use the word *tee* and the letter *T* in place of *wye* and *Y*; also we shall use *pi* (*π*) in place of *delta* (Δ) when dealing with pads and attenuators.

It can be shown that a resistance pad in the form of a *T* network will maintain the match that exists between a source and a load *when the series arms have equal resistance values.*

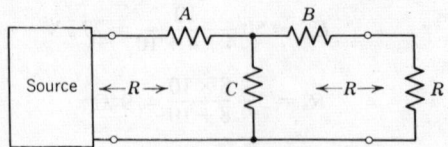

FIGURE 7.15   A resistance T pad matching a source to a load.

Consider the resistance network shown in Fig. 7.15. Looking into the network at the generator end, we must "see" a total resistance equal to R. Let the symbols A, B, and C represent ohms values:

$$R = A + \frac{(B+R)C}{B+R+C} \tag{7.2}$$

Looking back into the network at the receiver end, the source resistance being R, we get

$$R = B + \frac{(A+R)C}{A+R+C} \tag{7.3}$$

It is obvious that these two equations are identical if $B = A$.

We shall now show that a T pad with *series arms of equal resistance values* will maintain matched conditions while causing a desired reduction in voltage between a generator and its load. Assume that a 600-Ω transmission line is the "load" supplied by a generator that has 600 Ω of internal resistance. A reduction in voltage from 15 V at the generator output terminals to 3 V at the load is desired. The use of $R_1/2$ to denote the resistance of each series arm of a symmetrical T network is normal practice.

We have 15 V applied to a circuit with 600-Ω total equivalent resistance, and 3 V applied to a load resistance of 600 Ω (Fig. 7.16). Therefore

$$I = \frac{15}{600} = 0.025 \text{ A} \quad \text{and} \quad \frac{I}{5} = 0.005 \text{ A}$$

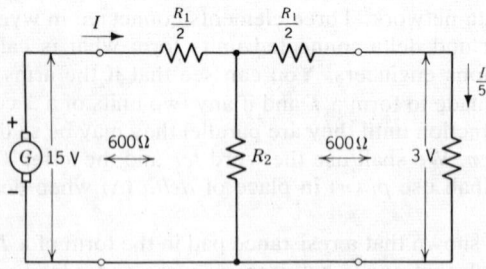

FIGURE 7.16   A T pad matching a generator to a load and causing a voltage reduction of 5:1.

We can now find the value of R_1 easily. Summing the voltages around the outside loop gives

$$15 - 0.025\frac{R_1}{2} - 0.005\frac{R_1}{2} - 0.005 \times 600 = 0$$

from which

$$\frac{R_1}{2} = 400\,\Omega, \quad R_1 = 800\,\Omega$$

The current in R_2 is $0.025 - 0.005 = 0.02$ A:

$$0.02R_2 = 15 - 0.025 \times 400$$

from which

$$R_2 = 250\,\Omega$$

Problem

9. Using the equations for T to π network transformation, determine the resistances of a π network that will replace the T in Fig. 7.16. Use $R_1 = 800, R_2 = 250$, as we just determined. (Ans: $R_A = 900\Omega = R_c, R_B = 1440\Omega$).

EXAMPLE 7.6   Let us now look into the problem of designing a resistance pad to match a generator to a load when their impedance values are not the same. The attenuation of voltage or of power may be a required amount also. A T network will be employed.

Assume that the pad is to cause a voltage reduction from generator output to load output such that the ratio is $4:1$.

Solution. The equation for *power ratio*, N, with the symbols in Fig. 7.17a is

$$\frac{V_1^2/R_1}{V_2^2/R_2} = N = \frac{V_1^2 R_2}{V_2^2 R_1}$$

From this,

$$\frac{V_1}{V_2} = \sqrt{\frac{R_1}{R_2}N} \qquad (7.5)$$

In Figure 7.17b, we see a particular condition in which a T pad is to be designed to match an 800-Ω source to a 500-Ω load, with a *power-reduction* ratio of $10:1$. The load receives power at 5 V, which means that the load current is $5/500 = 0.01$ A. The value of V_1 must then satisfy Equation (7.5) and the voltage ratio is

$$\frac{V_1}{V_2} = \sqrt{\frac{800 \times 10}{500}} = 4$$

The value of V_1 is $4 \times 5 = 20$ V.

The total input current is given by $20/800 = 0.025$ A. The current in C,

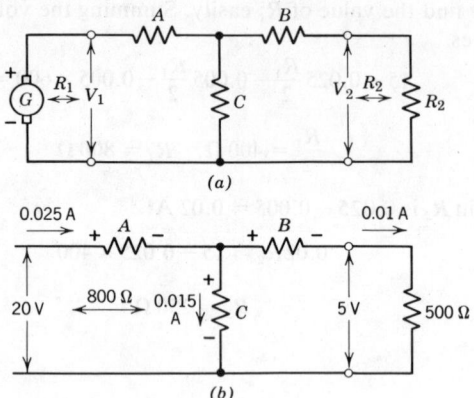

FIGURE 7.17 T pads matching generator to load. (a) A T pad matching two unequal resistances. (b) A particular matching problem.

the shunt arm of the pad, is $0.025 - 0.01 = 0.015$ A. These values are shown in Fig. 7.17b.

We may write voltage equations, as follows:

$$20 - 5 = 0.025A + 0.01B \tag{7.6}$$

$$20 = 0.025A + 0.015C \tag{7.7}$$

One more equation may be written, but it will be found to be identical with what we get when we subtract Equation (7.6) from Equation (7.7), namely,

$$5 = 0.015C - 0.01B \tag{7.8}$$

This means we must find another relationship among the three unknowns.

We recall that the *reciprocity theorem* allows us to interchange a voltage source in one branch of a passive circuit with an ammeter (assumed to have zero resistance) in another branch and to know that the ammeter reading will be the same in its new position as it was in the old. Accordingly, we draw the circuit in Fig. 7.18.

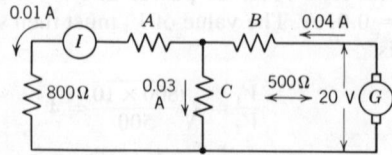

FIGURE 7.18 Reciprocity theorem applied to Figure 7.17.

The generator "sees" 500 Ω in its new location, so its current is $20/500 = 0.04$ A. The shunt arm has $0.04 - 0.01 = 0.03$ A.

We now have a third equation

$$20 = 0.04B + 0.03C \tag{7.9}$$

Rewriting Equation (7.9),

$$5 = -0.01B + 0.015C$$

Solving these two equations yields

$$B = 166.7 \,\Omega, \qquad C = 444.4 \,\Omega$$

And from Equation (7.7) we get $A = 533.3 \,\Omega$.

The power transfer ratio is $(20 \times 0.025)(5 \times 0.01) = 10:1$. This means that nine-tenths of the input power is dissipated in the matching network.

This T network may be transformed into an equivalent π network by means of the conventional transformation equations. It will be recalled that networks so related are equivalent insofar as their external voltages and currents are concerned.

Attenuators with unbalanced T, and bridged T, H, and O configuration arms, are common in practice. See Fig. 7.19. Design equations are available in handbooks. The engineer at the controls in a radio or television broadcast studio uses attenuators (pads with variable resistances) to change voltage ratios

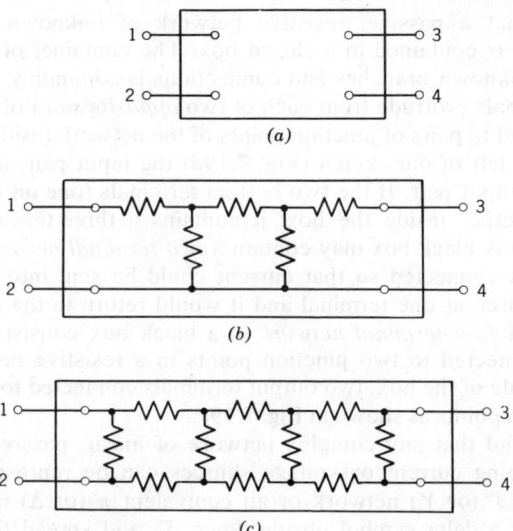

FIGURE 7.19 "Black box" and "hidden" networks. (a) Black box. (b) Three-terminal network. (c) Four-terminal network.

in communication networks of systems that feed an antenna or a recording device. The resistance values of components A, B, and C in Fig. 7.15 represented by $R_1/2$, $R_1/2$, and R_2 in Fig. 7.16, are changed by turning a single knob. This is called a gang-control arrangement. With proper design, the three-component can be made to "track" quite well, that is, maintain correct individual resistance values for a large number of voltage ratios obtained at corresponding positions of the control shaft. When very close tracking is required, an arrangement is employed that switches a large number of pad groups, each designed for a slightly increased voltage ratio. This gives the effect of a continuous, smooth variation.

Problem

10. Use the resistance values found for the arms A, B, and C (533.3, 166.7, and 444.4 Ω respectively) in the network of Fig. 7.17b, and compute the input impedance at the terminals, as seen by the generator. The load is 500 Ω. (Ans: 800 Ω.)

7.8   Hidden Passive Networks: Black Box

Assume that a passive, resistive network of unknown configuration and complexity is contained in a closed box. The container of this "hidden" network of unknown branches and connections is commonly called a *black box*. Two terminals protrude from each of two *opposite sides* of the box. Each pair is connected to pairs of junction points of the network inside. We shall call the pair at the left of our sketch (Fig. 7.19a) the input pair, and the pair on the right the output pair. If the two bottom terminals (one on each side) are connected together inside the box, it contains a three-terminal network as in Fig. 7.19b. A black box may contain a *two-terminal network* such as a group of resistors connected so that current could be sent into the group from an outside source at one terminal and it would return to the source at the other terminal. A *four-terminal network* in a black box consists of two input terminals connected to two junction points in a resistive network and, on the opposite side of the box, two output terminals connected to two entirely different junction points as shown in Fig. 7.19c.

We found that any complex network of linear, passive, circuit elements not containing current or voltage sources can be represented by either an equivalent T (or Y) network or an equivalent π (or Δ) network. Note that if we draw a delta symbol upside down, ∇, and spread the two lower lines apart we get two vertical lines supporting a horizontal line, which resembles the letter π. In Fig. 7.20 are shown T, π, H, and O resistive networks.

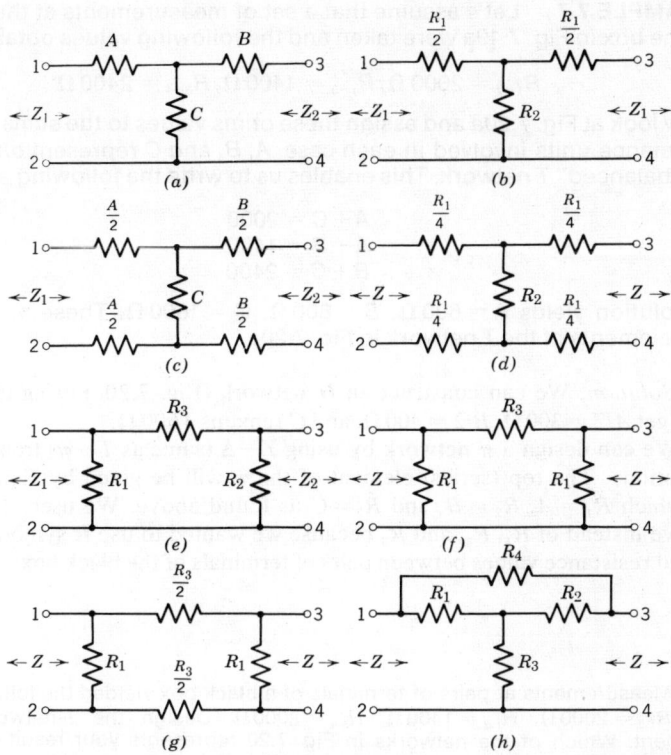

FIGURE 7.20 T, π, H, O networks. (a) General T. (b) Symmetrical T. (c) General H. (d) Symmetrical H. (e) General π. (f) Symmetrical π. (g) Symmetrical O. (h) Bridged T (c), (d), and (g) are balanced networks. Input and output resistances are represented by the letter Z because pads and attenuators are used mostly in alternating-current networks where ohms of impedance (Z) are encountered, instead of pure resistance, in many applications. Impedance will be explained when we get to alternating-current theory.

We shall now show how we can design a T or π network that will be the equivalent of the unknown circuit in the black box *in the respect that it will take the same input current and deliver the same output current as will the black box when supplied with the same input voltage and load resistance.* This means the equivalent T or π network will present the same resistance to its input terminals as that of the hidden network in the box, and the two networks will have the same output resistance. By output resistance we mean the resistance an ohmmeter would measure when connected to the output terminals.

EXAMPLE 7.7   Let's assume that a set of measurements at the terminals of the box in Fig. 7.19a were taken and the following values obtained:

$$R_{1-2} = 2000\ \Omega, R_{1-3} = 1400\ \Omega, R_{3-4} = 2400\ \Omega$$

Now look at Fig. 7.20a and assign these ohms values to the sums of pairs of resistance units involved in each case. A, B, and C represent *ohms* in that "unbalanced" T network. This enables us to write the following equations:

$$A + C = 2000$$
$$A + B = 1400$$
$$B + C = 2400$$

A solution yields $A = 600\ \Omega$, $B = 800\ \Omega$, $C = 1600\ \Omega$. These are evidently the elements of the T network in Fig. 7.20a.

Solution. We can construct an H network (Fig. 7.20c) using these values and get $A/2 = 300\ \Omega$, $B/2 = 400\ \Omega$, and C remains $1600\ \Omega$.

We can design a π network by using $Y - \Delta$ (same as $T - \pi$) transformation equations. The top (series) element of the π will be given by Equation (6.9) in which $R_1 = A$, $R_2 = B$, and $R_3 = C$ as found above. We used A, B, and C above instead of R_1, R_2, and R_3 because we wanted to use R symbols for measured resistance values between pairs of terminals of the black box.

Problems

11. Measurements at pairs of terminals of a black box yielded the following data: $R_{1-2} = 2000\ \Omega$, $R_{1-2} = 1400\ \Omega$, $R_{3-4} = 2000\ \Omega$. Design the T-network equivalent. Which of the networks in Fig. 7.20 represents your result? (Ans: Fig. 7.20b: $A = B = 700\ \Omega = R_1/2$, $C = 1300\ \Omega = R_2$.)
    $C = 1300\ \Omega = R_2$.)

12. After doing Problem 11 evaluate the elements of an H network that is equivalent to the one obtained in that problem. Is this a balanced network? (Ans: Fig. 7.20d : $R_1/4 = 175\ \Omega$, $R_2 = 1300\ \Omega$.
    $R_2 = 1300\ \Omega$.)

7.9   Review Questions

1. What are the characteristics of a linear circuit?
2. Circuits have active elements and passive elements. Name one element of each kind.
3. What quantities may be called *responses* of a network to an input (or applied) voltage?
4. A given network has two sources E_1 and E_2, and several resistances including one called the load resistance, R_L. Explain the idea used in solving for the load current by application of the superposition principle.

5. To apply Thevenin's theorem it is necessary to evaluate two quantities. What are they and what do you do with them after you get them?

6. How are the two quantities referred to in Question 5 evaluated?

7. In applying Norton's theorem, one of the two quantities required for Thévenin's is needed. Which is it?

8. What is the other quantity needed for Norton's theorem application, and how do we evaluate it?

9. In what kinds of circuit is the production of maximum power in the load an important consideration?

10. An amplifier for a loud speaker has an internal resistance of 8Ω. What should be the resistance of the loud speaker?

11. Assume the amplifier of Question 10 has an EMF of 4 V. How much power would the speaker receive if it gets maximum power and how much would be lost in the amplifier?

12. State the reciprocity theorem. Does it hold when there is a second active element, such as a voltage or current source inside the network?

13. What is an attenuator? Name two functions it can perform.

14. What is required of the series arms of a T pad of resistances that must match a generator to a load which has the same resistance as the generator? Answer this question with reference to the shunt arms of a π pad of resistances.

15. Explain what electrical engineers mean by the term "black box."

16. Outline the procedure of designing a T-network equivalent of the network in a black box. What must be the nature of the network in the box in order that your procedure will yield correct results?

17. What is the structure of a symmetrical H pad?

18. What is the structure of a symmetrical O pad?

19. What is the structure of a bridged T pad?

20. How could an adjustable pad be constructed so that it would provide practically continuous gradual control of attenuation?

7.10 Problems

Group A

13. (a) How much current flows in the 10Ω load resistance in Fig. 7.21? Can you solve this problem mentally? (b) Suppose the battery voltage is increased 50 percent to 90 V. How much current would flow in the load resistor? (c) Suppose the load resistance had been 20Ω when the battery supplied only 60 V, as in part (a). Would the *load current* have been half as much as we have with $R_L = 10 \Omega$? Calculate its value.

14. Solve for the current in the 8-Ω resistor in Fig. 7.22 by applying the principle of superposition.

15. Using Thévenin's theorem, determine the current in the 2-Ω resistance in Fig. 7.23.

16. Verify your answer to Problem 15 by applying Norton's theorem to the same network.

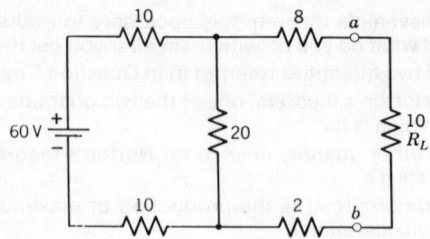

FIGURE 7.21   Resistances are in ohms for Problem 13.

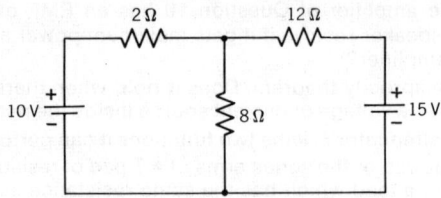

FIGURE 7.22   For Problem 14.

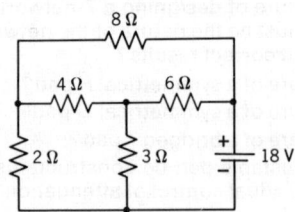

FIGURE 7.23   Circuit for Problem 15.

17. What should be the value of the load resistance in Fig. 7.24 so it can draw maximum power?

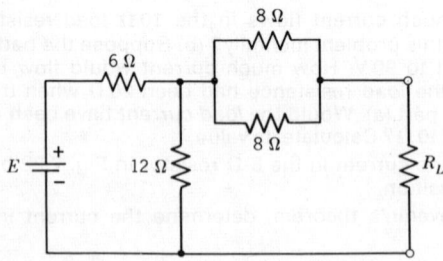

FIGURE 7.24   For Problems 17 and 18.

18. Let the load resistance in Fig. 7.24 be 8 Ω and let $E = 36$ V. Draw the Norton equivalent circuit and calculate the load current.

19. Calculate the amount of current that would be indicated by a zero-resistance ammeter in series with the 40-Ω resistance in Fig. 7.25. Then show that the reciprocity theorem holds.

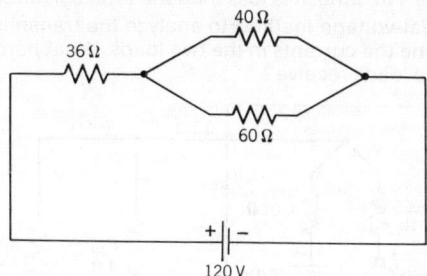

FIGURE 7.25   For Problem 19.

20. Show that the reciprocity theorem holds for Figure 7.23 by putting an ammeter in place of the battery and connecting the battery in series with the 2-Ω load resistor. The original current in the 2-Ω resistor is the answer to Problem 15.

Group B

21. Add a constant 20-V source in series with the 2-Ω resistor in Fig. 7.23 with such polarity that it will tend to send current upward through that resistor. Then use Thévenin's theorem to solve for the amount of current in the 4-Ω resistor.

22. To what value must the 2-Ω resistor in Fig. 7.23 be changed so that its branch will receive maximum power?

23. Transform the T section in Fig. 7.26 to a π section and calculate the battery current. (b) How much current is in each 3-Ω resistor?

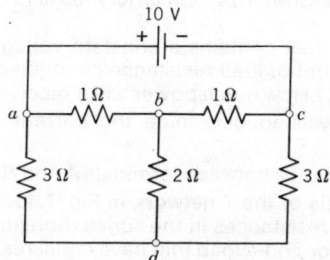

FIGURE 7.26   For Problem 23.

24. Problem 19 in Chapter 6 was to be solved for the current in the 4000-Ω resistor by one of the methods we had learned. Now solve for it by means of Thévenin's theorem. Consider the 4000 Ω to be the load resistance.

25. Apply Norton's theorem to solve Problem 7.24. (Look for a pitfall in this one. The short-circuit current flows from A to D in a zero-resistance conductor; let us say it is 1 in. long. It is less than the total current.)

26. Use the nodal-voltage method to analyze the transmission system of Fig. 7.27 and determine the currents in the two loads. What percent of the station output power do the loads receive?

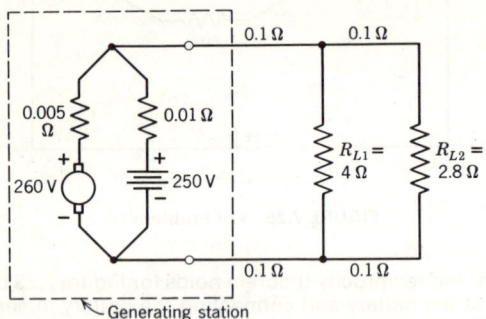

FIGURE 7.27   Circuit for Problem 26.

27. Determine I_L in the network of Fig. 7.28 by first assuming I_L to be 1 A and calculating the battery voltage required for this. Then calculate the correct value of I_L.

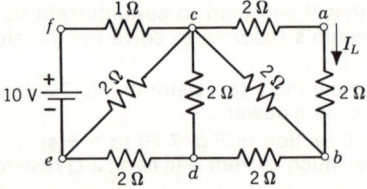

FIGURE 7.28   Circuit for Problem 27.

28. The box in Fig. 7.29 contains a constant-voltage generator and a delta network. What amount of load resistance connected to terminals a, b will receive maximum power? How much power will it receive?

29. Use nodal analysis to determine the current in the 2000-Ω resistance of Fig. 7.30.

30. What is the resistance between terminals A and B in Fig. 7.31?

31. Study the analysis of the T network in Fig. 7.15, which shows that such a network with equal resistances in the series (horizontal) arms can be designed to match a generator and a load that have equal resistances. By a similar analysis prove that a resistance pad of π form which matches a generator resistance to an equal load resistance has shunt (parallel) arms of equal resistance.

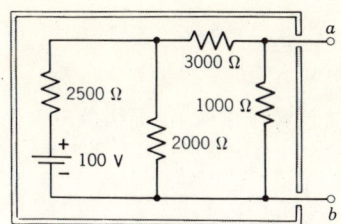

FIGURE 7.29   Circuit for Problem 28.

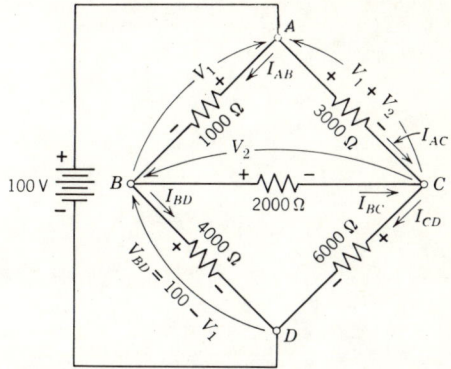

FIGURE 7.30   For Problem 29.

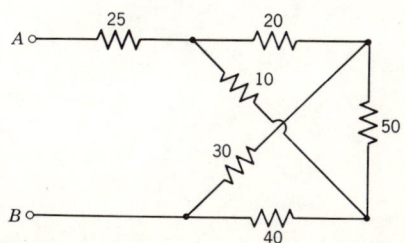

FIGURE 7.31   Resistances are in ohms. For Problem 30.

32. Sketch a π network with the series arm labeled R_B and each shunt arm labeled R_A, and show it connected in place of the T network of Fig. 7.16. Analyze the circuit and evaluate R_A and R_B. The 600-Ω generator and load must be matched and the 5 : 1 voltage ratio maintained.

33. Measurements at the terminals of a black box yielded the following ohm values: $R_{1\text{-}2} = 2550$, $R_{1\text{-}3} = 1650$, $R_{3\text{-}4} = 2700$. Design a balanced H pad that will be equivalent to the network in the box.

CHAPTER EIGHT

Magnetism and Magnetic Circuits

8.1 Magnetic Fields

The ability of a permanent magnet, such as the small horseshoe type, to pick up small iron objects (nails, tacks, iron filings) is quite familiar. Some mechanical pencils, flashlight cases, memorandum-board tabs, and even padded cloth potholders are fitted with small permanent magnets that serve to hold them in position on iron or steel surfaces. The small permanent magnets have the ability to retain the magnetic effect put into them during manufacture. This is done by means of passing an electric current, not through the magnet material, but through a coil of wire wound around a *magnetic path* of which the material of the small magnet is a part.

In this chapter we shall become acquainted with the magnetic effects of electric current, essentially *direct current*. (See Table 8.1 for terms we shall be using.) We have become quite familiar with electric circuits, so now we shall study the properties of *magnetic circuits*. These are *closed paths* in which *unbroken magnetic flux lines* exist.

Figure 8.1 shows some magnetic flux lines that exist inside a *bar magnet* and continue in *closed-loop fashion* in the space around the bar. Some additional lines that pass through the bar near its axis extend farther out in space and cannot be shown as complete loops in the illustration. Note that the

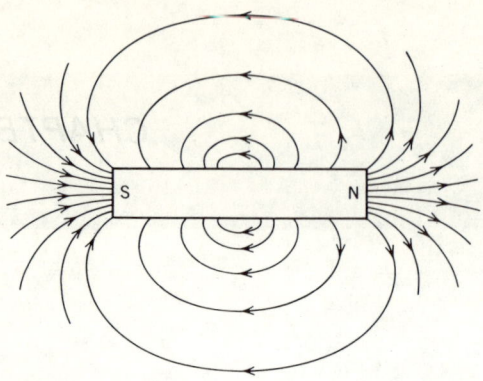

FIGURE 8.1   A bar magnet and its magnetic field.

magnetic flux lines come out at the N-pole end of the bar magnet and reenter it at the S-pole end. This arbitrary choice of *direction* of magnetic field flux was made by early experimenters. The assignment of magnetic flux as an identi-fication of the field is a convenience that enables us to analyze magnetic behavior not only in words but also *quantitatively*, that is, with numbers. The magnetic field cannot be detected by any of our personal senses: for example, sight, smell and feel, but its *effects* can be observed in various ways.

If we place a sheet of paper or cardboard on top of a bar magnet and sprinkle iron filings on it, they will align themselves in the manner shown in Fig. 8.2. Tapping the cardboard gently results in a well-defined pattern of the magnetic field. Magnetic flux lines are also called *lines of magnetic force* as suggested by Michael Faraday.

When two magnets are placed so that the north pole of one is near the south pole of the other, as in Fig. 8.3*a*, they experience a *force of attraction*. They

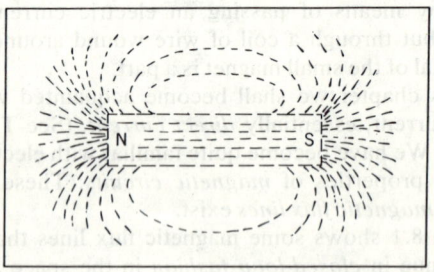

FIGURE 8.2   Iron filings line up on a piece of paper placed over a bar magnet.

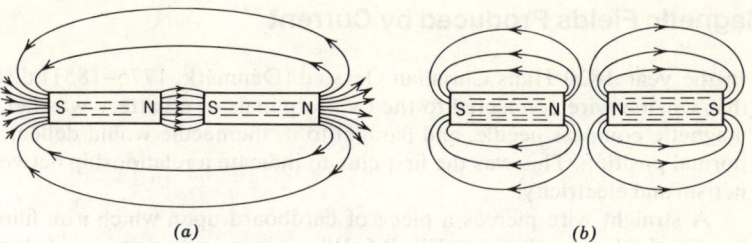

(a)   (b)

FIGURE 8.3   Magnetic fields of pieces of a bar magnet (a) unlike poles adjacent, (b) like poles adjacent. Flux lines do not cross one another.

will come together if friction or other forces are not strong enough to keep them apart. When two like poles are brought near each other, as in Fig. 8.3b, they experience a *force of repulsion* that tends to push them apart.

Additional important properties of magnetic lines of force are: (a) they always tend to shorten themselves, acting like stretched rubber bands: (b) they never cross one another: (c) they are always ready to pass through iron or other magnetic materials nearby, in preference to passing through air, even though their closed-loop paths are made longer thereby (Fig. 8.4); and (d) they always arrange their positions so that the maximum number of them can be accommodated in the region.

In our discussions we shall often use the term *magnetic field* instead of magnetic lines of force, and refer to magnetic flux as existing in a magnetic circuit in the same sense that electric current exists in an electric circuit. We shall become acquainted with a so-called *Ohm's law for the magnetic circuit* which will have the same form as Ohm's law for the electric circuit, with magnetic flux taking the place of electric current in the formula.

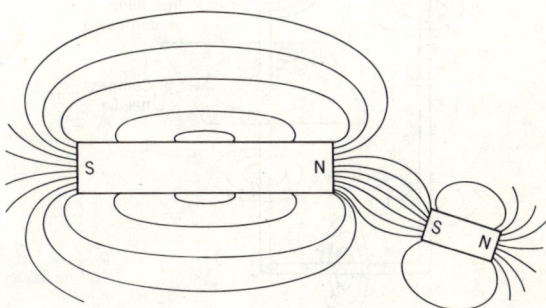

FIGURE 8.4   Soft-iron piece in field of a permanent magnet.

8.2 Magnetic Fields Produced by Current

In the year 1820 Hans Christian Oersted (Denmark, 1775–1851) discovered that when a wire connected to the two terminals of a battery was held over a magnetic compass needle, and parallel to it, the needle would deflect from its normal position. This was the first clue to indicate a relationship between magnetism and electricity.

A straight wire pierces a piece of cardboard upon which iron filings have been sprinkled, as shown in Fig. 8.5. When electric current is sent through the wire, the filings arrange themselves in concentric circles around the wire. To make a most distinctive pattern they usually need some encouragement in the form of tapping by a pencil or a finger on the edge of the cardboard. This pattern indicates the presence of magnetic lines of force. A compass needle placed on the cardboard will turn so that its north end will point *in the direction of the magnetic field* according to an established convention of direction. It is found that the cardboard may be placed anywhere along the length of the straight wire, and the filings and compass needle will line up in the identical manner if the current direction remains unchanged. This shows that the magnetic field exists along the full length of the current-carrying wire. The magnetic lines of force are in circular paths concentric with the axis of the wire, and the planes of the paths are always perpendicular to the wire axis.

The direction of magnetic force exerted on a north pole of a compass needle in the field in Fig. 8.5 must be along a concentric circle. To determine

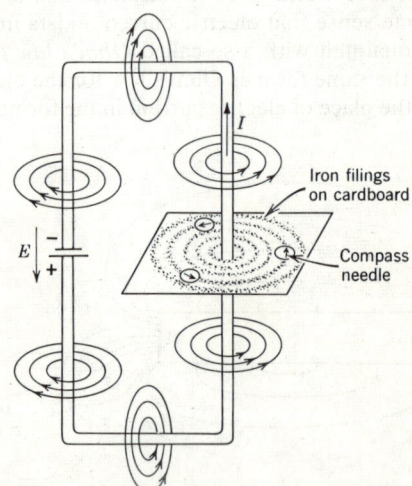

FIGURE 8.5   Magnetic field produced by current flowing in a loop of wire.

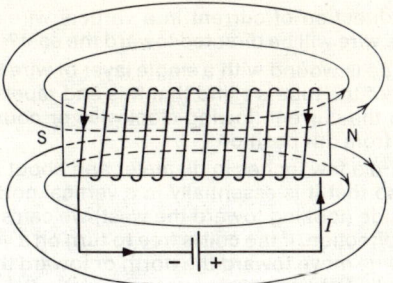

FIGURE 8.6   A coil of wire wound on a cardboard tube. Current in the wire establishes a magnetic field. Only four of the very large number of streamlines of the magnetic field are shown.

whether it will be clockwise or counterclockwise one may use a *right-hand rule: If a current-carrying conductor is grasped with the right hand with the thumb pointing along the conductor in the conventional direction of the current flow, the fingers encircling the conductor will point in the direction of the magnetic lines of force.* If we imagine that this is done near the cardboard in Fig. 8.5, the thumb would point upward along the wire and the fingers would point counterclockwise as we look down on the cardboard from above.

A coil of wire in which current is flowing produces a magnetic field that passes through the coil and returns on the outside as shown in Fig. 8.6. It is evident that, with the magnetic lines of force coming out the right-hand end, the current has established a north magnetic pole on that end and a south magnetic pole on the other end. A *right-hand rule for coils carrying current* is useful, and it says that, *if we grasp the whole coil with the right hand so that the fingers point in the direction of the current and if we extend the thumb parallel to the axis of the coil, it will point in the direction of the magnetic lines of force which are set up inside the coil and this will indicate the north-pole end of the coil.*

Both of the foregoing right-hand rules may be applied in reverse. If we know the direction of the magnetic field encircling a single wire carrying current, we can imagine that we may grasp the wire with the right hand so that the fingers point in the direction of the field. The thumb would then indicate the direction of current flow in the wire. In a somewhat similar manner we can determine the direction of current in a coil if we know which end (or, in a flat coil of very short length, which of its faces) is the one through which the lines of magnetic force emerge.

Problems

1. (a) A straight horizontal wire carries current toward the east. What are the directions of the magnetic field it produces above and below the wire? (b) What should be the direction of current in a vertical wire so that its magnetic field just north of the wire will be directed toward the east?

2. A cardboard tube is wound with a single layer of wire that carries current. As we look at the end of the tube we find that this end repels the north pole of a compass needle. Is the current flowing clockwise or counterclockwise around the tube as viewed from our position?

3. A flat coil of wire a few inches in diameter and about an eighth of an inch thick is suspended so that it is essentially in a vertical north-south plane. As viewed from the east side (looking toward the west) we can say that it carries current in the clockwise direction. If the coil is free to turn on a vertical axis, will the side of the coil toward us move toward the north or toward the south? What characteristics of the coil's field and the earth's magnetic field lead to an answer to this question?

8.3 Faraday's Law of Induced EMF

Michael Faraday (England, 1791–1867) made significant contributions to electrical science. Here we are interested in *Faraday's law of electromagnetic induction*. He often performed the experiment we shall describe, using various lengths of copper wire (hundreds of feet) and indicating instruments such as magnetic needles and a galvanometer.[1]

Figure 8.7 illustrates how an EMF can be induced in a coil of wire wound on a cardboard tube simply by holding a bar magnet so one of its poles is near an end of the coil and then moving the magnet toward, or away from, the coil.

[1]A galvanometer is a sensitive electric meter. Its pointer will deflect from the zero position, in the middle of its scale, either way depending on the direction of the very minute current flowing through its coil.

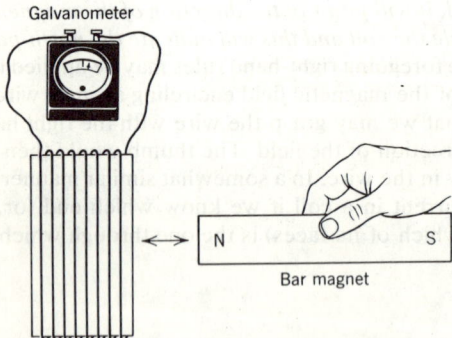

FIGURE 8.7 Inducing an EMF by changing the amount of magnetic flux linking a coil. The bar magnet is moved toward or away from the coil of wire. The induced EMF causes current to flow in the coil and through the galvanometer.

While the magnet is moving toward the end of the coil, the *number of magnetic flux lines* that pass through the turns of wire of the coil is increasing. We say the flux lines *link* with the turns of the coil and the *flux linkages* are increasing. *While this is going on*, that is, while there is *relative motion* between the magnet and the coil, *an EMF is induced* in the coil. This induced EMF will send current through the wire of the coil if we have a closed circuit. In the illustration, a galvanometer connected to the two ends of the coil closes the circuit. The presence of current in the closed circuit is indicated by a deflection of the galvanometer needle. If the wire happens to be wound around the cardboard tube in such a direction that the needle deflects to the right while the north pole approaches the coil, the needle will deflect toward the left if the north pole is drawn away from the coil. This change in deflection is the result of a reversal of current in the coil caused by a reversal of polarity of the induced EMF.

Faraday wound over 200 ft of copper wire in the form of a helix around a large block of wood and interposed between the turns a second helix of the same length. Turns of twine were inserted to prevent metallic contact. He connected a galvanometer to the two ends of one helix and to the ends of the other helix "a battery of one-hundred pairs of plates four inches square, with double copper, and well charged."[2] He noted that at the instant contact to the battery was made there was a slight deflection of the galvanometer needle in one direction. The needle soon returned to zero, however, and remained there while the battery continued sending a steady current through *its* helix. He knew current was flowing because that helix became heated. At the instant Faraday disconnected the battery, the galvanometer deflected about an equal amount *but in the reverse direction.* Later he used a helix wound on a glass tube and observed that a steel needle inserted inside the tube became magnetized when current was sent through the helix.

Faraday concluded that an *induced current* in the second helix (the one not connected to the battery) caused the galvanometer needle to move. This current was produced by an *induced voltage*, he reasoned. Experiments of this type, which were repeated many times by him, and independently during the same year, about 1830, by Joseph Henry (United States, 1797–1878) at the university which is now Princeton, formed the basis for his *law of electromagnetic induction.* One should recognize that lines of magnetic force produced by battery current in one wire helix also passed through, or *linked*, turns of wire in the other helix which had no metallic contact anywhere with the first helix. The amount of the voltage induced in the second helix depended on (*a*) the number of "flux linkages" and also (*b*) the rate at which the flux linkages changed with time. This involves how rapidly the field of lines of magnetic force (called flux) was established by the current when contact was made, or how rapidly the field collapsed and disappeared when contact was broken.

[2]William F. Magie, *Source Book in Physics*, New York: McGraw-Hill, 1935.

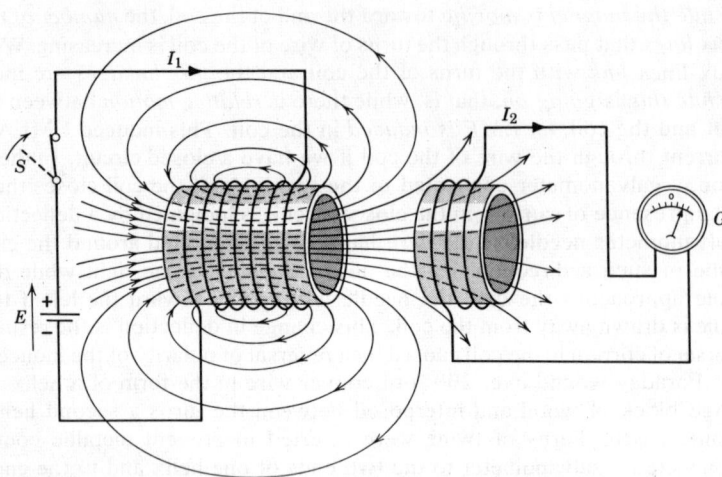

FIGURE 8.8 Illustrating electromagnetic induction. Switch S has just been closed and current I_1 is increasing, and so is the magnetic field it produces. Voltage is being induced in the smaller coil, and this induced voltage is sending the induced current I_2 through the closed circuit containing the galvanometer G.

Figure 8.8 shows how a changing magnetic field in one coil links turns in another coil, thereby inducing a voltage (also called an *electromotive force*, EMF) in it. This voltage will be induced even though the other coil may not be connected to anything, i.e., its wires may be left hanging loose. The *induced current*, I_2 flows in this illustration because the circuit containing the coil is closed. The galvanometer is used merely to indicate the presence of this current, which would flow if the ends of the wires were connected together rather than to the terminals of the galvanometer. In that case, a magnetic compass placed inside the coil, to serve as an "indicator," would be deflected while the current I_2 was building up.

The current I_1 will rise to a fixed, steady value in this example, if the switch is left closed. By that time I_2 will have decreased to zero, because induced currents flow only while the inducing magnetic field is *changing*. If, now, the switch is opened, the magnetic field produced by I_1 will collapse and very quickly disappear. *While this is happening*, the voltage induced in the other coil will be in the opposite direction to that which accompanied the *increasing* I_1 (just after the switch was closed) and the current I_2 induced in the other coil will reverse in direction of flow, causing the galvanometer to deflect to the left of center.

An induced voltage (EMF) exists also in the coil containing I_1 while I_1 is increasing, because its increasing magnetic field links its own turns, and this change in flux linkages induces the voltage. In fact, this *voltage of self-induction* slows the rate of rise of current I_1. If the wire of the left-hand coil is completely unwound after the switch has been opened and left lying along a straight line, the current sent through it at the close of the switch will rise almost instantly to its maximum and steady value. There would be practically no voltage of self-induction to cause the current I_1 to rise at a slower rate because the number of flux linkages in the *single-turn path* of I_1 would be very small. Furthermore, the flux linking the other coil would be so small that I_2 would be negligible in amount.

We may now state Faraday's law of electromagnetic induction: *When the magnetic flux linking a coil changes in amount, a voltage is induced in the turns of the coil which is proportional to the rate at which the flux linkages are changing with time.* This law holds for only one coil as well as for two. If only one coil is present, a voltage (or EMF) of *self-induction* is produced by the change in its own flux linkages. We shall leave further discussion and quantitative analysis of electromagnetic induction for a later chapter.

The MKS unit of magnetic flux is the weber (Wb), named after Wilhelm Edward Weber (Germany 1704–1791). The letter symbol for magnetic flux is the Greek letter phi: Φ.

You may well ask how much is one weber of magnetic flux. Let's imagine that the amount of flux that links a *one turn coil is changing* and thus inducing an EMF in the wire of the coil. If the induced EMF is to be 1 V, the magnetic flux must be *changing at the rate of 1 Wb/s.* We can define the weber in another way that is, perhaps, easier to understand. If magnetic flux passing through a single turn conductor increases at a constant rate from zero to 1 Wb in 1 s, it will induce an EMF of 1 V in the conductor. This means, of course, that the flux through the single turn will be 1 Wb at the end of the 1-s time interval. We must also emphasize that the flux need not be zero at either the beginning or the end of the 1-s interval. It is only necessary that the *change in the amount of magnetic flux* be 1 Wb. The flux may be 2 Wb and change to 3 at a constant rate during the 1-s interval, or it may change from 2 Wb to 1 in that time. Or, it may change from any initial value to a different final value that may be 1 Wb more or 1 Wb less at the end of the 1-s time interval.

The direction of an induced EMF (that is, its polarity) changes when the direction of relative motion between the flux and the coil changes. If we label the two ends of a coil *a* and *b*, we may find that *a* will be positive with respect to *b* while the flux linkages are increasing. In that case, *b* will be *positive* with respect to *a* while the flux linkages are decreasing.

EXAMPLE 8.1   The north pole of a permanent magnet is thrust into a coil. In 0.01 s the flux linkages changed from 0.02 to 0.05 Wb turns. How much EMF was being induced?

Solution. We must assume that the flux linkages were changing *at a constant rate* during the 0.01-s time interval. Then the value of induced EMF was constant and our answer will be the EMF at every instant during the interval:

$$\text{EMF} = \frac{N\Phi}{t} = \frac{0.05 - 0.02}{0.01} = 3 \text{ V}$$

Problems

4. The north pole of a permanent magnet has 0.005 Wb of flux coming out of it. It is thrust into a 10-turn coil. (a) How many flux linkages existed when all of the flux linked all of the turns of the coil?

   (b) A time interval of 0.004 s elapsed from the instant half the flux linked all of the turns to the instant all of the flux linked all of the turns. What was the average induced EMF during that time interval?

5. The magnet of Problem 4 was withdrawn completely, from its all-the-way-in position, in 6 ms. What was the average induced EMF?

8.4   Magnetomotive Force

We have seen that electric current produces a magnetic field and that when current flows in a coil of wire a magnetic field is established such that lines of magnetic force, which we also call magnetic flux, pass through the coil. It seems almost obvious that if we increase the amount of electric current the amount of magnetic flux will increase, and if the same amount of current flows in two different coils of the same shape and dimensions, except that one coil has more turns of wire than the other, more magnetic flux will be established through the coil that has the more turns. We say that magnetic flux is established by a *magnetomotive force*. The product amperes × turns gives us the amount of magnetomotive force.

The ampere-turn is the unit of magnetomotive force in the MKS system.

We shall use the script letter $\mathscr{F}$ to represent magnetomotive force, abbreviated MMF:

$$\mathscr{F} = NI \tag{8.1}$$

Where $\mathscr{F}$ is the magnetomotive force in ampere-turns, N is the number of turns of wire in the coil and I is the current in the coil in amperes.

8.5   Ohm's Law for the Magnetic Circuit

Now that we have a magnetomotive force producing magnetic flux in a magnetic circuit (corresponding loosely to electromotive force producing current in an electric circuit), is there something in magnetic units that corresponds to resistance in ohms? Yes, it is called *reluctance*. And, interestingly enough, reluctance is the name given to the magnetic circuit's opposition to the establishment of magnetic flux. It increases with the length of the path through which the magnetic flux must pass, it decreases when the cross-section area of the path is increased, and it depends on the kind of material of which the path is made. All of these things that determine the amount of reluctance of a closed magnetic circuit have counterparts (let's call them that) in the closed electric circuit. We shall soon develop a formula for reluctance that involves the dimensions of the path and a constant for the material of the path. The letter symbol for reluctance is the script letter $\mathscr{R}$.

Ohm's law for the magnetic circuit becomes

$$\text{Flux} = \frac{\text{magnetomotive force}}{\text{reluctance}}, \qquad \Phi = \frac{\mathscr{F}}{\mathscr{R}} \tag{8.2}$$

From this we get

$$\mathscr{R} = \frac{\mathscr{F}}{\Phi} = \frac{\text{ampere-turns}}{\text{webers}} \tag{8.3}$$

Evidently the *units of reluctance* are *ampere-turns per weber*.[3]

Reluctances of sections of a magnetic circuit are combined in the same manner that resistances in an electric circuit are combined. When sections are in series their reluctances are added; when in parallel the reciprocal relation holds. Reluctance is discussed further in Section 8.9.

Permeance is the name given to the reciprocal of reluctance. Permeances in parallel are added just as conductances in parallel are added. You will remember that conductance is the reciprocal of resistance.

8.6 Magnetic Flux Density

Suppose an amount of magnetic flux, Φ, exists in a path that has a cross-section area A perpendicular to the direction of the flux lines. The *number of flux lines per unit area* is then given by Φ divided by A. The letter **B** is used to represent magnetic flux density (see Fig. 8.9).

$$\mathbf{B} = \frac{\Phi}{A} \qquad \frac{\text{webers}}{\text{square meter}} \tag{8.4}$$

Webers per square meter are the units of flux density in the Rationalized MKS system. In the English system the units of flux density are *lines per square inch*, usually.

[3]It would be convenient if we had a one-word name for the unit of reluctance instead of having to use four: *ampere-turns per weber* (At/Wb). Someone suggested that *rel* be the name of the unit of reluctance, but it did not catch on. However, it should be easy to remember the units of reluctance by converting flux = MMF ÷ reluctance into reluctance = MMF ÷ flux ampere-turns/weber.

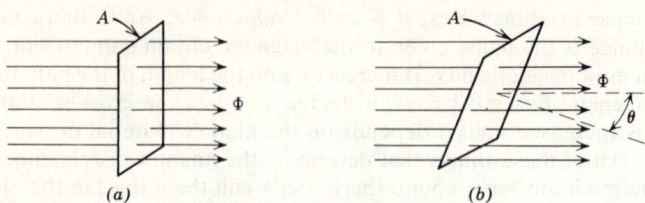

FIGURE 8.9   *(a)* Flux ϕ is perpendicular to area *A*. Flux density is *B* = ϕ/A. *(b)* Flux lines make an angle θ with the perpendicular to the area *A*. Flux density is *B* = (ϕ/A) cos θ.

1 Wb of magnetic flux = 10^8 lines of magnetic flux

The *line of magnetic flux* is called the *maxwell* (Mx) in honor of James Clerk Maxwell, a brilliant Scottish physicist and mathematician (1831–1879). The name *gauss* (G) is also used for magnetic flux density. It is a unit in the *centimeter-gram-second* (CGS) system and is named after Johann Karl Gauss (1777–1855), a German professor of mathematics:

1 G = 1 line of magnetic flux per square centimeter

EXAMPLE 8.2   A bar of iron 1 cm² in cross section has 10^{-4} Wb of magnetic flux in it. What is the flux density?

Solution

$$\mathbf{B} = \frac{\Phi}{A} = \frac{10^{-4}}{10^{-4}} = 1 \text{ Wb/m}^2$$

8.7   Magnetic Field Intensity

Now we know that ampere-turns produce magnetic flux. Increasing the number of ampere-turns of a coil will produce more flux and the magnetic field is said to become stronger. The *strength of a magnetic field* is also called the *field intensity* and its letter symbol is **H**.

   Magnetic field intensity is magnetomotive force per unit length. The MKS units are *ampere-turns per meter*.

   Let's look at Fig. 8.10, where *I* amperes flow in *N* turns wound on a toroidal ring whose length of magnetic path (average circumference) is $2\pi R$ meters. The number of *ampere-turns per unit length* is NI/l.

$$\mathbf{H} = \frac{NI}{2\pi R} \qquad \text{ampere-turns per meter} \qquad (8.5)$$

H is also called magnetizing force, although it is clearly not a force in the normal sense.

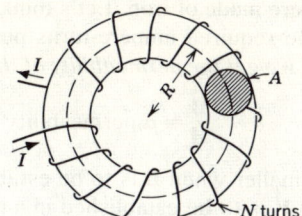

FIGURE 8.10   A toroidal magnetic circuit energized by *NI* ampereturns.

We must understand that when we calculate **H**, using *NI/l*, the length *l* is that of the flux path (or its part) over which *NI* is effective. There it is the full circumference of the toroid.

EXAMPLE 8.3 What is the magnetic field intensity in the iron bar of Example 8.1 if the relative permeability of the iron is 200?

Solution

$$\mathbf{H} = \frac{\mathbf{B}}{\mu} = \frac{\mathbf{B}}{\mu_r \mu_0}$$

$$\mathbf{H} = \frac{1}{2000 \times 4\pi \times 10^{-7}}$$

$$\mathbf{H} = 398 \text{ At/m}$$

8.8 Permeability

We know that iron and steel become magnetized quite easily. If you hold a steel bar horizontally so that one end points toward the earth's north pole, then tilt its north end downward about 70° below the horizontal and strike it a few sharp blows with a hammer it will become magnetized. If you bring the end of the bar that was dipped downward near a compass needle you will find that it is a north pole. It will attract the south-pole end of the compass. The earth's magnetic field slopes downward at an angle between 70° and 75° to the horizontal in the United States. The south magnetic pole of the earth is located well below the earth's surface in the north-central Canada.

Cast iron is not as easily magnetized as steel or wrought iron. Nonferrous metal—those without iron in them, such as aluminum and copper—cannot be magnetized. Neither can wood. There should be a factor that expresses *a measure of the ease with which materials can be magnetized*. There is, and it is called *permeability*.

If the ring in Fig. 8.10 is made of wood, a very much larger number of ampere-turns is required to establish a desired flux density than would be required if the ring were made of iron. Let's think of the relation of the desired flux density (**B**) to the required ampere-turns per unit length (**H**) to establish it. *If we divide **B** by **H** we get the permeability of the material:*

$$\frac{\mathbf{B}}{\mathbf{H}} = \mu \text{ permeability} \tag{8.6}$$

We said **H** is much smaller when **B** is to be established in a magnetic material (iron) than it is when **B** is to be established in a nonmagnetic material (wood), didn't we? This means that μ is much larger for magnetic materials than it is for nonmagnetic materials. One important nonmagnetic material in which magnetic flux is established is air. There is a special value of permeability for air and other nonmagnetic materials, and it is represented by μ_o:

$$\mu_o = 4\pi \times 10^{-7} \text{ for nonmagnetic materials and for empty space (vacuum)}$$

The units of permeability are henrys per meter.

Relative permeability is the ratio of the permeability of a material to the permeability of air (or free space) and it is denoted by μ_r:

$$\mu_r = \frac{\mu}{\mu_o} \tag{8.7}$$

so that permeability is determined numerically by multiplying μ_o by the relative permeability for the material to be used:

$$\mu = \mu_r \mu_o \tag{8.8}$$

Values of relative permeability of some materials are given in Table 8.1.

TABLE 8.1 **Maximum Relative Permeabilities of Ferromagnetic Materials**

Material	μ_r at Highest Densities
Cobalt	60
Nickel	50
Cast iron	90
Silicon iron	7,000
Transformer iron	5,500
Pure iron	8,000
Machine steel	450
Permalloy (Ni, Fe)	25,000 to 80,000

8.9 Reluctance Formula

Since reluctance in a magnetic circuit may be said to correspond to resistance in an electric circuit, it must increase with the length of path and decrease as the cross-section area increases. The permeability of the material of the circuit is involved also. The formula for reluctances is

$$\mathcal{R} = \frac{l}{\mu A} = \frac{l}{\mu_r \mu_o A} \text{ ampere-turns per weber} \tag{8.9}$$

In the MKS system, l is in meters, A is in square meters, $\mu_o = 4\pi \times 10^{-7}$ henrys per meter, and μ_r is a constant without units.

EXAMPLE 8.4   The ring in Fig. 8.10 is made of wood, the radius $R = 3$ in. and it is 1 in.² in cross-section area. There are 2 A of current in 500 turns of wire wrapped around the ring. How much magnetic flux is produced in the ring, and what is the flux density in webers per square meter?

Solution

$$\text{Flux} = \frac{\text{MMF}}{\mathcal{R}}, \qquad \mathcal{R} = \frac{l}{\mu_r \mu_o A} \frac{\text{At}}{\text{Wb}}$$

$$\mathcal{R} = \frac{2\pi \times 3 \times 2.54 \times 10^{-2}}{1 \times 4\pi \times 10^{-7} \times 2.54^2 \times 10^{-4}}$$

The length must be in meters (cm $\times 10^{-2}$) and the area in square meters, (cm² $\times 10^{-4}$). One inch $= 2.54$ cm:

$$\mathcal{R} = 59 \times 10^7 \frac{\text{At}}{\text{Wb}}$$

$$\Phi = \frac{2 \times 500}{59 \times 10^7} = 1.694 \times 10^{-6} \text{ Wb}$$

$$\mathbf{B} = \frac{\Phi}{A} = \frac{1.694 \times 10^{-6}}{(2.54)^2 \times 10^{-4}} = 26.2 \times 10^{-4} \text{ Wb/m}^2$$

Note that 1000 A (2A in 500 turns) are required to produce 1.694 μWb of flux in the wooden ring in Fig. 8.10. Now let's see how many ampere turns are required to produce *ten times that amount of flux* in the ring if it is made of high quality electrical sheet steel such as that used in the manufacture of transformers.

EXAMPLE 8.5   It is desired to produce 16.94 μWb of flux in a ring made of electrical sheet steel that has the same dimensions as the wooden ring in Fig. 8.10 and will replace it. The 1 in.² cross-section may be square instead of circular. How many ampere-turns will be required?

Solution. We shall need a value for the permeability. A good practical value of the *relative permeability* when this kind of steel is used efficiently may be taken as 2500. The new flux density will be ten times the flux density in Example 8.3:

$$\mathcal{R} = \frac{l}{\mu_r \mu_o A} = \frac{2\pi \times 3 \times 2.54 \times 10^{-1}}{2500 \times 4\pi \times 10^{-7} \times (2.54)^2 \times 10^{-4}}$$

$$\mathcal{R} = \frac{59.7 \times 10^7}{2500} = 23.88 \times 10^4 \text{ At/Wb}$$

$$\mathcal{F} = NI = \Phi \mathcal{R}$$

$$NI = 10 \times 1.694 \times 10^{-6} \times 23.88 \times 10^4$$

$$NI = 4 \text{ At}$$

This shows how the use of high-quality sheet steel in the core of a magnetic circuit *results in a very large reduction* of the ampere-turns required to establish a desired amount of flux. The ampere-turns required to produce *ten times as much flux* is only *one two hundred fiftieth* the amount required when the core is made of nonmagnetic material.

In this case we could have arrived at the correct answer (4 At) by a much shorter route. The solution we shall use is applicable only when the magnetic circuit is made of just one kind of material and has no air gaps.

Let us express the flux density in general form for both the air-core case ($\mathbf{B}_1$) and the steel core case ($\mathbf{B}_2$):

$$\mathbf{B}_1 = \frac{\Phi_1}{A} = \frac{(NI)_1 \mu_1}{l} \tag{8.10}$$

$$\mathbf{B}_2 = \frac{\Phi_2}{A} = \frac{(NI)_2 \mu_2}{l} \tag{8.11}$$

$$\mathbf{B}_2 = 10 \, \mathbf{B}_1$$

$$\Phi_2 = 10 \, \Phi_1, \, \mu_2 = 2500 \, \mu_1$$

Dividing Equation (8.11) by Equation (8.10),

$$\frac{(NI)_2}{(NI)_1} = \frac{B_2 \mu_1}{B_1 \mu_2} = \frac{10 \, B_1 \mu_1}{B_1 \times 2500 \, \mu_1}$$

$$(NI)_2 = \frac{(NI)_1}{250}$$

$$NI_2 = \frac{1000}{250} = 4 \text{ At}$$

Problems

6. (a) A 100-turn coil carries 2 A. How much MMF is produced?
    (b) The coil is wound on a wooden ring to form a toroid. What is the reluctance of the path of the magnetic field if 0.5 Wb of flux is present?

7. The magnetic field inside a solenoid has 2 μWb of flux. The diameter of the circular turns of wire of the solenoid is 2 cm. What is the average magnetic flux density inside the solenoid?

8. At a certain place on the earth's surface the horizontal component of the earth's magnetic field accounts for a maximum of 1.4 μWb of flux passing through a vertical area of 1000 cm^2. What is the total flux density of the earth's magnetic field if it makes a 30° angle with the vertical at this location?

9. What is the intensity (strength) of the earth's magnetic field in air where the flux density is 28 μWb per square meter?

10. If the earth's field, given in Problem 9, makes an angle of 70° with the earth's surface, how much flux passes through a square mile of the surface?

11. The flux density in a magnetic material is 0.8 Wb/m^2 and the field intensity is 640 NI/m. What is its relative permeability?

12. A transformer core has a cross-section area of 4 in.^{2} and a length of 32 in.^{2}. It has an air gap of $\frac{1}{32}$ in. and the flux density is 1.2 Wb/m^2. The relative permeability of the steel core is 2450. Calculate the reluctance of (a) the air gap, (b) the steel path.

13. How many magnetizing ampere-turns are required for the air gap and for the steel of the transformer of Problem 12?

We can easily set up an equation for the calculation of the ampere-turns needed to produce a desired flux density in an air gap. From "Ohm's law for the magnetic circuit,"

$$\Phi = \frac{\text{MMF}}{\mathcal{R}} = \frac{NI}{l/A\mu_0} = \frac{NIA\,\mu_0}{l}$$

For an air gap:

$$NI = \frac{\Phi l}{A\,\mu_0} = \frac{Bl}{\mu_0} \qquad\qquad (8.12)$$

Problems

14. A brass ring has 800 turns of wire uniformly distributed around it and they carry a current of 25 μA. The magnetic flux exists in the ring and the average length of the closed path is 48 cm. The cross section of the brass core has a radius of 1.45 cm. The permeability of brass is the same as that of air. (a) What is the reluctance of the magnetic path? (b) How much flux exists in the core?

15. Continue with Problem 13 and find (a) the flux density, (b) the magnetic field intensity. Find **H** by two methods [Equations (8.5) and (8.6)].

16. A magnetic circuit, 20 cm average length and 4 cm^2 in area, is made entirely of cast iron. A magnetomotive force of 440 At is required to produce 2.6×10^{-4} Wb of flux in the circuit. What is the relative permeability of the cast iron at that flux density?

8.10 Para-, Dia-, and Ferromagnetic Materials

Electrons in an atom are considered to be revolving in orbits around the nucleus (positively charged center) of the atom and, since such electron motion constitutes an electric current, this electron motion produces a magnetic field. There are billions of such motions and field contributions in a small piece of metal.

An electron has been found to be made up of a "cloud of negative electricity" spinning on its own axis similar to our earth's motion around its polar axis. This *additional current* (electric charge in motion) produces its own magnetic field.

First consider the situation in which the two relatively weak magnetic fields (one caused by orbital revolutions, the other by axial rotations) are in opposite directions and cancel each other. A material in which this takes place is termed *diamagnetic*. When diamagnetic material is placed in an external magnetic field, that is, in one that has been created by a source *external* to the material, the magnetic field inside the material is the same as the external magnetic field. According to advanced theory, as the applied magnetic field builds up in the diamagnetic material, the velocity of the oribiting electron is changed with the result that the magnetic field produced by the orbiting motion is changed in such a way as to *oppose* the externally applied field.

The permeabilities of diamagnetic materials are very slightly less than μ_o, the *decrease* is seldom greater than 0.01 percent. Wood and paraffin are two diamagnetic materials. Their permeabilities, respectively, are 0.99999950 and 0.99999942 *times that of free space*.

Now we consider a material with atoms in which the magnetic effects of electron rotation and spin do not quite cancel. This means there is a net magnetic effect in each atom. The effect is slight, however, and, since the orientation of atoms throughout the material is in all possible directions (called random orientation), the net effect is that the material shows no magnetic properties when it is placed in a region where a magnetic field with flux density **B** has already been established, there is an *increase* in **B**, the material is termed *paramagnetic*.

The permeability of paramagnetic material is slightly greater than that of free space. Aluminum is an example of this kind of material with a permeability 1.00000065 *times that of free space*.

A third, and by far most important, kind of material has a relatively large magnetic contribution from each of its atoms. The opposing magnetic effects of electron orbital motion and electron spin do not eliminate each other in an atom of this material. There is a relatively large contribution from each atom which aids in the establishment of an internal magnetic field, so that when the material is placed in a field the value of **B** is increased to many times the value that was present in the free space before the material was placed there. This kind of material is termed *ferromagnetic*. Such materials include, iron, cobalt, nickel, and many alloys (mixtures) of these with other materials, *Alnico*, an alloy of aluminum, nickel, and cobalt, is a ferromagnetic material used to make strong permanent magnets. Forces in these materials act to maintain alignment of electron current loops after an external magnetizing field has been removed. Permeabilities of these materials extend over a range from several hundred to tens of thousands of times greater than the permeability of free space (vacuum).

8.11   Magnetization Curves

When the current in a coil, which produces magnetic flux in an iron ring, is increased, the increase in the flux is not in direct proportion to the increase in current. Increasing the current increases **H** because **H** = NI/l, but the resulting increase in **B**, which is Φ/A, is not in direct proportion to the increase in **H**. Another way to say this is that the ratio of **B** to **H** is not constant as **H** is changed. This means the permeability of the iron changes with the flux density. Figure 8.11 shows how the permeability of a ferromagnetic material changes with flux density.

Magnetization curves, called **BH** curves, are furnished by manufacturers of ferromagnetic materials. (Incidentally, the ferro part of the word ferro-magnetic comes from the Latin word *ferrum* meaning iron. Wrought iron, cast iron, and alloys containing iron are classified as ferromagnetic materials.) Magnetization curves have the *magnetizing force* **H** on their horizontal scale. (**H** is also given the sophisticated name *magnetic potential gradient*.) On their vertical scale are corresponding values of flux density, **B**, that are produced by each particular value of **H**. See Fig. 8.12.

Magnetization curves are almost indispensable when ferromagnetic circuits are to be analyzed or designed. An example will show how useful they are.

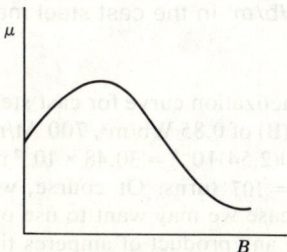

FIGURE 8.11   Showing how the permeability of a ferromagnetic material changes with the changes in flux density in the material.

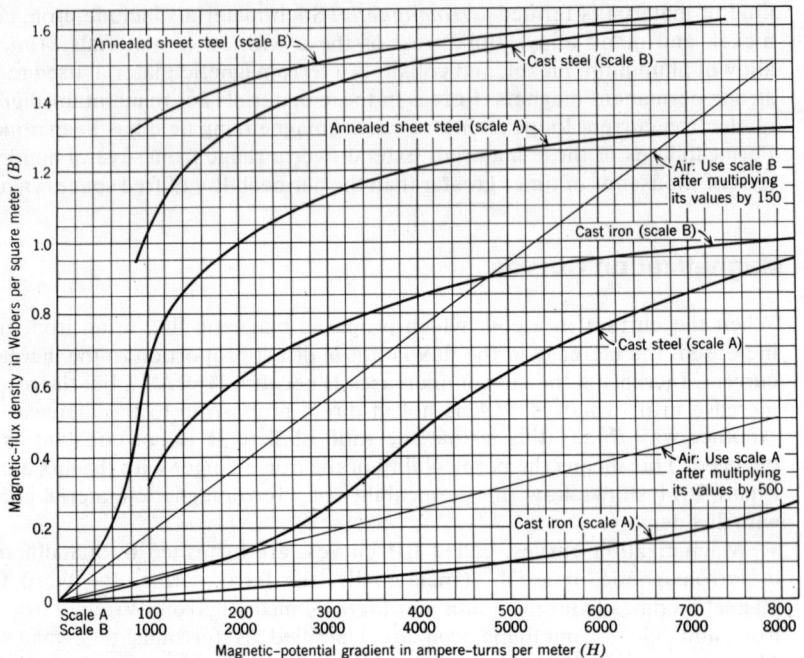

FIGURE 8.12 **Magnetization curves for commercial metals.**

EXAMPLE 8.6 Suppose we have a cast-steel ring shaped like our familiar one in Fig. 8.10. The special name for a donut-shaped ring is *torus*. Assume an average length of 12 in. and the cross section of the steel to be 0.5 in.². Let's determine, with the aid of the magnetization curve for cast steel, how many turns the magnetizing coil must have to produce a flux density of 0.85 Wb/m² in the cast steel magnetic circuit if we use 2 A of current.

Solution. The magnetization curve for cast steel (Fig. 8.12) shows that for a magnetic flux density (**B**) of 0.85 Wb/m², 700 At/m are required. The length of path in meters is $(12)(2.54)10^{-2} = 30.48 \times 10^{-2}$ m. $NI = 700 \times 30.48 \times 10^{-2} = 213.36$. $N = 213.36/2 = 107$ turns. Of course, we may have room for many more turns, in which case we may want to use only $\frac{1}{2}$ A and $213.36/0.5 = 428$ turns. We may choose any product of amperes times turns that will make 214 At. The use of the magnetization curve saved us a lot of calculations, didn't it?

Problems

17. Determine the ampere turns necessary to produce the same flux density specified in Example 8.6 if a cast iron ring with the same dimensions is used instead of the cast steel ring.

18. A magnetic circuit composed completely of annealed sheet steel has a magnetic field intensity (magnetic potential gradient) $H = 200$ At/m produced in it. How much flux density exists in the steel? How many ampere-turns are needed to maintain this flux density if the closed circuit of steel is 65 cm in length?

19. Using the magnetization curves of Fig. 8.12, determine the values of H for 1.2 Wb/m² flux density in (a) annealed sheet steel, (b) cast steel, (c) air.

The material in a well-designed magnetic circuit is operated at a flux density that is near the "knee" of the magnetization curve. The knee of the cast steel curve (Fig. 8.12) is located when the flux density B is about 1.2 Wb/m². Operating the core at densities somewhat higher requires more current and the resulting densities increase more slowly because *magnetic saturation* is setting in. *Saturation* is reached in a ferromagnetic material when small equal increases in the applied external magnetic field intensity produce equal increases in flux density. When this state exists, the magnetization curve becomes a straight line. Operating at densities somewhat lower does not give proportionate advantages in reduced current. Therefore, operation of ferromagnetic circuits at flux densites that fall in the region of the knee of the magnetization curve is desirable.

The normal permeability of a magnetic material is the ratio of values of B to corresponding values of H. That is,

$$\mu = \frac{B}{H}$$

When direct current in a magnetizing coil fluctuates, as in the case of an audio transformer that feeds a loudspeaker of a radio or television receiver, the flux density and corresponding values of H vary between certain limits. If the variations are called ΔB and ΔH, their ratio is called *incremental permeability* (Fig. 8.13a):

$$\mu_\Delta = \frac{\Delta B}{\Delta H} \tag{8.13}$$

where μ_Δ is the incremental permeability, ΔB is the difference between the highest and lowest values of B, and ΔH is the difference between the corresponding highest and lowest values of H.

If the magnetizing coil is supplied with alternating current, the magnetic flux oscillates between a maximum value in one direction and an equal maxi-

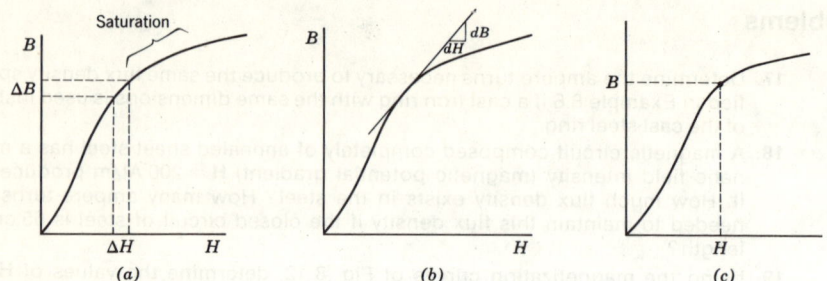

FIGURE 8.13 Permeabilities of a ferromagnetic material. (a) Incremental permeability: $\Delta\mu = \Delta B/\Delta H$. **(b) Differential permeability:** $\mu_d = dB/dH$. **(c) Permeability at constant flux density:** $\mu = B/H$.

mum value in the opposite direction. Inasmuch as μ is the slope of the **BH** curve, it can be written

$$\mu_d = \frac{d\mathbf{B}}{d\mathbf{H}} \tag{8.14}$$

and its value can be taken at a point P in the region of the knee of the curve (Fig. 8.13*b*). This is called the differential permeability. The slope of a curve at a point on it is the slope of a tangent to the curve. Since the tangent is a straight line, the slope is given by the $d\mathbf{B}/d\mathbf{H}$ value on the line.

When a ferromagnetic path has a flux density that is constant, as is the case with electromagnets supplied with direct current, one can use the ratio of **B** to **H** at a point chosen as the center of the knee region on the magnetization curve (Fig. 8.13*c*). This makes the solution of certain kinds of problems easier than they would be if the ferromagnetic part of a magnetic circuit is treated as a nonlinear reluctance. When μ is constant, the reluctance is constant in a given design. See Equation (8.9).

8.12 Magnetic Flux in Air Gaps

The fact that $\mu = \mu_0$, a constant, in air makes the determination of the number of ampere-turns required to produce a given amount of flux in the gap rather easy to understand. The basic formulas for reluctance and total flux apply. Let's look at an example.

EXAMPLE 8.7 An air gap 1 mm long and 2 cm² in cross section exists between a lifting magnet and its load. Most of the ampere-turns on the magnet are needed to establish the flux in the air-gap part of the circuit. If the total flux is 100 μWb, how many of the total ampere-turns are needed to supply the magnetomotive force across the air gap?

Solution

$$\mathscr{R} = \frac{1 \times 10^{-3}}{1 \times 4\pi \times 10^{-7} \times 2 \times 10^{-4}} = \frac{10^8}{8\pi} \text{ At/Wb}$$

$$\Phi = \frac{NI}{\mathscr{R}}, \qquad NI = \Phi \mathscr{R}$$

$$NI = 200 \times 10^{-6} \times \frac{10^8}{8\pi} = 796 \text{ At}$$

The flux density is

$$B = \frac{200 \times 10^{-6}}{2 \times 10^{-4}} = 1 \text{ Wb/m}^2$$

In this example, the flux density in the air gap was chosen to be a good value for an annealed sheet steel path that must carry the same flux. Note from the magnetization curve for annealed sheet steel in Fig. 8.12 that the steel path requires only 200 At/m. Even a 20-cm length of steel path would require only $0.2 \times 200 = 40$ At. For an electromagnet with those dimensions, the total ampere-turns required would be $796 + 40 = 836$. The air gap "grabs" 95 percent of the applied ampere-turns.

The reverse problem for an air gap is solved just as easily. That is the kind in which the ampere-turn value is given and you want to know how much flux will be produced in the air gap.

EXAMPLE 8.8   The magnetomotive force that is to establish flux in an air gap 5 mm long is 5000 At. What *flux density* will it establish?

Solution

$$\mathbf{B} = \frac{\Phi}{A} = \frac{NI\mu_r\mu_0 A}{lA}$$

$$\mathbf{B} = \frac{5000 \times 1 \times 4\pi \times 10^{-7}}{5 \times 10^{-3}}$$

$$\mathbf{B} = 1.256 \text{ Wb/m}^2$$

Of course, if we know the cross-section area of the gap we can calculate the total flux produced by $\Phi = \mathbf{B}A$.

8.13   Series Magnetic Circuits

Magnetization curves are used to help analyze ferromagnetic circuits made up of sections in series. Fig. 8.14*a* shows such a circuit. The cross sections of the steel and iron sections are square and their lengths are closely approximated. There is crowding of the magnetic flux at all four inside corners, so that the exact path lengths cannot be known. Let us assume that a flux density of 1.4 Wb/m² is desired in the cast-steel portion, and then compute the number

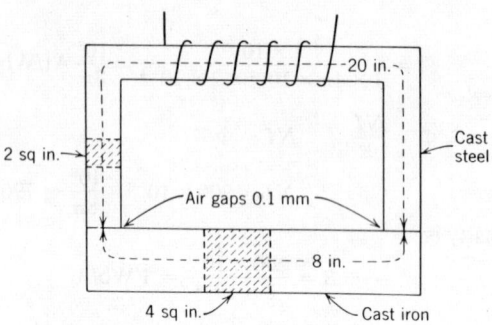

FIGURE 8.14a   A series ferromagnetic circuit.

of ampere-turns required to maintain the flux in the whole circuit. In a practical case there will be small air gaps where the steel and cast iron meet. An average of $\frac{1}{10}$ mm each is a reasonable assumption.

For the cast steel, the magnetization curve in Fig. 8.12 shows $H = 2600$ At/m are required to produce 1.4 Wb/m², so the MMF required by the 20-in. path is

$$MMF = 2600 \times 20 \times 0.0254 = 1320 \text{ At}$$

In the cast iron, $\mathbf{B} = \Phi/A$. The flux is the same as that in the cast-steel section, so $\Phi = 1.4[2(2.54)^2 \div 10^4] = 0.001806$ Wb.

$$\mathbf{B} = 0.001806 \div (4 \times 2.54^2 \times 10^{-4}) = 0.7 \text{ Wb/m}^2$$

This result could have been obtained by simply using the inverse ratio of the areas, since the area in cast iron is twice that in steel.

From the magnetization curve for cast iron, the required value of H is 2550 At/m. The 8-in. length of cast-iron path requires

$$MMF = 2550 \times 8 \times 0.0254 = 518 \text{ At}$$

In the air gaps, B is the same as in the cast steel, that is, 1.4 Wb/m². From the upper curve for air gaps. $H = 7480 \times 150 = 1.122 \times 10^6$ At/m. The two air gaps, with a total length of 0.2 mm, will require a MMF of $1.122 \times 10^6 \times 0.2 \times 10^{-3} = 225$ At. The total ampere-turns required, according to these computations, is $1320 + 518 + 225 = 2063$.

It was easier to determine $\mathbf{H}$ for the air gaps by using the straight-line graph than to calculate it. However, let's calculate the ampere-turns required.

$$\mathbf{H} = \frac{\mathbf{B}}{\mu_0} = \frac{1.4 \times 10^7}{4\pi}$$
$$\mathbf{H} = 1.115 \times 10^6 \text{ At/m}$$
$$NI = 1.115 \times 10^6 \times 0.2 \times 10^{-3}$$
$$NI = 223 \text{ At}$$

Problem

20. Refer to the series magnetic circuit of Fig. 8.14a and *assume* that all of the ferro-magnetic material is *cast steel* and that it has a flux density of 1.25 Wb/m² in the 20-in. length. If the air gaps are each 0.10 mm in length, find the MMF required.

We have learned how to determine the total magnetomotive force (ampere-turns) required to establish a desired flux density in a series magnetic circuit composed of two or more different materials (one of which may be air). The inverse problem, which is the determination of how much magnetic flux is produced in a series magnetic circuit containing two or more different kinds of material is not so straightforward. The reason is that the permeability of each ferromagnetic path depends on the flux density which is unknown. There is a straightforward graphical method[4] of solving such a problem, but we shall not present it here.

We can solve such a problem by calculation if we are willing to use a cut-and-try approach. This may involve three or four approximate solutions in order to obtain the accuracy desired. Let's run through a simple example.

EXAMPLE 8.9   The magnetic circuit in Fig. 8.14*b* is energized by a MMF of 2000 At. How much flux is produced in the air gap?

Solution. We choose what we think is a reasonable amount of flux density in the steel and calculate the number of ampere-turns it would require. Then we calculate the additional number that the air gap will require in order that it may have the same flux density. The sum of the two ampere-turn amounts is then compared with the amount available. Unless it comes within a desired

[4]See H. Alex Romanowitz, *Electrical Fundamentals and Circuit Analysis*, John Wiley & Sons, Inc., 1966, pp. 179–181, for an analysis of nonlinear magnetic circuits.

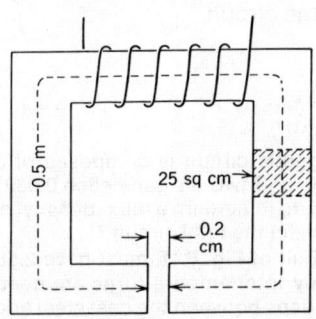

FIGURE 8.14*b*   For Example 8.9. Annealed sheet-steel magnetic circuit with air gap.

accuracy (shall we say only a few percent more or less than the given amount — 2000 At in this case) we adjust our assumed flux density and calculate it again.

Let's say we shall start with $\mathbf{B} = 1.2$ Wb/m². From the curve for annealed sheet steel we get $\mathbf{H} = 390$ At/m.

For the steel part, $\mathscr{F}_s = 390 \times 0.5 = 195$ At.

For the air gap,

$$\mathbf{H} = \frac{1.2}{4\pi \times 10^{-7}} = 9.55 \times 10^5 \text{ At/m}$$

$$\mathscr{F}_a = 9.55 \times 10^5 \times 2 \times 10^{-3} = 1910 \text{ At}$$

$$\text{Total } NI = 195 + 1910 = 2105 \text{ At}$$

Evidently the estimate of 1.2 Wb/m² is a little too high. Our result is nearly 5 percent more than the given number of ampere-turns. Let's reduce the estimated flux density about 4 percent and see what happens.

$$\text{For } \mathbf{B} = 0.96 \times 1.2 = 1.152 \text{ Wb/m}^2 \text{ (call it 1.15),}$$

the corresponding

$$\mathbf{H} = 320 \text{ At/m, and for the steel,}$$
$$\mathscr{F}_s = 320 \times 0.5 = 160 \text{ At}$$

For the air gap,

$$\mathbf{H} = \frac{1.15}{4\pi \times 10^{-7}} = 9.15 \times 10^5 \text{ At/m}$$

$$\mathscr{F}_a = 9.15 \times 10^5 \times 2 \times 10^{-3} = 1830 \text{ At}$$
$$\text{Total } \mathscr{F} = 160 + 1830 = 1980 \text{ At (close enough)}$$

The total magnetomotive force required to produce a desired flux in a series magnetic circuit is the sum of the magnetomotive forces required for each section of the circuit.

Problems

21. A series ferromagnetic circuit is composed of 32 in. of annealed sheet steel, 32 in. of cast steel, and two air gaps each 0.0394 in. long. How many ampere-turns are needed to maintain a flux density of 1.2 Wb/m² in the (a) sheet steel, (b) cast steel, (c) the total circuit?

22. The magnetic circuit of Fig. 8.15 must have 2.58×10^{-4} Wb of flux circulating in it. (a) How many total ampere-turns are needed? Neglect the effect of the two very short air gaps between the cast steel and sheet steel pieces. (b) Calculate the ampere-turns required for the 0.5-in. air gap by a second method that does not make use of the graph in Fig. 8.12.

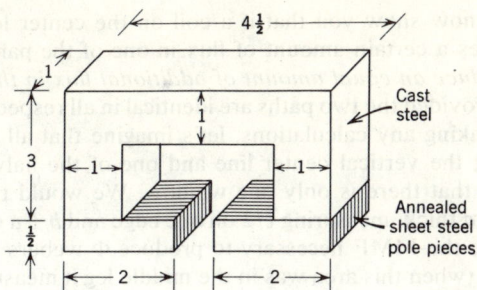

FIGURE 8.15   For Problem 21. Dimensions are in inches.

8.14   Parallel Magnetic Circuits

In order to improve the performance of a ferromagnetic circuit a *shell-type* core is used instead of one that has a single closed path. A shell-type core has two "windows" instead of one (Fig. 8.16a) and sheet-steel *E* and *I laminations*, which are thin flat pieces. There are as many *E*-shaped as there are *I*-shaped pieces. They are assembled in staggered fashion so they will stay together. By this it is meant that the next layer behind the one seen at the front has the *E* piece at the bottom with its three legs pointing upward and the *I* piece is across the top. The middle leg has twice the area of an outside leg. After the "exciting coil" is wound on a form that has the dimensions of the inside leg ($2a \times d$ in the illustration) the laminations are slipped into and over it, and the assembly looks as shown in Fig. 8.16b.

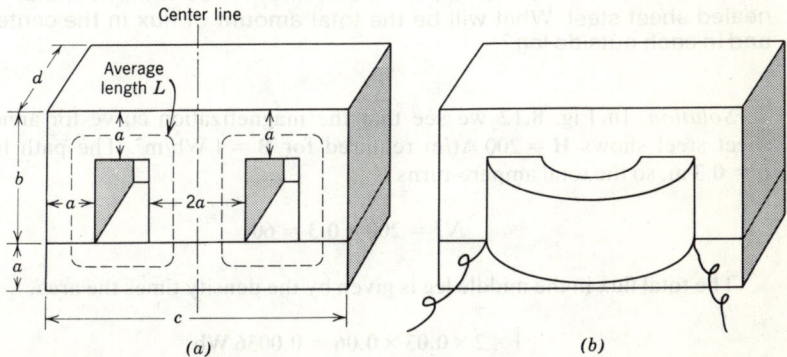

FIGURE 8.16   Shell-type core construction. (a) Two identical magnetic circuits in parallel. (b) coil on center leg of shell-type core.

We shall now show you that if a coil on the center leg with NI ampere-turns produces a certain amount of flux in one of the parallel paths of length L, it will produce an equal amount of additional flux in the other path, whose length is L, provided the two paths are identical in all respects.

Before making any calculations, let's imagine that all E and I laminations are cut along the vertical center line and one of the halves is placed on the other half so that there is only one window. We would then have a stack of laminations $2d$ thick, measuring $c/2$ on one edge and $b + a$ on the other. We can now calculate the MMF necessary to produce Φ webers of flux in the $a \times 2d$ cross section (when this area was in the middle leg it measured $2a \times d$, which is the same area) of a single path L units long. We'll have to admit that half of the flux will be in the steel that makes up half the total area ($a \times d$) and the other half of the flux will be in the other half ($a \times d$) of the area. *We have two identical paths in parallel.* With this configuration, a given amount of ampere-turns of a single coil will produce the same total flux in the steel surrounded by the coil as it will in the center section of the shell-type configuration in Fig. 8.16a. Keep in mind that we are working with flux densities. Since the required number of ampere-turns for a desired *flux density* in a path of length L can be determined, the *cross-section area* does not enter into the determination. This means it doesn't matter whether the area is $2a \times d$ or just $a \times d$ in Fig. 8.16a so long as the flux density is constant throughout the paths of length L. Let's calculate the magnetomotive force required to produce a desired flux density in the middle leg of the core in Fig. 8.16 assuming there are no air gaps.

EXAMPLE 8.10   Assume that there are no air gaps in the parallel magnetic circuits in Fig. 8.16a and determine the ampere-turns of the coil required to produce a flux density of 1 Wb/m² in the middle leg. The dimensions are $a = 3$ cm, $b = 9$ cm, $d = 6$ cm, and $L = 30$ cm. The metal is annealed sheet steel. What will be the total amount of flux in the center leg and in each outside leg?

Solution. In Fig. 8.12 we see that the magnetization curve for annealed sheet steel shows $\mathbf{H} = 200$ At/m required for $\mathbf{B} = 1$ Wb/m². The path length $L = 0.3$ m, so the total ampere-turns is

$$NI = 200 \times 0.3 = 60$$

The total flux in the middle leg is given by the density times the area,

$$1 \times 2 \times 0.03 \times 0.06 = 0.0036 \text{ Wb}$$

The flux in each outside leg is $0.0036/2 = 0.0018$ Wb

Problem

23. A parallel magnetic circuit of annealed sheet steel, shaped like that in Fig. 8.16a, has dimensions $a = 2$ cm, $b = 8$ cm, $d = 5$ cm, and $L = 25$ cm. Each of the two flux paths has two 1-mm air gaps in series. How much current is required in a 250-turn coil to produce 0.002 Wb of flux in the middle leg?

TABLE 8.2   Quantities, Units, and Symbols Related to Magnetic Fields

Flux Φ, webers; Area A, Square meters

Flux density **B**, Wb/m², $\mathbf{B} = \dfrac{\Phi}{A}$

Magnetomotive forces, $\mathscr{F}$, Ampere turns, At or NI

Magnetic field intensity **H**, ampere turns per meter $\dfrac{\text{AT}}{\text{m}}$

Permeability of free space (vacuum) $\mu_0 = 4\pi \times 10^{-7}$ henrys per meter

Relative permeability $\mu_r = \dfrac{\mu}{\mu_0}$, no units, just a ratio.

Ohm's law $\Phi = \dfrac{\mathscr{F}}{\mathscr{R}}$, webers $= \dfrac{\text{ampere turns}}{\text{ampere turns per weber}}$

Reluctance $\mathscr{R} = \dfrac{l}{\mu_r \mu_0 A}$, ampere turns per weber

Useful Formulas for Computation

In ferromagnetic material: At $= \mathbf{H} \times l$ Get **H** from curve in Fig. 8.12.

In air gap: At $= NI = \Phi \mathscr{R} = \dfrac{\mathbf{B} A l}{\mu_0 A} = \dfrac{\mathbf{B} l}{4\pi \times 10^{-7}}$ by calculation.

Alternate: Get **H**, at given flux density, from straight line in Fig. 8.12.

Series magnetic circuit: $\mathscr{F}$ is distributed throughout the closed path as applied voltage is in an electric circuit. Total $\mathscr{F}$ is the sum of all of the separate MMFs in the series sections.

Parallel magnetic circuits: One $\mathscr{F}$ applied in a common section produces flux which divides at a junction of the parallel paths, just as applied voltage produces current that divides at a junction of parallel branches of a network.

The sum of the MMFs around each parallel closed path equals the MMF of the applied ampere-turns.

8.15 Ampere's Circuital Law

Look at Equation (8.5). **H** is the magnetic field intensity produced in the circular magnetic path by NI ampere-turns. Note that this circular path *encloses* NI ampere-turns. Also note that $2\pi R \mathbf{H} = NI$.

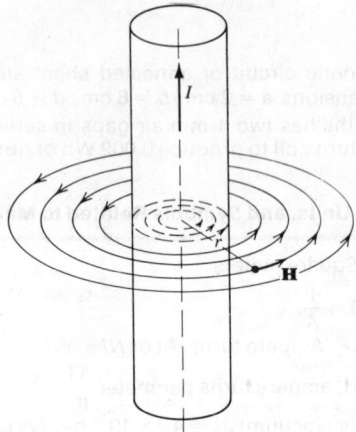

FIGURE 8.17   Magnetic flux produced by current flowing in a straight, solid conductor.

Now look at Fig. 8.17. It represents a section of a long, straight, solid conductor (somewhat magnified for purposes of illustration) carrying a current *I* amperes. If we multiply **H** by the circumference of the circular path along which **H** is constant ($2\pi r$), we get *NI* in which *N* is one turn. If we choose a path for which *r* is smaller, **H** on that circle will be larger and $2\pi r$**H** will still equal *NI* in which *N* is one turn. This leads to *Ampere's Circuital law* which says:

The product **H** times the distance around a closed path is equal to the magnetomotive force that produces **H**.

This statement needs a qualification[5] in order that the law may be precisely stated. When the path is one along which **H** is not constant and not tangent to the path at all points, the *component of H must be used*. In the general case the integral must be taken around the closed path, as explained in the footnote. Where *N* is more than one turn, *NI* ampere-turns is the same as if *NI* amperes exist in only one turn.

If we know the value of *I* in Fig. 7.16, we can get **H** at any distance *r* as follows:

$$\mathbf{H} = \frac{I}{2\pi r} \text{ amperes per meter} \tag{8.15}$$

[5]Ampere's circuital law: The integral around a closed path of the component of **H** along the path times the differential length (*d***L**) of the path is equal to the current enclosed, i.e., $\oint \mathbf{H} \cdot d\mathbf{L} = I$ enclosed. This is the same as $\oint \mathbf{H} d\mathbf{L} \cos \theta$, in which θ is the angle between vectors **H** and *d***L** at every point on the closed path.

This equation is correct only if r is not less than the radius of the conductor. Also, r must be small compared with the length of the straight conductor in which I flows. That is, r must be less than one-tenth of the length of the conductor.

If r is less than the radius to the surface of the conductor, the current enclosed by the circular path *inside the conductor* will be less than I. The area of the small circle will be πr^2 and the enclosed current will be

$$I_{\text{encl}} = \frac{\pi r^2}{\pi R^2} I = \frac{r^2}{R^2} I$$

Using this,

$$2\pi r \mathbf{H}_i = \frac{r^2}{R^2} I$$

and $\mathbf{H}_i$ inside the conductor is

$$\mathbf{H}_i = \frac{Ir}{2\pi R^2} \text{ amperes per meter} \tag{8.17}$$

Problem

24. A long straight cylindrical conductor 1 cm in diameter carries 80 A. Calculate values for, and plot a curve of **H** vs r from $r = 0$ to $r = 2.5$ cm.

8.16   Magnetic Shielding

There are times when it is desirable to prevent lines of magnetic force from passing through a space occupied by a sensitive device such as a meter movement or other mechanism that should not be magnetized. Sometimes stray magnetic fields exist in these spaces.

It is common practice to use a shield made of high-permeability ferromagnetic material to surround the instrument or device. The lines of magnetic force take up paths in the low-reluctance material in preference to passing through the high-reluctance air paths in which the device is located. See Fig. 8.18. A hollow metal enclosure made of steel laminations of rectangular or circular shape may be used, or the metal may be soft iron.

Problems

25. A long straight circular conductor carries 50 A of current. What is the magnetic field intensity at a point 3 ft from the center of the conductor?
26. How far from the axis of the conductor of Problem 25 does the field have an intensity of 25 At/m?
27. The conductor of Problem 8.25 is $\frac{1}{2}$ in. in diameter. What is the field intensity at all points $\frac{1}{8}$ in. from the axis of the conductor?

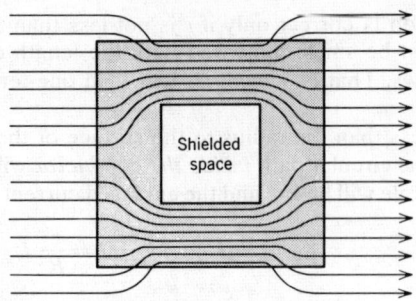

FIGURE 8.18   Magnetic shielding by a high-permeability enclosure. The enclosure may be a section of a hollow cylinder, as well.

8.17   Magnetic Hysteresis

Consider an iron ring, or torus, as shown in Fig. 8.10 on which a coil of wire is wound. To produce a magnetic field that is uniform in the ring, the turns of wire should be uniformly distributed around the length of the ring.

If no flux exists in the ring at the outset, an increase in coil current from zero to a maximum value I_m will cause increasing flux densities to a maximum value $\mathbf{B}_m$ and the **B–H** curve will be the magnetization curve for the kind of iron of which the ring is made. This we already know. If the current value is decreased, however, causing the magnetic-potential gradient to diminish from its maximum value $\mathbf{H}_m$, the corresponding values of flux density will be larger than those on the magnetization curve, and the **B–H** curve will follow the path from M to R in Fig. 8.19. At R there is a *residual-flux density*, which is retained by the iron after the magnetomotive force has been removed (the coil current is zero). Apparently the magnetic axes of some of the atoms remain in the alignment forced upon them by the strong magnetic field produced by the exciting ampere-turns, with the result that the ring retains some "residual magnetism." To reduce the flux density to zero it is necessary to reverse the current in the coil and thus apply a magnetic potential gradient $-\mathbf{H}_c$. $\mathbf{H}_c$ has been called the "coercive force," an MMF that "coerces" the residual field to disappear.

As the reversed current increases until $-\mathbf{H}_m$ is reached, the curve follows MCM' until $-\mathbf{B}_m$ exists in the ring. Reduction of the current to zero, reversal, and then an increase to produce $\mathbf{H}_m$ again is accompanied by lagging **B** values and corresponding **B–H** relations along the portion $M'R'C'M$. This magnetic behavior is called "hysteresis" and the closed curve is called a "hysteresis loop." Owing to the nature of ferromagnetic materials, the loops resulting from a second, or even a third, "cycle" of current changes would not lie exactly on

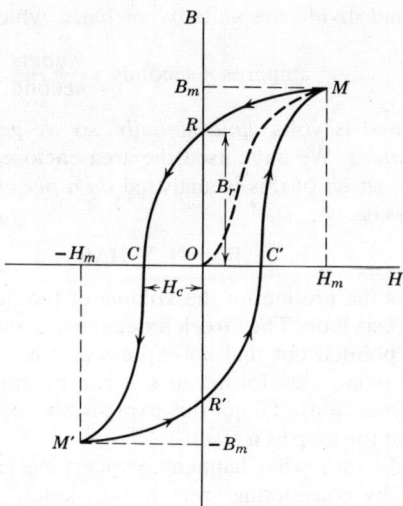

FIGURE 8.19   A hysteresis loop for iron.

top of this first one. After a relatively few cycles, the successive loops would follow a fixed path, which the loop in Fig. 8.19 may well represent.

8.18   Hysteresis Loss

Inasmuch as *energy* is required to align the magnetic poles of the atoms of a ferromagnetic material, to break up their "residual alignment" existing at points R and R' on the hysteresis loop, and to realign them again, we should be willing to admit that there must be some energy lost during every cycle of current change in the coil. This energy is taken, of course, from the source of the current.

In alternating current equipment, such as transformers used in power transmission, sixty of these cycles of magnetic action occur *every second*. An important quantity in such practical applications is the amount of energy loss, *per cycle*, in a given ferromagnetic circuit. It is obtained by multiplying the *volume* of the magnetic material (in cubic meters) by the *area* of the hysteresis loop (in ampere-turns per meter times webers per square meter). The units will be *joules per cycle*:

$$\text{(meters)}^3 \times \frac{\text{amperes}}{\text{meter}} \times \frac{\text{webers}}{\text{(meter)}^2} \rightarrow \text{amperes} \times \text{webers}$$

If we multiply and divide the units by *seconds*, which is permissible, we get

$$\text{amperes} \times \text{seconds} \times \frac{\text{webers}}{\text{second}}$$

Webers per second is volts *dimensionally*, so we get volts × amperes × seconds, which is *joules*. We have used the area enclosed by only *one excursion around the loop*, so all of this is analyzed *on a per cycle basis*. The equation for hysteresis loss is

$$W_h = V \sum \mathbf{H} \Delta \mathbf{B}$$

which symbolizes the product of the volume of the material multiplied by the area of the hysteresis loop. The Greek letter sigma Σ means *summation of*.

It should be pointed out that an expression for the energy loss per cycle *per unit volume* is most useful because it can be applied to a magnetic path of any overall dimensions. To get this expression we simply divide by volume and get the area of the loop as a result.

Further insight into what happens, concerning energy, during one cycle can be obtained by considering areas in the sketch of the hysteresis loop in Fig. 8.19.

The area enclosed by $C'M\mathbf{B}_mOC'$ represents the energy *per cubic meter* given to the magnetic material while $\mathbf{H}$ goes from zero to $+\mathbf{H}_m$. The area $M\mathbf{B}_mRM$ represents the energy per cubic meter returned to the electric circuit (from which it came). The area $RCOR$ represents energy put in while $\mathbf{H}$ goes from zero to $\mathbf{H}_c$, which is in the opposite direction. Additional energy put in is represented by the area $CM'(-\mathbf{B}_m)R'OC$, and a second amount returned is represented by the area within $M'(-\mathbf{B}_m)R'M'$. The net effect is that the area inside the loop, measured in $\mathbf{BH}$ units, gives the energy *per cubic meter per cycle*. This is converted into heat within the material and increases its temperature.

A close approximation of the area of a loop, in $\mathbf{BH}$ units, can be obtained by plotting the loop to scale on coordinate paper, which has small divisions, and then adding all of the squares enclosed by the closed path.

Dr. Charles P. Steinmetz (Germany, 1865–1923), who came to this country in 1889 and became known as an "electrical wizard" with the General Electric Company, discovered many theoretical relationships as well as practical principles in electrical engineering. Among them is the dependence of magnetic hysteresis loss on the *maximum instantaneous flux density* in ferromagnetic materials. This is an important and useful expression. In a symmetrical cycle of magnetization, the hysteresis loss *per unit volume per cycle* is

$$W_h = \eta \mathbf{B}_m^{1.6} \text{ joules} \tag{8.18}$$

The coefficient η (eta) is in *joules per cubic meter per weber per square meter*. It is determined by experiment for each kind of ferromagnetic material used.

Apparently materials with the best magnetic qualities have the smallest values of η. For example, the best transformer steels have η values around 130, for cast steel they are around 2500, and for cast iron about 3750. An alloy called *Hypernik* of 50 percent nickel and 50 percent iron has an η value of 25. It is used in current-transformer cores in instrumentation.

8.19   Energy Loss Caused by Eddy Currents

When alternating current flows in an exciting coil, such as a winding on the ferromagnetic core of a transformer, voltages are induced *in the core material* by the changing magnetic flux. The currents that flow in the core material (because these induced voltages are present) are called *eddy currents*, and they contribute an energy loss because their paths have resistance. The *power* loss is proportional to i^2R, but current and resistance values cannot be determined directly. Experiments have shown that the eddy-current power loss is proportional to $\mathbf{B}^2$ and to the square of f, the number of complete magnetization cycles per second. The expression for the *eddy-current loss per cycle per cubic meter of magnetic material* is

$$W_e = k\mathbf{B}_m^2 d^2 f^2 \text{ joules per cubic meter per cycle} \qquad (8.19)$$

In this equation k is a constant, called the *eddy-current coefficient*, for the kind of magnetic material used, and f is the number of cycles per second. For transformer steel, k is about 7.5×10^5. The symbol d represents the thickness of a sheet of the steel, so one may conclude correctly that this equation is to be used for computations for laminated (layers of sheets) magnetic paths. A typical value of d for a sheet of transformer steel is 0.3556 mm, which is 3.556×10^{-4} m.

To get an idea of how much loss occurs per cycle and what the power loss is in a transformer core, let us consider the following example.

EXAMPLE 8.11   A transformer that operates on a 60 Hz frequency has a laminated sheet-steel core for which the eddy-current coefficient k is $(7.5)10^5$ and the thickness of a sheet is 3.556×10^{-4} m. The maximum instantaneous flux density is 1.5 Wb/m². The total volume of the magnetic core is 2000 cm³. Compute the energy loss per cycle attributed to eddy currents.

Solution. Using Equation (8.19) we have

$$W_e = 7.5(10)^5(1.5)^2 3.556^2(10^{-8})60^2 = 769 \text{ J/m}^3$$

The total loss per cycle is

$$W = (769)(2000)10^{-6} = 1.538 \text{ J/cycle}$$

The total eddy-current *power loss* is

$$P_e = 1.538(60) = 92.2 \text{ W}$$

Problems

28. A power transformer has a core material for which the constant η is 130 J/m³/ Wb/m². Its volume is 8000 cc and the maximum flux density is 1.25 Wb/m². What is the hysteresis loss in watts if the frequency of the alternating current is 60 Hz?

29. The steel laminations of the transformer core of Problem 28 are 0.5 mm thick and the eddy-coefficient $k = (7.5)10^5$. What is the eddy-current loss in watts?

8.20 Permanent Magnets

Relatively recent developments of magnetic alloys have made possible the manufacture of permanent magnets of greatly increased strength. Longer air gaps with larger energy storage are now practical, although high cost must be taken into account and therefore design must utilize the material with maximum effectiveness. The new alloys are physically hard and present difficult problems in manufacture. Permanent–magnet materials require comparatively high magnetic field intensity—approximately 150,000 At/m or more—to become magnetized and they have large values of residual flux density when the applied **H** is removed.

After a permanent magnet has been removed from the closed magnetic circuit in which it was magnetized, it loses some of its strength owing to the reluctance of the air gap through which it must maintain magnetic flux. Before removal, however, it may have a *residual flux* density of more than 0.9 Wb/m² if it is made of cobalt steel and even more than that if it is one of the Alnico alloys which contain aluminum, nickel, and cobalt.[6]

8.21 Pull of a Magnet

Although electromagnets are used for commercial lifting jobs, such as loading scrap iron into a conveyance or raising an armature to a higher position, the equation that gives the force, or pull P, in newtons is

$$\mathbf{P} = \frac{\mathbf{B}^2 \mathbf{A}}{2\mu_o} \text{ N} \tag{8.19}$$

---

[6]For a discussion of permanent magnets and their design, see H. Alex Romanowitz, *Electrical Fundamentals and Circuit Analysis*, New York: John Wiley & Sons, 1966. *op. cit.* pp. 189–194.

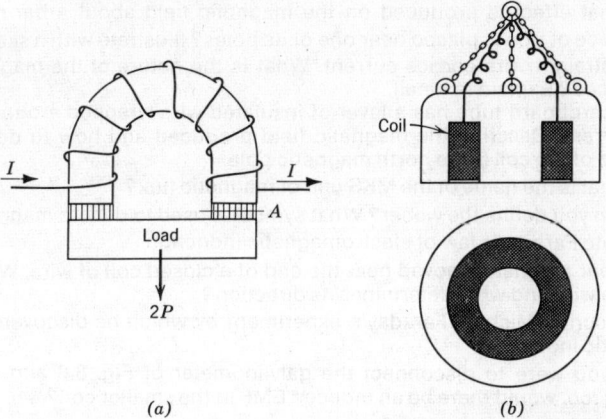

FIGURE 8.20 Lifting electromagnets. (a) Half-toroid lifting magnet. (b) Enclosed-coil circular electromagnet.

B is the flux density (webers per square meter) of the air gaps, two of which are present in practical electromagnet applications. A is the effective cross-section area of one of two identical gaps. See Fig. 8.20a.

EXAMPLE 8.12 As an example of relative magnitudes, let us assume **B** = 0.5 Wb/m² and A = 0.01 m².

Solution

$$P = \frac{(0.5)^2(0.01)}{(2)(4\pi)(10^{-7})} = 995 \text{ N}$$

The object lifted weights $2P$ N, which is $(2)(995)(0.225) = 447.8$ lb.

Because of fringing of flux at the pole faces and nonuniformity of **B** in the gap, certain approximation must be introduced in design computations for electromagnets.

8.22 Review Questions

1. Name four important characteristics of magnetic flux.
2. What is the name given to the end of a permanent magnet from which lines of flux are said to emerge?
3. What are the laws related to attraction and repulsion of magnetic poles?
4. How could you demonstrate the existence of a magnetic field?

5. What effect is produced on the magnetic field about a bar magnet when a piece of iron is placed near one of its poles? Illustrate with a sketch.
6. A straight wire carries current. What is the nature of the magnetic field produced? Explain in detail.
7. A cardboard tube has a layer of insulated wire wrapped around it that carries current. Describe the magnetic field produced and how to determine which end of the coil is the north magnetic pole.
8. What is the name of the MKS unit of magnetic flux?
9. Can you define the weber? What symbol is used to denote magnetic flux?
10. State Faraday's law of electromagnetic induction.
11. A bar magnet is moved near the end of a closed coil of wire. What happens in the wire and what determines its direction?
12. Describe Michael Faraday's experiment by which he discovered electromagnetic induction.
13. If you were to disconnect the galvanometer of Fig. 8.8 *and then close the switch*, would there be an induced EMF in the smaller coil?
14. What is magnetomotive force, and what symbol represents it?
15. State "Ohm's law for the magnetic circuit."
16. Define magnetic flux density. What are its MKS units?
17. What is the name of the unit of magnetic flux in the (a) MKS system, (b) CGS system?
18. Answer Question 17 for magnetic flux density.
19. Distinguish between magnetic flux density and magnetic field intensity. Define the latter mathematically.
20. An amount of flux Φ passes horizontally through an area A which is tilted at an angle of θ degrees with respect to the perpendicular to the flux. Express the flux density mathematically.
21. Define permeability as the ratio of two magnetic quantities.
22. What is the permeability of free space, and what is its symbol?
23. Define relative permeability.
24. To what electrical quantity is reluctance similar?
25. State the formula by means of which reluctance can be calculated.
26. Why could we use $\mu = 4\pi \times 10^{-7}$ in calculating the reluctance of a path in either paramagnetic or diamagnetic material?
27. What are the quantities represented by a saturation curve? Which is usually the ordinate quantity?
28. Distinguish between magnetomotive force and magnetizing force.
29. A magnetic path 10 cm long in a material is subjected to 100 At of magnetomotive force. What is the field intensity in the path?
30. A flux density of 0.5 Wb/m² is to be established in an air gap 1 mm long. As much field intensity is required for this as for a cast steel path 95 m (311 ft) long. Why this big difference?
31. Compare the method of calculating ampere-turns required for establishing flux in parallel magnetic circuits with voltage required for establishing current in parallel electric circuits.
32. State Ampere's circuital law.

33. How can stray electric fields be prevented from passing through a space in which it is not wanted?

34. What is magnetic hysteresis?

35. How does hysteresis loss vary with flux density?

36. What quantities are plotted to form a hysteresis loop?

37. What are eddy currents in a magnetic circuit and what is done to minimize the energy loss involved?

38. An electromagnet like that represented in Fig. 8.19a has a certain number of ampere-turns energizing it. Although Equation (8.19) does not have length of air gap as a factor that determines the lifting force, it is a fact that as the air gap gets shorter the pull increases very materially. Why?

39. Imagine two parallel wires 1 cm apart, each carrying current in the same direction. Sketch the magnetic fields produced by both currents. How does the field between the wires compare in strength with the field outside the wires? Does flux from each current encircle the other wire? If magnetic lines of force act like stretched rubber bands, would you expect the wires to be drawn toward each other by equal forces?

40. Assume the currents in the wires of Question 39 to flow in opposite directions. Is the field stronger between the wires than it is in the outside regions? Would you expect the wires to be forced apart?

8.23 Problems

30. Show that $B = \mu_r \mu_0 H$ starting with $\Phi = \mathscr{F}/\mathscr{R}$.

31. The relative permeability of a certain ferromagnetic alloy is 1500 when $B = 0.95$ Wb/m². Where H is doubled the flux density is increased 20 percent. What is the permeability at the higher density?

32. A transformer core made of annealed sheet steel has the form and dimensions shown in Fig. 8.21. It is assumed there are no air gaps. A coil of 1000 turns

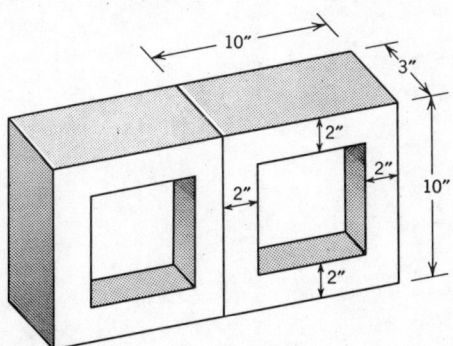

FIGURE 8.21   Annealed sheet-steel core of a shell-type transformer. The two sections have identical dimensions.

surrounds the inside sections. How much current is required to produce 5610×10^{-6} Wb of flux in each outer section?

33. A 1000-turn coil carrying 1.6 A surrounds the inside sections of the transformer core of Fig. 8.21. How much flux does it produce in each outer section?

34. Calculate data and plot a curve showing how relative permeability of annealed sheet steel changes with flux density in the steel for values of **B** from 0.1 Wb/m² to 1.4 Wb/m². Obtain coordinates of ten points.

35. A ferromagnetic material has a hysteresis coefficient $\eta = 200$. (a) Calculate the hysteresis loss per cubic meter per cycle when the maximum flux density is 1.2 Wb/m². (b) How much power would be lost in hysteresis in 0.02 m³ volume at a frequency of 60 Hz?

36. Calculate the energy loss per cycle in 0.02 m³ of the material of Problem 35 which is made of steel laminations 0.0355 cm thick. The eddy-current coefficient $k = 7.5 \times 10^6$ and $f = 60$ Hz. What is the power loss?

37. A magnetic potential gradient **H** = 390 At/m is required to maintain 1.2 Wb/m² flux density in annealed sheet steel. A series magnetic circuit has an annealed sheet steel path 1 m long and an air gap *l* m long. The cross-section area of both the steel and air gap is 1 cm². If the reluctance of the air gap is the same as the reluctance of the steel path, how long is the air gap?

38. An electromagnet of the type shown in Fig. 8.20 is used to raise a load of 3000 lb. Each air-gap area, *A*, through which the flux passes is 25 in.² How much flux density is required?

39. The equivalent series magnetic circuit of the electromagnet and its load in Problem 8.38 has 40 in. of cast steel and two 0.2-in. air gaps. If we assume that the effective cross-section area of the steel path is 25 in.² and the air gaps have the same effective areas, how many amperes of current must be sent through 400 turns of wire on the electromagnet? What is the dc resistance of the coil if it is wound with 700 ft of No. 12 AWG wire? How much dc voltage is required?

40. A core material for use in an electromagnet should not have a flux density greater than 1.25 Wb/m². How much area should each of two pole faces have if the magnet is to lift 5000 lb?

CHAPTER NINE

Direct — Current Meters and Measurements Instruments

It is essential that students preparing for a career in electrical engineering become familiar with the construction and use of instruments that measure voltage, current, power, resistance, and all other basic quantities with which they must work. Quantitative data are needed in almost every form of engineering practice; accuracy and reliability are of paramount importance.

In this chapter we shall study instruments that are used in direct-current practice. Later, after we have become familiar with details of alternating-current circuit behavior, we shall study instruments and measurements employed in ac practice.

9.1   Turning-Coil Instruments

The pointer of a direct-current meter measuring electric current or voltage moves up-scale because it is rigidly attached to a small coil of wire that turns on a vertical axis when current is sent through it. The magnetic field produced by the current in the coil reacts with another magnetic field in the meter in such a way that a turning force is exerted on the coil.

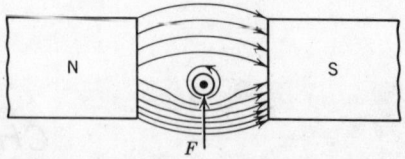

FIGURE 9.1   Wire carrying current directed out of the page. Flux produced by the current adds to the field below the wire, making it stronger, and subtracts from the field above the wire, making it weaker. Force *F* acts on the wire on the strong side. "Motor action" is illustrated.

Figure 9.1 shows a single wire carrying current out of the paper while it lies in a magnetic field which is between two poles of a permanent magnet. The magnetic flux produced by the current in the wire is added to the flux of the permanent magnet below the wire and it is directed in opposition to the flux of the permanent magnet above the wire. The net effect is a sideways force on the wire in the upward direction. Remember that one of the characteristics of magnetic lines of force is that they act like stretched rubber bands. Those below the wire try to shorten themselves and, in so doing, they exert an upward force on the wire. They actually exert upward force *on the current in the wire*, but the current cannot get out of the wire, so the force is transmitted to the wire itself.*

If we reverse the current in the wire in Fig. 9.1 the force will reverse direction and push the wire downward. The current *I* in the single turn in Fig. 9.2

*The electron current in a television picture tube passes through a crosswise magnetic field. Forces exerted on the electron stream move it sidewise and up-and-down to produce the picture.

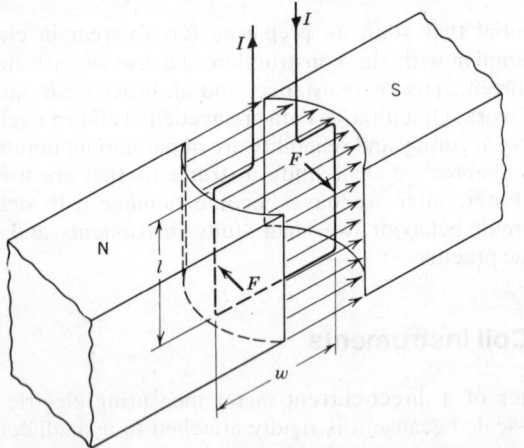

FIGURE 9.2   Forces *F* acting on a one-turn coil carrying current in a magnetic field. In practice a solid-iron cylinder is mounted between the pole faces, leaving a short air gap for the coil sides to turn in and causing the flux lines to be radial in direction.

flows downward in the side near the S pole and upward in the side near the N pole. This accounts for the opposite directions of the forces F. This pair of forces creates a *torque* that causes the single-turn coil to rotate about its vertical axis. If the design and construction were such that a horizontal pointer were attached to the coil, it would turn clockwise as viewed from above.

When a current I is directed at right angles to a magnetic field of flux density B, the force on length l of the current (or on length l of the conductor carrying the current) is given by

$$F = BIl \text{ newtons} \tag{9.1}$$

If the angle between the direction of the current and the direction of the magnetic field is θ, the force is given by $F = BIl \sin \theta$. The turning moment (torque) on the coil in Fig. 9.2, since $\theta = 90°$, is

$$T = 2BIlr \text{ newton-meters} \tag{9.2}$$

in which r is the "moment arm" of each force because it is the radius of the path in which the coil sides move. The width of the coil is $w = 2r$, and wl is the area A enclosed by the coil, so it follows that

$$T = BAI = \Phi I \tag{9.3}$$

When the flux Φ is constant, the torque is directly proportional to the current.

EXAMPLE 9.1   A meter coil of 30 turns carries 20 mA of current at right angles to a magnetic field where $B = 0.25$ Wb/m². The length of one side of the coil is 1 in. and the coil width is 1.5 in. How much torque in newton-meters and in ounce-inches is produced?

  Solution
$$T = 0.25(1 \times 1.5) \, 2.54^2 \times 10^{-4} \times 20 \times 10^{-3} \times 30$$
$$T = 1.45 \times 10^{-4} \text{ N-m}$$
$$T = 1.45 \times 10^{-4} \times 0.2248 \times 16 \times 39.37$$
$$T = 0.0205 \text{ oz in.}$$

Problems

1. A conductor parallel to the ground carries 5 A toward the east. It lies in a magnetic field whose density is $B = 0.75$ Wb/m² directed north. How much force is exerted on each meter of conductor length, and in what direction?
2. Solve Problem 1 assuming that the wire is perpendicular to the earth, current flows vertically upward, and the field flux is directed (a) toward the northeast, (b) 30° south of east.
3. A straight wire is carrying 10 A of current vertically upward in a magnetic field and a 1-m length of it experiences a force of 0.25 N, toward the west. What is the direction of the magnetic field and what is its flux density?

4. A square loop of wire carrying 2 A of current is parallel to the earth. Two of its sides are 0.5 m long and extend east to west; its other two sides extend north and south. How much torque is exerted on the loop if it lies in a horizontal magnetic field the flux density of which is 0.5 Wb/m² directed toward the west? Does the axis of the moving loop run north-to-south or east-to-west? If the current flows in a clockwise fashion as observed when we look down at the loop from above, does the side of the loop toward the east move up or down?

A familiar property of a spiral spring is that its angular deflection is proportional to the torque applied. The deflection of the pointer of a dc meter is restrained by a spiral spring mounted around the meter shaft so that the movement of the coil and pointer winds up the spring a small amount. This makes possible the use of a *linear scale* on the meter dial because the deflection of the spring is proportional to the torque applied, according to Hooke's law in mechanics.

9.2   The D Arsonval Movement

The movement shown in Fig. 9.3 is by far the most widely used type of direct-current meter movement. Its special merits, in addition to employing a linear scale (division lines are equally spaced throughout), are high sensitivity, dependability, minimum interference from stray magnetic fields, and ruggedness.

The D'Arsonval movement is used in both ammeters and voltmeters of the direct-current type. To obtain adequate torque with very small current, a coil with many turns of very fine wire is used. It is comparatively light in

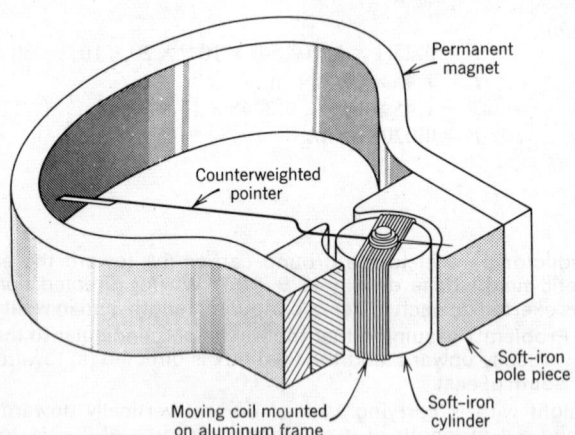

FIGURE 9.3   Cut-away view of D'Arsonval movement in a direct-current meter.

weight, and this is required to minimize friction in the bearings that support the shaft, coil, springs, and pointer.

If the current through the meter varies so rapidly that the coil and pointer cannot follow the variations, the pointer will come to rest on the scale at a position determined by the average torque. The meter reading is then the average current or average voltage.

9.3   DC Ammeters and Shunts

Ammeters for measuring direct-current employ the D'Arsonval principle because of its many advantages. The current in the moving coil is kept small by the use of a *shunt* which carries most of the current to be measured. Meters that measure up to about 25 A usually have shunts mounted inside the meter case. Such meters are connected directly in a line containing the current to be measured, that is, *in a series connection.* When an external shunt is used, only the shunt is connected in the line and especially designed leads (wires) serve to connect the shunt to the two terminals of the meter.

EXAMPLE 9.2   A moveable coil has a resistance of 50 Ω and a normal current capacity of 1 mA. It is to be used in an ammeter which has a full-scale reading of 10 A. We shall show the connections and calculate the resistance of the internal shunt.

Solution. Figure 9.4a shows the connections. I, the current to be measured, divides at the junction into I_c and I_s. When I is 10 A, $I_c = 1$ mA, and I_s is 9.999 A. The voltage drop V_c through the coil is equal to the voltage drop E_s across the shunt:

$$V_c = I_c R_c = 0.001 \times 50 = 0.050 \text{ V} = V_s$$

The resistance of the shunt is

$$R_s = \frac{V_s}{I_s} = \frac{0.050}{9.999} = 0.005 \ \Omega$$

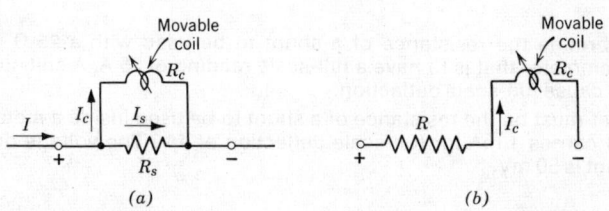

(a)   (b)

FIGURE 9.4   Internal elements and connections of (a) ammeter, (b) voltmeter using same moving element.

Another way to determine R_s is to use $I_c R_c = I_s R_s$:

$$0.001 \times 50 = 9.999\,R_s$$
$$R_s = 0.005\ \Omega$$

This is a very small amount of resistance, but it can be obtained with the aid of precision instruments.

Usually, an adjustable resistance is connected in series with the moving coil of a meter in order to provide a means for adjusting the "coil resistance" to the correct value. Suppose you want the coil to have exactly 50 Ω resistance. It is practical to design it to have less than 50 Ω, say, approximately 40 Ω, and then connect in series with the coil an adjustable resistance that may be varied from perhaps 50 to 15 Ω. If the coil, after manufacture, has 40.7 Ω, then 9.3 Ω could be added by the adjustable unit. The path for the coil current would then have the required 50 Ω and a 50 mV shunt would send 1 mA through it when a 50 mV drop exists across the shunt.

·Suppose a meter *with an internal shunt* has 10 A full-scale capacity. To use it on a 100 A line would require an external shunt. Calling the meter resistance R_m and the shunt resistance R_s, we have

$$I_m R_m = I_s R_s \tag{9.4}$$

from which R_s can be determined.

EXAMPLE 9.3   The 10-A meter with an internal resistance of 0.005 Ω is to be used to measure current that may reach a maximum of 100 A. What should be the resistance of an external shunt for use with this meter?

Solution. From Equation (9.4).

$$R_s = \frac{10 \times 0.005}{100 - 10}$$

$$R_s = 0.000555\ \Omega$$

Problems

5. Determine the resistance of a shunt to be used with a 25-Ω moving coil in an ammeter that is to have a full-scale reading of 25 A. A coil current of 10 mA will cause full-scale deflection.

6. What must be the resistance of a shunt to be used inside a meter if its moving coil carries 1 mA for full-scale deflection of 1A? The voltage drop across the shunt is 50 mV.

The measuring capacity (full-scale ampere value) of dc meters may be changed by changing the shunt. Shunts are rated in millivolts drop at maximum current: 50 mV-100 A shunt, 100 mV-200 A shunt, and so on. The one in Example 9.2 is a 50 mV-10 A shunt.

9.4 DC Voltmeters

The pointer of the ammeter of Example 9.2 deflects to the full-scale position when the current in the moving coil is 1 mA. If the meter can be supplied with an internal resistance unit R, of correct ohms value, *connected in series with the moving coil*, the meter can be connected between two lines that have a potential difference of 150 V and its coil will carry 1 mA. The pointer will again deflect to the full-scale position where the reading can be 150 V instead of 10 A. That is, the meter can be fitted with a *voltage scale* (instead of a current scale) and become a *voltmeter*. This is exactly how dc voltmeters are constructed.

EXAMPLE 9.4   Suppose the movable coil of the meter in Example 9.2 is to be used in a voltmeter with a full-scale reading of 150 V. A series resistance R must be used. The arrangement is shown in Fig. 9.5a. The coil current (0.001 A) flows through R. Determine the value of R.

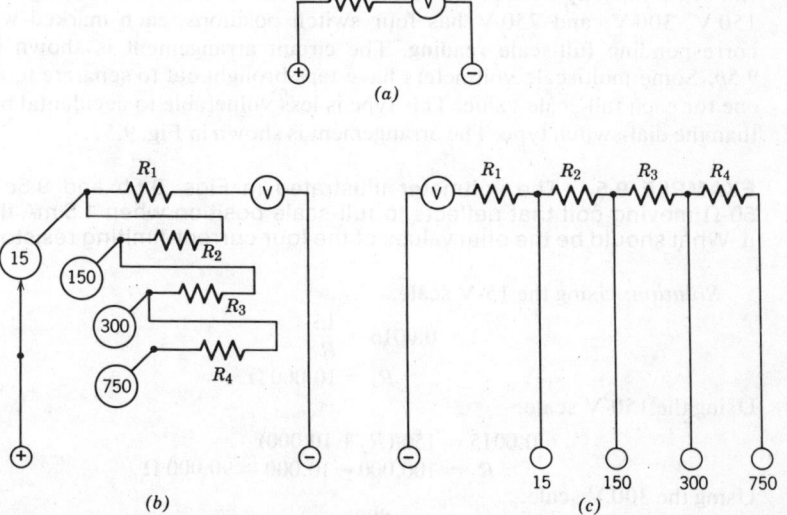

FIGURE 9.5 Internal connections of voltmeters. (a) Simple one-scale meter. (b) Multi-Lange voltmeter, dial-switch type. (c) Separate-terminal-post type.

Solution

$$0.001 = \frac{150}{R + 10}$$

Solving,

$$R = 149,990 \ \Omega$$

Problems

7. A voltmeter has a 20-Ω moving coil that carries 5 mA when the deflection is full-scale at 300 V. How much resistance must be connected in series with the coil?

8. Suppose the voltmeter in Problem 7 is to have a second scale 0–150 V. What should be the amount of series resistance? Why is R not exactly half of the amount needed in Problem 7?

9.5 Multirange Meters

The range of a voltmeter may be changed by switching to a different value of series resistance. A dial switch mounted on the panel of a multirange voltmeter may be set at one of several specific positions in accordance with the full-scale capacity desired. A voltmeter capable of full-scale readings of 15 V, 150 V, 300 V, and 750 V has four switch positions, each marked with the corresponding full-scale reading. The circuit arrangement is shown in Fig. 9.5*b*. Some multiscale voltmeters have taps brought out to separate terminals, one for each full-scale value. This type is less vulnerable to accidental burnout than the dial-switch type. The arrangement is shown in Fig. 9.5*c*.

EXAMPLE 9.5 The voltmeter illustrated in Figs. 9.5*b* and 9.5*c* has a 50-Ω moving coil that deflects to full-scale position when 1.5 mA flows in it. What should be the ohm values of the four current-limiting resistors?

Solution: Using the 15-V scale,

$$0.0015 = \frac{15}{R_1}$$
$$R_1 = 10,000 \ \Omega$$

Using the 150-V scale,

$$0.0015 = 150/(R_2 + 10,000)$$
$$R_2 = 100,000 - 10,000 = 90,000 \ \Omega$$

Using the 300 V scale,

$$0.0015 = \left(\frac{300}{R_3 + 100,000} \right)$$
$$R_3 = 200,000 - 100,000 = 100,000 \ \Omega$$

Using the 750-V scale,

$$0.0015 = \left(\frac{750}{R_4 + 200,000}\right)$$

$$R_4 = 500,000 - 200,000 = 300,000 \, \Omega$$

Multiscale ammeters are also fitted with dial switches for use in selecting different, shunts to change the capacity of the meter.

EXAMPLE 9.6 The meter coil in Fig. 9.6a has 50 Ω resistance and deflects to full scale when it carries 1 mA. What should be the ohms value of the three shunt resistances if the full-scale readings are 1 A, 5 A, 10 A?

Solution. When the full-scale reading is to be 1 A, 999 mA must flow through the shunt.
When

$$I_c R_c = I_s R_s; \qquad R_s = \frac{I_c}{I_s} R_c$$

$$R_{s1} = \frac{1}{999} \times 50 = 0.05005 \, \Omega$$

When the 5 A range is to be used,

$$R_{s2} = \frac{1}{4999} \times 50 = 0.01000 \, \Omega$$

and when the 10 A range is to be used,

$$R_{s3} = \frac{1}{9999} \times 50 = 0.005000 \, \Omega$$

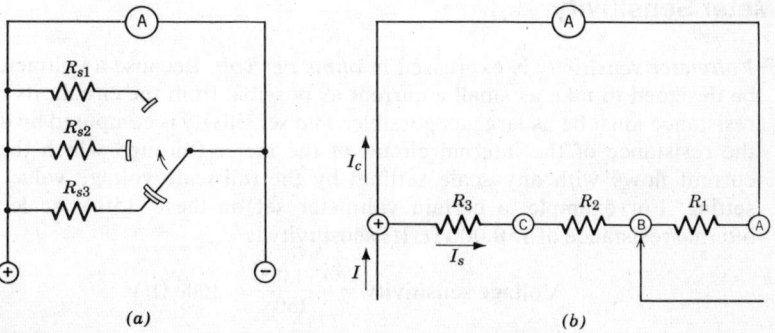

FIGURE 9.6   Internal connections of a multirange ammeter. (a) dial-switch type. (b) External terminal-post type.

Although the next figure beyond the last zero in the R_{s2} and the R_{s3} value is not zero, it is impractical to expect more than the indicated accuracies in the manufacture of such low-resistance elements. Furthermore, other errors in the construction and in the reading of the meter would be greater than that introduced by the shunt. Variation in contact resistance contributes to inaccuracy of this type of meter.

An important requirement of multirange ammeter construction is that the sliding contact of the scale-changing switch must not break contact with the terminal of one shunt before it makes contact with the next terminal. If it did, the circuit would be *open*, which normally would be undesirable. Furthermore, an arc may form at the time of breaking the circuit, which would cause the contact surfaces to deteriorate and soon result in trouble and costly repairs. In contrast, the dial switch of a multirange voltmeter must make only one contact at a time. The student should understand, with a little thought, why this is true. What would happen if the switch did not completely clear the 150-V contact when the user moved the knob to the 750-V position and then applied 750 V to the meter?

In precision work, the resistance of *connecting leads* from the shunt to the ammeter terminals must be taken into account. Leads are furnished with each shunt, and only these should be used with it.

The arrangement of shunts in Fig. 9.6b does not require a scale-changing switch and assures that some resistance will always be in parallel with the moving coil of the instrument. The ohms values of R_1, R_2, and R_3 can be selected so that the instrument can have full-scale capacities of 1, 5, 10 A, or 0.5, 2, 5 A, or other sets of full-scale values. When R_1, R_2, and R_3 are constructed with their proper resistance values, the terminal posts A, B, C may be designated 1, 5, 10 or 0.5, 2, 5 according to the full-scale values of the meter.

9.6   Meter Sensitivity

Voltmeter sensitivity is expressed in *ohms per volt*. Because a voltmeter must be designed to take as small a current as possible from the circuit, its internal resistance must be as large as possible. The sensitivity is computed by dividing the resistance of the internal circuit of the meter (through which the meter current flows with any scale setting) by the full-scale voltage value at that setting. For example, a certain voltmeter set on the 0–150 V scale has an internal resistance of 150,000 Ω. Its sensitivity is

$$\text{Voltage sensitivity} = \frac{150,000}{150} = 1000 \ \Omega/\text{V}$$

Suppose this meter has three voltage ranges: 0–15, 0–150, 0–300. Any one of these voltage values, used with the internal resistance of the meter *at that voltage setting*, would give the same result for sensitivity. When set on the 15-V scale the internal resistance would have to be 15,000 V so that the moving-coil current would be the required value for full-scale deflection, The voltage sensitivity is thus $15,000/15 = 1000 \, \Omega/V$.

The moving coil of a voltmeter, or, in other words, the meter movement, may be thought of as a milliammeter that could be used alone. One with a full-scale current of 10 mA could be put in a series circuit, in which the current is known not to exceed that value, and used to measure current. Or it could serve as a low-resistance millivoltmeter if used between two points that have a potential difference not larger than $I_m R_c$ which is the product of the moving-coil resistance and the maximum current it should receive, that is, the current that gives full-scale deflection.

It is interesting to note that, because the full-scale current in the meter element is the ratio of the voltage drop across it to its resistance and since the units can be expressed as *volts per ohm*, the *ohms-per-volt* sensitivity must then be given by *the reciprocal of the full-scale current value*. If a meter requires 1 mA for full scale deflection (called *current sensitivity*), it follows that $1/0.001 = 1000$ must be the ohms-per-volt sensitivity. This is $R_c \div V_c$, the coil resistance divided by the voltage drop across the coil.

Problem

9. What is the sensitivity of a voltmeter that: (a) has 300 volts full-scale and a resistance of 300,000 Ω? (b) has a full-scale coil current of 50 μA? (c) has 10-MΩ resistance when used with the 0–100 V scale?

9.7   Loading effect of Voltmeters

The larger the ohms-per-volt sensitivity of a voltmeter, the smaller will be the current it will require for a given measurement. Electronic-circuit currents are frequently found to be in the milliampere or microampere range, and series resistors of thousands of ohms may be in the circuit branch where a voltage measurement is to be taken. It is quite understandable that relatively large current values taken by a voltmeter will cause erroneous results, owing to added voltage drops in the series resistances: A voltmeter with 15,000-Ω/V sensitivity used to measure a 15-V drop, between points A and B in a series circuit, would take 1 mA of current. This meter current will add a 5-V drop, which was not there before, across a 5000-Ω *series* resistor. A voltmeter with 20,000-Ω/V sensitivity would take only 0.05 mA, which would cause a 0.25-V drop. This is an error of about 1.7 percent. A *vacuum-tube voltmeter*, with a constant resistance of, say, 100 MΩ would take an insignificant amount of current and so *would not load the circuit.* Some vacuum-tube voltmeters are built to have infinite input impedance.

Problems

10. A 50,000-Ω resistor is in series with a 150,000 Ω resistor and they are connected to a 200-V source. (a) calculate the voltage across the 15,000-Ω resistor. (b) What would a 0-150 scale voltmeter (sensitivity 1000 Ω/V) read when connected across the resistor? If you assume the voltmeter reading is the voltage across the resistor when the meter is not connected, what would the error be in volts and in percent?

11. Do Problem 10 but assume you measure the voltage with a 0-150 scale, 20,000 Ω/V meter.

9.8   Practical Considerations

As you must remember, ammeters are connected *in series* with the unit whose current is to be measured. This means having the current turned off, opening the line that carried the current, and connecting the two open ends to the input (+) and output (−) terminals of the meter. The current must enter the meter at its (+) terminal.

The internal resistance of an ammeter is very small, as we have observed, and this is necessary. If the meter resistance were appreciable, its insertion would change the amount of the current. There are times when the meter resistance must be taken into account. This is especially true when using *milliammeters.*

Occasionally we feel obliged to put a fuse in series with a milliammeter to protect it from overloads. Unfortunately, very small capacity fuses have resistances that are not negligible. This is true of fast-action instrument fuses in the range 2 mA to $\frac{1}{32}$ A. Their resistance may be as much as 30 to 50 Ω. This must be taken into account if a change in current caused by the addition of this much resistance into the circuit may be more than can be tolerated.

We must be very careful in using ammeters because they are susceptible to burn-out. It is good practice to connect the meter, at first, so that it is set at its largest current capacity. We must also know that the current to be measured does not exceed the capacity of the meter. It is good practice to *connect a short-circuiting switch* across the terminals of the ammeter as an added protection. Then when the line is energized, the switch will carry most of the current. Opening the switch will result in all of the current going through the ammeter. If the pointer flies off scale the switch can be closed instantly and, generally, the meter will not have time to burn out. If an ammeter has a scale-changing switch, the switch should be set at *maximum current capacity* before the meter is put away for future use.

Voltmeters are not as vulnerable as ammeters. Excess voltage will throw the pointer off scale and may bend it. A multirange voltmeter is susceptible to burn-out when the user is careless. If the scale-change switch is set too low, or if connections are made to binding posts on the meter that are for too small a scale, the meter may be burned out. For example, if you try to measure 150 V with the leads (wires) connected to the 15-V scale you will, in all probability, burn out the meter. When put away, multirange voltmeters should have their scale-changing switches set at maximum voltage capacity.

ACCURACY OF METER READINGS

Suppose a 10 A meter has 50 divisions on its scale. Each division represents a current increment (change) of 0.2 A. If the meter accuracy is given as 2 percent, this means that any reading from 10 to 100 percent of full scale is accurate to ±2 percent of the full-scale reading, or ±0.2 A. Readings that fall in the first 10 percent of the total deflection are not reliable because of bearing friction and the ±0.2 A inherent error. The least current that should be read on this meter is 1.0 A. At this reading the actual current may be 0.8 or 1.2 A. If readings in this range are to be taken and better accuracy expected, a meter with smaller capacity (0–2.5 or 0–3 A) should be used. Too much error may be introduced when taking readings in the lowest third of a meter scale.

9.9   Galvanometers

A galvanometer is a dc instrument with a D'Arsonval movement designed to carry current in the microampere or very low milliampere range. The coil is capable of turning either clockwise or counterclockwise. When there is no current in the coil it is in a neutral position and its pointer rests at zero *in the middle of the scale*. Current in one direction deflects the pointer to the right, and current in the other direction deflects it toward the left.

A useful application of the galvanometer is in a bridge circuit by which resistance is measured. When the control resistance of the bridge is set at too large a value, the needle deflects to one side of zero; when the control resistance is too small, the deflection is to the other side of zero. When the bridge is balanced, no current flows through the galvanometer and this is indicated by a zero reading. We shall study the theory of measurement with a bridge in Section 9.14.

9.10   Resistance Measurement by Voltmeters and Ammeters

Figure 9.7 shows two circuits used for measuring an unknown resistance with the aid of Ohm's law. Because the resistance value is unknown it is wise to

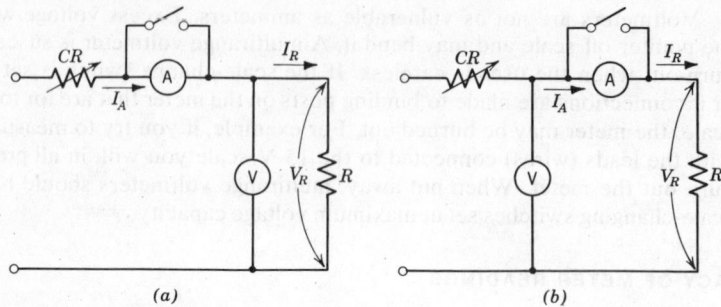

FIGURE 9.7   Voltmeter-ammeter method of measuring unknown resistance R.

connect a short-circuiting switch across the ammeter. Should the unknown resistance be so low as to draw more current than the ammeter can handle, the switch will carry it and the meter will not be burned out.

Good quality voltmeters have a "push-to-read" button on their top so that the voltmeter is energized only while a reading is being taken. If the voltmeter does not have this convenience the ammeter can be read before the voltmeter is connected. The current limiting resistor can be used to control the amount of current employed in the measurement.

The connections in Fig. 9.7a are used when R is small. The ammeter carries voltmeter current which is often negligible in comparison with I_R so that the ammeter reading can be taken as I_R. If this is not the case, as when measuring resistances that are not considered small in comparison with R, correction for voltmeter current in the ammeter must be made. Let subscripts A and v refer to ammeter and voltmeter respectively. In Fig. 9.7a,

$$V = V_R \tag{9.5}$$

$$I_R = I_A - V_R/R_v \tag{9.6}$$

and the unknown resistance is

$$R = \frac{V_R}{I_R} = \frac{V}{I_R} \tag{9.7}$$

EXAMPLE 9.7   An unknown resistance is measured using the circuit of Fig. 9.7a by means of a voltmeter that has 15,000 Ω resistance and a milli-ammeter. The resistance, R, of the unknown is 3000 Ω, but we are not supposed to know this. The fact remains, however, that the voltmeter will take one-fifth as much current as the unknown resistance, and this is revealed in the readings. Assume that $I_A = 0.060$, $V_R = 150$ V, $R_A = 0.1$ Ω.

Solution. Let's calculate the value of the unknown resistance R.

From Equation (9.6),

$$I_R = 0.060 - 150/15,000 = 0.05 \text{ A}$$

and, from (9.7),

$$R = \frac{150}{0.05} = 3000 \ \Omega$$

The voltage drop across the ammeter due to 0.06 A is only 0.006 V.

If we had not made the correction in the current and had used the un-corrected ammeter reading, we would have obtained a value for R that is much too low:

$$R = \frac{V}{I_A} = \frac{150}{0.06} = 2500 \ \Omega$$

It turns out that if the resistance of the voltmeter is as much as four times the unknown resistance, the determination of the amount of the unknown R, without making a correction for voltmeter current, is only about 4.5 percent lower than the actual value. When using the connections in Fig. 9.7b, the voltage across R is the line voltage V minus the voltage drop across the ammeter:

$$V_R = V - I_A R_A \tag{9.8}$$

and

$$R = \frac{(V - I_A R_A)}{I_A} \tag{9.9}$$

These connections should be used when the current taken by the unknown resistance is about the same as, or less than, the voltmeter current. It is obvious that, in Fig. 9.7b, V/I_A gives the sum of the ammeter resistance and the resistance R. Since the ammeter resistance is always very small, it may be neglected when R is very much larger than R_A.

EXAMPLE 9.8 Assume, in Fig. 9.7b, that $V = 150$ V, $R_V = 15,000 \ \Omega$, $I_R = 0.15$ A and $R_A = 0.2 \ \Omega$. Determine R with and without correction.

Solution

Without correction:

$$R = 150/0.15 = 1000 \ \Omega$$

With correction:

$$I_A R_A = 0.15 \times 0.2 = 0.03 \text{ V}$$
$$R = (150 - 0.03)/0.15 = 999.80 \ \Omega$$
$$\text{Error: } = (0.2/1000)100 = 2/100 \text{ of 1 percent}$$

Problems

12. A resistance was measured by the voltmeter-ammeter method. The readings were 115 V, 5 A. What was the resistance value? No corrections were required.

13. An unknown resistance was measured by the voltmeter-ammeter method, with the voltmeter connected across the resistance and the ammeter carrying the total current, which was 2 mA. The ammeter resistance was 25 Ω; the voltmeter resistance was 150,000 and its reading was 120 V. Calculate the unknown resistance.

14. An unknown resistance was measured by the voltmeter-ammeter method with the voltmeter connected across the supply lines so that it measured the voltage drop across the ammeter and unknown resistance in series. The readings were 9 V and 3 A. Meter resistances were 15,000 Ω and 0.5 Ω. Calculate the unknown resistance.

9.11   Ohmmeters

An ohmmeter is a resistance-measuring instrument which contains a dry-cell battery that furnishes current to send through an unknown resistance. This same current causes the deflection of the pointer of a current meter, in the ohmmeter case, over a scale graduated in ohms. Figure 9.8a shows a simple circuit of an ohmmeter that should be used to measure medium and large values of resistance. R_1 is a constant resistance of 24,000 Ω; R_0 is a variable resistance called the "zero adjust" unit. When the test prods are touched to each other at their tips, the meter pointer deflects to its full-scale position. This is the zero-

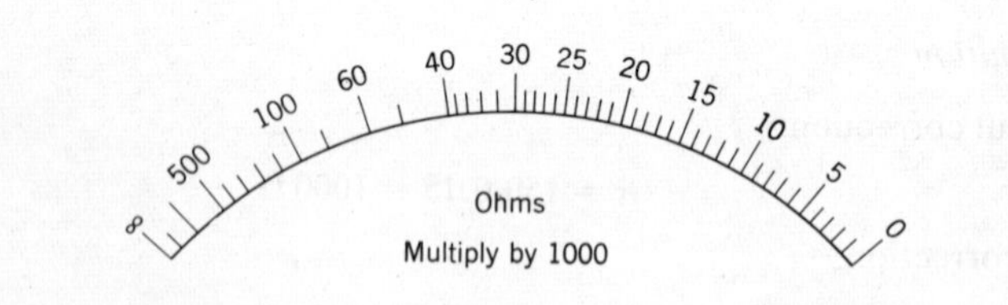

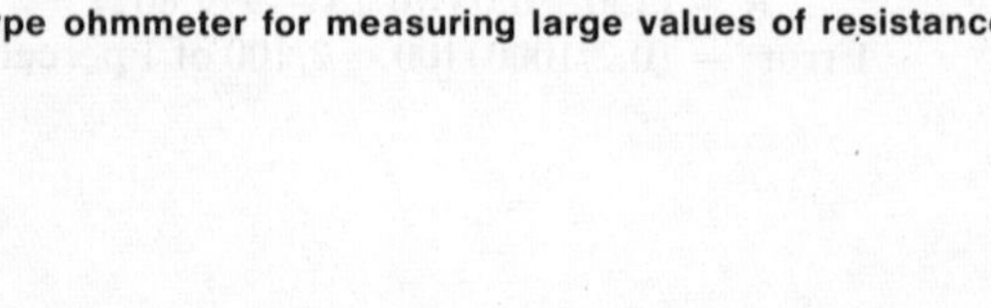

(a)

(b)

FIGURE 9.8   Series-type ohmmeter for measuring large values of resistance. (a) Internal circuit. (b) Scale.

ohms point on the scale, so marked because there are zero ohms connected between the prods. If the pointer is not exactly on zero, the zero-adjust resistance is changed until it is.

The total resistance in the circuit must be $1.5/(50 \times 10^{-6}) = 30,000 \ \Omega$ in order that the current will be $50 \ \mu A$ when the probes are shorted. This means $6K$ of the $10K$ in R_0 will be in the circuit. The zero-adjust unit is indispensable because after the ohmmeter has been in use for some time the battery voltage will drop below 1.5 V. R_0 must then be set for a little less than $6000 \ \Omega$ in order to get the meter to read zero ohms with the probes shorted.

With the probes not touching anything the resistance between them (air path) is infinite and the current through the meter is zero. Therefore the infinite-ohms point on the scale is at the extreme *left*.

The midpoint on the scale is reached by the pointer when the meter carries $25 \ \mu A$. In this case the resistance, R_x, being measured is between the probes and its value may be computed as follows:

$$\frac{1.5}{(30,000 + R_x)} = 25 \times 10^{-6}$$

from which

$$R_x = 30,000 \ \Omega$$

A value of R_x smaller than this would result in a meter current larger than $25 \ \mu A$ and the pointer would be between the center and the zero-ohms position, that is, in the right-hand half of the scale.

Figure 9.8*b* shows a typical scale for this kind of ohmmeter. The meter resistance is ignored when this circuit is used. Why?

LOW-RANGE OHMMETER

When unknown resistances with small ohms values are to be measured, a circuit simplified to the form of Fig. 9.9*a* gives much more accurate measurements. When the prods are touched together the meter is short circuited and it gets no current. There is, of course, no deflection of the pointer. Therefore the zero-ohms point is at the extreme left of the scale. Full-scale current flows when the prods are separated, so the position of *full-scale deflection* is the *infinite ohms point*.

If the prods are not connected to anything and the variable resistance R is set at $5 \ \Omega$ there will be $2.5 \ \Omega$ in series with $147.5 \ \Omega$. The 3 V source will deliver $3/150 = 0.020 \ A$, 10 mA of which will flow through the meter, making it show a full-scale reading. This point on the meter dial is marked infinite ohms (∞).

If $R_x = 2.5 \ \Omega$ the meter current will be slightly more than half of the full-scale value. In this case the total resistance of the circuit is $147.5 + (2.5 \times 2.5)/5 = 148.75 \ \Omega$. The battery current is $3/148.75 = 0.020168 \ A$ and the meter

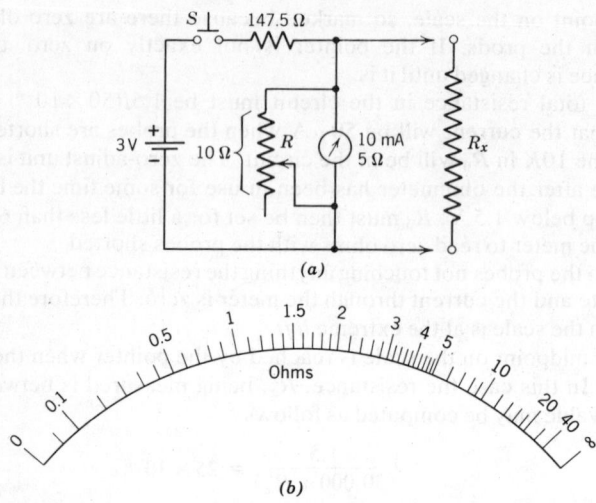

FIGURE 9.9   Parallel-type ohmmeter for measuring small values of resistance. (a) Internal circuit. (b) Scale.

current is $0.02068/4 = 0.0504 = 5.042$ mA. This means that the 2.5-Ω mark on the scale is only slightly above the middle point.

The compensating resistance R is used to set the full-scale deflection at the infinite-ohms point. To set the pointer at the proper place when the meter carries no current, a screw at the pivot of the pointer can be turned slightly by a screw driver.

Problems

15. An unknown resistance was measured by means of the circuit of Fig. 9.8a and found to be 45,000 Ω. How much current passed through the meter?

16. An unknown resistance was measured by means of the circuit of Fig. 9.9a and found to be 5 Ω. How much current passed through the meter?

17. Assume the battery voltage in Fig. 9.8a decreased 5 percent owing to age. Calculate the value of R_0 required for full-scale deflection to be obtained with the test prods touching each other.

18. In Fig. 9.8a how much resistance is being measured when 10 μA is going through the meter? Assume $R_0 = 6000 \Omega$.

You may wonder how unknown resistances in the 40 to 5000-Ω range can be measured accurately, since measurement in these ranges is not covered by the meters illustrated here. The answer is that there are *multirange ohmmeters* that have a scale-change switch which adjusts the internal circuit so that the meter reading may be multiplied by 10, 100, 1000, or 10,000 according to the position of the switch. The range that will result in a needle deflection near the middle of the scale is chosen in the measurement process.

We might add that single instruments are available which will measure not only a full range of resistances but also dc volts from zero to 5000 in 7 ranges, ac volts from zero to 5000 in 7 ranges, and direct current from zero to 80 μA and also from zero to 800 mA in 4 ranges. The accuracies of such meters are of the order of ± 2 to ± 3 percent of full scale. This is good enough for normal laboratory practice.

Vacuum tube voltmeters have very high internal resistance (infinite in some cases) so that they will not draw enough current to load the circuit.

9.12 Measurement of High Resistance with a Voltmeter

If you have a voltmeter whose internal resistance is known, you can measure an unknown resistance with it alone. A requirement for reasonable accuracy is that the unknown resitance value should be not less than about one-tenth, nor more than about ten times, the resistance of the voltmeter.

Look at Fig. 9.10. R_x may be a resistor or it may even represent the insulation resistance of a high-voltage cable. The procedure is to apply high enough voltage, V_0, to give a good reading on the voltmeter, perhaps half-way up the scale. Then record the reading and call it V. Also record the resistance of the voltmeter used for this reading. Next, read the amount of applied voltage, V_0. You may even use a higher range setting on the voltmeter to do this, or use another meter. Although the equations we will use involve the current I through the meter, we do not need its value because our algebraic solution of the equations eliminates it. In Fig. 9.10 we have

$$V_0 - V = I R_x$$
$$V = I R_m$$

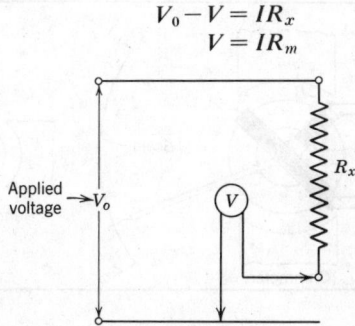

FIGURE 9.10   Resistance measurement with only a voltmeter.

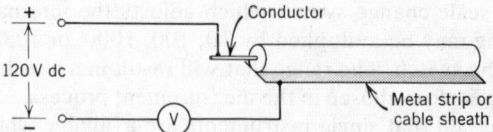

FIGURE 9.11   Measuring insulation resistance by means of a voltmeter. Circuit for Example 9.9.

Solving for R_x,

$$R_x = \frac{V_0 - V}{V} R_m \qquad (9.10)$$

EXAMPLE 9.9   Let R_x be the insulation resistance of an electric cable. The connections are shown in Fig. 9.11. The meter resistance is 150,000 Ω and it reads 12 V. How much is the insulation resistance of the cable?

Solution

$$R_x = \frac{120 - 12}{12} \, 150,000 = 1,350,000 \ \Omega = 1.35 \ \text{MΩ}$$

9.13   Megger Insulation Tester

To obtain enough current in an ohmmeter coil to cause readable deflections when the resistance of electrical insulation is to be measured, it is necessary to have a source of relatively high voltage. Insulation resistance is in the range of kilohms to thousands of megohms, so a 500-V, dc source is common. A device used for measuring the resistance between windings and the frame of electric machinery, the insulation resistance of cables, and of insulators is called a *Megger* (Fig. 9.12). The Megger Insulation Tester will also produce as much

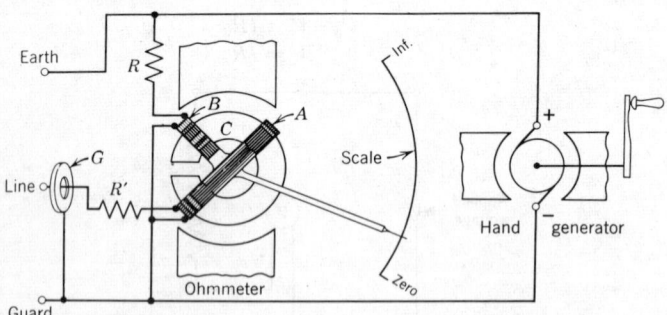

FIGURE 9.12   Basic diagram of electrical connections for the Megger Insulation Tester. (Courtesy James G. Biddle Co., Plymouth Meeting, Pa.)

as 2500 V direct current, which is used for applying high-test potential to the insulation of high-voltage equipment.

The Megger Insulation Tester has a rectangular magnetic circuit with two air gaps, one in each end. A shaft containing a deflecting coil, a potential coil, and a pointer is free to turn in one air gap, and a voltage-generating armature is mounted in the other air gap. The armature is driven through gears operated by a hand crank. Through a unique design, the pointer position is arbitrary when the armature is not being rotated, and the pointer "floats" over the scale and may remain in any position. But it will stop at the infinity position when the crank is rotated if the test terminals are open-circuited or connected to a resistance that is many times larger than the largest values used in practice. But, when a resistance is connected to the terminals, current flows in the deflecting coil which will turn and draw the pointer down-scale from the infinity position. The scale is calibrated in terms of resistance, a substantial part of it in megohms.[2]

9.14   Wheatstone Bridge

The Wheatstone[3] bridge is the basic circuit used in laboratories to obtain precise measurements of elements used in electric circuit analysis.

In Fig. 9.13 the unknown resistance to be measured, R_x, is balanced against known resistances. Although R_1, R_2, and R_3 may be variable in small amounts, it will be seen that R_1 and R_2 need be variable only in powers of 10. Each of

[2]For more details on the Megger Insulation Tester circuit, the student is referred to Chester L. Dawes, *Electrical Engineering*, Vol. I, New York: McGraw-Hill, 1957, or to the manufacturer.

[3]Sir Charles Wheatstone (1802–1875) was an English physicist and engineer who experimented in many branches of physics and obtained many patents as a result of his work and discoveries.

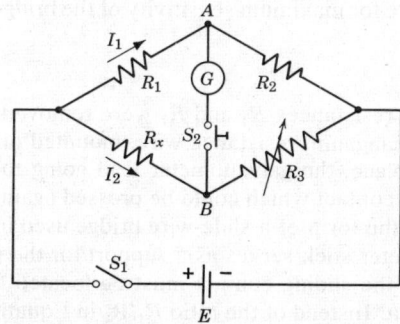

FIGURE 9.13   Basic circuit of Wheatstone bridge. During use, switch S_1 is left closed. Switch S_2 is held open by a spring until closed by the operator.

these may be designed to have only the possible values of 1, 10, 100, 1000 Ω. *Their ratio* may then be set, decimally, between 0.001 and 1000. It can be seen that this dictates a possible range of measurement from 0.001 times the smallest setting of R_3 to 1000 times the largest setting of R_3.

The Wheatstone bridge is said to be *balanced* when the galvanometer reads zero with both switches closed in the circuit of Fig. 9.13. This is accomplished by choosing the proper ratio for R_1/R_2 and the required setting of the variable resistance R_3. With no current in the closed branch containing the galvanometer, the voltage drop across R_1 equals the voltage drop across R_x. Also, the *IR* drops across R_2 and R_3 are equal:

$$I_1 R_1 = I_2 R_x$$
$$I_1 R_2 = I_2 R_3$$

Solving these two equations for R_x gives

$$R_x = \frac{R_1}{R_2} R_3 \qquad (9.11)$$

Evidently, only the ratio of R_1 to R_2 is needed, and not their individual resistance values. This suggests the design of a three-terminal unit representing R_1 and R_2 which will provide a series of resistance ratios. Dial-type resistance units are generally used, with steps of equal resistance in each unit. The ohms values of steps of adjacent units are related in powers of 10.

A caution is necessary here to warn against a possible faulty reading that may be obtained with a bridge that has three separate decade boxes for use as R_1, R_2, and R_3. Suppose R_x turned out to be small—in the range 1–2 Ω and the value of R_1/R_2 was obtained using $R_1 = 1\,\Omega$, $R_2 = 10,000\,\Omega$. Assuming $R_x = 1.5\,\Omega$, a balance would be obtained when R_3 is set at 15,000 Ω. The bridge would not have as reliable sensitivity as would be available if R_2 were set at 100 Ω with $R_1 = 1\,\Omega$. The ratio arms should be kept as close to each other in ohm value as possible for maximum sensitivity of the bridge.

SLIDE-WIRE BRIDGE

If, in Fig. 9.13, the resistances R_1 and R_2 were removed and a 1-m length of German silver, or manganin, resistance wire, mounted on a meter stick, were connected in their place, the galvanometer lead going to junction A could be fastened to a sliding contact which could be pressed against the wire to make a connection. This is the form of a slide-wire bridge used in elementary-physics laboratories. The meter stick serves as a support for the wire, and the reading on the stick where the sliding contact must be located for bridge balance is recorded as test data. Instead of the ratio R_1/R_2 in Equation (9.10) the ratio of the lengths of the parts of the wire on the two sides of the contact point are used. For example, if we let *l* represent the length of wire on the R_1 side of the circuit and $100 - l$ represent the length on the R_2 side (l is in centimeters), we may then write

$$R_x = \frac{l}{100 - l} R_3 \tag{9.12}$$

This relation is obtainable from Fig. 9.13 where R represents R_3 used here.

Because the wire has uniform cross section and constant resistivity throughout, the resistances of sections are to each other as their lengths. Suppose the sliding contact is at 10.5 cm for balance and $R_3 = 3565 \ \Omega$. R_x is then $(10.5/89.5)3565 = 418 \ \Omega$. The accuracy of the measurement may be such that the third digit is questionable.

Problems

19. A voltmeter that has $150K \ \Omega$ internal resistance was used to measure a resistor that has a large ohms value. When a 120-V source was applied to the two in series, the meter read 36 V. Calculate the resistance of the resistor.

20. A voltmeter with $10,000 \ \Omega/V$ sensitivity has scales 0–15, 0–150, 0–300 V. A piece of electric cable 5 ft long is used as a sample on which to measure insulation resistance. Readings were taken in the same manner as those of Fig. 9.11 and found to be 115 V, and then 10 V (using the 0–15 V scale). Calculate the resistance of the insulation in the 5-ft sample.

21. The cable of Problem 20 had a lead sheath covering on its complete outside surface. How much current flowed through the voltmeter (set on its 0–15 V scale) during the test? How much current flowed through each 1-ft section of the insulation? What was the resistance *per foot* of the insulation?

22. When an unknown resistance was measured on the Wheatstone bridge of Fig. 9.13, the ratio arms were set at $R_1/R_2 = 100/1$ and R_3 has to be set at 123.5 to balance the bridge. What was the value of the unknown resistance?

23. An unknown resistance was measured on the Wheatstone bridge and found to be $28.5 \ \Omega$. R_3 had to be set at $285 \ \Omega$ when R_1 was $10 \ \Omega$. What was the value of R_2 in this test?

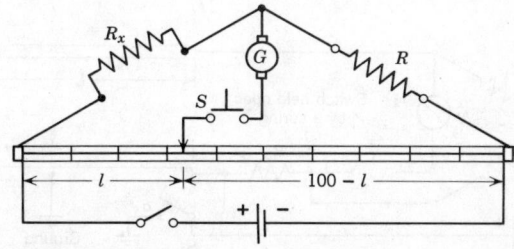

FIGURE 9.14 Slidewire bridge. Bare wire made of high resistivity material supported at the ends of a meter stick. Resistances of sections of the wire are directly proportional to their lengths.

24. To measure R_x, the unknown resistance, using the slide-wire bridge of Fig. 9.14, the following readings were taken. $I_G = 0$, $R = 1250\ \Omega$, $l = 34$ cm. Calculate R_x.

25. Suppose R_x is calculated to be 360 Ω when $R = 1000\ \Omega$ in Fig. 9.14. What must be the observed value of l?

9.15 The Varley Loop

Application of the Wheatstone bridge is very effective in the determination of where current leaks "to ground" through a fault in the resistance of a telephone-line wire. A section of the two-conductor line is connected to the R_x terminals of the bridge circuit. See Fig. 9.15.

R_1 and R_2 are the ratio arms of the bridge and R_3 is the known variable resistance, usually in the form of a "decade box" with dial switches. The decade box may have four dials, the first with 1-Ω steps, the second with 10-Ω steps, the third and fourth with 100-Ω and 1000-Ω steps, respectively.

The bare conductors are connected to the terminals of R_1 and R_3 in the telephone company headquarters. The other bare ends are connected together at a point beyond the suspected location of the fault. The switch S is placed at position b and a balance obtained by adjusting R_3. R_3 and d in series form one arm of the bridge. If r is the ohms resistance per foot of conductor, the total resistance of this arm is $R_3 + rd$ ohms. The relations for balance are

$$\frac{R_1}{R_2} = \frac{r(L+L-d)}{R_3+rd} \tag{9.13}$$

The value of r is known if the wire size and material are known and a wire table is available. The distance d to the ground fault can then be computed. If r is not known, it can be determined from a measurement with switch S connected to terminal a. A bridge balance for this case permits computation of the resistance of length $2L$ of the conductor. Calling this R, it is evident that

$$r = \frac{R}{2L}$$

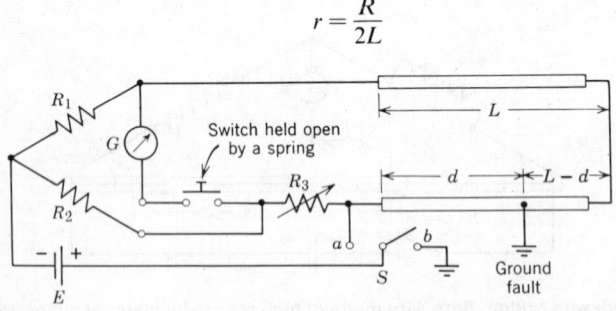

FIGURE 9.15 A Varley loop connected to a two-wire line with a ground fault.

Substituting this into Equation (9.13)

$$\frac{R_1}{R_2} = \frac{(R/2L)(2L-d)}{R_3 + Rd/2L}$$

Solving for d,

$$d = \frac{2L}{R}\left(\frac{RR_2 - R_3R_1}{R_1 + R_2}\right) \text{ feet} \qquad (9.14)$$

This is the distance to the ground fault.

If $R_1 = R_2$ in Equation (9.14),

$$d = \frac{L}{R}(R - R_3) \qquad (9.15)$$

EXAMPLE 9.10   The Varley loop is connected, as shown in Fig. 9.15, to a two-wire line with conductors 3200 ft long. With the switch S in position b, a balance is obtained when $R_1 = 10\ \Omega$, $R_2 = 1000\ \Omega$, $R_3 = 145\ \Omega$. When S is in position a with R_1 and R_2 unchanged, $R_3 = 275\ \Omega$ for balance. Find the distance to the fault.

Solution. First obtain R, the resistance of the whole line, from the second balance test.

$$R = \frac{R_3}{R_2/R_1}$$

$$R = \frac{275}{100} = 2.75$$

$$d = \frac{2 \times 3200}{2.75} \cdot \frac{(2.75 \times 1000) - (145 \times 10)}{1010}$$

$$d = 2995 \text{ ft}$$

It should be stated here that electronic instruments are available which operate on alternating current and on which direct readings are obtainable for distances to ground faults, to short circuits, and to disturbances on lines which result in changes in input impedance.

9.16 Review Questions

1. A rectangular loop of wire carries I amperes of current while suspended in a magnetic field as shown in Fig. 9.16a. The magnetic field has flux density B and is parallel to the plane of the loop and perpendicular to the vertical sides. Looking down on the loop from above, and disregarding the horizontal sections, view (b) shows current going downward on the left and upward on the right. How does the flux produced by the current affect the magnetic field below and above the wires in view (b)? In which directions will forces act on the wires, as viewed in (b)? In this view, will the loop turn clockwise or counterclockwise?

2. What is the formula that gives the amount of force on one side of the loop in Fig. 9.16? What is the formula for the torque on the loop?

3. Your *left hand* may be used, with the thumb, forefinger, and middle finger extended so they are mutually perpendicular to one another, to show the direction of force on a conductor that carries current in a magnetic field. If the forefinger points in the direction of the flux and the middle finger points in the direction of the current, the thumb will point in the direction of the force on the conductor. Try this for the force on both sides of the coil. Taking into account the fact that you must remember to use your left hand and also which way you must point your fingers, decide for yourself whether you would rather use this rule or the concept called for in Question 1 to determine the direction of the force.

4. Describe the D'Arsonval meter movement.

5. What is the characteristic of a spiral spring that makes possible a *linear* scale for a D'Arsonval type meter? (Hooke's law)

6. Explain the purpose of a shunt in an ammeter.

7. How is an ammeter connected to measure current in a line that supplies power to a motor?

8. Make a sketch showing an ammeter with an external shunt connected properly to measure current supplied to a motor. Is there much voltage drop across the meter? Why?

9. If the movement of an ammeter is to be used in a voltmeter, what changes and additions are necessary? Illustrate with a sketch.

10. Add a voltmeter, properly connected, to your sketch for Question 8.

11. More current than is needed to cause the needle to deflect flows through an ammeter that has an internal shunt. Is this also true of a voltmeter and its internal resistance?

12. Is the current taken by a voltmeter from a line usually negligible? Why?

13. How are the elements in a multirange ammeter connected? Draw a diagram.

14. How are the elements in a multirange voltmeter circuits connected? Draw a diagram.

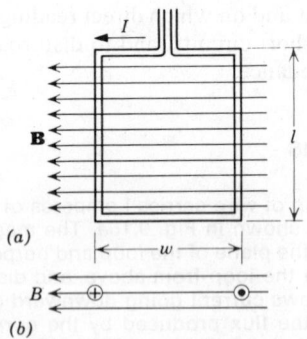

FIGURE 9.16 A loop carrying current in a magnetic field. View *b* shows the two sides as seen from above. The current goes down on the left side and up on the right.

15. Define voltage sensitivity of a voltmeter. How is it related to the voltmeter's full-scale current value?

16. What is meant by the loading effect of a voltmeter?

17. (a) What will happen if the + terminal of a voltmeter is connected to the − polarity line and the − terminal to the + polarity line of a supply source? (b) What will happen if an ammeter is connected between the two lines coming from a supply source? (c) Calculate the initial current through the ammeter of (b) using $R_m = 0.01\ \Omega$, $E = 120$ V.

18. In taking a reading of 10 V, which meter would give the better accuracy: a 0–150, 0–15, or 0–50 scale instrument?

19. What is a galvanometer? Where, on its scale, is the pointer located when its current is zero?

20. Sketch two possible circuit connections of the voltmeter when an unknown resistance is to be measured by the voltmeter-ammeter method. What determines which circuit to use when an unknown resistance is to be measured?

21. What is the current source when an ohmmeter is used?

22. How does the circuit of an ohmmeter for measuring high resistances differ from the circuit of one for measuring low resistances? Where is the zero-ohms on each meter?

23. How would you bring the pointer of an ohmmeter to the exact zero-ohms position if it were not there when you started to use the meter?

24. Explain how to measure a high-valued resistance using a voltmeter alone. Make a sketch and derive the formula for the resistance desired.

25. What is a Megger?

26. Draw the basic circuit of a Wheatstone bridge, label all parts, and write the formula by means of which the unknown resistance can be calculated.

27. How does a slide-wire bridge differ from a Wheatstone bridge? Which of the two is the more accurate?

28. What is the Varley loop used for?

9.17 Problems

Group A

26. A D'Arsonval meter movement has 40 Ω resistance and requires 2.5 mA for full-scale deflection. Explain how to use it in the design of an ammeter that reads 5 A full scale, and compute necessary values.

27. Explain how to use the meter movement of Problem 24 in a voltmeter that reads 150 V full scale.

28. A milliammeter has a 2000-Ω movement coil that deflects to the full-scale position when it carries 100 μA. What shunt resistance is required when the meter is to measure 100 mA at full scale?

29. The meter movement in Problem 26 is to be used in a multirange voltmeter *with a separate resistor* and binding post for each range. (This is not to be a tapped arrangement like that in Fig. 9.5. Such an arrangement is called for in the next problem.) Sketch the internal circuit and compute the ohms value of resistor for each of the following ranges: 0–10, 0–50, 0–150, 0–300, 0–750 V.

30. A voltmeter with the same movement and scales as those of Problem 29 is designed with rotary dial-switch operation for changing scales. Leads are brought from junctions of a set of resistors *connected in series*. Draw the circuit. Calculate the correct ohms values of the individual resistors.

31. (a) Calculate the power in each of the series resistors in the design of Problem 30. (b) Use a catalog and pick out a set of precision resistors for use in each of the five range settings. None should have smaller wattage ratings than the calculated value.

32. The moving coil in a 5 A meter has a resistance of 50 Ω. Its internal shunt resistance is 0.0005 Ω. Calculate the full-scale deflection current.

33. The insulation resistance of a piece of insulated wire was measured using a 300-V, 300,000-Ω voltmeter. A 240-V supply source was used and one reading taken with the voltmeter was 30 V. What was the resistance of the insulation?

34. The insulation resistance of the field coil of a dc motor was measured with the aid of a millivoltmeter that had an internal resistance of 50 Ω and a reading of 2.5 mV. The supply voltage was 120 V. What was the insulation resistance? *Note.* Using a millivoltmeter in this way, without making preliminary tests, is hazardous. Can you tell why? Problem 35 should be considered along with this one.

35. Suppose a motor coil like the one in Problem 34 is suspected of having a defect in its insulation. (a) Would there be a possibility of damage to a 0–150 V, 150,000-Ω voltmeter if it were connected in series with the coil insulation and 120 V applied to the series pair? (b) Suppose this voltmeter read 60 V when so connected. Calculate the insulation resistance. (c) If a 0–50 milivoltmeter having 50 Ω internal resistance were used in place of the voltmeter in part (a), what would it read? (d) What would be the result if a 30 mV, 50-Ω meter were used?

Group B

36. A Wheatstone bridge circuit like that in Fig. 9.13 has available values of 1, 10, 100, 1000 for R_1 and also for R_2. R_3 represents a known resistance value obtained from a *four-dial unit*. Each dial of R_3 is on a decade with the following possible settings *in addition to zero*: units (1–9 Ω), tens (10–90), hundreds (100–900), thousands (1000–9000). This means R_3 can be set at any value between zero and 9999 ohms in 1-Ω steps. (a) What combination of settings of R_1, R_2, and R_3 will produce a balance of the bridge when the "unknown" resistance R_x is 1234 Ω? (b) What situation would result if the ratio 10:1 were used for R_1/R_2? If it were decided that adequate balance is obtained with $R_3 = 123 \Omega$, what would be the percent error in the result? (The "unknown" resistance R_x was measured accurately by an instructor and found to be 1234 Ω.)

37. A 1-mi length of telephone line made of two conductors of 18-gage copper wire was tested with a Varley loop to locate a fault. R_1 was made equal to R_2 (Fig. 9.15) and R_3 had to be set at 2.35 Ω to produce a balance of the bridge. Calculate the distance to the fault.

38. A 150-V voltmeter has 147,950 Ω in series with a 50-Ω moving coil that requires 1 mA for full-scale deflection. With only 148 V across the meter it happens to read 150 V. The pointer goes off scale when 150 V are applied. What should be done inside the meter to reduce the error to zero when 150 V are applied? After the correction, will it read correctly when 148 V are applied?

39. A 100 mA milliammeter has a 50-Ω movement that should deflect to full scale while carrying 1 mA. When used to measure a current that is known to be exactly 91 mA, the meter pointer comes to rest at the full scale position and thus gives a reading of 100 mA. If you had a supply of resistors for use to correct the meter reading, how would you proceed and what value of resistance would you use ,

CHAPTER TEN

Inductance and Electromagnetic Induction

After studying resistance in detail, it is logical to investigate a second phenomenon that has an important effect on circuit behavior *when the current in the circuit is changing in value.* It is called **inductance**. The effect is most prominent when the conductor carrying the changing current is wound in the form of a coil, although it is present under certain conditions in a straight conductor.

When current is constant during a time interval, inductance has no effect during that interval even though the current is flowing in a coil or other kind of path that encloses an area in which flux, produced by the current, exists. But let the current value change, even during a very short time interval, and *the result will be an induced EMF* caused by the linkage of the changing flux with the conductor. The amount of this induced EMF is expressible in terms of the quantity known as *inductance.*

10.1 Faraday s Law

A coil of wire, wound on a hollow cardboard cylinder, carries current as shown in vertical projection in Fig. 10.1. Magnetic flux produced by the current passes

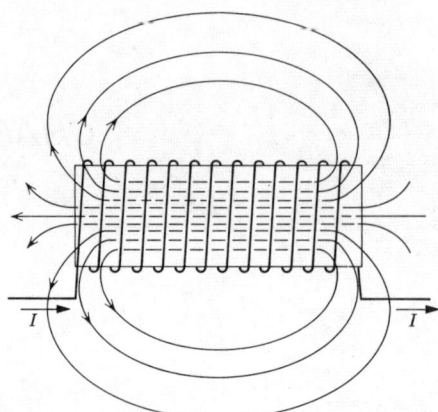

FIGURE 10.1   A coil of wire carrying a current that can increase and thus produce an increase in flux linkages, or decrease and cause a decrease in flux linkages.

toward the left through the coil and some of it links only part of the turns. None of the flux lines is actually broken (as we learned in Chapter 8); instead they continue through the air around the coil and enter at the right end of the cylinder.

Now suppose the current is increasing. The amount of magnetic flux is also increasing and the increasing flux linkages produce an induced EMF in accordance with Faraday's law, as we learned in Chapter 7. Let's say that during a small "increment" of time, Δt, the flux linkages increased an amount $\Delta(n\Phi)$. By Farady's law the induced EMF is

$$\text{EMF} = -\frac{\Delta(n\Phi)}{\Delta t} \text{ volts} \tag{10.1}$$

in which $\Delta(n\Phi)$ is a small change in flux linkages (weber-turns) that occurs in the short time interval Δt seconds. For example, if a slight change of current in a coil increases the flux linkages from 0.04 to 0.05 Wb-turns in one thousandth of a second, the EMF induced in the coil *while the current change is going on* is $-(0.05-0.04)/0.001 = -10$ V. The minus sign means that the induced EMF is *opposite in direction to the applied EMF which is responsible for the increase in current.*

Notice that we are describing action that is taking place in only one coil. The 10 V is an EMF of *self-induction*, and the coil is said to have *inductance*. By opposing the increase in *voltage* that caused the current to increase, the *EMF of self-induction can be said to oppose the increase in current* that takes place.

Let's apply logical reasoning to a case like the above but where the current in the coil is *decreasing*. In the short time interval, Δt, the change in flux linkages is negative and equal to $-\Delta(n\Phi)$. The induced EMF is then

$$\text{EMF} = -\frac{-\Delta(n\Phi)}{\Delta t} = +\frac{\Delta(n\Phi)}{\Delta t} \tag{10.2}$$

This EMF is in the same direction as the applied EMF [it has to be if $-\Delta(n\Phi)/\Delta t$ were in the opposite direction to the applied EMF] so this EMF *opposes the decrease in current*. In other words, it tries to prevent the current from decreasing.

10.2　Lenz's Law

The *opposition to a change in current* that takes place when inductance is present is the basis of a statement called *Lenz's law*.

We have seen how this phenomenon exists when current changes in a single coil. Later we shall learn a lot about what happens when flux produced by one coil links turns of a second coil located near the first coil. We can go into that situation a little bit even now and illustrate how Lenz's law applies when a second coil is involved. Look at Fig. 10.2.

If magnetic flux produced by a *changing* current, I_1, in a coil (No. 1) links turns of another coil (No. 2) that is part of a closed circuit, the EMF induced in the turns of No. 2 will produce an *induced current*, I_2, in No. 2. The direction of this induced current will be such that the flux it produces *will oppose the change in the flux that I_1 produced*.

STATEMENT OF LENZ'S LAW. An induced current will flow in such a direction that its own magnetic flux will be directed to oppose the change in flux that produced the induced current.

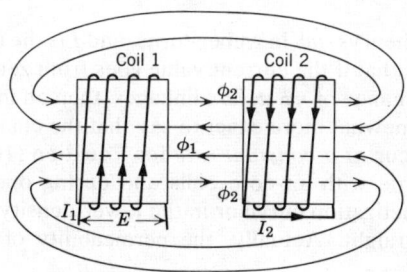

FIGURE 10.2　Increasing I_1, causes ϕ_1 to increase, thus inducing current I_2 in Coil 2. Direction of I_2 must be such that its flux ϕ_2 opposes the change that is taking place in I_1 by inducing an EMF opposite to the increase in E.

Note that Lenz's law, as just stated, specifies that a *current* is induced. The law applies whether an induced current actually flows or just *tends to flow*. If changing flux links turns of a coil that cannot carry current because it is part of an open circuit, there will still be an *induced EMF*. The *polarity* of this induced EMF can be determined easily by *assuming that it produces current* and assigning direction to the current, so that the flux it produces will oppose *the change in flux that produced the induced EMF*. This is not as complicated or confusing as it may sound. All you have to do is to imagine that an induced current *can flow* even though you know the circuit is open. Then *assign direction* to this imaginary current that will cause *its flux* to oppose *the change* in the original flux. That word *change* is very important. If the induced current is the result of *an increase* in the original flux, its own flux opposes the increase. Got it? If not, go back to the start of this section and read more slowly.

10.3 Self-Inductance

The unit of inductance is the *henry*, named for Joseph Henry (United States, 1799–1878).

DEFINITION: An element (coil, conductor, current-carrying region) in which an EMF of 1 volt is induced when current in it is changing at the rate of 1 amp per second has an inductance of 1 henry.

The symbol for henry is H

Inductance is often defined as *flux linkages per unit current*, and the equation for this is

$$L = \frac{n\phi}{I} \tag{10.3}$$

in which L is in henrys, $n\phi$ in weber-turns, and I is the current in amperes. This equation implies that if the current value goes from zero to I or from I to zero, causing a total change of $n\phi$ in flux linkages, then an inductance of L henrys is present. It is somewhat more exact to say that the changes in flux linkages and current must occur *at a constant rate* for Equation (10.3) to hold. This would be essentially true with air-core coils and during operation in the saturated region of a magnetization curve or in the lower-density region where the curve is practically straight. Actually, the permeability of the flux path must be constant.

L in the equation

$$L = \frac{\Delta(n\phi)}{\Delta i} \tag{10.4}$$

symbolizes an *incremental inductance* that is present when small changes in current (Δi) cause small changes in flux linkages.[1]

Figure 10.3 shows how the *same amount* of change in current produces widely differing amounts of change in flux linkages of an iron-core coil. Incremental inductance is the *slope of the curve* at the midpoint of the increment.

The self-inductance of a coil depends only on the number of turns, the physical dimensions of the coil, and the material of its core and its surroundings. The inductance of an air-core coil is proportional to the square of the number of its turns. When the number of turns is increased, not only is more flux produced by the same amount of current, but also there are more turns to be linked by all of the flux.

A one-layer coil of wire with a length several times its diameter is called a *solenoid*. For a solenoid with a length at least ten times its diameter, the following equation may be used to compute the inductance.

$$L = \frac{\mu_0 N^2 A}{l} \text{ henrys} \tag{10.5}$$

in which N = number of turns, A = area of solenoid in square meters, l = length of solenoid in meters.

Special equations may be used to compute the inductance of shorter solenoids, multilayer coils, and flat (pancake) coils used in radio and television

[1] In calculus form

$$L = \frac{d(n\phi)}{di} \quad \text{or} \quad L = \frac{d\lambda}{di}$$

in which $n\phi$ and λ represent flux linkages.

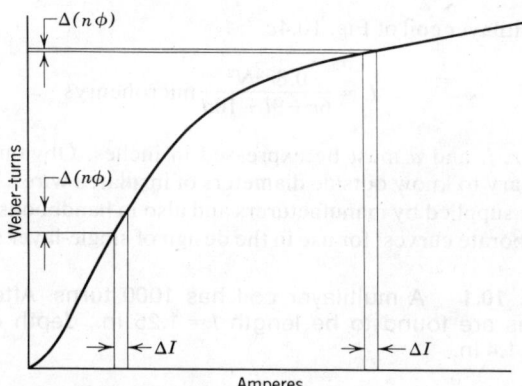

FIGURE 10.3   Incremental inductance $L = \Delta(n\phi)/\Delta I$ depends on the flux density. Equal values of ΔI are shown for unsaturated and saturated ferromagnetic paths.

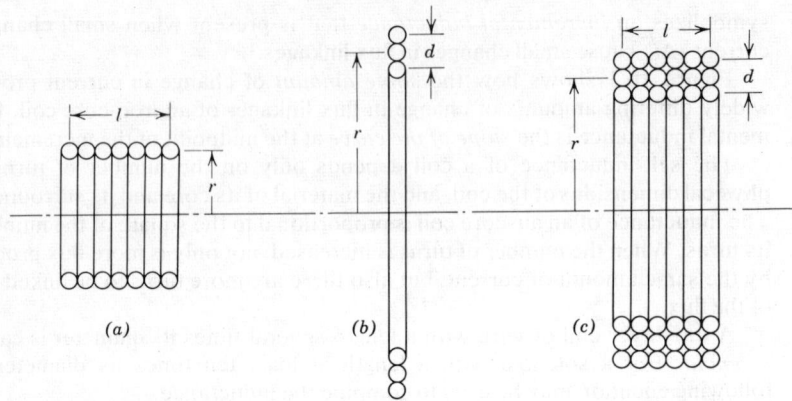

FIGURE 10.4 Forms of coils with *air cores* for which inductance can be computed. Views are sections through axes.

apparatus and illustrated in Fig. 10.4. For the short solenoid in Fig. 10.4*a*

$$L = \frac{r^2 N^2}{9r + 10l} \text{ microhenrys} \tag{10.6}$$

The dimensions are in *inches*.
For the pancake coil of Fig. 10.4*b*

$$L = \frac{r^2 N^2}{8r + 11d} \text{ microhenrys} \tag{10.7}$$

For the multilayer coil of Fig. 10.4*c*

$$L = \frac{0.8 r^2 N^2}{6r + 9l + 10d} \text{ microhenrys} \tag{10.8}$$

Note that r, l, and d must be expressed in inches. Obviously, in coil design it is necessary to know outside diameters of insulated wires. These are found in wire tables supplied by manufacturers and also in handbooks. Some handbooks furnish elaborate curves[2] for use in the design of single-layer solenoids.

EXAMPLE 10.1 A multilayer coil has 1000 turns. After it is wound its dimensions are found to be length $l = 1.25$ in., depth $d = 1.5$ in., mean radius $r = 1.4$ in.

[2]See *Reference Data for Radio Engineers*, 4th Edition, Chapter 5, International Telephone and Telegraph Corp., New York, 1956.

Solution. The computed inductance is

$$L = \frac{0.8 \times 1.4^2 \times 1000^2}{6 \times 1.4 + 9 \times 1.25 + 10 \times 1.5} = 45{,}300 \ \mu\text{H} = 45.3 \ \text{mH}$$

10.4 Counter EMF in Terms of L

Equation (10.4), which is a mathematical definition of self-inductance, is repeated here:

$$L = \frac{\Delta(n\Phi)}{\Delta i} \qquad (10.9)$$

and Faraday's law is recalled:

$$\text{EMF} = \frac{-\Delta(n\Phi)}{\Delta t} \qquad (10.10)$$

The minus sign signifies that this is an opposing EMF, which we can call a counter-EMF of self-induction in cases where $\Delta(n\Phi)$ is produced by changes in current in a single coil.
Solving (10.9) for $\Delta(n\Phi)$.

$$\Delta(n\Phi) = L\Delta i$$

Putting this into (10.1),

$$\text{EMF} = -L\frac{\Delta i}{\Delta t} \qquad (10.11)$$

Using usual calculus notation,

$$e_L = -L\frac{di}{dt} \ \text{volts}. \qquad (10.12)$$

where e_L is the EMF of self induction of a coil whose inductance is L henrys and whose current is changing at the rate di/dt because it represents the voltage across a pure inductance in a dc circuit in which the current is changing in value and also in an ac circuit where current is continually changing in instantaneous value.
Equation (10.12) is Lenz's law in mathematical form

Problems

1. A coil has 250 mH self-inductance and negligible resistance. What is the voltage drop across it at the instant its current is changing at the rate of 20 A/s?
2. The current in a coil that has negligible resistance is changing at the rate of 1 A/s and the voltage drop across it is 1 V. What is its inductance?

3. At what rate is current changing in a 1.5-H coil if the EMF of self-induction is 120 V?

4. Magnetic flux linking a coil is changing at a constant rate through an interval of 0.5 s. The total change in flux linkages is 150 Wb turns. (a) How much EMF is induced? (b) How much is induced at the instant that is half-way through the 0.5 s time interval? (c) How much at the instant 90 percent of the time interval has elapsed?

5. Magnetic flux linking a 1-turn coil is changing with time. At what rate is the flux changing when the induced EMF is $+10$ V? When the induced EMF is -10 V?

6. A current of 2 A flows in a coil of 100 turns. Assume that all the flux links all the turns. Calculate the amount, and specify the direction, of the induced voltage in each of the following cases. Let "forward direction" mean the direction in which the current is flowing.
   (a) The flux linkages change from 10^{-5} Wb-turns to 5×10^{-4} Wb-turns in 10^{-2} s. (*Note*. This is an increase in flux linkages.)
   (b) The flux linkages change from 10^{-5} Wb-turns to 10^{-6} Wb-turns in 5×10^{-3} s.

7. An air-core solenoid has a single layer of wire wrapped closely on a plastic tube. The length of the coil is 2 in., its diameter is 1.5 in., and it has 500 turns. What is its inductance in microhenrys?

8. An air-coil has the form shown in Fig. 10.4c. The inside diameter is 2 in., the length is 2 in., and there are 9 layers of the 288-turn winding with an average thickness of $\frac{1}{16}$ in. per layer. Calculate the inductance in microhenrys. Then express it in millihenrys and in henrys.

10.5 Energy in a Magnetic Field

When a current I is flowing in an inductance L, which has negligible resistance, voltage has been involved along with the current over a period of time. All of the energy has been stored in the magnetic field, since there is no resistance in which energy can be dissipated. The current takes time to build up from zero to any value I, and in the process the field builds up from zero to its value corresponding to the current I.

At any instant the product of the instantaneous voltage across the inductance and the instantaneous current is the instantaneous power:

$$p = e_L i \text{ joules per second (or watts)}$$

During a short time interval Δt, the energy ΔW put into the magnetic field is equal to the power times the time:

$$\Delta W = e_L i \, \Delta t \text{ joules} \tag{10.13}$$

The time interval Δt is so short that the *average value* of the current during that time is represented by i in this equation. Now, $e_L = L \, \Delta i / \Delta t$ is the voltage required across L to maintain the current. Substituting this in Equation (10.13)

$$\Delta W = L\frac{\Delta i}{\Delta t}i\Delta t = Li\Delta i \text{ joules} \tag{10.14}$$

This is the increase in field energy produced by Δi, the *incremental increase* in current.

It is customary to represent the adding up, or *summation*, of all the incremental contributions to a final value, such as will result in a case of this kind, by the following symbols:

$$W = \sum_{i=0}^{i=I} Li\Delta i \text{ joules} \tag{10.15}$$

This means that all contributions of energy due to an *extremely large number of increments of current*, each represented by Δi, are added together from $i = 0$ to $i = I$, the value at which the energy is to be calculated. The final result of this addition is

$$W_L = \tfrac{1}{2}LI^2 \text{ joules} \tag{10.16}$$

in the magnetic field of L. We now follow this method of deriving Equation (10.16), with a simple summation process in the calculus called *integration*.[3]

When an instantaneous current, i *amperes*, exists in a coil with inductance L henrys and no resistance is present, the instantaneous power is $e_L i$ watts when the units are volts and amperes. During a time interval dt, the energy put into the field is

$$dW = e_L i\,dt \text{ joules}$$

Now, from Equation (10.12) (the negative sign has no significance here),

$$e_L = L\frac{di}{dt} \text{ volts}$$

so that

$$dW = \left(L\frac{di}{dt}\right)i\,dt = Li\,di \text{ joules}$$

The total energy put into the field while i is increasing from zero to I is

$$W = \int_0^I Li\,di = \tfrac{1}{2}Li^2 \Big|_0^I = \tfrac{1}{2}LI^2 \text{ joules}$$

EXAMPLE 10.2   Four field coils of a motor with 10H of inductance each are in series, and carry 2 A of current. How much energy is stored in the magnetic field of the motor?

[3]If the simple calculus that follows bothers you, ignore this proof and take Equation (10.16) as a fact. We authors like to show off once in a while.

Solution

$$W = \tfrac{1}{2} \times 10 \times 4 \times 2^2 = 80 \, \text{J}$$

10.6 Mutual Inductance

In Section 10.2 we discussed the situation when we had a second coil near, but independent of, a first coil in which current is changing. *Mutual inductance* is said to exist between the two coils when some magnetic flux of one coil can link with turns of the other coil. We shall now examine this phenomenon in some detail.

Figure 10.5 shows this situation. The switch has just been closed and current i_1 is increasing in Coil 1. Voltage induced in Coil 2 is causing i_2 to flow, and you will notice that flux ϕ_{21} produced by i_2 is in opposition to ϕ_{11} produced by i_1. The remainder ϕ_{22} of the flux produced by i_2 is that which links only the turns of Coil 2.

The flux responsible for the induction of voltage in Coil 2 is ϕ_{12}, that part of ϕ_{11} which reaches over to Coil 2. The mutual inductance from Coil 1 to Coil 2 is M_{12}, expressed as flux linkages in Coil 2 per unit current in Coil 1:

$$M_{12} = \frac{N_2 \phi_{12}}{i_1} \text{ henrys} \tag{10.17}$$

The mutual inductance from Coil 2 to Coil 1 is M_{21}, expressed as flux linkages in Coil 1 per unit current in Coil 2:

$$M_{21} = \frac{N_1 \phi_{21}}{i_2} \tag{10.18}$$

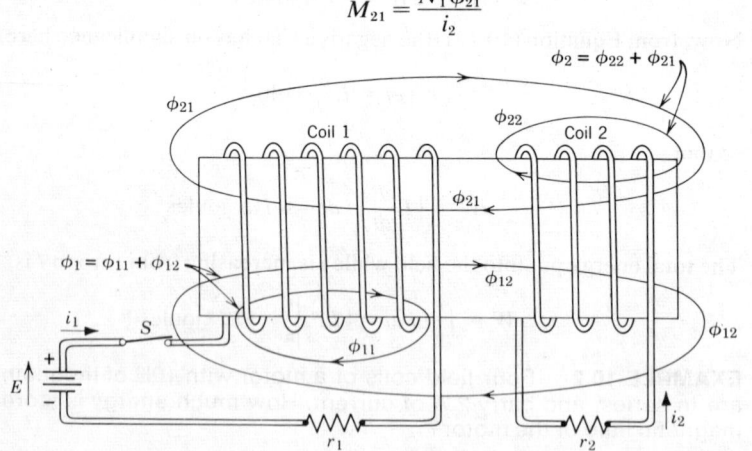

FIGURE 10.5 Magnetically coupled coils. The switch has just been closed. Primary current i_1 and secondary current i_2 are increasing and their fluxes oppose each other.

Let's recall that the self-inductances of the coil are

$$L_1 = \frac{N_1 \phi_1}{i_1} \text{ henrys}, \qquad L_2 = \frac{N_2 \phi_2}{i_2} \text{ henrys}$$

First, consider an ideal case in which all the flux produced by each current links all the turns of both coils. Then $\phi_{12} = \phi_1$ and $\phi_{21} = \phi_2$. We may then correctly write

$$L_1 L_2 = M_{21} M_{12} = M^2 \tag{10.19}$$

and

$$M = \sqrt{L_1 L_2} \text{ henrys} \tag{10.20}$$

which is the *theoretical maximum possible* mutual inductance.

There is always some *leakage flux* produced by each current that does not link any turns of the other coil. These are ϕ_{11} and ϕ_{22} in Fig. 10.5. Nevertheless, it is always true that $M_{12} = M_{21}$ if the magnetic permeability of the mutual flux path remains constant.

The ratio of the actual mutual inductance between two coils to the theoretical maximum is called the *coefficient of coupling*, and its symbol is k:

$$k = \frac{M}{\sqrt{L_1 L_2}}$$

The *coefficient of coupling* may also be defined in terms of the self- and mutual fluxes:

$$k = \sqrt{\frac{\phi_{12}}{\phi_1} \cdot \frac{\phi_{21}}{\phi_2}} \tag{10.22}$$

To prove this, we obtain expressions for these fluxes in terms of M and N from the inductance formulas:

$$\phi_{12} = \frac{M_{12} i_1}{N_2}; \quad \phi_1 = \frac{L_1 i_1}{N_1}; \quad \phi_{21} = \frac{M_2 i_2}{N_1}; \quad \phi_2 = \frac{L_2 i_2}{N_2}$$

and substitute them into Equation (9.20),

$$k = \sqrt{\frac{(M_{12} i_1 / N_2)}{(L_1 i_1 / N_1)} \cdot \frac{(M_{21} i_2 / N_1)}{(L_2 i_2 / N_2)}} = \sqrt{\frac{M_{12}}{L_1} \cdot \frac{M_{21}}{L_2}}$$

$$k = \sqrt{\frac{M^2}{L_1 L_2}} = \frac{M}{\sqrt{L_1 L_2}} \quad \text{since } M_{12} = M_{21}{}^4$$

---

[4]This is proved on pp. 219 and 220 of *Electrical Fundamentals and Circuit Analysis*, H. Alex Romanowitz, Wiley, New York, 1966.

EXAMPLE 10.3   Two coils for which $L_1 = 100$ mH, $L_2 = 400$ mH are near each other. What would be the maximum possible amount of mutual inductance?

Solution

$$\text{Maximum } M = \sqrt{100 \times 10^{-3} \times 400 \times 10^{-3}} = 200 \times 10^{-3} \text{ H}$$

EXAMPLE 10.4   If the actual value of the coefficient of coupling in Example 10.3 is 0.8, what is M?

Solution

$$M = 0.8 \times 200 \times 10^{-3}$$
$$= 160 \times 10^{-3} \text{ H}$$

10.7   Inductors in Series

Figure 10.6a represents two inductors (inductance coils) in series with the axis of one coil perpendicular to the axis of the other. In this case there is no mutual inductance: $M = 0$. The dots at the ends of the coil symbols (called polarity marks) have a special meaning, which can be explained with reference to Fig. 10.6b. If the coils are connected in series, represented as L_1 and L_2 in Fig. 10.6b, so that when the current enters the dotted end of L_1 and leaves, it must enter L_2 at its dotted end, the fluxes of the two coils will add. If the connections to L_2 are reversed so that the current must enter its undotted end, their fluxes will subtract (i.e., oppose each other). We are assuming that the axes of the inductance coils are on the same straight line.

The total inductance of a series pair that are connected such that their fluxes aid each other, or add to each other, is given by

$$L_T{}^+ = L_1 + L_2 + 2M \tag{10.23}$$

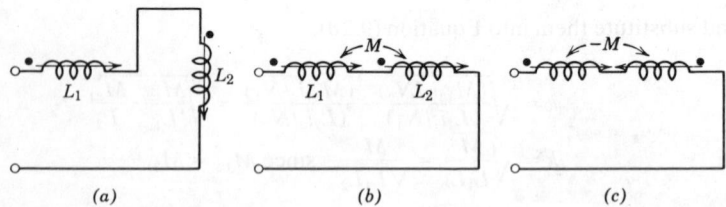

(a)   *(b)*   *(c)*

FIGURE 10.6   Coils connected in series. (a) No mutual inductance because the net effect of mutual flux linkages is zero. (b) Mutual inductance present, fluxes add. (c) Mutual inductance present, fluxes are in opposition.

The third term is $2M$ because each coil contributes the same amount to the total of flux linkages per unit current. When the fluxes oppose each other, the total inductance is

$$L_T^- = L_1 + L_2 - 2M \qquad (10.24)$$

Subtracting L_T^- from L_T^+,

$$L_T^+ - L_T^- = 4M$$

$$M = \frac{L_T^+ - L_T^-}{4} \text{ henrys} \qquad (10.25)$$

We must further discuss the situations in Fig. 10.6b and c. In the case where the coil fluxes add, we represent the inductance related to each coil separately in this mutual inductance situation and add them together to get Equation (10.23):

$$L_T^+ = (L_1 + M) + (L_2 + M) \qquad (10.26)$$

and in the case of opposing fluxes (c)

$$L_T^- = (L_1 - M) + (L_2 - M) \qquad (10.27)$$

This way of grouping together all henry values involving the same coil is necessary when we have a more complicated mutual inductance situation as in Fig. 10.7.

EXAMPLE 10.7   The fluxes of A and B in Fig. 10.7 are in opposition so we say that they *subtract*. Their mutual inductance $(-M_{AB})$ is therefore negative. Fluxes of A and C add, and fluxes of B and C subtract. The three coils are in series because they all carry the same current. The mutual inductance M_{BC} is negative and M_{AC} is positive. The total inductance is given by

Solution

$$\begin{aligned}
L_T &= (L_A - M_{AB} + M_{AC}) + (L_B - M_{BC} - M_{BA}) + (L_C + M_{CA} - M_{CB}) \\
&= (10 - 6 + 5) + (20 - 8 - 6) + (15 + 5 - 8) \\
&= 9 + 6 + 12 = 27 \text{ H}
\end{aligned}$$

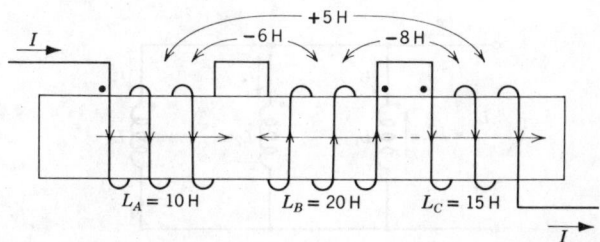

FIGURE 10.7   Three inductances, A, B, and C in series with mutual inductance between every pair.

10.8 Inductors in Parallel

When two or more self-inductances are connected in parallel and there is no mutual inductance between them, their henry values are related in reciprocal fashion in the same manner as parallel resistances are related. Figure 10.8 shows three such inductors.

The total current, i_T, is the sum of the branch currents, no matter how many branches are in parallel:

$$i_T = i_1 + i_2 + i_3 + \cdots$$

Representing their rates of change with respect to time, we have

$$\frac{di_T}{dt} = \frac{di_1}{dt} + \frac{di_2}{dt} + \frac{di_3}{dt} + \cdots$$

Remember that voltages across parallel branches are all equal, and that $e = L(di/dt)$ across each branch and across the total network. From this

$$\frac{di}{dt} = \frac{e}{L}$$

and we have

$$\frac{e}{L_T} = \frac{e}{L_1} + \frac{e}{L_2} + \frac{e}{L_3} + \cdots$$

from which, for parallel inductances

$$\frac{1}{L_T} = \frac{1}{L_1} + \frac{1}{L_2} + \frac{1}{L_3} + \cdots \tag{10.28}$$

Based on this analysis, *we may reflect back to the series connection of inductances* and conclude that *their henry values must be added* because their EMFs of self-induction (terminal to terminal) are in series and must therefore be added and their sum must be equal to the EMF across the terminals of the whole series circuit.

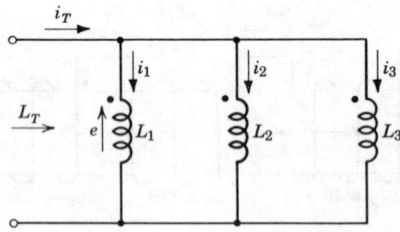

FIGURE 10.8 Inductances in parallel.

If only two inductors are in parallel, and no mutual inductance is present, their equivalent inductance is given by their product divided by their sum, as in the case of two parallel resistances. Thus, if only L_1 and L_2 in Fig. 10.8 are present,

$$L_T = \frac{L_1 L_2}{L_1 + L_2} \tag{10.29}$$

Caution. Neither Equation (10.26) nor Equation (10.27) can be used when mutual inductance is present. The following explanation applies to such a case.

PARALLEL INDUCTORS WITH MUTUAL INDUCTANCE

EXAMPLE 10.6 When mutual inductance is present in a parallel-connected group of inductances, one must know whether each pair of mutual fluxes add to each other or subtract. The polarity marks on the coils of Fig. 10.9 reveal this information. The mutual inductance between L_1 and L_2 is negative because their dotted ends are not of the same polarity. If terminal *a* is positive, terminal *b* is negative and the dotted end of L_A is positive and the dotted end of L_B is negative. In order for their fluxes to add, the dotted end of L_2 would have to be connected to the line from terminal *a*.

Solution. We may say, then:

$$M_{AB} = -4H, \ M_{AC} = 2H, \ M_{BC} = -3H$$

In order to use Equation (10.26) we need to obtain values for L_1, L_2, and L_3 which include the appropriate mutual inductances. Using given values,

$$L_1 = 10 - 4 + 2 = 8 \text{ H}$$
$$L_2 = 20 - 4 - 3 = 13 \text{ H}$$
$$L_3 = 15 - 3 + 2 = 14 \text{ H}$$

From Equation (10.28),

$$\frac{1}{L_T} = \frac{1}{8} + \frac{1}{13} + \frac{1}{14} = 0.273$$
$$L_T = 3.66 \text{ H}$$

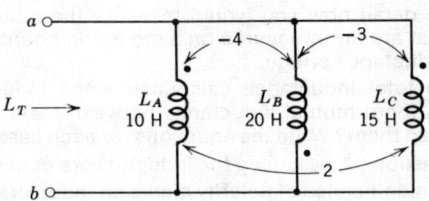

FIGURE 10.9 Inductances in parallel. Fluxes of L_A and L_C add, fluxes of L_A and L_B are in opposition. So are fluxes of L_B and L_C.

10.9   Review Questions

1. A coil of wire carries a steady current which establishes a magnetic field of 100 μWb, which is constant. Is an EMF induced? Why?

2. State Faraday's law of electromagnetic induction.

3. Two amperes flow in a 100-turn coil. If the current increases so rapidly that the flux linkages increase at the rate of 5 Wb-turns/s, what else is happening in the coil? Does this effect aid or oppose the increase in the current?

4. Answer Question 3 after assuming the current *decreases* at the rate indicated. State the effect on the decreasing current.

5. How much EMF is induced in the case of Questions 3 and 4?

6. Does the induced EMF in the above examples produce an induced current? Why?

7. Two coils are close together but their wires are not connected to each other. Coil 1 has 250 turns; Coil 2 has 100 turns. Part of the flux of Coil 1 links turns of Coil 2 and vice-versa. As I_1 increases, current I_2 in Coil 2 begins to flow. Why? What relation exists between the flux of I_2 and the flux of I_1?

8. In Question 7, what is the effect on I_1 of the EMF induced in Coil 1 by the flux produced by I_2? What law applied here?

9. State Lenz's law.

10. Define self-inductance.

11. How is EMF of self-induction related to self-inductance?

12. What is incremental inductance?

13. Compare the incremental inductance of an air-core coil with that of a coil of a closed iron ring. How does saturation affect the incremental inductance of the coil on the ring?

14. How does the self-inductance of an air-core coil vary with the number of turns?

15. When do two coils have mutual inductance?

16. How would you arrange two coils, that must be near each other, so they will not have mutual inductance?

17. Define mutual inductance between Coil 1 and Coil 2.

18. What do M_{12} and M_{21} mean, and how do they compare in value?

19. Define coefficient of coupling between two coils.

20. Two coils have self-inductances L_1 and L_2. How could you calculate their maximum possible mutual inductance? Is this obtainable? Why?

21. Describe in detail how you would measure the mutual inductance between two coils that are firmly secured on a mounting board. Assume you know how to use an inductance bridge.

22. How is the total inductance calculated when inductors are connected in series (a) without mutual inductance between them, (b) with mutual inductance between them? Write the equations for each case.

23. Answer Question 22, assuming the inductors are connected in parallel.

24. Explain the significance of polarity marks on inductors.

25. The magnetic circuit of an electric motor is supplied with energy in order to establish a strong magnetic field. With a given amount of current, the strength

of the field is increased when the number of turns on the field poles is increased. How is the energy in the field related to the inductance of the series-connected field coils and the current they carry?

10.10 Problems

Group A

9. Two air-core coils have inductances $L_1 = 100$ mH, $L_2 = 250$ mH. What is the theoretical maximum possible mutual inductance that can be obtained with these coils?

10. The mutual inductance between two coils in a radio receiver is 100 mH. One coil has 100 mH of self-inductance. What is the self-inductance of the other if $k = 0.5$?

11. Two coils for which $L_1 = 387.2$ mH, $L_2 = 125$ mH, have $M = 110$ mH. What is the coefficient of coupling?

12. The self-inductances of two coils are $L_1 = 150$ mH, $L_2 = 250$ mH. When they are connected in series with their fluxes adding, their total inductance is 620 mH. When the connections to one of the coils is reversed (they are still in series), their total inductance is 180 mH. How much mutual inductance exists between them?

13. What is the coefficient of coupling between the coils of Problem 12 when they are connected so that their fluxes (a) add, (b) subtract?

14. A field coil on a dc motor has $L = 15$ H and $I = 1.5$ A. (a) How much energy is stored in the magnetic fields of four of these field coils when connected in series? (b) How far above the earth would an object weighing 25 lb (11.34 kg) have to be lifted so it would have the same amount of potential energy?

15. An air-core coil has 780 turns and an inductance of 460 μH. What will be its inductance if 180 turns are removed?

Group B

16. What is the total inductance of the circuit in Fig. 10.10 if $L_1 = 5$ H, $L_2 = 8$ H, $k = 0.45$?

17. What would be the total inductance of the circuit in Fig. 10.10 if the connection to L_2 were reversed?

18. What is the total inductance of the circuit in Fig. 10.11 if $L_1 = 10$ H, $L_2 = 15$ H, $L_3 = 12$ H, $M_{12} = 5$ H, $M_{23} = 3$ H, $M_{13} = 1$ H?

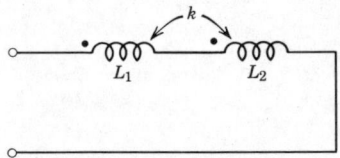

FIGURE 10.10   For Problems 16 and 17.

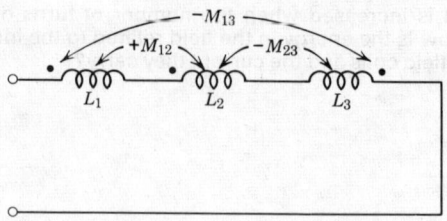

FIGURE 10.11   For Problem 18.

19. An air-core coil has 250 turns and an inductance of 1 mH. Current passing through it has a maximum instantaneous value of 10 mA. What is the maximum instantaneous value of the flux produced?

20. Seventy percent of the flux in Problem 19 links all 200 turns of a second coil. Calculate the mutual inductance between the two coils.

21. What is the coefficient of coupling between the two coils of Problem 20 if the self-inductance of Coil 2 is 555 μH?

CHAPTER ELEVEN

Inductance and Resistance in a Circuit

We obtained much practical and useful information from Chapter 10 about inductance and how an inductor behaves while its current is changing. But we did not take into account the fact that since a coil of wire must have resistance, there must be an IR drop present as well as an induced EMF (voltage) of self-induction.

In this chapter we shall confine our study of how an inductance coil that has resistance behaves while carrying a direct current that varies exponentially with time. A good example is the inductance coil of an automobile ignition system. A changing direct current from the battery and generator produces a changing flux in one coil and this induces a high voltage in a second coil which is connected by the distributor to the spark plugs in rapid succession.

Later, when we begin a study of alternating-current theory, we shall learn a lot more about the behavior of a coil that has resistance as well as inductance.

11.1 Rise of Current in a Coil

The inductance and resistance of a coil of wire are represented separately, and in series, in Fig. 11.1a. If we start counting time at the instant the switch is closed, instantaneous values of current will be represented by points on the

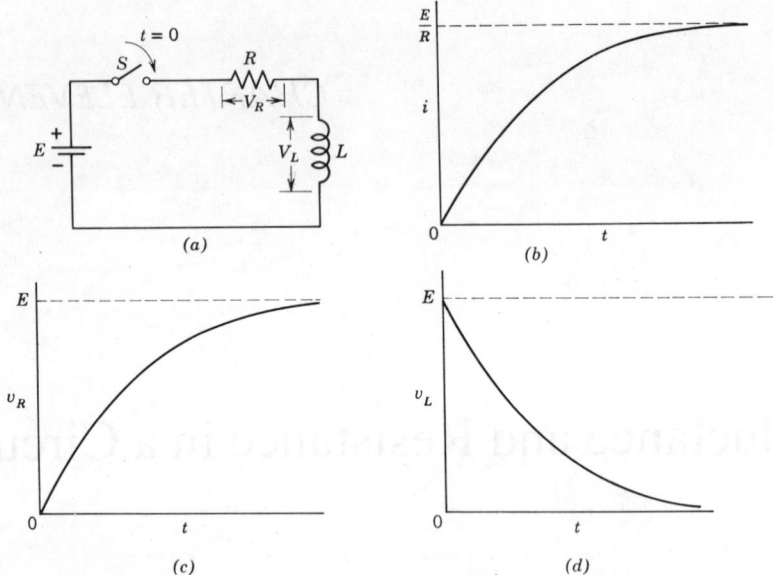

FIGURE 11.1   Constant voltage applied to R and L in series. (a) Circuit. (b) Instantaneous current versus time. (c) Voltage drop across R (iR) versus time. (d) Voltage drop across L (Ldi/dt) versus time.

curve in (*b*) and instantaneous value of voltage across the resistance in the coil by points on the curve in (*c*). The reason the current does not build up instantly to its final fixed value is that the *counter EMF of self induction is opposite in polarity to the applied voltage* and v_R, *which is the difference* between E and the counter EMF, is all that is left to send current through the coil. This voltage must be equal to iR because the inductive effect neutralizes the remainder of the total voltage.

We should pause here and get a few basic ideas firmly in mind. Before the switch is closed the current is zero. The current is zero at the instant the switch blade makes contact because it must start from zero. *It is impossible to change the amount of current in an inductance from one value to another in zero time.*[1]

[1] In an inductance, the voltage across its terminals is

$$v_L = -L\frac{di}{dt}, \quad \frac{di}{dt} = \frac{-v_L}{L}$$

If the current could change from i_1 to i_2 *in zero time* the rate of change of current with respect to time (di/dt) would be *infinite*. L is real, so v_L would have to be infinite. This is impossible. So *it is impossible for the current in an inductance to change from one value to another in zero time.*

As the current increases in value in Fig. 11.1*b* the *voltage* v_R *across the resistance increases* and this reduces *the rate at which the current increases.* We can easily prove this, and thus show why the current-versus-time curve becomes *less and less steep.* Since $E = v_L + v_R$ and, as we learned in Chapter 10, $v_L = L(di/dt)$, in which di/dt is the *rate of change* of current,

$$E = L\frac{di}{dt} + v_R \tag{11.1}$$

$L(di/dt)$ must be *decreasing* if v_R is *increasing* because E is constant. From $v_L = L(di/dt)$ we get

$$\frac{di}{dt} = \frac{1}{L} \times v_L \tag{11.2}$$

Since $1/L$ is a constant, di/dt must be decreasing as v_L decreases. Therefore the current is increasing at a slower rate as time goes on, and the current curve flattens out. The final value of the current is

$$I_f = \frac{E}{R} \tag{11.3}$$

That is, the current must be limited by the resistance only, since di/dt becomes zero and so does the counter EMF, which is given by $L(di/dt)$.

In this circuit the initial current (at $t = 0$) is zero:

$$I_0 = 0 \tag{11.4}$$

and, at $t = 0$,

$$v_L = L\frac{di}{dt} = E \tag{11.5}$$

Perhaps we should pause here and anticipate your logical question. "I thought the EMF of self-induction was *minus* $L(di/dt)$. What happened to the minus sign?" We'll have to admit that you are right, but we can get ourselves out of this apparent contradiction by means of a simple analogy. Suppose we imagine two batteries connected in parallel instead of only the one as a source in Fig. 11.1, and we say its positive terminal is connected to the hinge post at the left side of the switch and its negative terminal is connected to the negative terminal of the existing battery. Call the voltage of this second battery E also. The two terminal voltages oppose each other insofar as producing any circulating current through themselves alone is concerned. Summing voltages around the batteries only, $E + (-E) = 0$. The second E is considered negative only when we write a Kirchhoff voltage equation around *a closed circuit which goes through* the second battery. Otherwise, the voltage of the second battery is just as positive as the voltage of the existing battery. Going back to v_L in the circuit, its polarity is $+$ on the top and $-$ on the bottom *while the current is increasing.* The curve of Fig. 11.1*d* shows v_L starting out equal to E at $t = 0$ and falling off to zero. v_L does oppose the change (increase) in current and the minus sign indicates this when we add the voltages around the closed circuit.

$$E - L\frac{di}{dt} - Ri = 0$$

which is Equation (11.1). Let's now change v_R to Ri in Equation (11.1) and rewrite it:

$$E = Ri + L\frac{di}{dt} \tag{11.6}$$

The expression for i that satisfies this equation is[2]

$$i = \frac{E}{R}(1 - \epsilon^{Rt/L}) \tag{11.7}$$

In this equation t is in seconds, and its value is zero at the instant the switch makes contact. This equation gives us the value of i at that instant. We called this I_0, and we then can say

$$I_0 = \frac{E}{R}(1 - \epsilon^0)$$

Since any number raised to the zero power $= 1$, we get

$$I_0 = \frac{E}{R}(1 - 1) = 0$$

and this verifies Equation (11.4).

EXAMPLE 11.1   A coil that has $L = 1$ H and $R = 5\,\Omega$ is connected to a constant 10-V source at $t = 0$. Analyze the performance at $t = 0$ to $t = \infty$.

Solution

$$i = \frac{E}{R}(1 - \epsilon^{-Rt/L})$$

for all values of t.
At $t = 0$:

$$i = \frac{10}{5}(1 - 1) = 0$$

The initial rate of change of current is, from Equation (11.6),

$$\left.\frac{di}{dt}\right|_{t=0} = \frac{10}{1} = 10 \text{ A/s}$$

[2]This equation is solved for i in the Appendix, Section A.2.

The final value of current is

$$I_f = \frac{10}{5}(1 - \epsilon^{-\infty}) = \frac{10}{5} = 2 \text{ A}$$

At $t = 0$, $v_R = 0 \times 5 = 0$, $v_L = 10 - 0 = 10$ V
At $t = \infty$, $v_R = 2 \times 5 = 10$ V, $v_L = 10 - 10 = 0$ V

Now let's get expressions for v_R and v_L that will hold as time goes on and also at $t = 0$.

Since $v_R = iR$, we can change Equation (11.7) to

$$v_R = E(1 - \epsilon^{-Rt/L}) \tag{11.8}$$

From this we see that $v_R = E(1-1) = 0$ at $t = 0$. This must be true since $I_0 = 0$ and, at $t = 0$, $v_R = RI_0$.

Since

$$v_L = E - v_R$$
$$v_L = E - E(1 - \epsilon^{-Rt/L}) \tag{11.9}$$
$$v_L = E\epsilon^{-Rt/L}$$

And it is evident from this that, at $t = 0$, $v_L = E$.

The next value of t that we must consider is that which will give us the amounts of I, V_R, and V_L which we call the final, or steady-state, values. In practice we need only wait from several seconds to even a small fraction of a second (in many cases) for the current to reach its final, steady value. But we let $t = \infty$ (infinity) in our equations, instead of some very large value, and we find it convenient that

$$\epsilon^{-\infty} = \frac{1}{\epsilon^{\infty}} = 0$$

The final value of current, I_f, is then found using Equation (11.7),

$$I_f = \frac{E}{R}(1 - 0) = \frac{E}{R} \tag{11.10}$$

We predicted this [Equation (11.3)] when we were discussing how the current curve flattens out as time goes on — note the curve in Fig. 11.1c. From Equation (11.8) we get the final value of the voltage across the resistance part of the circuit:

$$\text{Final } V_R = E(1 - 0) = E \tag{11.11}$$

and, from Equation (11.9),

$$\text{Final } V_L = E\epsilon^{-\infty} = E/\epsilon^{\infty} = 0 \tag{11.12}$$

Note the curve in Fig. 11.1d.

Problems

1. A coil which has $L = 500$ mH, $R = 1\,\Omega$ is suddenly connected to a constant-voltage source of 3 V. What are the initial and final values of current?
2. At what rate is the current in the coil of Problem 1 changing at (a) $t = 0$, (b) $t = \infty$?
3. What are the values of the voltage drop due to resistance at $t = 0$ and at $t = \infty$?
4. What are the instantaneous EMFs of self-induction at $t = 0$ and at $t = \infty$?
5. What is the instantaneous current at $t = 500$ ms?
6. What are the values of v_R and v_L at $t = 500$ ms?

11.2 Time Constant

The circuit in Fig. 11.2a contains a coil that has inductance L and resistance R, both values of which are known. Therefore the value of L/R is known. In the term $\epsilon^{-Rt/L}$, the exponent becomes -1 when $L/R = t$. This is easy to see, if we agree to let time go on from $t = 0$ to $t = L/R$ while the current is increasing. Where $t = L/R$, $R/L = 1/t$ and the term becomes $\epsilon^{-t/t} = \epsilon^{-1}$. Note that in the time interval $t = L/R$, the current will be

$$i = \frac{E}{R}(1 - \epsilon^{-1}) = \frac{E}{R}(1 - 0.367) = 0.633\frac{E}{R} \qquad (11.13)$$

The steady-state value of the current, which is its final and maximum value in this case, *is given by* E/R. Equation (11.12) says that at the end of the time

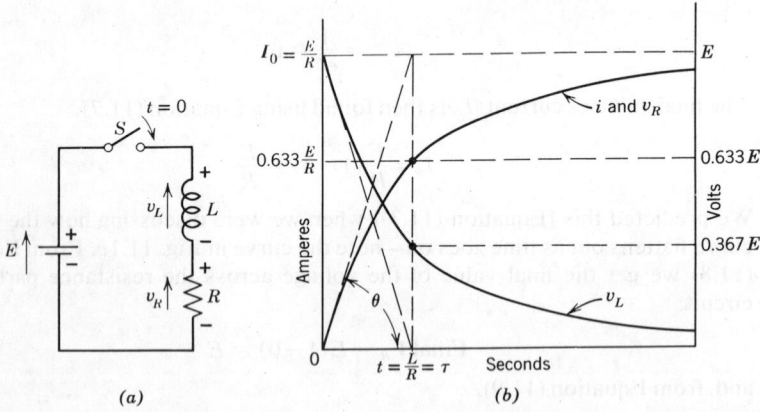

(a)   (b)

FIGURE 11.2 Constant voltage applied to an inductance of L henrys in series with R ohms. (a) Circuit schematic. (b) Curves of instantaneous voltage and current. Voltage scale chosen so that curves of i and V_R coincide. Tan $\theta = E/L$.

interval L/R the current has reached 63.3 percent of its maximum value. A little thought reveals to us that we could apply a constant voltage E to *any coil* we may have, or procure, and not matter what its values of L or R happen to be, the current will rise to 63.3 percent of its final value in L/R seconds. Since this holds for any pair of L and R in series, the time interval L/R has been given the special name *time constant*. The Greek letter *tau* (τ) is the symbol for time constant in this book.

Let's check the units involved and see if L/R is measured in seconds.

$$L \rightarrow \frac{\text{webers}}{\text{amperes}} = \frac{\text{volts} \times \text{seconds}}{\text{amperes}}$$

$$R \rightarrow \frac{\text{volts}}{\text{amperes}}$$

$$\frac{L}{R} \rightarrow \frac{\text{volts} \times \text{seconds}}{\text{amperes}} \times \frac{\text{amperes}}{\text{volts}} = \text{seconds} \quad ,$$

DEFINITION: *The time constant of a series R,L circuit is the time required for the changing current to accomplish 63.3 percent of the total change it will undergo. It is given by* $\tau = L/R$ *seconds. L is in henrys, R is in ohms.*

Based on our limited experience so far, we could have defined time constant as the time required for the current to reach 63.3 percent of its final value. But this will not hold in many cases, as we shall see, because we shall encounter cases where the changing current is decreasing instead of increasing, and 63.3 percent of its final value would be zero if its final value is zero. However, even in this case the time constant is the length of time required for the current *to go through 63.3 percent of the total change it will undergo.* If the instantaneous value of a *decreasing current* is dependent on the exponential factor $\epsilon^{-t/\tau}$ it will drop from the initial value (at $t = 0$) to $(100 - 63.3)$ percent = 36.7 percent of the initial value in τ seconds. Thus in τ seconds it will go through 63.3 percent of the total change it will undergo.

Observe in Fig. 11.2b that the dash line drawn tangent to the current curve at the origin (when $t = 0$) intersects the horizontal dash line, representing the final value of current, vertically above the $t = L/R$ point on the the seconds scale. This dash line makes an angle θ with the time axis and its slope (amperes/second) is the *initial rate of change of the current with respect to time.* Of course, the rate of change decreases after $t = 0$. But, if the rate of change could continue at its value at $t = 0$, the current would reach the maximum value in L/R seconds, which is the time constant interval. This *initial rate of change* of current can easily be shown to be equal to E/L amperes per second. Since the initial current is zero, let $i = 0$ in Equation (11.6),

$$E = 0 + L \frac{di}{dt}, \quad \text{at } t = 0$$

Therefore

$$\left.\frac{di}{dt}\right|_{t=0} = \frac{E}{L} \text{ amperes per second} \tag{11.14}$$

EXAMPLE 11.2 (a) What is the time constant of the coil in Example 11.1? (b) What is the current value in the coil of Example 11.1 at the end of the time-constant interval?

Solution

$$\text{(a) } \tau = \frac{L}{R} = \tfrac{1}{5} \text{ s}$$

$$\text{(b) } i \text{ at } t = \tau = \frac{E}{R}(1 - \epsilon^{R\tau/L})$$

$$i = 2(1 - \epsilon^{-(RL/LR)}) = 2(1 - \epsilon^{-1})$$
$$i = 2(1 - 1/\epsilon) = 2(1 - 1/2.718)$$
$$i = 2(1 - 0.367) = 1.266 \text{ A}$$

Note that whenever we want to know current at values of t between 0 and ∞, we need to evaluate $\epsilon^{-(R/L)t} = \epsilon^{-t/\tau}$. Table 11.1 has data that will be useful in solving problems in which t/τ is given in the table.

TABLE 11.1 **Exponentials for Selected Values** t/τ

t/τ	$\epsilon^{-t/\tau}$	$1 - \epsilon^{-t/\tau}$	t/τ	$\epsilon^{-t/\tau}$	$1 - \epsilon^{-t/\tau}$
0	1.000	0.060	1.0	0.367	0.633
0.1	0.904	0.096	1.2	0.301	0.699
0.2	0.818	0.182	1.4	0.246	0.754
0.3	0.740	0.260	1.5	0.223	0.777
0.4	0.670	0.330	2.0	0.135	0.865
0.5	0.606	0.394	2.5	0.082	0.918
0.6	0.549	0.451	3.0	0.049	0.951
0.7	0.496	0.504	4.0	0.018	0.982
0.8	0.449	0.551	5.0	0.007	0.993
0.9	0.406	0.594	∞	0.000	1.000

Knowing the current at any chosen instant after the switch is closed, we can determine the voltage drop due to the resistance and the voltage drop due to the EMF of self-induction by making use of Ohm's law. In Example 11.2, the voltage v_R at the end of the time-constant interval is

$$v_R = iR = 1.266 \times 5 = 6.33 \text{ V}$$

and the EMF of self-induction is v_L:

$$v_L = E - v_R = 10 - 6.33 = 3.67 \text{ V}$$

We may want to know how long it would take for the voltage drop across R in a series R–L circuit to reach a certain value. An example will show how this may be determined.

EXAMPLE 11.3   A coil that has $L = 2.40$ H, $R = 4\,\Omega$ is connected to a constant 100-V supply source at $t = 0$. How long does it take the voltage across the resistance to reach 50 V?

Solution. Since we want to know the value of t in Equation (11.7), we shall first solve for it in general terms and then put in our known quantities:

$$i = \frac{E}{R}\left(1 - \epsilon^{-Rt/L}\right)$$
$$iR/E = 1 - \epsilon^{-Rt/L}$$
$$\epsilon^{-Rt/L} = 1 - iR/E$$

Note, here, that we need not solve for the current because we can replace iR by 50 V:

$$\epsilon^{-Rt/L} = 1 - \frac{50}{100} = 0.5$$

$$\epsilon^{Rt/L} = \frac{1}{0.5} = 2$$

$$Rt/L = \ln 2 = 0.693$$

$$t = 0.693\left(\frac{2.4}{4}\right) = 0.416\ \text{s}$$

EXAMPLE 11.4   Our choice of desired voltage across R in Example 11.3 lets us know that the current must be $50 \div 4 = 12.5$ A at that instant. Let's use the basic equation and calculate the current at the end of the 0.416 s time interval.

Solution

$$i = \left(\frac{100}{4}\right)\left[1 - \epsilon^{-4(0.416)/2.4}\right]$$
$$i = 25\left[1 - \epsilon^{-0.693}\right]$$

Now

$$\epsilon^{-0.693} = 1/\epsilon^{0.693} = \tfrac{1}{2}$$
$$i = 25(1 - 0.5) = 12.5\ \text{A}$$

as we predicted.

Problems

7. Assume an inductance coil with $L = 8$ H, $R = 10\,\Omega$ is suddenly connected to 100 V that remain constant. How long will it take the voltage drop across R to reach (a) 80 V? (b) 40 V?

8. In Problem 7, calculate the (a) rate of change of current at $t = 0$, (b) rate of change of current when $v_R = 80$ V.
9. In Problem 7, calculate the (a) EMF of self-induction at the instant the current is 5 A. Can you do this mentally?
10. (a) How long will it take for the current to rise to 5 A in Problem 7? (b) Using the answer to (a), calculate the EMF of self-induction at the instant the current reaches 5 A and thus verify your answer to Problem 9.
11. Calculate the instantaneous value of current at the end of two time-constant intervals for Problem 7.

In the solution of Example 11.2 we noted that the exponential term $\epsilon^{-Rt/L}$ of the current equation can be expressed as $\epsilon^{-t/\tau}$. This enables us to use the value of L/R, which is constant for a given coil (or whenever both L and R are constant), and thus simplify the equation for current and voltage:

$$i = (E/R)\,(1 - \epsilon^{-t/\tau}) \qquad\qquad (11.15)$$
$$v_R = E(1 - \epsilon^{-t/\tau}) \qquad\qquad (11.16)$$
$$v_L = E\epsilon^{-t/\tau} \qquad\qquad (11.17)$$

Recall that the initial rate of rise of current depends only on the value of L when E is constant: $di/dt = E/L$. This means that *the current will take a longer time to reach its final value if we increase L.* The curves in Fig. 11.3 show this. The final value of current, I_f, is shown to be constant, which means that R is constant, in these cases of increasing L. Because $\tau_1 = L/R$, branches or coils with large values of L can be said to have *long time constants* and those with small values of L can be said to give *short time constants*. Note, also, that if the value of L is fixed, long time constants can be obtained when R is relatively small and short time constants result when R is relatively large.

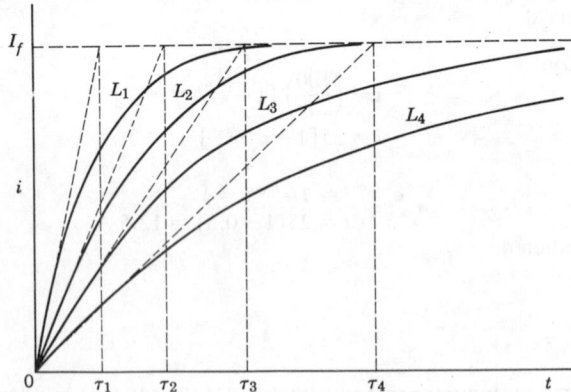

FIGURE 11.3 Current versus time in series R–L circuit. Applied voltage E and resistance R are constant. $L_4 > L_3 > L_2 > L_1$. τ_1 is a short time constant; τ_4 is a long time constant.

11.3   Decreasing Current in R–L Circuits

We shall now study the action of an inductance in a series R–L circuit while the current in it is *decreasing*. Assume the current I_0 in Fig. 11.4a to be at its steady-state value, which is E/R, and then let the switch be closed at $t = 0$. At that instant three important things happen simultaneously. The fuse blows, v_L instantly increases from zero and becomes equal to I_0R, and the potential difference between the left end of R and the bottom end of L drops from its value E (just before $t = 0$) to zero. *And the current*, which was flowing downward through L, *keeps on going*. The self-induced voltage $-L(di/dt)$ opposes the tendency for the current to decrease and thus delays the fall of the current to zero. This is another manifestation of Lenz's law.

At the instant the switch was closed, v_R was equal to E. It decreases along with the decrease in current, as shown in (c). Note that v_L is opposite in direction (polarity) to v_R, so we show its curve with negative values of voltage.

The Kirchhoff voltage equation around the loop is

$$Ri + L\frac{di}{dt} = 0 \tag{11.18}$$

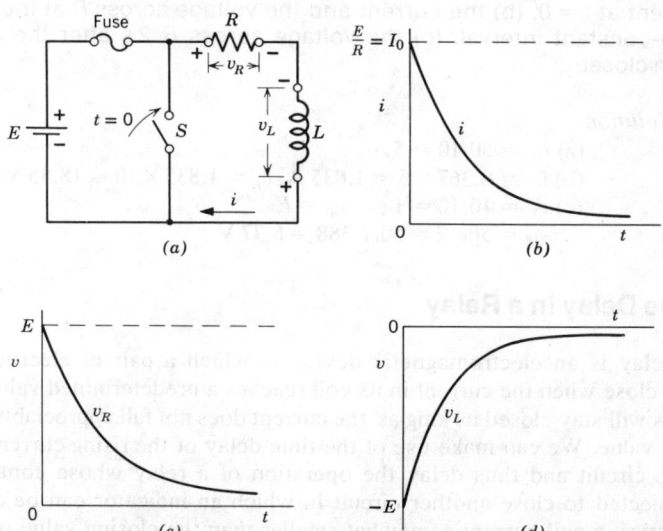

(a)

(b)

(c)

(d)

FIGURE 11.4   Current decreasing in a coil that has resistance R and inductance L. (a) Schematic of the circuit. At $t = 0$, $i = E/R = I_0$ and switch S is closed. Current immediately starts to decrease as shown in (b). Voltages v_R and v_L are equal and opposite in polarity. $Ldi/dt = v_L$ and this acts as a source of decreasing EMF.

Solving[3] for i,

$$i = I_0 \epsilon^{-Rt/L} \qquad (11.19)$$

in which I_0 is the current value at $t = 0$.

In the circuit of Fig. 11.4a, $I_0 = E/R$, so at any value of t after the switch is closed,

$$i = \frac{E}{R} \epsilon^{-Rt/L} \qquad (11.20)$$

The voltage across R is then

$$v_R = iR = E\epsilon^{-Rt/L} \qquad (11.21)$$

The voltage v_L is equal, but opposite in polarity, to v_R:

$$v_L = -E\epsilon^{-Rt/L} \qquad (11.22)$$

The time constant is, obviously, L/R, whether the current is decreasing or increasing. The rate of change of current at $t = 0$ is

$$\left.\frac{di}{dt}\right|_{t=0} = \tan \theta = -\frac{E}{L} \text{ amperes per second} \qquad (11.23)$$

EXAMPLE 5 Let $E = 50\,\Omega$, $R = 10\,\Omega$, $L = 10\,H$ in Fig. 11.4a. Find (a) the current at $t = 0$, (b) the current and the voltage across R at the end of the time-constant interval, (c) the voltage across R 2 s after the switch has been closed.

Solution

(a) $I_0 = 50/10 = 5\,A$

(b) $i_\tau = 0.367 \times 5 = 1.835\,A$, $v_R = 1.835 \times 10 = 18.35\,V$

(c) $\tau = 10/10 = 1$ s. $v_R = E\epsilon^{-t/\tau}$

$v_R = 50\epsilon^{-2} = 50/7.388 = 6.77\,V$

11.4   Time Delay in a Relay

A relay is an electromagnetic device in which a pair of electrical contacts will close when the current in its coil reaches a predetermined value. The contacts will stay closed as long as the current does not fall appreciably lower than that value. We can make use of the time delay of the rising current in a series R–L circuit and thus delay the operation of a relay whose contacts can be connected to close another circuit in which an indicator can be actuated. In practice, a coil current somewhat smaller than the closing value will keep the contacts closed because the reluctance of the magnetic circuit is less than it was when the contacts were open.

[3]Solved in the Appendix, Section A.3.

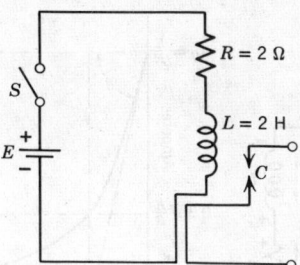

FIGURE 11.5   Relay closes contact C when current reaches a certain value.

EXAMPLE 6   The relay circuit in Fig. 11.5 has $L = 2$ H, $R = 2\,\Omega$, $E = 10$ V. The relay can carry 5 A continuously, but its contacts will close when the current reaches 4.5 A. How much time delay occurs after the switch is closed?

Solution. The time constant is $L/R = 1$ s. Using Equation (11.14),

$$4.5 = \frac{10}{2}(1 - \epsilon^{-t/1})$$
$$1 - \epsilon^{-t} = 0.9, \quad \epsilon^{-t} = 0.1$$
$$\epsilon^{t} = 10, \, t = \ln 10 = 2.3 \text{ s}$$

If this is too long to wait, we can add some resistance in series with the relay and increase the applied voltage. For example, if we add $6\,\Omega$ resistance in series with the relay and increase the applied voltage to 40 V, the time constant will be 0.25 s ($L/R = 2/8$) and the relay will close in 0.575 s.

When we study capacitors and their operation in series with a resistance we shall find that the R-C circuit is more practical for use as a timing circuit because it is much more convenient to change the value of C in an R-C circuit than the value of L in an R-L circuit. The time constant of an R-C circuit will be found to be the product of R and C, that is $\tau = RC$ seconds.

Problems

12. In Fig. 11.6a let $R = 100\,\Omega$, $L = 20$ H, and $E = 200$ V. Assume that the switch changes from the top position to the bottom position in zero time. (a) What is the current before the switch is actuated? This is also the current at $t = 0$. (b) What is the rate of change of current at $t = 0$? (c) What is the time constant? (d) What is the current value at $t = 200$ ms?

13. In Fig. 11.6a, how long does it take for the current to decrease to 0.25 A?

14. (a) In Problem 12, what is the voltage of self-induction when $i = 0.25$ A? (b) How much energy is in the magnetic field at this instant? (c) At what rate is the energy being dissipated in R?

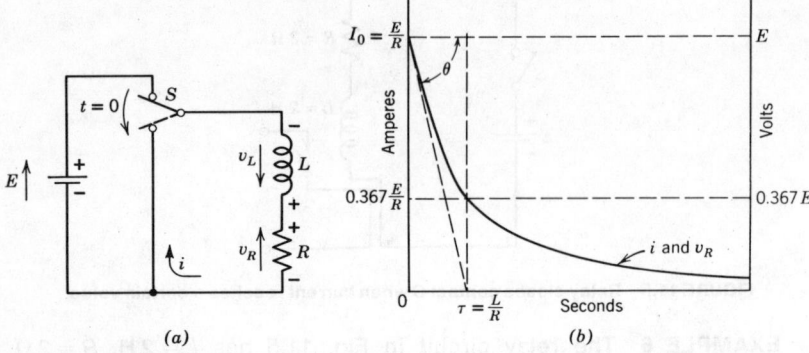

(a)   *(b)*

FIGURE 11.6 **A coil with inductance L henrys and R ohms carrying current E/R amperes is suddenly short-circuited. (a) Circuit schematic. (c) Curves of instantaneous current and voltage. The voltage scale was chosen so that the curves coincide. Note that $\tan \theta = E/L$ as in the case of increasing current (Fig. 11.2b).**

15. A coil with $L = 0.5$ H, $R = 10\,\Omega$ is carrying a steady current of 800 mA. It is suddenly short-circuited at $t = 0$. At $t = 0.05$ s: (a) What is the instantaneous current?; What is it at $t = 100$ s? (b) What is the power loss in the coil? (c) What is the voltage of self-induction?

16. Verify the fact that if $E = 40$ V and $6\,\Omega$ resistance are added in series with the relay of Fig. 11.5, the current will rise to 4.5 A in 0.575 s.

11.5   Energy Dissipation in a Magnetic Field

Field coils of motors have large values of inductance and, therefore, they store large amounts of energy when they carry direct current. If the current is caused to reduce the zero very quickly, as is the case when a switch is opened, a very large induced EMF appears at the terminals of the coil. This may result in the voltage between layers of the coil windings puncturing the insulation. It is certain that the induced voltage would force enough current through a voltmeter, connected to the terminals, to burn it out. Figure 11.7a shows such a circuit.

In order to protect the winding insulation and a meter that may be connected, a discharge resistance is arranged so that it will be connected across the coil terminals before the circuit is entirely broken by the disconnecting switch, as shown in Fig. 11.7b.

An easy approach to an understanding of what happens is to recall that the current in an inductive circuit cannot change in value in zero time. We have here a useful application of this principle, which, incidentally, we have emphasized so much that you may have become tired of hearing about it.

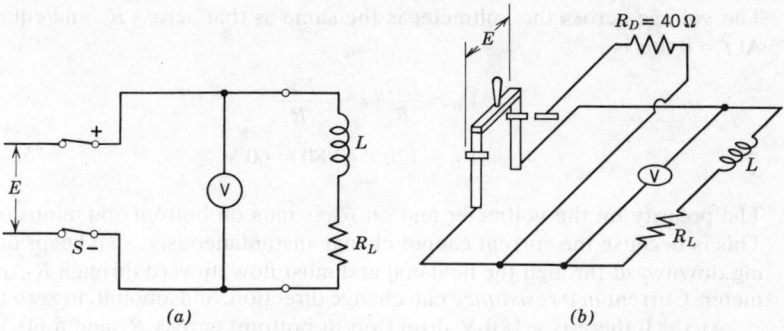

FIGURE 11.7 (a) Field coil with voltmeter connected across its terminals $L = 10\,H$, $R_L = 80\,\Omega$, $E = 120\,V$. Meter resistance $= 150{,}000\,\Omega$. (b) Field-discharge resistor is connected in parallel with meter before switch blades break contact. Supply voltage, same as in (a), is $E = 120\,V$.

Using the numerical values supplied with the illustration, we see that before the switch in Fig. 11.17a is opened the current is $120/80 = 1.5\,A$. If we open the switch suddenly and assume there will be no arcing at the switch contacts, the 1.5 A would be forced to exist (at $t = 0$) in the 150,000-ohm meter.

Look what the voltage at the meter terminals would have to be:

$$V = 1.5 \times 150{,}000 = 225{,}000 \text{ V}$$

Even if only 1 A were forced through the voltmeter it would burn out, but that would require $V = 1 \times 150{,}000 = 150{,}000$ V. Ridiculous! So — what does happen, anyway?

Well, arcing occurs at the switch blades, and the arc at each blade may stretch out to be an inch or two long. And the meter goes up in smoke while this is happening. We shall now show how the added discharge resistance, R_D, prevents this and protects the meter.

EXAMPLE 11.7   Assume the switch in Fig. 11.7b is opened quickly and $R_D = 40\,\Omega$ is connected across the field coil that has the voltmeter on its terminals. Neglecting arcing at the switch blades, determine the amount of voltage across the meter at $t = 0$, the instant R_D is connected and the supply voltage E simultaneously removed.

Solution. With the field fully energized, $I = E/R_L = 120/80 = 1.5\,A$. Using Equation (11.20), and noting that the current must now flow through $R_D = 40\,\Omega$ in parallel with the 150,000-Ω voltmeter, we may ignore the meter resistance. The equivalent resistance of 40 and 150,000 Ω in parallel is 39.988 Ω. The current is

$$i = \frac{E}{R_L}\left(1 - \epsilon^{-(R_L + R_D)t/L}\right)$$

The voltage across the voltmeter is the same as that across R_D and equals iR_D At $t = 0$ this is

$$V_D = \frac{ER_D}{R_L}\epsilon^0 = \frac{ER_D}{R_L}$$

(11.24)

$$V_D = 120 \times 40/80 = 60 \text{ V}$$

The polarity on the voltmeter and on R_D is plus on bottom and minus on top. This is because the current cannot change instantaneously, so it keeps on flowing downward through the field coil and must flow upward through R_D and the meter. Current *in a resistance* can change direction, and amount, in zero time.

At $t = 0$ there is a 120-V drop (top to bottom) across R_L and a 60-V drop (bottom to top) across the meter and R_D. This means the induced EMF in L is 180 V. The induced EMF must also be $I_0(R_L + R_D) = 1.5 (80 + 40) = 180$ V. This is a 180 V rise (top to bottom) through L from which 180-V drop is subtracted to agree with Kirchhoff's voltage law around the clockwise path.

Equation (11.24) is important in a case of this kind. You will be asked to apply it in Problems.

11.6 Review Questions

1. A coil carrying no current is suddenly connected to a constant direct-voltage source. Why is the initial current (at $t = 0$) zero?
2. What is the effect of the EMF of self-induction in the coil mentioned in Question 1?
3. Why is the steady-state current in the coil of Question 1 determined by the coil resistance only, if the applied voltage is kept constant?
4. Define the time constant of a circuit whose current changes exponentially with time. To what is the time constant equal, for a circuit of inductance L henrys and resistance R ohms?
5. Assume the final value of current in the coil of Question 1 to be 10 A. What was its value at the end of one time constant interval?
6. If the voltage applied to the coil in Question 1 is E, the coil inductance is L, and the resistance is R, what relation of two of these quantities gives the rate of change of current at $t = 0$?
7. Prove, using units, that the units of L/R is seconds.
8. What would be the rate of change of current in an inductance if the current could change from one value to another in zero time?
9. A coil carrying $I_0 = 10$ A is suddenly short-circuited (at $t = 0$). What is the current value at the end of the time constant interval?
10. If R_L in Question 9 is 5 Ω, what is the voltage across R at the end of τ seconds? What is the induced EMF at that instant? Do these two voltage aid or oppose each other?
11. What happens to the energy in the magnetic field of the coil in Question 9?

12. Would the current in the coil of Question 9 decrease to zero faster or slower if the coil resistance were more than 5 Ω?

13. A voltmeter is connected to the terminals of the field coil of a motor. Explain what would happen to the meter if the switch through which current passes to the coil and meter were suddenly opened. Why would this happen?

14. What precautions might be taken to prevent damage to the meter in Question 13?

15. A coil has a certain time constant. Do you think it may be easier to decrease the time constant than to increase it? Assume additional resistance is easier to obtain and costs less than additional inductance. Explain.

11.7 Problems

17. The voltage applied to a series circuit of 240 Ω and 12 H is 120 V. Find how long it takes for the current to increase to (a) one fourth and (b) three fourths of its final value.

18. Continuing with Problem 17, find the voltage across the resistance at the end of each of the following total time intervals: 25, 50, 75, 100, 250 ms.

19. How much energy exists in the magnetic field of the coil of Problem 17 (a) at the end of one time-constant period, and (b) when steady state is reached?

20. Assume that, in Fig. 11.4a, $E = 12$ V, $R = 3\,\Omega$, $L = 6$ H. What is the voltage across R at 1 s after the switch is closed?

21. In Problem 20, what is the rate of change of current at $t = 0$?

22. In Problem 20, how long will it take for the induced EMF to drop to one half of the value it has at $t = 0$?

23. What is the energy in the magnetic field of L at the end of the time period specified in Problem 22?

24. The field winding of an electric motor has 80 Ω resistance and 12 H inductance. It is connected to a 240-V source. A field-discharge resistance of 80 Ω is suddenly connected in series with the winding as it is simultaneously removed from the 240-V source like the source is removed in Fig. 11.7b. (a) Calculate the rate of change of current at the instant the discharge resistance is connected. (b) What is the voltage across the field winding terminals at that instant? This amount of voltage would also be applied to a voltmeter if one were connected to the motor terminals.

25. (a) In Problem 17, how much energy remains in the magnetic field τ seconds after the current starts to decrease? (b) At what rate is the energy being dissipated in the total resistance in the circuit at that instant? Both the resistance of the winding and the external 80 Ω dissipate energy.

Group B

26. A coil for which $L = 0.5$ H, $R = 10\,\Omega$ is carrying a steady current of 800 mA. It is suddenly short-circuited at $t = 0$. (a) Find the current values at $t = 0.05$ s and 100 s. (b) What is the initial rate of change of current? (c) What is the power loss in the coil at $t = 0.05$ s? (d) What is the voltage of self-induction at $t = 0.05$ s?

27. In Problem 26, how long does it take for the current to fall to 40 mA?

28. Assume, in Fig. 11.7, that $E = 240$ V, $L = 12$ H, $R_l = 120$ Ω, and the voltmeter resistance is 300,000 Ω. What would be the voltage across the voltmeter if R_D is not connected when the switch is opened?

29. In Problem 28, assume $R_D = 40$ Ω is reconnected before the switch is opened and calculate the voltage across the voltmeter at the instant the switch is opened.

30. In Fig. 11.7, $E = 120$ V, $L = 10$ H, $R_l = 80$ Ω. We found in Section 11.5 that the voltmeter would be burned out unless a protective field-discharge resistor R_D were provided. Calculate the minimum value that R_D must have in order to prevent the voltmeter from receiving more than 150 V.

CHAPTER TWELVE

Electric Fields and Capacitance

We are familiar with the fact that an object held in the hand falls to the ground when it is released. This is because a *gravitational field* exists above the earth's surface and this field exerts a force on the object causing it to move toward the earth. The force is exerted on the object while we hold it and we must exert an equal force opposite in direction to the "force of gravity" to prevent the object from moving downward.

Imagine a small sphere (*A*) charged with positive electricity and suspended by an insulating thread. It produces an electric field in space all around it. If we bring a second sphere (*B*), suspended by a thread charged with negative electricity, near it, the second sphere will be acted on by a force that urges it to move toward the positively charged sphere. We say the electric field produced by the positive sphere exerts a force on the negative sphere in a manner similar to the action of the gravitational field on the object mentioned above. If the sphere is positively charged, the electric field will exert a force that urges it to move away from sphere (*A*).

It is also true that the second sphere (*B*) produces an electric field of its own, and it is correct to say that its field exerts a force on the first sphere (*A*). This force will urge *A* to move *toward B* if *B* is negatively charged and *away* from *B* if *B* is positively charged.

12.1   Coulomb's Law

The first thing we must do in our study of electric fields is to express mathematically the amount of force that exists between two charged bodies. Charles-Augustin Coulomb (France, 1736–1806) expressed this force by a mathematical relation known as *Coulomb's law*:

$$F = \frac{Q_1 Q_2}{4\pi\epsilon d^2} \text{newtons} \tag{12.1}$$

In this equation, Q_1 and Q_2 represent the charges in coulombs and d represents the distance between them in meters. The factor epsilon, ϵ, is a constant that has a particular value for each kind of medium (air, insulating oil, or other material) in which the charges are located. This quantity will be discussed in detail in Section 12.6.

We must remember that, as given by Coulomb's law, the force between two charged bodies (or charges) *is directly proportional* to the product of the amounts of the charges *and inversely proportional* to the square of the distance between them.

If the charges are both positive or both negative, the force is positive and one of *repulsion*. If the charges differ in sign (one + and the other −) the force is one of *attraction*. It is better to say that the electric fields produce forces on the charges that push them toward each other than to say that each charge attracts the other.

EXAMPLE 12.1   A small metal sphere charged with 100 μC of positive electricity is suspended in a vacuum 3.281 ft from another sphere charged with 50 μC of negative electricity. The value of ϵ for vacuum is 8.854×10^{-12}. How much force exists between the charges?

Solution. The value of d, in meters, is 1 because there are 3.281 ft in a meter.

$$F = \frac{(100 \times 10^{-6})(-50 \times 10^{-6})}{4\pi \times 8.854 \times 10^{-12} \times 1^2} = -45 \text{ N}$$

Multiplying this by 0.2248 lb/N gives

$$F = -10.116 \text{ lb}$$

The minus sign signifies a force of attraction.

12.2   Electric Flux Density and Field Intensity

Imagine a small isolated sphere with charge Q_1 uniformly distributed over its surface (Fig. 12.1). It sends out electric flux radially in all directions, and the number of flux lines through a small area of constant size decreases as

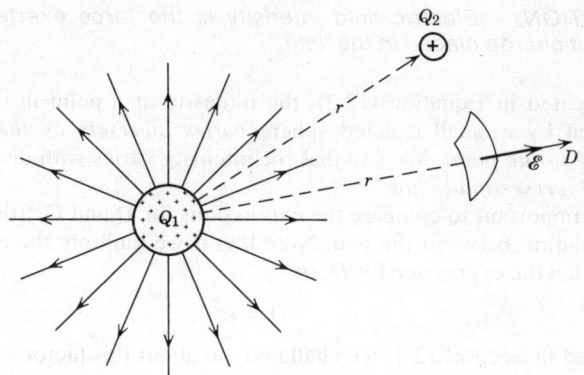

FIGURE 12.1   A *positure* **charge** Q_1 **sends out electric flux radially in all directions. At a point** *r* **meters away, the flux density** $D = Q_1/4\pi r^2$. **The electric field intensity** $\mathscr{E} = Q_1/4\pi\epsilon r^2$. **The force on charge** Q_2 **in the field is** $F = \mathscr{E} Q_2$. $\mathscr{E}$ **is the force** *per unit charge* **and this is called electric field intensity or electric field strength.**

the area moves radially outward. It is easy to understand that the field gets weaker as distance from the sphere increases.

Each unit of positive charge sends out one unit of electric flux, so that

$$\psi = Q$$

is true *by definition*.

The symbol for electric flux is ψ(psi).

If we imagine a sphere with *r* meters radius whose center is at the center of our charged body, it is easy to see that the amount of *flux per unit area on the imagined sphere* is the total flux ψ divided by the area *S* through which it passes:

$$\text{Flux density } D = \frac{\psi}{4\pi r^2} = \frac{Q}{4\pi r^2} \tag{12.2}$$

The units of *D* are coulombs per square meter.

Looking at Fig. 11.1, let's assume that a second charge $Q_2 = 1$ C is located in the field produced by Q_1 and *r* meters away. Using Coulomb's law for the force on this *unit charge*, we have

$$\frac{F}{Q_2} = \frac{Q_1}{4\pi\epsilon r^2} = \mathscr{E} \text{ newtons per coulomb} \tag{11.3}$$

Here we have a new quantity, whose symbol is $\mathscr{E}$, and we call it *electric field intensity*.

DEFINITION: *Electric field intensity is the force exerted by the field on a unit charge placed in the field.*

As indicated in Equation (12.3), the intensity at a point in the electric field produced by a small isolated sphere *varies inversely as the square of the distance to the point.* We say that the intensity varies with distance according to the *"inverse-square law."*

It is important to compare the expressions for D and E, from which we can get a relation between the two. Note that if we multiply the expression for $\mathscr{E}$ by ϵ we get the expression for D, so

$$D = \epsilon \vec{\mathscr{E}} \qquad (12.4)$$

As stated in Section 12.1, we shall find out about this factor ϵ in Section 12.6. The symbol for ϵ when the medium is *free space*, i.e., *a vacuum*, is ϵ_0 and its value is

$$\epsilon_0 = \frac{1}{36\pi \times 10^9} = 8.854 \times 10^{-12}$$

We shall see later that its units are *farads per meter.*

Both **D** and $\vec{\mathscr{E}}$ have *direction* as well as magnitude, and are *vector quantities* because of this. We should express them in vector notation, using $\mathbf{a}_r$ as a unit vector in the radially outward direction. The use of the unit vector does not alter the numerical value. It merely *indicates direction.* In general,

$$\mathbf{D} = \frac{Q}{4\pi r^2}\, \mathbf{a}_r$$

$$\vec{\mathscr{E}} = \frac{Q}{4\pi \epsilon r^2}\, \mathbf{a}_r$$

$$\mathbf{D} = e\vec{\mathscr{E}}$$

are *vector forms* of Equations (12.2), (12.3), (12.4).

EXAMPLE 12.2   A positive charge of 250 μC is located in free space at the origin of coordinates in the XY plane. Determine the flux density and field intensity at the point $x = 0, y = 4$ m. How much force does the field exert on a point charge $Q = -25$ μC at the point, and what is its direction?

Solution

$$D = \frac{250 \times 10^{-6}}{4\pi \times 4^2} = 1.244 \times 10^{-6} \text{ C/m}^2$$

$$\mathscr{E} = \frac{D}{\epsilon_0} = \frac{1.244 \times 10^{-6}}{8.854 \times 10^{-12}} = 14.05 \times 10^4 \text{ N/C}$$

$$F = Q\mathscr{E} = 14.05 \times 10^4(-25)\,10^{-6} = -3.51 \text{ N toward the origin}$$

EXAMPLE 12.3 Assume the medium in which the charges in Example 12.2 are located is a gas whose relative permittivity is 1.5. Find the quantities requested. Actual permittivity is equal to relative permittivity times ϵ_0. See Section 12.6.

Solution. Since D is independent of ϵ, $D = 1.244 \times 10^{-6} \, C/m^2$:

$$\mathscr{E} = \frac{D}{\epsilon_r \epsilon_0} = \frac{1.244 \times 10^{-6}}{1.5 \times 8.854 \times 10^{-12}} = 9.36 \times 10^4 \, N/C$$

$$F = (-25 \times 10^{-6})(9.36 \times 10^4) = -2.34 \, N \text{ toward the origin}$$

Problems

1. A small sphere, charged with $150 \, \mu C$ of positive electricity, is located in free space. How much force does its field exert on (a) a $10 \, \mu C$ point charge 2 m away, (b) a $-25 \, \mu C$ point charge 5 m away? (c) What are the directions of the forces on the charges?
2. (a) Calculate the flux density of the field produced at $r = 2$ m by the $150 \, \mu C$ charge in Problem 1, and specify the direction of the "flux density vector." (b) Repeat (a) for $r = 5$ m.
3. What are the intensities of the field at the locations of the two smaller charges in Problem 1. Are the directions of the fields the same at the two locations?
4. A positive charge of $100 \, \mu C$ is acted on by a force of 10 N when placed at a point in an electric field. What is the field intensity there?
5. Calculate the amount of concentrated charge required if it is 2 m from the $100 \, \mu C$ charge of Problem 4 and produces the field intensity indicated.

12.3 Capacitors

Two metal plates form a capacitor when located near each other (they are usually parallel to each other) so that an electric field will be established between them when they are charged with opposite kinds of electricity. Capacitors are used to store energy. The time of this storage is very short in many applications — even as short as one nanosecond (10^{-9} s) or less, while in other cases it may be measured in milliseconds, seconds, or minutes. Some capacitors used in power circuits have their amounts of charge varying slightly, but they remain in a charged condition many hours at a time. The capacitor that is connected across the "points" in an automobile ignition system absorbs energy that otherwise would produce arcing between the points and cause rapid pitting and wear on the surfaces of the points.

One popular type of small capacitor is made of two long, narrow sheets of metal foil separated by a layer of insulating paper (paraffined or oiled) and rolled into tubular form. Some are mounted in metal cans filled with insulating oil.

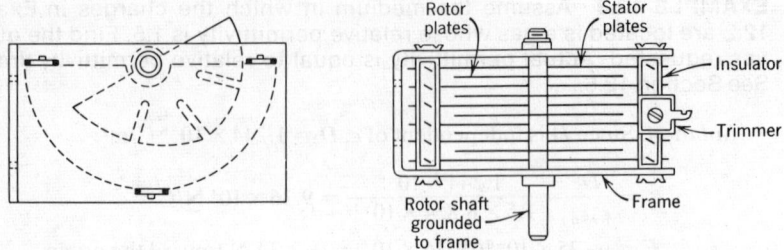

FIGURE 12.2   A variable capacitor with 5 movable and 6 stationary plates.

Another type of capacitor that can be made *electrically large* (able to store a relatively large amount of charge for its physical size) is the electrolytic capacitor. It utilizes a very thin layer of nonconducting chemical on one of its plates to serve as the insulating medium. As a result, the plates may be brought very close to each other mechanically, which contributes to an increase in charge-storage capacity.

Miniature *ceramic capacitors*, which resemble small buttons without holes in appearance, are used in large quantities in communications equipment and other electronic circuit applications. Each has two wire stems, called leads, extending from it for making connections.

A popular type of high-voltage capacitor has mica as the insulating medium separating the plates. Mica capacitors may be used on alternating current and have practically zero leakage current.

Variable capacitors have a set of metal plates that can be moved in and out of the spaces between another set of stationary plates. The capacitance values of variable capacitors used in radio and television receivers are very small. A variable capacitor is shown in Fig. 12.2. A *trimmer*, shown mounted at the edge of the insulating post, is used to make fine adjustments in calibrating the tuning dial on the rotor shaft.

The "electrical size" of a capacitor is measured in units of *capacitance*; the basic unit is the *farad* and the symbol is *F*. The amount of capacitance is determined by the plate area facing the opposite-polarity plate area, by the separation distance between the plates, and by the kind of material between the facing surfaces of the plates.

12.4   Charging a Capacitor

Figure 12.3 represents two parallel metal plates separated a relatively short distance—say, 1 mm—in air. The resistance of the connecting wires is represented by *r*. *E* is a constant-voltage source.

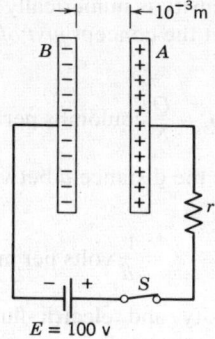

FIGURE 12.3   Two parallel conducting plates form a capacitor. The resistor r represents the total resistance in the circuit.

At the instant the switch was closed, electrons began moving from plate A through the connecting wire to the source, through it, and to plate B. This continued for the brief period of time required to establish a full 100-V potential difference between the plates: A is $+100$ V with respect to B. No charge moved between the plates in the charge-free, leak-proof space between them. Energy was built up, however, in the *electric field* between the plates. A charge, which we may call $+Q$, resides on A and an equal and opposite charge $-Q$ resides on B. The charges must be equal in amount because the source E supplies 1 electron to plate B for every one that is removed from plate A. If this were not true there would be two different current values at the same instant in a series circuit, which is impossible.

If the connecting wires are disconnected at the source and brought together, a discharge current will flow, and the energy in the field will be dissipated in the form of heat in the resistance of the circuit.

12.5   The Electric Field of a Parallel-Plate Capacitor

The charge on a capacitor plate divided by the area facing the other plate gives the *surface-charge density*. If the charge is Q and the area is S the surface charge density is

$$\rho_s = \frac{Q}{S} \text{ coulombs per square meter} \qquad (12.5)$$

Every coulomb of positive charge gives off a coulomb of electric flux. This is a very important basic concept. *Every coulomb of negative charge receives one coulomb of flux.*

The surface-charge density is numerically equal to the electric-flux density, D *just off the surface*. Let the concept *just off the surface* mean 10^{-100} mm (or less) away:

$$D = \rho_s = \frac{Q}{S} \text{ coulombs per square meter}$$

The voltage V divided by the distance d between the plates is the *electric-field intensity* $\mathscr{E}$:

$$\mathscr{E} = \frac{V}{d} \text{ volts per meter} \qquad (12.6)$$

Both electric-field intensity and electric-flux density are *vector quantities* because they have *direction* as well as magnitude. They are usually printed in boldface type $\vec{\mathscr{E}}$, $\mathbf{D}$ when expressed in vector form and in normal type $\mathscr{E}$, D when they represent only the magnitudes of the quantities.

12.6 Permittivity

The ratio of D to $\mathscr{E}$ is a property of the medium in which these field quantities exist. It is called the permittivity (ϵ) of the medium. It is also called the *dielectric constant*. Different media (meaning materials) used between the plates of capacitors include paraffined paper, oiled paper, oil alone, ceramic insulation, and mica (see Table 12.1). These materials are called *dielectrics*. The basic definition of permittivity ϵ is

$$\epsilon = \frac{D}{\mathscr{E}} \text{ farads per meter} \qquad (12.7)$$

We shall now show that the units of ϵ, the permittivity, are *farads per meter*. In doing this the shortest way, we use *volts per meter* instead of *newtons per coulomb* as the units of ϵ. We shall show this first:

$$\mathscr{E} \rightarrow \frac{\text{newtons}}{\text{coulomb}} \times \frac{\text{meters}}{\text{meters}} = \frac{\text{joules}}{\text{coulomb} \times \text{meters}}$$

Now joules/coulomb is work per unit charge = volts. Therefore

$$\epsilon = \frac{D}{\mathscr{E}}, \quad \epsilon \rightarrow \frac{\text{coulombs}}{\text{meter}^2} \div \frac{\text{volts}}{\text{meter}}$$

$$\epsilon \rightarrow \frac{\text{coulombs}}{\text{meter}^2} \times \frac{\text{meters}}{\text{volts}} = \frac{\text{coulombs}}{\text{volts}} \times \frac{1}{\text{meter}}$$

Since coulombs divided by volts = capacitance in farads,

$$\epsilon \rightarrow \text{farads per meter}$$

TABLE 12.1   Relative Permittivities and Dielectric Strengths
              at Frequencies less than 2 kHz)

Dielectric	Relative Permittivity	Dielectric Strength (Kv/mm)
Air	1	3
Amber	2.86	55
Asphalt	2.7	1–2
Bakelite	4.5	10–25
Ebonite	2.8	30–100
Fiber	2.5	4
Glass (Drown)	6.9	10–30
Glass (Pyrex)	4.8	50
Mica	5–9	60–230
Oil	2–5	5–20
Paper	2–4	30–40
Paraffin	3–5	32
Porcelain	6–7.5	1.5–10
Rubber	2–3.5	10–35
Slate	6–7.5	0.2–2.5
Water (distilled)	81	15

From Equation (12.7) we obtain

$$\mathscr{E} = \frac{D}{\epsilon} \text{ volts per meter} \tag{12.8}$$

And from Equations (12.5) and (12.6), we have

$$\frac{V}{d} = \frac{D}{\epsilon} = \frac{p_s}{\epsilon} \tag{12.9}$$

The voltage between the plates may now be expressed in terms of surface-charge density:

$$V = \frac{p_s d}{\epsilon} \text{ volts} \tag{12.10}$$

The student must memorize the following definition of capacitance:

$$C = \frac{Q}{V}; \qquad \text{Farads} = \frac{\text{coulomb}}{\text{volts}}$$

Note how useful it is in obtaining the expression for the capacitance of a parallel-plate capacitor.

Since the total charge on a parallel-plate capacitor is given by $Q = p_s S$, the *capacitance, defined as charge per volt,* is then

$$C = \frac{Q}{V} = \frac{p_s S}{p_s d/\epsilon} = \frac{\epsilon S}{d} \text{ farads} \tag{12.11}$$

The student must understand that this formula for capacitance holds for *parallel-plate capacitors only.* The *permittivity of free space*, ϵ_0, is the constant:

$$\epsilon_0 = \frac{1}{36\pi \times 10^9} = 8.854 \times 10^{-12} \text{ farad per meter}$$

This value of dielectric constant is used when the medium is a *vacuum* or *air.* A quantity called *relative permittivity*, ϵ_r, is given in reference books for all kinds of dielectric materials. It is the *numerical ratio* of the actual permittivity of the material to the permittivity of free space (vacuum):

$$\epsilon_r = \frac{\epsilon}{\epsilon_0} \text{ a dimensionless numeric} \tag{12.12}$$

This means that, in order to obtain the permittivity of a dielectric to use in calculating capacitance, the relative value ϵ_r obtained from the reference must be multiplied by ϵ_0. As an example, the relative permittivity of paraffined paper is about 2.2, so the value that would be used for ϵ in Equation (12.11) is 2.2 ϵ_0.

EXAMPLE 12.4 A capacitor with paraffin-paper dielectric has an effective area per plate of 1 m² and a plate separation of 10^{-4} m. Determine the capacitance.

Solution

$$C = \frac{\epsilon S}{d} = 2.2 \times \frac{1}{36\pi \times 10^9} \times \frac{1}{10^{-4}} = 0.1947 \times 10^{-6} \text{ F}$$

This is very nearly 0.2 μF, a popular size of fixed capacitor.

EXAMPLE 12.5 A variable capacitor with air dielectric ($\epsilon_r = 1$) has 11 movable and 12 stationary plates. When they are completely meshed for maximum capacitance, the area of each plate that faces the dielectric is 0.0015 m² and the separation between opposite plates is 0.001 m. Compute the maximum capacitance of this variable capacitor.

Solution

$$C = 1 \times \frac{1}{36\pi \times 10^9} \times 0.0015 \times 11 \times 2 \times \frac{1}{0.001} = 292 \times 10^{-12} \text{ F}$$

This is 292 pF, a popular size for a variable capacitor.

In multiple-plate capacitors, charges collect on both sides of the plates that are enclosed by dielectric material and oppositely charged plates. In this example, the 11 movable plates are enclosed by air dielectric and oppositely charged plates; therefore the total area is given by multiplying the area of one plate (0.0015 m²) by 11 times 2.

Problem

6. A parallel-plate capacitor (the dielectric is air) is charged with 10^{-8} coulomb of electricity. What is the surface-charge density if the area of the positive plate is 0.01 m²? What is the electric-flux density just off the inside surface of the positive plate? What is it just off the negative plate?

7. (a) Calculate the electric-field intensity between the plates of the capacitor in Fig. 12.3 if the area of each plate is 0.0025 m² and they are 1 mm apart. (b) What is the flux density in the air between the plates? (c) Calculate the capacitance and the charge on each plate.

12.7   Electric Fields in Simple Geometries

PARALLEL-PLANE CONDUCTORS

Figure 12.4 represents two parallel-plane conductors placed close together in a medium with relative permittivity $\epsilon_r = 1$. Electric-flux lines not shown here are horizontal; they *originate on the positive charges* of the conductor on the right, which is V_1 volts higher in potential than the one negatively charged conductor on the left. The separation distance is d meters. Planes parallel to

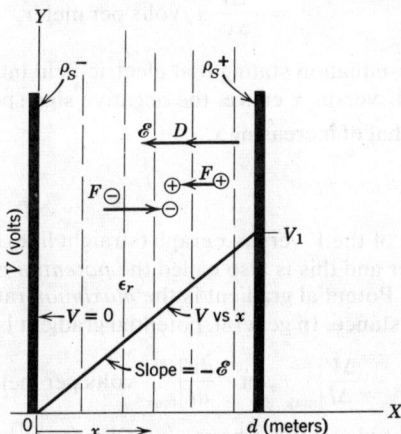

FIGURE 12.4   Electric field in dielectric between parallel plates with V_1 volts of potential difference. $V = 0$ (assumed) on negative plate.

the conductors, at various constant values of x, are indicated by dash lines. These planes are regarded as *equipotential surfaces* because every point in a chosen one is at the same potential with respect to one of the conductors as is every other point in that chosen plane. Furthermore, since the value of the field intensity is constant ($\mathscr{E} = V_1/d$) throughout the dielectric material between the plates, we may write for the potential at x meters from the negative plate;

$$V_x = \mathscr{E}x = \frac{\text{volts}}{\text{meter}} \times \text{meters} = \text{volts} \qquad (12.13)$$

The $\mathscr{E}$ and D fields are constant and directed toward the left at every point between the charged plates that is x meters from the negative plate.

A positively charged particle such as an ion of a gas, which may be present between the plates, will be driven toward the left by a *force proportional to the amount of its charge*, q_+:

$$\mathbf{F}_+ = \vec{\mathscr{E}} q_+ \qquad (12.14)$$

A negatively charged particle, for example, an electron (charge q_e), will be driven toward the right by a force

$$\mathbf{F}_- = -\vec{\mathscr{E}} q_e \qquad (12.15)$$

in which q_e is the electronic charge (-1.602×10^{-19} C).

Equation (11.13) is acutally the equation of the straight line in Fig. 12.4, drawn from 0 to V_1. $\mathscr{E}$ is the negative of the slope of that line. In fact, we could write

$$\mathscr{E} = \frac{-\Delta V}{\Delta x} \, \mathbf{a}_x \text{ volts per meter} \qquad (12.16)$$

which is a vector equation stating that electric-field intensity is the negative of the slope of the V versus x curve, the negative sign specifying the direction of $\vec{\mathscr{E}}$ as opposite to that of increasing x.

POTENTIAL GRADIENT

The slope $\Delta V/\Delta x$ of the V versus x graph (straight line) in Fig. 12.4 is expressed in volts per meter and this is also called the *potential gradient* in the space between the plates. Potential gradient is the *maximum* rate of change of potential with respect to distance. In general, potential gradient is given by

$$\left.\frac{\Delta V}{\Delta l}\right|_{\text{max}} \quad \text{or} \quad \left.\frac{dV}{dl}\right|_{\text{max}} \text{ volts per meter.}$$

This is the magnitude. It must have a unit vector symbol attached to make it complete, i.e. a vector expression.

Obviously, we could have many Δl's in the space between the plates in Fig.

12.4. One could be parallel to the plates, in which case the rate of change of V with respect to distance would be zero, wouldn't it? Because $\vec{\mathscr{E}}$ is necessarily directed toward the negative plate, $\vec{\mathscr{E}}$ is given by *minus* the potential gradient.

ELECTRON VOLTS

Consider that an electron leaves the negative plate in Fig. 12.4 and is driven by the electric field to the positive plate. Its energy is increased because the force of the field acts on it through the distance between the plates. When it reaches the positive plate its energy is V_1 *electron volts* higher than it was at the instant it left the negative plate. Assume its velocity was zero then. Since *voltage* is *work per unit charge*, the energy given to the electron is

$$W = V_1 Q_e$$

in which q_e is the charge of the electron.

An *electron volt of energy* is added each time the electron passes through a potential difference of 1 V in the direction of increasing potential. Therefore,

1 electron volt $= 1.602 \times 10^{-19}$ J of energy

FARADAY SCREEN

Sometimes it is desirable to prevent electric charges that exist on a circuit component, such as a coil of wire or a metal plate, from inducing opposite charges on another component. This can be done by placing a wire mesh or a solid metal sheet, called a *Faraday screen*, between the charged body and the other component that should not have a charge induced on its surface. The screen is connected to earth, that is, to a point on the apparatus called "ground" because a wire has been put in place that connects that point to a rod or a pipe that enters the earth.

Electrons pass between the earth and the screen to provide the induced charge that otherwise would shift to the side of the protected component if the screen were not in place.

An area that should be kept free from electromagnetic waves and their disturbing effects can be isolated by surrounding it with a grounded continuous metal screen built into the walls, ceiling, and floor, or added on their surfaces.

EXAMPLE 12.6   Two parallel plates, each 10 cm square, and 5 mm apart in air, are perpendicular to the Y axis. One plate lies in the *XZ* plane and the other cuts the Y axis at y = 0.005 m. The plate in the *XZ* plane is 100 V negative with respect to the other plate. (a) What is the capacitance of this capacitor? (b) What would be the capacitance if the dielectric were bakelite?

Solution

$$\text{(a) } C = \frac{\epsilon S}{d} = \frac{1}{36\pi \times 10^9} \times \frac{(0.10)^2}{0.005}$$

$$C = 17.7 \times 10^{-12} \text{ F} = 17.7 \text{ pF}$$

$$\text{(b) } C = \frac{\epsilon_r \epsilon_0 S}{d} = 4.5 \times 17.7 = 79.65 \text{ pF}$$

EXAMPLE 12.7   What are the electric field intensity and flux density in the air capacitor of Example 12.6?

Solution

(a) $\mathscr{E} = 100/0.005 = 20,000$ V/m
(b) $D = \epsilon_r \epsilon_0 \mathscr{E} = 1 \times 8.854 \times 10^{-12} \times 2 \times 10^4 = 17.708 \times 10^{-8}$ C/m²

EXAMPLE 12.8   Repeat Example 12.7 assuming the dielectric is bakelite.

Solution

(a) $\mathscr{E} = 100/0.005 = 20,000$ V/m
(b) $D = 4.5 \times 17.708 \times 10^{-8} = 79.68 \times 10^{-8}$ C/m²

EXAMPLE 12.9   Calculate the amount of charge on each plate of the capacitor of Example 12.8 when the dielectric is (a) air, (b) bakelite.

Solution

$$\rho_s = D, \quad Q = \rho_s S = DS$$
(a) $Q = 17.708 \times 10^{-8} \times 10^{-2} = 17.708 \times 10^{-10}$ C
(b) $Q = 79.68 \times 10^{-8} \times 10^{-2} = 79.68 \times 10^{-10}$ C

It is important to observe that when we choose a dielectric material with a larger pemittivity, $\mathscr{E}$ is not changed but D and the amount of charge on the plates are increased. Recall that $\mathscr{E} = $ volts $\div$ distance; $D = \epsilon\mathscr{E}$.

AN INFINITELY LONG, UNIFORM LINE CHARGE

Consider a long filamentary conductor (Fig. 12.5a) with a uniform-charge density of ρ_L coulombs per meter. The expressions we shall use for **D** and $\mathscr{E}$ at any point r meters (at right angles) from the line were derived assuming the line extends from $-\infty$ to $+\infty$. However, when r *is very much smaller* than the distance from the point P to the nearest end of a line charge *of finite length*, the expressions for **D** and $\mathscr{E}$ are sufficiently accurate.

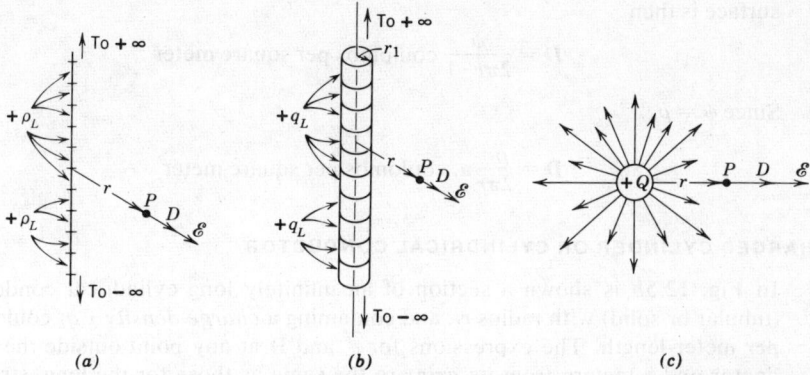

FIGURE 12.5 (a) An infinitely long filamentory conductor with uniform line-charge density of ρ coulombs per meter. (b) An infinitely long, uniformly charged cylindrical conductor. Charge density is q_L coulombs per meter length. (c) A concentrated positive charge produces radial D and ϵ fields in all directions.

By calculus integration[1] it has been found that the electric-field intensity at r meters from an infinitely long line of positive charge is given by

$$\vec{\mathscr{E}} = \frac{\rho_L}{2\pi\epsilon r}\,\mathbf{a}_r \text{ volts per meter} \tag{12.17}$$

As we explained before, you shouldn't let $\mathbf{a}_r$ bother you because it is simply *a pointer* to indicate that *the direction* of the $\vec{\mathscr{E}}$ field *is radially outward* from the wire. If the line-charge density had been assumed negative $(-\rho_L)$ then $-\mathbf{a}_r$ would appear in Equation (12.16). We *must not neglect direction* in specifying quantities that have direction. For example, if a rod man in a surveying party were told to move 10 ft, he would naturally ask in what direction? Distance (or length) is a *vector quantity*, and it is not fully defined for use in a problem or in many other situations until both *magnitude and direction* are given.

The electric-flux density at point P in Fig. 12.5a is

$$\mathbf{D} = \epsilon\vec{\mathscr{E}} = \frac{\rho_L}{2\pi r}\,\mathbf{a}_r \text{ coulombs per square meter} \tag{12.18}$$

This result is easily obtained when we know that all of the electric flux coming from the *charge on 1 meter length* of the infinitely long conductor will will be directed *radially outward* and confined between two planes 1 m apart and perpendicular to the filamentary conductor. The total flux is then $\psi = \rho_L \times 1$ and it passes through the *curved surface only* of a cylinder r meters in radius with its axis on the line containing the charge. The flux density at the

[1]W. H. Hayt, Jr., *Engineering Electromagnetics*, 2nd Edition, New York: McGraw-Hill, 1967.

surface is then

$$D = \frac{\psi}{2\pi r \cdot 1} \text{ coulombs per square meter}$$

Since $\psi = \rho_L$,

$$\mathbf{D} = \frac{\rho_L}{2\pi r} \mathbf{a}_r \text{ coulombs per square meter}$$

A CHARGED CYLINDER OR CYLINDRICAL CONDUCTOR

In Fig. 12.5b is shown a section of an infinitely long cylindrical conductor (tubular or solid) with radius r_1, and containing a *charge density* $+ q_L$ coulombs per meter length. The expressions for $\vec{\mathscr{E}}$ and $\mathbf{D}$ at any point outside the conductor and r meters *from its axis* are the same as those for the long, straight filamentary charge [Equations (12.6) and (12.7)]. Actually $q_L = \rho_L$ for a line charge located back inside the conductor at its axis. One should note that the surface-charge density is related to q_L as follows:

$$\rho_s = \frac{q_L}{2\pi r_1 \times 1} = \frac{q_L}{2\pi r_1} \text{ coulomb per square meter}$$

CONCENTRATED CHARGE

The electric-field intensity at any point r meters from the center of a concentrated charge $+ Q$ is directed radially outward, as shown in Fig. 12.5c. Its value is obtainable directly from Coulomb's law, and also from the expression for D which is easily found.

First let us assume that a small *positive charge q* is located r meters from Q. The force on this "test" charge is

$$\mathbf{F} = \frac{Q(q)}{4\pi\epsilon r^2} \mathbf{a}_r \tag{12.18}$$

The expression for $\vec{\mathscr{E}}$, *the force per unit charge*, is

$$\frac{\mathbf{F}}{q} = \vec{\mathscr{E}} = \frac{Q}{4\pi\epsilon r^2} \mathbf{a}_r \tag{12.19}$$

Also $\mathbf{D} = \epsilon\vec{\mathscr{E}}$

$$\mathbf{D} = \frac{Q}{4\pi r^2} \mathbf{a}_r \tag{12.20}$$

Let us get D another way. The total flux put out by Q coulombs of charge is $\psi = Q$ coulombs of flux. The magnitude of the flux density on an imaginary sphere r meters in radius and centered at Q is

$$D = \frac{Q}{S} = \frac{Q}{4\pi r^2} \text{ coulombs per square meter} \qquad (12.21)$$

This is the *scalar magnitude* of the *vector* quantity **D** in Equation (12.20).

EXAMPLE 12.10   A long straight conductor in air is charged with $+10$ μC/m of length. Calculate the field intensity at a point on a line perpendicular to the conductor and 20 cm from its axis.

     Solution

$$\mathscr{E} = \frac{10 \times 10^{-6}}{2\pi \times \dfrac{1}{36\pi \times 10^9} \times 0.2}$$

$$\mathscr{E} = 900,000 \text{ V/m, directed radially outward.}$$

EXAMPLE 12.11   How much electric flux comes from a 1 cm length of the conductor of Example 12.10?

     Solution

$$Q \text{ per cm} = 10^{-5} \text{ C/m} \div 100 = 10^{-7} \text{ C/cm}$$

This flux radiates outward and all of it passes through a "band" 1 cm wide located at whatever distance we may choose. Let's choose $r = 0.2$ m, as in Example 12.10. The flux *per square meter* of band area at that distance will be $D = \psi/S$. Now

$$S = 2\pi \times 0.2 \times 0.01 = 0.004\pi \text{ m}^2$$
$$D = 10^{-7} \div 0.004\pi = 79.6 \times 10^{-7} \text{ C/m}^2$$

Now let's get D from $\mathscr{E}$ in Example 12.10:

$$D = \epsilon_0 \mathscr{E} = 8.854 \times 10^{-12} \times 9 \times 10^5 = 79.6 \times 10^{-7} \text{ C/m}^2$$

Problems

8. A long straight conductor is tubular in form and has a radius of 0.5 cm. Its axis coincides with the Z axis, which means it is perpendicular to the XY plane. It has a charge of -40 μC/cm length. How much electric flux in its field passes through a cylindrical surface 1 m in diameter and 2 m long, if the axis of the cylindrical surface coincides with the Z axis? What is the direction of the flux?

9. What is the flux density, in vector form, just off the surface of the conductor of Problem 8?

FIELD PRODUCED BY MORE THAN ONE SOURCE

Because D and $\vec{\mathscr{E}}$ have direction in addition to magnitude, the flux density and and field intensity produced by two or more sources is the resultant (vector sum) of the separate contributions made by the sources considered separately. Instead of using vector methods, which is an easy way to obtain the resultant, we shall use simple trigonometry.

Let us consider two sources of field flux, one a line charge and the other a concentrated charge. If we had only concentrated charges, each one would be used in the solution just as the single one is in the following example.

EXAMPLE 12.12 A long straight conductor coincides with the Z axis with its center at the origin. It has a distributed charge of $\rho_L = 10^{-5}$ C/m. A concentrated charge $Q = 50 \times 10^{-6}$ C is located at the point $x = 0$, $y = 4$ m. The medium is a gas for which the relative permittivity is 2. What are the flux density and field intensity at the point $x = 3$, $y = 4$? See Fig. 12.6.

Solution. Contribution to D by the line charge: $D_L = 10^{-5}/2\pi \times 5 = 0.318 \times 10^{-6}$ C/m². It makes an angle $\theta = \cos^{-1} 0.6 = 53.1°$ with the x axis. Contribution by the concentrated charge $= D_Q = 50 \times 10^{-6}/4\pi \times 3^2 = 0.442 \times 10^{-6}$ in the $+x$ direction. We must add the two contributions. We get the component (effect) of D_L along the x axis and add it to D_Q. This will be D_x. The x component of D:

$$D_{Lx} = 0.318 \times 10^{-6} \cos \theta = 0.318 \times 10^{-6} \times 0.6 = 0.1908 \times 10^{-6}$$
$$D_x = 0.1908 \times 10^{-6} + 0.442 \times 10^{-6} = 0.6328 \times 10^{-6}$$

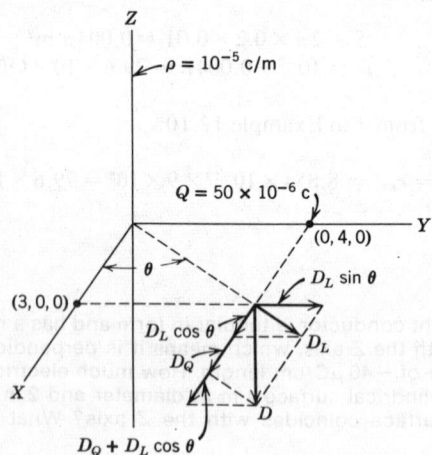

FIGURE 12.6   For Example 12.12.

The Y component of D is obtained by using the sine of θ:

$$D_y = 0.318 \times 10^{-6} \sin \theta = 0.318 \times 10^{-6} \times 0.8 = 0.2544 \times 10^{-6}$$

The net flux density is the resultant:

$$D = \sqrt{0.6328^2 + 0.2544^2} \times 10^{-6}$$
$$D = \sqrt{0.4004 + 0.0647} \times 10^{-6}$$
$$D = 0.682 \times 10^{-6} \text{ C/m}^2$$

The angle α that D makes with the x axis is

$$\alpha = \cos^{-1}(0.6328/0.682)$$
$$\alpha = \cos^{-1} 0.928 = 21.9°$$

The field intensity is $\mathscr{E} = D/\epsilon = D/\epsilon_r\epsilon_0$

$$\mathscr{E} = 0.682 \times 10^{-6}/2\epsilon_0$$
$$\mathscr{E} = 0.341 \times 10^{-6}/8.854 \times 10^{-12}$$
$$\mathscr{E} = 38.5 \times 10^3 \text{ N/C}$$

in the same direction as **D**.

Problems

10. Two concentrated charges in air are: $Q_1 = 5 \times 10^{-5}$ C at $(4, 0, 0)$ and $Q_2 = -10^{-4}$ C at $(0, 8, 0)$. Find the magnitude of the flux density produced at the orign by these two charges acting together.

11. Find the magnitude and direction of the field intensity produced by the charges of Problem 10.

12. Suppose a positive charge $Q_3 = 25 \times 10^{-6}$ is placed at the point $x = 3$, $y = 0$ and that the line charge and the point charge of Example 12.11 are present also. Find the magnitude and direction of the field intensity at the point $x = 3$, $y = 4$ produced by all three charges.

12.8 Elastance

The reciprocal of capacitance is called *elastance*. Its symbol in this book is $\mathscr{S}$ and the unit is the *daraf*, which is the word *farad* spelled backward. It is a very small unit, so the megadaraf (10^6 darafs) is more practical.

$$\mathscr{S} = \frac{1}{C} \text{ darafs}, \qquad \text{when } C \text{ is in farads} \qquad (12.22)$$

Since $V = Q/C$, and Q coulombs of charge produce an equal number of coulombs of electric flux, represented by ψ,

$$V = \psi \mathscr{S} \text{ volts} \qquad (12.23)$$

Elastance in a capacitor is similar to resistance in a wire. Therefore the last equation indicates a relationship similar to Ohm's law. *When capacitors are in series, the flux has the same value in all of them.* This is because the same amount of charge is transferred through all of them.

12.9   Capacitors in Series

Figure 12.7 shows three capacitors in series across a dc voltage. The reason the electric flux ψ has the same value in all three is easily understood if the capacitors are regarded as *uncharged* before E is applied. During the charging process the source applying the potential difference E will take as many electrons from the left-hand plates of C_1 as it supplies to the right-hand plates of C_3. The net positive charge on C_1 is thus equal to the net negative charge on C_3. Each capacitor has *equal and opposite charges* on its plates. This means each dielectric has the same amount of flux passing through it. Using Equation (12.23) we can write

$$V_1 = \psi \mathscr{S}_1, \quad V_2 = \psi \mathscr{S}_2, \quad V_3 = \psi \mathscr{S}_3 \qquad (12.24)$$

Since

$$V = \psi(\mathscr{S}_1 + \mathscr{S}_2 + \mathscr{S}_3)$$
$$V = V_1 + V_2 + V_3 \qquad (12.25)$$

so that the equivalent elastance is

$$V/\psi = \mathscr{S}_1 + \mathscr{S}_2 + \mathscr{S}_3 = \mathscr{S}_e \qquad (12.26)$$

Therefore the equivalent elastance $\mathscr{S}_e$ of a number of capacitors in series is equal to the sum of the individual elastances. Since $C = 1/\mathscr{S}$, the *equivalent*

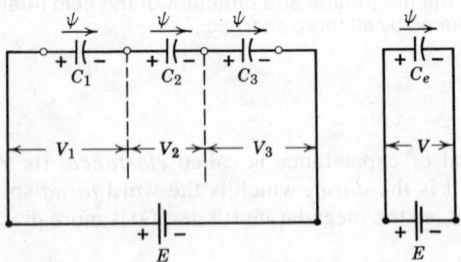

FIGURE 12.7   Three capacitors in series and their equivalent C_e. Equation (12.27) defines C_e.

capacitance of the three in series is C_e

$$C_e = \frac{1}{\dfrac{1}{C_1} + \dfrac{1}{C_2} + \dfrac{1}{C_3}} \text{ farads} \qquad (12.27)$$

It is seen that capacitances in series combine to give the equivalent capacitance as resistances in parallel combine to give the equivalent resistance.

Using the first two equations of (12.24),

$$\frac{V_1}{V_2} = \frac{\psi \mathscr{S}_1}{\psi \mathscr{S}_2} = \frac{C_2}{C_1}$$

which shows that when two capacitors are in series, their voltage drops are inversely proportional to their capacitances and directly proportional to their elastances. When three capacitances are in series, the voltage across *any one of them* is to the total voltage as its elastance is to the sum of the three elastances. This can be shown by using any one of the three equations of (12.24) and dividing it by Equation (12.25),

$$\frac{V_x}{V} = \frac{\mathscr{S}_x}{\mathscr{S}_1 + \mathscr{S}_2 + \mathscr{S}_3} \qquad (12.28)$$

Where the subscript x can be 1, 2, or 3, and V is the total voltage. This last equation is the same as

$$\frac{V_x}{V} = \frac{1/(\mathscr{S}_1 + \mathscr{S}_2 + \mathscr{S}_3)}{1/\mathscr{S}_x} = \frac{C_e}{C_x} \qquad (12.29)$$

This gives V_x on any one of three capacitors connected in series in terms of their equivalent capacitance and the capacitance of the one chosen.

EXAMPLE 13   A 1-μf, a 2-μf, and a 4-μf capacitor are connected in series across 250-volts direct current. Calculate the value of C_e, the equivalent capacitance of the group. Calculate the voltages V_1, V_2, and V_3 across the corresponding capacitors.

Solution. The elastances are 1, $\frac{1}{2}$, $\frac{1}{4}$ megadarafs and the elastance of the equivalent capacitor is:

$$\mathscr{S}_e = 1 + 0.5 + 0.25 = 1.75 \text{ megadarafs}$$

$$C_e = \frac{1}{\mathscr{S}_e} = \frac{4}{7} \mu\text{F}$$

$$\frac{V_1}{250} = \frac{1}{1.75}, \qquad V_1 = 142.9 \text{ V}$$

$$\frac{V_2}{250} = \frac{0.5}{1.75}, \qquad V_2 = 71.4 \text{ V}$$

$$\frac{V_3}{250} = \frac{0.25}{1.75}, \qquad V_3 = 35.7 \text{ V}$$

$$V_1 + V_2 + V_3 = E = 250.0 \text{ V}$$

12.10　Capacitors in Parallel

Figure 12.8 shows three capacitors in parallel and their equivalent C_e. One can reason that their equivalent is a single capacitor whose plate area is the sum of the areas of the plates of all three, because it must accommodate the *total charge* on all three at the *same voltage*. Since $C = Q/E$, C must increase the same amount Q does, if E remains unchanged. To derive the expression for C_e, the relationships learned in deriving the equivalent capacitance for the series connection are used.

The flux in each capacitor is equal to its voltage divided by its elastance: therefore,

$$\psi_T = \frac{E}{\mathscr{S}_1} + \frac{E}{\mathscr{S}_2} + \frac{E}{\mathscr{S}_3} = E\left(\frac{1}{\mathscr{S}_1} + \frac{1}{\mathscr{S}_2} + \frac{1}{\mathscr{S}_3}\right)$$

$$\frac{\psi_T}{E} = \frac{1}{\mathscr{S}_1} + \frac{1}{\mathscr{S}_2} + \frac{1}{\mathscr{S}_3} = \frac{1}{\mathscr{S}_e}$$

But $1/\mathscr{S}_x = C_x$ and $1/\mathscr{S}_e = C_e$: therefore

$$C_1 + C_2 + C_3 = C_e \text{ farads} \tag{12.31}$$

Problems

13. The capacitances in microfarads of three capacitors are $C_1 = 0.01$, $C_2 = 0.05$, and $C_3 = 0.10$. Compute the total capacitance when they are connected (a) all in series; (b) all in parallel; and (c) C_1 and C_2 in parallel and C_3 in series with the pair.

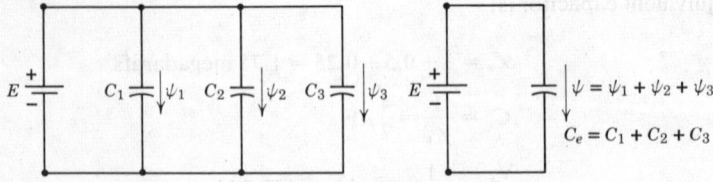

FIGURE 12.8　Three capacitors in parallel and their equivalent C_e.

14. A constant potential difference of 100 V is applied to the group of capacitors in part *a* of Problem 13. What is the voltage across each capacitor? Solve this problem when it applies to group *c* of Problem 13.

An interesting situation develops when two or more charged capacitors of different sizes are connected together, as when the switch is closed in Fig. 12.9*a*. Because there is no loss of charge, $Q_1 + Q_2$ remains the total charge on the equivalent capacitor for which $C_e = C_1 + C_2$. The voltages V_1 and V_2, usually different, will change to a common value, V, after the switch is closed.

EXAMPLE 12.14   In Fig. 12.9*a*, let $C_1 = 0.02 \, \mu F$, $V_1 = 100 \, V$, $C_2 = 0.03 \, \mu F$, $V_2 = 200 \, V$. Determine V_e and the charge on each capacitor after the switch has been closed.

Solution

$$Q_1 = 0.02 \times 10^{-6} \times 100 = 2 \times 10^{-6} \, C$$
$$Q_2 = 0.03 \times 10^{-6} \times 200 = 6 \times 10^{-6} \, C$$

The equivalent capacitance is

$$C_e = (0.02 + 0.03)10^{-6} = 0.05 \times 10^{-6} \, F$$

The equivalent voltage is

$$V_e = (Q_1 + Q_2)/C_e = 8 \times 10^{-6}/0.05 \times 10^{-6} = 160 \, V$$
$$Q_1' = 0.02 \times 10^{-6} \times 160 = 3.2 \times 10^{-6} \, C$$
$$Q_2' = 0.03 \times 10^{-6} \times 160 = 4.8 \times 10^{-6} \, C$$

EXAMPLE 12.15   A tricky situation exists when the polarities of the voltages are opposite, as in Fig. 12.9*b*. We then take $Q_2 = -6 \times 10^{-6} \, C$. The net charge that prevails after the switch has been closed is $2 \times 10^{-6} - 6 \times 10^{-6} = -4 \times 10^{-6} \, C$:

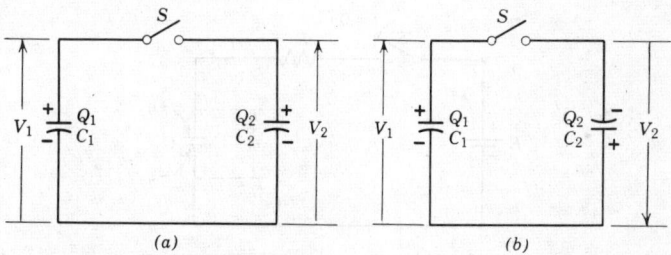

(a)   (b)

FIGURE 12.9   Two charged capacitors to be connected together. (a) Like-polarity terminals to be joined. (b) Unlike-polarity terminals to be joined.

Solution

$$C_1V_1 + C_2V_2 = C_eV_e$$
$$0.02 \times 10^{-6} \times 100 + [0.03 \times 10^{-6}(-200)] = 0.05 \times 10^{-6} V_e$$
$$V_e = (2-6)/0.05 = -80 \text{ V on both capacitors}$$

Their polarities are the same: minus on bottom, plus on top.
Let's check the individual charges:

$$Q_1' = 0.02 \times 10^{-6}(-80) = -1.6 \times 10^{-6} \text{ C}$$
$$Q_2' = 0.03 \times 10^{-6}(-80) = -2.4 \times 10^{-6} \text{ C}$$
$$Q_1' + Q_2' = -4 \times 10^{-6} \text{ C}$$

12.11 Charging a Capacitor from a DC Source

From the definition of capacitance, $C = Q/V$, it is known that the charge q on a capacitor at any instant is equal to the capacitance times the voltage at that instant (refer to Fig. 12.10):

$$q = Cv_c \qquad (12.32)$$

For a very small change in voltage, Δv_c, the change in the charge is

$$\Delta q = C\Delta v_c$$

If that change is accomplished in a very small time interval Δt, we may write $\Delta q/\Delta t = C(\Delta v_c/\Delta t)$ coulombs per second (amperes). But rate of flow of charge is current (Δq coulombs flowed in the time Δt), so

$$\frac{\Delta q}{\Delta t} = i_c = C\frac{\Delta v_c}{\Delta t} \text{ amperes} \qquad (12.33)$$

$$q = Cv_c, \qquad dq = Cdv_c, \qquad \frac{dq}{dt} = i_c = C\frac{dv_c}{dt} \qquad (12.34)$$

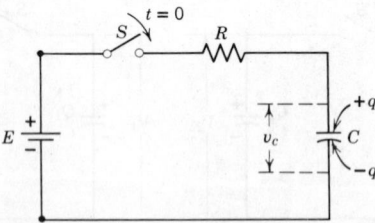

FIGURE 12.10 A capacitor charging through a series resistance. Charges $+q$ and $-q$, equal in amount, build up on the capacitor. The maximum value becomes $Q = CE$ coulombs. C is in farads, E in volts.

This shows that capacitor current depends on the size of the capacitor (i.e., its capacitance C) and the *rate at which its voltage is changing with time.* It is very important to note that when the voltage across a capacitor is not changing, the current is zero, and that the faster the voltage is changing the larger is the current. Maximum current flows when the rate of change of the capacitor voltage, with respect to time, is a maximum. In the circuit of Fig. 12.10 the capacitor C has no charge ($q = 0$) at $t = 0$, the instant the switch is closed, and therefore the voltage on it starts at zero:

$$v_c|_{t=0} = \frac{q}{C} = 0$$

At the instant the switch makes contact, the *full value of E is impressed across R* and the initial value of current is

$$i_c|_{t=0} = \frac{E}{R} \text{ amperes}$$

Using this for i_c in Equation (12.33) we can write for the instant $t = 0$:

$$\frac{E}{R} = C \frac{\Delta v_c}{\Delta t}\Big|_{t=0}$$

$$\frac{\Delta v_c}{\Delta t}\Big|_{t=0} = \frac{E}{RC} \text{ volts per second} \tag{12.35}$$

which is the *initial rate of rise* of the capacitor voltage. If it were possible for the capacitor voltage to continue rising at *this constant rate*, it would reach the full voltage E of the battery in RC seconds. This is evident when one observes that a constant rate of E/RC (volts per second) times RC (seconds) would give a rise of E in that time. RC is the *time constant* of the circuit. This will be verified in the exponent of ϵ when the equation of the instantaneous voltage is obtained, just as it was in the exponent in the equation of current in an inductance. Since we mentioned time constant, let's show that the *units* of RC *are seconds*. R = volts/amperes, C = coulombs/volts

$$RC = \frac{\text{volts}}{\text{amperes}} \times \frac{\text{coulombs}}{\text{volts}} = \frac{\text{coulombs}}{\text{amperes}} = \text{seconds}$$

At the instant the capacitor voltage reaches the battery voltage E, the current stops flowing, that is, it becomes zero. The current decreases during the charging process because the voltage building up on the capacitor opposes the flow of current.

An important fact about the changing of voltage on a capacitor should be brought out here. At the instant the switch in Fig. 12.10 is closed, all of the voltage E appears instantly across the resistance R. None of it appears instantly across C. *It is not possible for the voltage across a capacitor to change in zero time.* For that to be possible, the capacitor current would have to be infinitely large, as will now be shown. Rewriting Equation (12.33),

$$i_c = C \frac{\Delta v_c}{\Delta t} \text{ amperes}$$

In order for a change (Δv_c) in capacitor voltage to take place *in zero time*, the time interval Δt would have to be zero. This would make the quotient $\Delta v_c / \Delta t$ infinite. Since C has some finite value (i.e., C is not zero or infinite) the right-hand side would become infinite. This means the current would be infinite, which is impossible. Therefore the voltage across a capacitor cannot change in zero time.

Using Kirchhoff's voltage law after $t = 0$ and because current is flowing, the following equation holds at any instant.

$$E = i_c R + v_c = i_c R + \frac{q}{C} \tag{12.36}$$

Remembering that E is a *constant voltage*, any incremental increase in voltage due to an increase in the charge q must be balanced by an equal decrease in iR,

$$\frac{\Delta q}{C} = -\Delta i_c R \tag{12.37}$$

These changes occur simultaneously in the time increment Δt. Dividing both sides by Δt

$$\frac{1}{C} \frac{\Delta q}{\Delta t} = -\frac{\Delta i_c}{\Delta t} R$$

But the time rate of change of charge is *current*.

$$\frac{1}{C} i_c = -\frac{\Delta i_c}{\Delta t} R, \quad \frac{\Delta i_c}{\Delta t} = -\frac{i_c}{RC} \text{ amperes per second} \tag{12.38}$$

This rate of change of current with respect to time evidently depends on the amount of current flowing at the instant the rate of change is considered. At the instant the switch is closed ($t = 0$) the current is a maximum because there is no voltage drop across the capacitor and all of the applied voltage is available for sending current through the resistance.

$$\left. \frac{\Delta i_c}{\Delta t} \right|_{t=0} = -\frac{E/R}{RC} = -\frac{E}{R^2 C} \text{ amperes per second} \tag{12.39}$$

This is the *maximum value of* $\Delta i_c / \Delta t$. Notice that it is negative, which means that the current is *decreasing*.

The curves of instantaneous current and voltage of the capacitor while it is charging through the resistance R are shown in Fig. 12.11. The equation for instantaneous current may be obtained by continuing from Equation (12.38):

$$\frac{\Delta i_c}{i_c} = \frac{-\Delta t}{RC}$$

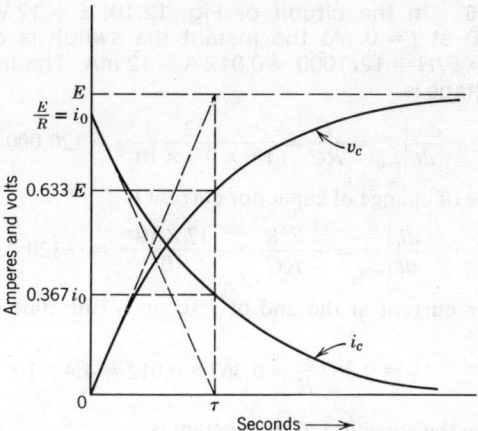

FIGURE 12.11 Curves showing instantaneous values of current and voltage of a capacitor charging through a resistor.

If you are acquainted with elementary calculus you have recognized before this that incremental expressions such as Δi and Δt may well be written as *differentials di and dt*.

The last expression is a simple equation in calculus to solve when expressed as follows:

$$\frac{di_c}{i_c} = -\frac{dt}{RC} \tag{12.40}$$

The solution, given in the Appendix in Section A4, is

$$i_c = \frac{E}{R} \epsilon^{-t/RC} \text{ amperes} \tag{12.41}$$

When $t = \tau = RC$,

$$i_c = \frac{E}{R} \epsilon^{-1} = 0.367 \frac{E}{R} \text{ amperes}$$

Therefore, at the end of the time-constant interval the current has gone through 63.3 percent of the total change it will make. Also, at $t = \infty$, $i_c = 0$.

The equation for the instantaneous voltage v_c on the capacitor is readily found using Equation (12.41) because v_c is the applied voltage minus the voltage drop v_R across the resistor, the latter being i_cR.

$$v_c = E - i_cR = E - E\epsilon^{-t/RC}$$
$$v_c = E(1 - \epsilon^{-t/RC}) \text{ volts} \tag{12.42}$$
$$v_R = E\epsilon^{-t/RC} \text{ volts}$$

When $t = 0$, $v_c = 0$, and when $t = \infty$, $v_c = E$.

EXAMPLE 16   In the circuit of Fig. 12.10, $E = 12$ V, $R = 1000 \, \Omega$, $C = 0.1 \, \mu$F, $v_c = 0$ at $t = 0$. At the instant the switch is closed ($t = 0$): The current is $i = E/R = 12/1000 = 0.012$ A $= 12$ mA. The initial rate of rise of capacitor voltage is

$$\frac{dv_c}{dt}\Big|_{t=0} = \frac{E}{RC} = \frac{12}{10^3 \times 0.1 \times 10^{-6}} = 120{,}000 \text{ V/s}$$

The initial rate of change of capacitor current is

$$\frac{di_c}{dt}\Big|_{t=0} = -\frac{E/R}{RC} = -\frac{12 \times 10^{-3}}{10^{-4}} = -120 \text{ A/s}$$

The capacitor current at the end of τ seconds (the time-constant interval) is

$$i_c = 0.367 \frac{E}{R} = 0.367 \times 0.012 = 4.4 \times 10^{-3} \text{ A}$$

The voltage on the capacitor at that instant is

$$v_c = 12 - 4.4 \times 10^{-3} \times 1000 = 7.6 \text{ V}$$

The instantaneous value of current at the end of any time interval t is obtained from Equation (12.41) when expressed as follows

$$i_c = 12 \times 10^{-3} \, \epsilon^{-10^4 t}$$

and the capacitor voltage at that instant is given by $E - iR_c$. Suppose $t = 2\tau = 2 \times 10^{-4}$ s. At that instant

$$i_c = 12 \times 10^{-3} \, \epsilon^{-2} = 1.624 \times 10^{-3} \text{ A}$$

and the capacitor voltage at that instant is

$$12 - 1.624 \times 10^{-3} \times 1000 = 10.376 \text{ V}$$

Problems

15. A capacitor ($C = 2 \, \mu$F) and a resistance ($R = 100$ kΩ) are connected as in Figure 12.10 and $E = 100$ volts is suddenly applied. What are (a) the initial and final values of current, (b) the time constant, and (c) the current at the end of the time constant interval?

16. In Problem 15, find (a) the voltages across C and R at $t = 0$ and $t = \infty$, (b) the voltages across C and R at $t = \tau$.

17. (a) What is the current in the circuit of Problem 15 at the instant $t = 0.3$ s? (b) What are the voltages across C and R at that instant?

18. What are the values of the rate of change of current in the circuit of Problem 15 when (a) $t = 0$ and (b) $t = \infty$?

19. How long will it take the current in Problem 15 to decrease to half its initial value?

12.12   Energy Storage in a Capacitor

When a capacitor is in the process of being charged, small amounts of charge are continually added until the total charge Q coulombs is present. During the addition of the tiny increments of charge Δq, the capacitor voltage undergoes a corresponding incremental increase Δv_c such that

$$\Delta q = C\Delta v_c \text{ coulombs}$$

in which C is the capacitance in farads.

During the addition of Δq, the instantaneous current i is the time rate of change of charge.

$$i = \frac{\Delta q}{\Delta t} = C\frac{\Delta v_c}{\Delta t}$$

Instantaneous power is voltage times current:

$$p = v_c i = Cv_c\frac{\Delta v_c}{\Delta t}\text{watts}$$

Energy is power times time, so the contribution to the energy in the electric field between the capacitor plates during the time interval Δt is

$$\Delta w = p(\Delta t) = Cv_c\frac{\Delta v_c}{\Delta t}\Delta t = Cv_c\Delta v_c$$

Summing up all of the Δw contributions that occur while the capacitor voltage is being raised from zero to V_c during the charging process, we use the same symbolism as with the magnetic-field energy case in Section 10.5.

$$W_C = \sum_{v_c=0}^{v_c=V_c} Cv_c\Delta v_c = \tfrac{1}{2}CV_c^2\text{ joules} \tag{12.43}$$

In calculus form, the more precise derivation is as follows: The capacitor-charging current is

$$i = \frac{dq}{dt} = \frac{d}{dt}(Cv_c) = C\frac{dv_c}{dt}$$

The instantaneous power is

$$p = v_c i = v_c C\frac{dv_c}{dt}$$

The incremental energy during time dt is

$$dW = v_c C\frac{dv_c}{dt}\cdot dt = v_c C\,dv_c$$

The total energy delivered to the capacitor while its voltage increases from zero to V_c volts is

$$W = \int_0^{V_c} Cv_c \, dv_c = \tfrac{1}{2}CV_c^2 \text{ joules} \qquad (12.43)$$

This is integrated in Appendix A.5.

This is the equation for calculating the energy in the electric field of a capacitor of C farads with V rms volts between its plates *in the steady-state case*.

Since $\qquad\qquad CV = Q, W = \tfrac{1}{2}QV = \tfrac{1}{2}Q^2C \qquad (12.45)$

12.13   Discharge of a Capacitor

The capacitor in the circuit shown in Fig. 12.12 has an initial voltage V_0 at t_0, the instant the switch is closed. There is no way for V_0 to be kept constant, so it will start to decrease at the instant current starts flowing.

The incremental change in capacitor voltage will equal the incremental change in the iR drop. The capacitor voltage is decreasing, so its incremental change is negative.

$$\Delta v_c = -\Delta i_c R \qquad (12.46)$$

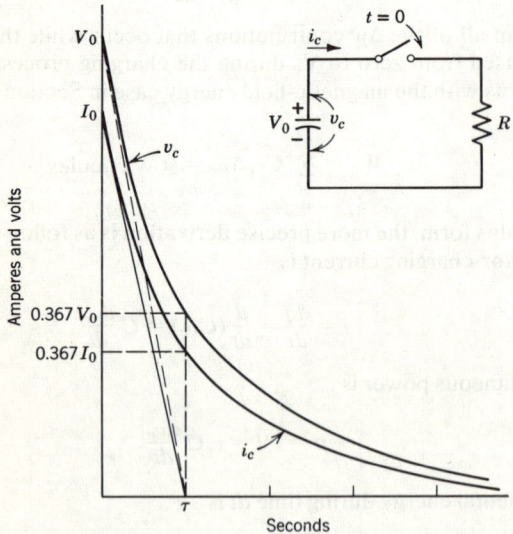

FIGURE 12.12   Capacitor discharge through a resistance. Circuit and curves.

Several times we have used the relation $C = \Delta q / \Delta v_c$, from which we get

$$\Delta v_c = \frac{\Delta q}{C}$$

Using Equation (12.46),

$$- \Delta i_c R = \frac{\Delta q}{C}$$

$$i_c R + \frac{\Delta q}{C} = 0$$

Dividing by Δt,

$$\frac{\Delta i_c}{\Delta t} R + \frac{\Delta q}{\Delta t} \cdot \frac{1}{C} = 0$$

The calculus form of this equation is

$$R \frac{di_c}{dt} + \frac{1}{C} i_c = 0 \tag{12.47}$$

The solution of this equation (given in Appendix A.6) is

$$i_c = \frac{E_0}{R} \epsilon^{-t/RC} \text{ amperes} \tag{12.48}$$

in which E_0 is the potential difference between the capacitor terminals at $t = 0$.

Comparison of this equation with Equation (12.41) shows that the instantaneous current at any instant during discharge is the same at the end of the same time interval as it is during charge if $E_0 = E$.

12.14   Timing Circuits

In our study of R-L circuits we learned that the time from the instant voltage is applied until the current, or the voltage across a resistor, reaches a definite value can be delayed a desired amount. This is accomplished by the choice of the time constant or the applied voltage. Use of the R-C circuit is more practical because it has the advantage of more convenient and less costly changes of C and the same options of selection of R.

Vacuum tubes and gas tubes, as well as transistors, can be caused to conduct when certain control voltages reach particular values. In some cases the voltage starts out at too large a value and immediately begins decreasing as does v_C which equals v_R in Fig. 12.12. If the application requires voltage to build up from zero or some other small value, the capacitor-charging circuit may be used and the voltage picked off R in Fig. 12.10.

EXAMPLE 12.17   When the double-pole switch is thrown to the left in Fig. 12.13 the 4 μF capacitor quickly charges to 100 V with the indicated

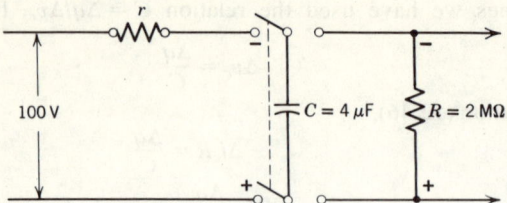

FIGURE 12.13   For Example 12.17. Discharge of *C* through *R* provides delay time during which voltage across *R* decreases to a desired critical value.

polarity. At the instant it is caused to make contact on the right, the voltage across R starts to decrease from 100 V and in 8 s (τ) it is down to 36.7 V. It is obvious that changing R to 0.5, 0.25, or 0.1 times its value will reduce the time constant correspondingly and thus the voltage will fall to 36.7 V in 4, 2, or 0.8 s.

Application. The capacitor-charging circuit is readily adapted to timing control where an increase in voltage is desired. The voltage across the capacitor is applied to the controlled device. In the case when 100 V is applied to $C = 4\,\mu\text{F}$ $R = 2\,\text{M}\Omega$ in series, the capacitor voltage will rise from zero to 36.7 V in 8 s. If shorter, or longer, time intervals are desired, R or C may easily be chosen to give the desired result. Figure 12.14 shows various time delays in an R-C circuit with short, medium, and relatively long time constants

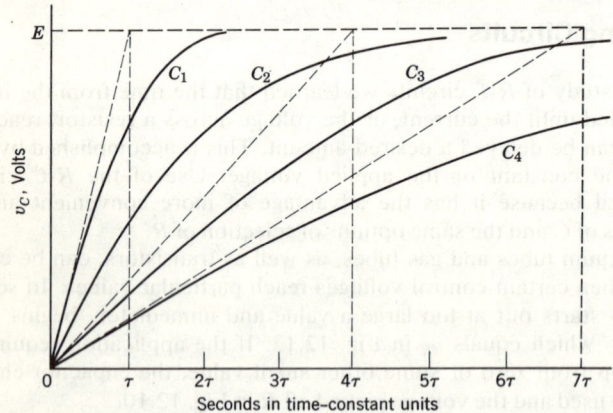

FIGURE 12.14   Time delays in rise of capacitor voltage in *R–C* circuits, like the one in Fig. 12.10, that have short (τ) time constants, medium (2τ), and relatively long (7τ) time constants. If *R* is constant, then $C_4 > C_3 > C_2 > C_1$.

Problems

20. A 5 μF capacitor, connected in series with a 2 MΩ resistor, is charged to 120 V. After the applied voltage is removed, the capacitor is connected across the resistor. Find (a) the voltage on the capacitor at the end of τ seconds (b) 2τ seconds, (c) the current at each instant.

21. How much energy was stored in the field of the capacitor of Problem 20 at the start of its discharge, and how much at the end of 2τ seconds?

22. How much time is required for the voltage of the 5 μF capacitor in Problem 20 to decrease to 50 V?

23. At the instant the voltage of the 5 μF is 50 V, what are (a) the energy in the field of the capacitor, (b) the current, (c) the rate at which energy is being dissipated in the resistor? This refers to Problem 22.

24. Prove that when a capacitor starts discharging through a resistance, the rate of change of current at $t = 0$ is given by V/R^2C in which V is the voltage on the capacitor at $t = 0$, and R and C represent the usual quantities.

25. (a) How much charge is on the plates of a 0.1 μ F capacitor when the potential difference between its terminals is 20 V? (b) At what rate does its discharge current change at the instant it is connected to a 50 k resistor? (c) How long will it take for the voltage across the resistor to decrease to 5 V?

12.15   Electron Motion in an Electrostatic Field

If a particle with a charge $+q$ coulombs is released at *zero velocity* in an electric field of intensity $\vec{\mathscr{E}}$, a force $\mathbf{F} = q\vec{\mathscr{E}}$ is exerted by the field on the charge This force causes the charge (the mass of which is represented by m) to move with an acceleration according to the familiar relation $F = ma$. The acceleration is then

$$\mathbf{a} = \frac{q\vec{\mathscr{E}}}{m} \text{ meters per second squared} \tag{12.49}$$

When the force is constant, the particle (starting from rest) acquires a velocity U in t seconds given by

$$U = at \text{ meters per second} \tag{12.50}$$

and in that time it will travel a distance d:

$$d = \tfrac{1}{2}at^2 \text{ meters} \tag{12.51}$$

The *kinetic energy* of the charged particle is then

$$KE = \tfrac{1}{2}mU^2 \text{ joules} \tag{12.52}$$

This energy has the same numerical value as is calculated using $W = \mathscr{E}q_e$, if the charge moved through a total potential difference of E volts.

EXAMPLE 12.18   An electron starts from rest in an electrostatic field between two parallel plates in a vacuum. It travels from a point just off the negative plate and is raised 100 V in potential by the time it reaches the positive plate, which is 0.005 m away. We shall determine important numerical values related to the electron's motion.

Solution. Force on the electron:

$$F = \mathscr{E}q_e = (100/0.005)(1.602 \times 10^{-19})$$
$$= 3.204 \times 10^{-15} \text{ N}$$

Acceleration: $a = \dfrac{F}{m} = (3.204 \times 10^{-15})/9.11 \times 10^{-31}$

$$= 3.52 \times 10^{15}$$

meters per second squared

Time of travel:

$$d = \tfrac{1}{2}at^2, \qquad t = \sqrt{\frac{2d}{a}} = \sqrt{\frac{2 \times 0.005}{3.52 \times 10^{15}}} = 1.685 \times 10^{-9} \text{ s}$$

Velocity at the positive plate:

$$U = at = 3.52 \times 10^{15} \times 1.685 \times 10^{-9}$$
$$= 5.93 \times 10^6 \text{ m/s}$$

Kinetic energy upon striking the plate: $KE = \tfrac{1}{2}mU^2$

$$KE = \tfrac{1}{2} \times 9.11 \times 10^{-31} \times (5.93 \times 10^6)^2 = 1.602 \times 10^{-17} \text{ J}$$

Check: $W = Vq_e = 100 \times 1.602 \times 10^{-19} = 1.602 \times 10^{-17} \text{ J}$

12.16   Electron Motion in a Steady Magnetic Field

A steady magnetic field does not exert a force on an electrically charged particle at rest. It does exert force, however, on a charged particle in motion, but the force is *always perpendicular* to the direction of motion of the particle. The force is always at right angles to the direction in which the particle is proceeding, and because of this the force cannot change the *magnitude* of the particle's velocity. This means the field cannot supply energy to a charged particle, so the kinetic energy of the particle remains unchanged. This is in contrast to the effect of an electric field on a charged particle, which, as we have seen, exerts a force that is independent of the direction of motion of the particle and causes an *energy transfer* between the field and the particle.

The magnitude of the force exerted by a steady magnetic field of flux density B (webers per square meter) on a moving positive charge Q (coulombs) with velocity U (meters per second) is given by

$$F = QUB \sin \theta \text{ newtons} \tag{12.53}$$

in which θ is the angle between the *direction of motion* of the particle and the *direction of the magnetic flux*. The direction of the force is perpendicular to the direction of motion (i.e., the direction of instantaneous velocity U) and also perpendicular to the magnetic-flux lines.

The force on a positive charge moving with a linear velocity U will act in the same direction as a force on a current I that flows in the direction in which the positive charge is moving. If the thumb, forefinger, and the next finger of the right hand are extended at right angles to one another, and so placed that the thumb is in the direction of the current and the forefinger in the direction of the magnetic field, the middle finger will point in the direction of the force.

When the moving charge is negative (an electron or a negative ion), one must consider the direction of the *current* to be opposite to the direction of the moving negative charge.

If the directions of charge motion and flux are not at right angles, the thumb and forefinger may still be pointed as specified above and the third finger pointed at right angles to both. One may not be able, physically, to force his thumb and forefinger more than 135° apart, but he can use his imagination if necessary.

Figure 12.15a shows a *positively charged* particle that started along the x axis in a magnetic field B which is parallel to the Y axis. The force F on the particle is shown with it in four different positions in its circular path in the X-Z plane. In this case, $\sin \theta = \sin 90° = 1$. The force is obviously larger than

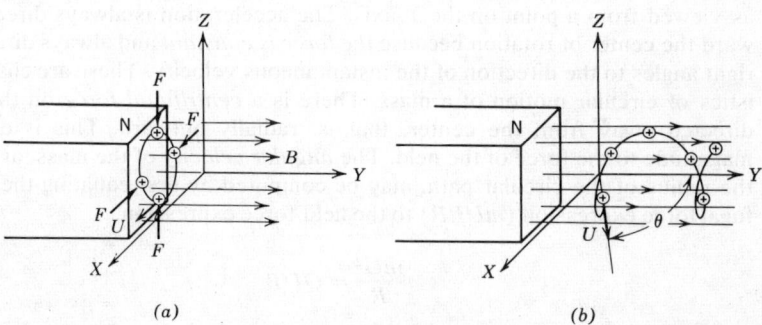

(a) (b)

FIGURE 12.15 (a) A positively charged particle started perpendicular to a uniform B field moves in a circular path in the field. (b) A positively charged particle started at an angle θ with a uniform B field moves in a helical path.

it would be for any other value of θ except 270°, in which case the particle would be rotating in the opposite direction.

In Fig. 12.15*b* the angle θ has been chosen to be 60°. This is readily seen to indicate that there is a component of the velocity U which is parallel to the Y axis and to the magnetic field. Since the field force cannot change the magnitude of the particle's velocity, it will progress in the Y direction while it rotates. Its path will describe a *helix*. The force will be given by

$$F = QUB \sin 60° = 0.866\, QUB \text{ newtons}$$

As the particle rotates, it will move in its helical path and progress parallel to the Y axis with a constant velocity of $U \cos 60° = 0.5\, U$ meters per second.

EXAMPLE 12.19   In Figure 12.15*b* let $Q = 1.602 \times 10^{-19}$ coulomb, $U = 5 \times 10^7$ meter per sec, $B = 0.5$ wb per sq meter, $\theta = 60°$.

(a) How much force is exerted by the field and how fast is the particle proceeding in the $+Y$ direction?

Solution

$$F = 1.602 \times 10^{-19} \times 5 \times 10^7 \times 0.5 \sin 60° = 3.46 \times 10^{-12} \text{ N}$$
$$U_y = 5 \times 10^7 \cos 60° = 2.5 \times 10^7 \text{ m/s}$$

(b) What is the acceleration produced by this force if the particle's mass is 167.2×10^{-29} kg (a hydrogen ion)?

Solution

$$a = (3.46 \times 10^{-12})/167.2 \times 10^{-29} = 2.07 \times 10^{15} \text{ m/s}^2$$

The force in this example is keeping the particle rotating in a circular path as viewed from a point on the Y axis. The acceleration is always directed toward the center of rotation because *the force is constant* and always directed at right angles to the direction of the instantaneous velocity. These are characteristics of circular motion of a mass. There is a *centrifugal force* on the mass directed away from the center, that is, radially outward. This is equal in magnitude to the force of the field. The *angular velocity* of the mass, as well as the radius of the circular path, may be computed by first equating the centrifugal force expression (mU^2/R) to the field-force expression.

$$\frac{mU^2}{R} = QUB \tag{12.54}$$

from this

$$\frac{mU}{R} = QB$$

The angular velocity is

$$\omega = \frac{U}{R} = \frac{QB}{m} \text{ radians per second} \tag{12.55}$$

The radius of the circular path is

$$R = \frac{Um}{QB} \text{ meters} \tag{12.56}$$

Let us compute the radius of the path of the hydrogen ion in our example.

$$\text{Radius of helix} = \frac{5 \times 10^7 \cos 30° \times 167.2 \times 10^{-29}}{1.602 \times 10^{-19} \times 0.5} = 0.904 \text{ m}$$

The kinetic energy of the particle is

$$KE = \tfrac{1}{2} \times 167.2 \times 10^{-29} (5 \times 10^7)^2 = 2.09 \times 10^{-12} \text{ J}$$

In this example we have assumed that the mass of the particle in motion is the same as the rest mass.

12.17   Practical Applications of Capacitors in DC Circuits

Although capacitors are used to a much greater extent in alternating-current systems, many of which will be studied later in this book, there are many dc applications. We have learned about the use of a capacitor to store energy that is to be released at a given time, and the important part it can play in a timing circuit.

The ability of a capacitor to resist sudden changes in its terminal voltage makes it useful in "smoothing out" ripples (small, fast variations in magnitude) in a dc voltage applied to its terminals. It is used for this purpose in filter circuits in dc power supplies in radio and television receivers and transmitters.

Very high dc voltages are obtained by charging a group of capacitors in parallel to their maximum safe working voltage and then changing their connections to put them in series. Thus, a million volts can be obtained by charging 20 capacitors in parallel to 50,000 V and then connecting them in series. In this way powerful electric fields become available, as well as large amounts of power (rate of supply of energy).

12.18   Review Questions

1. State Coulomb's law, first in words and then in mathematical form.
2. Define electric flux density and give its symbol.
3. Define electric field intensity and give its symbol.

4. How much electric flux is associated with Q coulombs of charge? How does the flux of $+Q$ differ from the flux of $-Q$?

5. What is the flux density at a point 1 m from a concentrated charge of 4π C?

6. Define permittivity. If we interpret Equation (12.7) as giving units of *flux density per unit field strength*, can we say that the permittivity of a material is a *measure* of how easy it is to establish an electric field in it? A given field intensity (strength) between two capacitor plates can establish twice the flux density in a slab of dielectric for which $\epsilon = 2\epsilon_0$ as it can if the dielectric is air.

7. Define relative permittivity.

8. What is the permittivity of free space (vacuum)?

The following 6 questions (9 to 14) pertain to a particular charged parallel-plate capacitor.

9. What is the capacitance if $Q = 10^{-8}$ C and $V = 100$ V?

10. What is the field intensity between the plates if they are 1 cm apart?

11. What is the surface-charge density of the area if the positive plate is 100 cm^2?

12. What is the permittivity of the medium, using the fact that surface-charge density equals flux density in a parallel plate capacitor?

13. What is the formula for the capacitance of a parallel-plate capacitor?

14. How is C changed if the (a) area of plates is doubled, (b) distance between plates is doubled, (c) the dielectric is replaced by one that has five times the permittivity of the original one?

15. Define the electron volt, and give its value in joules.

16. An electron in a vacuum tube leaves the cathode at zero velocity and is accelerated to the anode (plate) which is 100 V above the cathode. With how much energy, in electron volts, does the electron strike the plate? How much in joules?

17. Describe the nature of the electric field produced by (a) a concentrated charge, (b) a uniform line charge on a long, straight conductor, (c) long, straight tubular conductor uniformly charged.

18. How do we calculate the capacitance of a group of capacitors that are connected in parallel? How, when they are all in series?

19. When capacitors are all in series, why does the electric flux in all of them have the same value?

20. Define elastance and give its units.

21. How is the time constant of a series R-C circuit calculated?

22. A voltage V is suddenly applied to a series R-C circuit. Why is the initial current (at $t = 0$) determined by R only?

23. A capacitor C is being charged. What is the formula for the energy stored in the field of C at any value of terminal voltage? Name the units.

24. What is the formula for the instantaneous current while the capacitor in an R-C series circuit is being charged?

25. Answer Question 24 for the case while the capacitor originally charged to voltage V is being discharged through R.

26. Tell how a capacitor is useful as an element in a timing circuit.

27. An electron (charge $-q_e$) is released in an electric field. What is the formula for the force exerted by the field in the electron?

28. An electron arrives at the plate of a vacuum tube with a velocity of U m/sec. What is the formula for its kinetic energy on arrival?

29. Compare the forces and their results when a charged particle is placed at rest in (a) an electric field, and (b) a magnetic field.

30. Answer Question 29 when the particle is injected crosswise to the (a) magnetic field and, (b) electric field.

31. Name at least three useful applications of capacitors in dc circuits.

12.19 Problems

Group A

26. Two parallel flat plates in a vacuum are 5 mm apart and connected to a 250-V source. An electron is released with zero velocity at the negative plate. Calculate how long it takes the electron to reach the positive plate.

27. Calculate the following for the electron in Problem 26: (a) its velocity when it reaches the electrode, (b) the increase in energy given to it. Its mass is 9.11×10^{-31} kg.

28. An electron is injected at right angles into a magnetic field ($B = 0.6$ Wb/m^2) with a velocity of 10^8 m/s. (a) What is the amount of force exerted on the electron? (b) What is the amount and direction of the acceleration given to it?

29. What is the diameter of the circular path of the electron in Problem 28, and what is its kinetic energy?

30. Two identical charges, each 100 μC are located at two adjacent corners of a square 50 cm on a side. Each of the other two adjacent corners has a charge of -100 μC. Find the flux density at the center of the square area.

31. Two long parallel wires are 20 cm apart. One has $\rho_L = +10$ μcoulombs per meter of charge, the other has $\rho_L = -10$ μcoulombs per meter. Find the field intensity at a point midway between the wires. The medium is free space.

32. A long straight metal tube, 0.4 cm outside diameter, has a uniformly distributed surface charge of $20\pi\mu$C per meter length. (a) How much flux passes through an imaginary cylinder defined as follows: radius 0.5 m, length 2 m, axis coincides with the axis of the conductor. (b) Let the axis coincide with the Z axis and assume the existence of two additional point charges: $Q_1 = 500$ μC at $x = 5$ m, $y = 0$, $z = 0$, and $Q_2 = -500$ μC at $x = 0$, $y = 5$ m, $z = 0$. What is the flux density at the point $x = 5$, $y = 5$, $z = 0$ at the midpoint of the tube axis?

33. The capacitances in microfarads of three capacitors are $C_1 = 0.02$, $C_2 = 0.05$, $C_3 = 0.08$. Calculate the total capacitance when they are connected (a) all in series, (b) all in parallel, (c) C_1 and C_3 in parallel and C_2 in series with the pair.

34. A potential difference of 100 V is applied to the capacitors of Problem 33 when connected in series. What is the voltage across each capacitor?

35. Six capacitors of 8 μF each are mounted on a board with solid conductors and switches as shown in Fig. 12.16. (a) What are the switch positions for minimum capacitance, and what is its value? (b) Answer for maximum capacitance also. What are the switch positions for a capacitance of 18 μF?

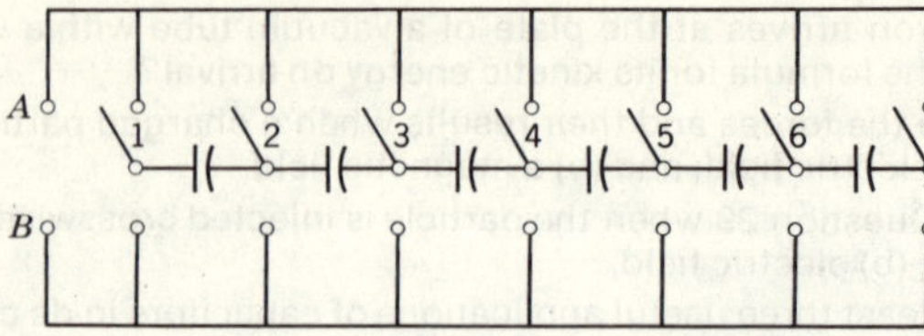

FIGURE 12.16   For Problem 35. An arrangement of capacitors that gives many possible values of capacitance.

36. Two capacitors are connected as shown in Fig. 12.17. After the switch is closed and a steady state is achieved, what are the following for each capacitor: (a) charge, (b) voltage, (c) stored energy after the steady state is reached?

37. Two parallel plates with 500 V between them are separated by a dielectric 4 mm thick for which $\epsilon_R = 5$. The plates have a total area of $4\pi \times 10^{-4}$ m². What are the (a) field intensity in the dielectric, (b) the flux density just off the positive plate, just off the negative plate and at all points in the dielectric (neglect fringing at the edges), (c) charge on each plate, (d) capacitance, (e) energy stored?

38. A 0.5 μF capacitor is in series with a 0.5 MΩ resistor and the pair is suddenly connected to 300 V. (a) at what rate is the current changing at $t = 0$? (b) What is the current value at $t = 0.25$ s? at 1 s? (c) How long does it take for the capacitor energy to become 0.02 J?

39. How much is the energy in the capacitor field of Problem 38 when its voltage becomes constant? At that instant what is the power in the resistor?

Group B

40. The capacitor of Problem 38 is charged to 300 V and then suddenly disconnected from the source and connected to two 0.5 MΩ resistors in series. What is the power in each resistor (a) 0.5 s after they have been connected, (b) 1 s after they have been connected?

41. By the time the current in the circuit of Problem 40 becomes zero, how much energy has been dissipated in each resistor?

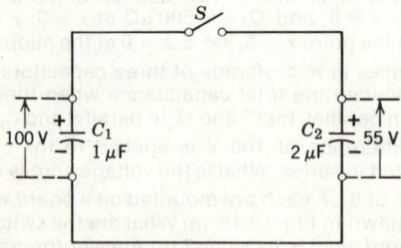

FIGURE 12.17   Two charged capacitors to be connected in series. For Problem 36.

42. An 8 μF capacitor is to be used in a timing circuit that utilizes the voltage across a resistor. A 12-V battery is the only source available for charging the capacitor. A delay of 1 s until the voltage on the resistor reaches 9 V is required. What value of resistance must be used?

43. An 8 μF capacitor charged to 100 V is suddenly discharged through a resistor in a timing circuit. Its voltage is to be used to allow the collector current of a transistor to increase and operate a relay. The critical voltage is 20 V and the time delay required is 10 s. What amount of resistance must be used?

44. Two capacitors are connected in series. One has $C_1 = 0.25 \mu$F, 200 V maximum, and the other has $C_2 = 0.1 \mu$F, 400 V maximum. What is the maximum safe potential difference that can be applied to the pair?

45. Two plates of a capacitor have zero volts potential difference between them. After 5×10^{10} electrons have been removed from one plate and added to the other, how many coulombs of flux are established between them?

46. The potential difference between the plates of Problem 45 became 200 V after the charge transfer. What is the capacitance of the capacitor?

47. A tubular capacitor is made by rolling up two sheets of aluminum foil 1.5 in. wide and 4 ft long separated by 2 sheets of 4 mm thick waxpaper 1.75 in. wide and sufficiently long to prevent the two aluminum sheets from touching each other. The relative permeability of the paper is 4 and it has a dielectric strength of 200 V/mm. What is (a) the capacitance of the capacitor, (b) its voltage rating?

48. The charge on each capacitor in Fig. 12.18 is 5 μC. What is their voltage after the switches have been closed?

49. Calculate the charge on each capacitor in Fig. 12.18 after the switches have been closed.

50. When a capacitor is charging in a series R-C circuit, the time required for the capacitor voltage to reach a desired fraction of its final value is called its "rise time." This is usually chosen to be some fixed value around 90 percent. Suppose $C = 12$ pF and the rise time (90 percent value) must not exceed 3 ns. What is the maximum value of resistance that may be used?

51. The resistance in an R-C circuit is 400 Ω and the rise time (90 percent value) is 5 ns. What value of capacitance must be used?

52. A 100 μF capacitor is charged to 5000 V through a resistance R that must limit the charging current to 100 mA. Calculate R and the time required to charge the capacitor to 95 percent of its maximum possible energy from this source. What is the value of this energy in watt-seconds?

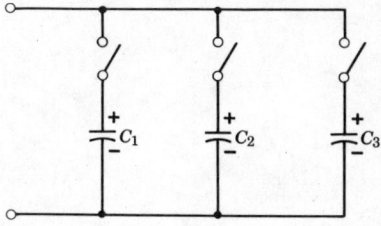

FIGURE 12.18   For Problem 48. $C_1 = 0.01 \mu F_1$, $C_2 = 0.02 \mu F$, $C_3 = 0.025 \mu F$.

Alternating Current and Voltage

We have learned something about time-varying current and voltage by studying the "transients" present when a constant dc voltage is applied to an *R-L* branch (resistance and inductance in series) or an *R-C* branch, which contains a resistance and a capacitance in series. We found that the current either becomes constant at some fixed value, or reduces to zero, after a period of several time constant's duration, and we called this the "steady state."

An alternating voltage and an alternating current of sine-wave form are in of several time constant's duration, and we called this the "steady state." their instantaneous maximum values remain constant. Under these conditions their effective values (which are generally the readings of meters measuring them) also remain constant.

It is interesting to learn that even though electric charges surge back and forth through a light bulb, a motor winding, or an electric heater, energy is given to the device in the process and heat or mechanical power, or both, are supplied to it. Electrical charges are not "used up." They serve as carriers of energy given to them at an energy "source." Most of the mechanical energy given to a generator is converted by it into electrical energy.

13.1 Generation of an Alternating Voltage

The two most widely used devices for generating an alternating voltage for power and light applications are rotating machinery and electrical oscillators. The voltage continuously changes in magnitude, going through negative as well as positive values that follow the ordinates of a *sine curve*[1].

The rotation of one wire loop, of which there are many, on the armature of a generator results in the generation of a voltage, the *instantaneous value* of which may be represented by the equation

$$e = E_m \sin \omega t \qquad (13.1)$$

in which ω (omega) is the angular velocity of the loop in *electrical radians per second*. The number of electrical radians per revolution of a coil on the armature is 2π times the number of *pairs* of magnet poles on the machine. Every time a loop of the armature winding turns a sufficient amount for each of its sides to pass completely under a pair of magnet poles (north and south) the armature has turned through 2π electrical radians. It is evident in Fig. 13.1 that the flux *through a loop* will change simultaneously from its full value in one direction to its full value in the opposite direction and then to its full value in the original direction again. It is also evident that, at the instant the flux through the moving coil is a maximum (positions $a, c,$ and d), the *rate of change*

[1]A brief discussion of sine, cosine, and tangent is given in the Appendix A.1. Some material on vector algebra, which is useful in phasor calculations, is also presented. Slide rule operations are included.

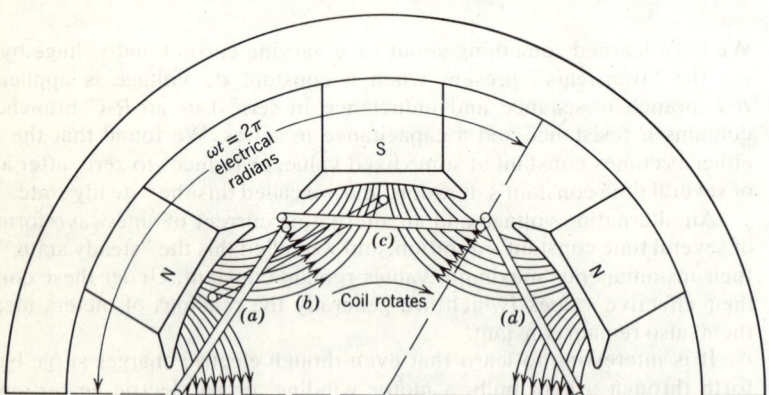

FIGURE 13.1   Half of the pole structure of a six-pole generator. A rotating armature coil is shown in four instantaneous positions. Generated voltage is zero in positions *a, c,* and *d* and a maximum in position *b* when the net flux through the coil is zero, but the rate of change of flux linkages is a maximum.

of flux linkages is zero. To understand this, one need only note that at this instant the flux through the coil has stopped increasing and is ready to start decreasing.

At the instant the sides of a coil are under the centers of the magnet poles, position *b*, the net flux *through* the coil is zero because as much comes out as goes in. The enclosed flux is in the process of changing from a small amount in one direction to a small amount in the opposite direction in an *extremely small increment* of time Δ*t*. This can be *a large rate of change*. One weber of magnetic flux is equal to 10^8 maxwells (or lines) of flux. To generate only one volt, a single turn must experience a *rate of change* of flux linkages in the amount of 10^8 lines per sec, which is 1 Wb/s.

13.2 Sine- and Cosine-Wave Forms

A plot of Equation (13.1) is shown in Fig. 13.2*a*. The abscissa is the angular displacement in electrical radians (radians per second times seconds), and inasmuch as a plot of the equation must be stopped somewhere, *one cycle* only of the generated voltage is shown. The count of time for the sine wave was begun at the instant the coil was in position *a* in Fig. 13.1.

If one wishes to start the count of time when the coil is in position *b*, one must wait until the generating coil advances $\pi/2$ electrical radians farther. This places the vertical axis of Figure 13.2*a* at the $\pi/2$ point on the ωt scale, and so at $t = 0$ the instantaneous voltage is a maximum. Figure 12.2*b* is drawn for this condition. The equation of this curve is $e = E_m \cos \omega t$.

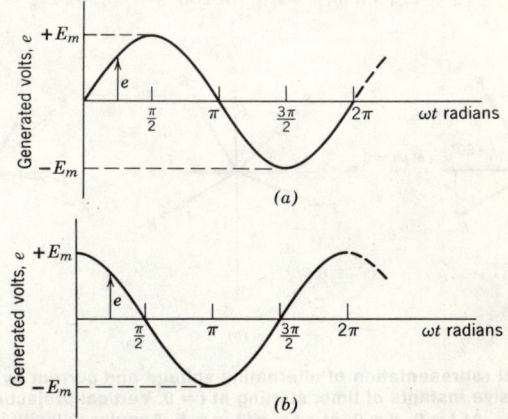

FIGURE 13.2   Curves representing instantaneous values of generated voltages. (a) Sine-wave form; $e = E_m \cos \omega t$ volts. (b) Cosine-wave form; $e = E \cos \omega t$ volts.

13.3 Phasors

Notice, in Fig. 13.2*a*, the vertical arrow labeled *e*. It could have been drawn vertically at any point on the horizontal (ωt) axis to meet a corresponding radial arrow E_m. It represents the *instantaneous value* of the generated voltage. At $\omega t = \pi/2$ radians (which is the angle turned through by the generator coil in *t* seconds), the instantaneous value *e* of the generated voltage is E_m, the *maximum*. This is indicated in Equation (13.1). If we substitute $\pi/2$ for ωt we get $e = E_m \sin (\pi/2) = E_m$. The subscript *m* means maximum.

We can draw a rotating arrow in the plane of the paper, as shown in Fig. 13.3*a*, label it E_m and call its *vertical projection e*. E_m arrows are shown at $t = 0$, and also at the end of five time intervals. Suppose t_1 is such that $\omega t_1 = 1.047$ radian = 60°. At that instant the generated voltage e_1 is

$$e_1 = E_m \sin \omega t_1 = E_m \sin 60° = 0.866\, E_m \text{ volts}$$

At $t = t_2$ assume $\omega t_2 = 90°$:

$$e_2 = E_m \sin \omega t_2 = E_m \sin 90° = E_m$$

At $t = t_3$ assume $\omega t_3 = 135°$:

$$e_3 = E_m \sin \omega t_3 = E_m \sin 135° = 0.707\, E_m \text{ volts}$$

When $360° > \omega t > 180°$, ωt is negative.
At $t = t_4$ assume $\omega t_4 = 200°$:

$$\text{Sin } 200° = -\sin 160° = -0.342$$

$$e_4 = E_m \sin \omega t_4 = E_m \sin 200° = -0.342\, E_m \text{ volts}$$

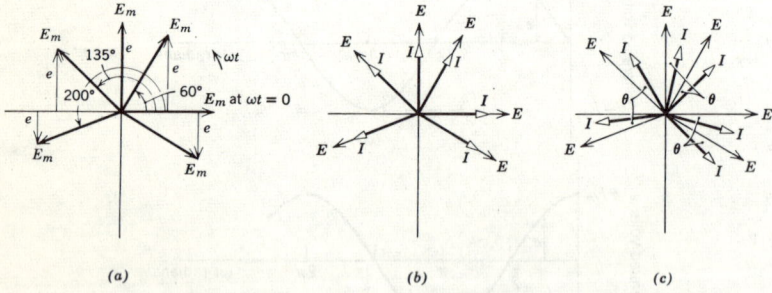

(a)   (b)   (c)

FIGURE 13.3 Graphical representation of alternating voltage and current by phasors. (*a*) Voltage phasor, E_m at 6 successive instants of time, starting at $t = 0$. Vertical projection represents the instantaneous value of E . At $t = 0$, $e = 0$, at $\omega t = \pi/2$, $e = E$. Angular velocity is ω radians per second. (*b*) Voltage E and current I are in phase. Effective values are represented. (*c*) Current I lags voltage, E, in phase, by angle θ as both go through all possible instantaneous values.

And at $t = t_5$ assume $\omega t_5 = 330°$:

$$\text{Sin } 330° = -\sin 30° = -0.500$$

$$e_5 = E_m \sin \omega t_5 = E_m \sin 330° = -0.500\, E_m \text{ volts}$$

Voltage is a phasor quantity. A phasor quantity changes instantaneous values as time changes usually as the sine or cosine of ωt changes with time. Phasors of voltage or current are usually expressed as effective values rather than as maximum values in electric circuit analysis.

13.4   Phase Angle

The angle the phasor makes with the reference line (which is taken to be at zero degrees or radians) is called the *phase angle* of the quantity the phasor represents. At $t = t_1$ the phase angle is $60°$, at $t = t_3$ it is $135°$, etc.

Alternating current is a phasor quantity also. We shall soon see that if the voltage happens to be applied to the terminals of a *pure resistance* (no L or C present) the current will be in phase with the voltage and the *current phasor* will always have the same phase angle as the voltage phasor. Current in a pure resistance is represented on the phasor diagram on the arrow for E_m that is on that resistance, but the arrow *head* is drawn at a different point. Because current and voltage usually have numerical values that are substantially different, the length of the arrows representing them differ in length.

An important point should be made here before we go any further. The arrows of phasor diagrams are labeled, preferably, with *effective values* of voltage, or current, and not with maximum instantaneous values. Effective values are *meter reading values*, in almost every case, and these are related to, but different from, maximum instantaneous values, as we shall see.

Now we can examine Fig. 13.3b. The current I, which flows in a resistor to which E is applied, is in phase with E at every instant of time. At $t = 0$, the E phasor is usually drawn horizontally to the right.

A different situation is represented in Fig. 13.3c. The voltage E is evidently applied to a coil that has both resistance and inductance, with the result that the current is delayed in going through its instantaneous values. Consequently, the current *lags the voltage*, in time-phase, by the angle θ. The current in an R-L circuit is always θ degrees (or radians) behind the applied voltage in the following sense. It reaches its positive maximum after the voltage has reached its positive maximum, and it is delayed the same amount in reaching other values. For example, by the time the instantaneous current value has fallen to zero, the instantaneous voltage has already done so. This is shown in Fig. 13.4.

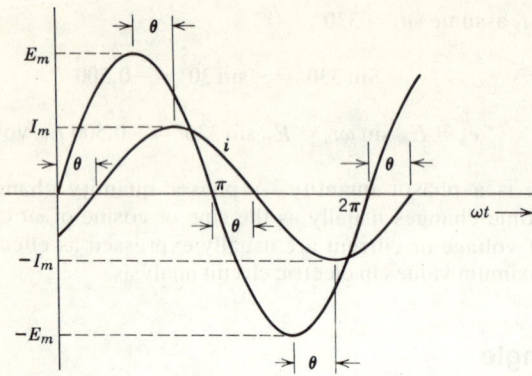

FIGURE 13.4 Instantaneous volts and amperes of an inductor with resistance. Current lags voltage in time phase by an angle θ.

CYCLE

When a varying quantity, such as voltage or current, repeats all of its instantaneous values in the same order, as is the case in Fig. 13.4, one complete set of all possible values is called a *cycle*. One cycle of the voltage wave spans the θ to $(\theta + 2\pi)$ points on the radian axis. The full cycle of current values starts θ radians after the voltage cycle starts (because the current *lags* the voltage) and continues for θ radians after the voltage is zero and ready to start another cycle.

RADIAN

An arc of a circle can be expressed as so many degrees of arc or radians of arc. An arc that is one fourth of the circumference is 90° of arc or $\pi/2$ radians of arc. The number of degrees of arc is the number of degrees in the "central angle" in the circle that subtends the arc. Likewise, the number of radians of arc is the number of radians in the central angle. The vertex (intersection point of the two sides) of a central angle is the center point of the circle.

A very important relation between length of arc and radius of a circle is

$$\text{radius} \times \text{radians of central angle} = \text{length of arc}$$
$$r\theta = l \tag{13.2}$$

in which r is the radius of a circle, θ is the central angle *in radians*, and l is the length of arc subtended by the two sides forming θ.

From this

$$\theta = \frac{l}{r} \tag{13.3}$$

We get the definition of *radian* from Equation (13.3). *An angle formed by two radii of a circle that subtend an arc of the circle equal in length to the radius is one* **radian**. We also note that the radian is a dimensionless quantity, since r and l are given in the same units of length.

$$2\pi \text{ radians} = 360°$$

$$1 \text{ radian} = \frac{360}{2\pi} = 57.3°$$

FREQUENCY

It has been customary, until recent years, to define *frequency* in electrical technology as the number of *cycles per second* of a varying, recurring quantity. The new name is *hertz*, pronounced *hurts*. The name was chosen to honor Heinrich Rudolph Hertz (Germany, 1857–1894) who pioneered in electro-magnetic wave research.

We used to say the frequency of the alternating current that lights our homes is 60 cycles per second. Now we use only one word, instead of three, and say 60 hertz.

ANGULAR FREQUENCY

The quantity ω in Equation (13.1) is equal to 2π times the frequency because in each revolution of the phasor, and therefore in every cycle, an angle of 2π radians is passed through. That is, $\omega = 2\pi f$ radians per second:

Angular frequency = $2\pi f$ radians/second

EXAMPLE 13.1   An ac voltage is defined as 169.7 in 377t volts. (a) What is the frequency? (b) What is the maximum instantaneous value and what is its value at $t = 1/240$ s and at $t = 1/720$ s?

Solution

(a) $2\pi f = 377, \quad f = 60 \text{ Hz}$
(b) At $t = 1/240$ s
$e = 169.7 \sin (377/240)$
$377/240 = 1.571 \text{ rad}$
$1.571 \text{ rad} \times 57.3 °/\text{rad} = 90°$
$e = 169.7 \sin 90° = 169.7 \text{ V} = E_m$

At $t = 1/720$ s, $\omega t = 0.524$ rad
$\omega t = 0.524 \times 57.3° = 30°$
$e = 169.7 \sin 30° = 84.85 \text{ V}$

EXAMPLE 13.2   An ac voltage with sine-wave form has a maximum instantaneous value of 30 V, which it reaches in 1/1600 s after going positive. What is the frequency?

Solution

$$30 = 30 \sin (2\pi f/1600)$$
$$\sin (2\pi f/1600) = 1$$
$$2\pi f/1600 \text{ is a } 90° \text{ angle } = 1.571 \text{ rad}$$
$$f = (1.571 \times 1600)/2\pi = 400 \text{ Hz}$$

EXAMPLE 13.3   A loop of wire is rotated in a magnetic field produced in a two pole-pole generator. (a) Through how many mechanical degrees must the loop be turned to generate one cycle of voltage? (b) Answer for the case of the six-pole machine in Fig. 13.1. (c) Write a formula for the number of electrical degrees for each mechanical degree in an *N*-pole generator.

Solution (a) Loop must pass completely under both poles. Therefore the loop must turn through 360 mechanical degrees to generate 1 cycle. (b) Loop must turn $\frac{1}{3}$ of a revolution to pass under one pair of poles:

$$\frac{1}{3} \text{ revolution} = 120 \text{ mechanical degrees for 1 cycle}$$

(c) A machine with *N* poles has *N*/2 pairs of poles, each pair representing 360 electrical degrees. A coil must make 1 rev in order to pass under *N*/2 pairs of poles. One revolution = 360 mechanical degrees. The number of electrical degrees for each mechanical degree is

$$(N/2) \times 360 \div 360 = N/2 \text{ electrical degrees per mechanical degree}$$

Note that $(N/2) \times 360 = $ total electrical degrees per revolution. Total electrical degrees per revolution ÷ total mechanical degrees per revolution gives the electrical degrees a coil passes through for each mechanical degree turned.

Problems

1. An alternating current is defined by $i = 20 \sin 377t$ amperes. What is its frequency and what are its maximum instantaneous positive and negative values?
2. What is the instantaneous value of the current in Problem 1 at $t = 1$ ms?
3. A current with sine-wave form has an instantaneous value of 10 A at the instant ωt has a radian value corresponding to 60°. What is the maximum instantaneous value?
4. How many revolutions per minute must a 4-pole generator make in order to generate a 60-Hz voltage?

5. A generated voltage $e = 169.7 \sin \omega t$ has a value of 90 V at $t = 0.05$ s. What is the angular frequency? What is the frequency in hertz?

6. Two radii of a circle subtend an arc 12 in. in length. The radius is 4 in. What is the angle between the radii in radians and in degrees?

7. Two radii of a circle, whose diameter is 30 cm, make an angle of 172° with each other. What length of arc do they subtend on the circumference?

8. Prove that radius × angular velocity of a particle rotating in a circle of radius R gives the linear velocity of the particle.

13.5   Average and Effective (Rms) Values

Equation (13.1) gives the instantaneous value of a generated voltage of sine-wave form. If the voltage is applied to a pure resistance R (one without inductance or capacitance included in its physical form), a current of sine-wave form flows, and it may be represented by

$$i = I_m \sin \omega_t \qquad (13.4)$$

in which $I_m = V_m/R$. When $\omega t = \pi/2$, $e = V_m$ and $i = V_m/R$; when $\omega t = 3\pi/2$, $e = -V_m$ and $i = -V_m/R$ and it flows in the reverse direction, which is indicated by the minus sign.

It would be impractical to use the maximum instantaneous value of alternating current to specify it as a current of that many amperes because it has that value only twice each cycle. Furthermore, the average value cannot be used because the current is positive as much as it is negative, so the average value is zero. See Fig. 13.5. The average value of the current during only a positive half-cycle may be computed, however. It is found to be $2I_m/\pi = 0.637I_m$.[2]

Although the average value over half a cycle might be used, it would not be as logical a choice as what we shall find to be an *effective value* which is related to the power developed in a resistance by the alternating current. Inasmuch as

[2]Proved in Appendix A.7.

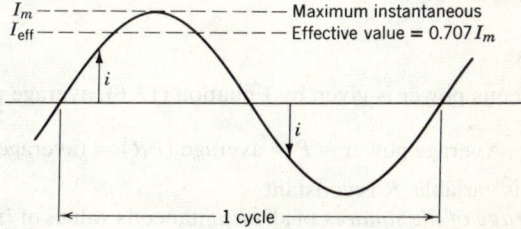

FIGURE 13.5   **Representing an alternating current. The instantaneous value i goes both positive and negative. The average value over a whole cycle is zero.**

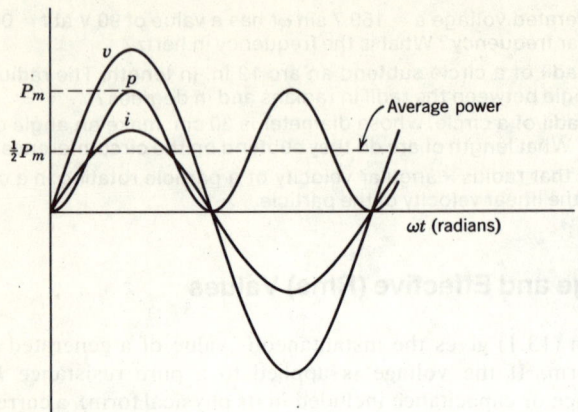

FIGURE 13.6 Instantaneous voltage, current, power, and average power in a resistance.

a direct current of I amperes develops I^2R watts in a resistance R, it has been found most logical to conclude that I_{eff} amperes of alternating current should produce the same power as the same number of amperes, I_{dc}, of direct current.

We shall first learn about instantaneous power and average power. This will lead to the determination of effective values of a-c voltage and current that are of sine-wave form.

Because instantaneous power in a resistor is obtained by multiplying instantaneous voltage by instantaneous current: $p = ei$, we may write

$$p = (iR)i = i^2R \tag{13.5}$$

and

$$p = v(v/R) = v^2/R \tag{13.6}$$

Figure 13.6 shows how the products of instantaneous voltage and current give positive power throughout each cycle. Its instantaneous values vary between $E_m I_m$ and zero.

AVERAGE POWER

If instantaneous power is given by Equation (13.6), average power P is defined as

$$\text{Average power} = P = \text{average } (i^2R) = (\text{average } i^2)R \tag{13.7}$$

since only i is variable, R is constant.

The *average of the squares* of all instantaneous values of i is always positive.

The effective value of an alternating current should be expressed in terms of its maximum instantaneous value, for then equations like 13.1 and 13.4 may be written. Figure 13.7 shows one cycle of a current wave (maximum instantaneous value = 1 A) and a curve of instantaneous current-squared values. The average of the sum of all i^2 values may be shown[3] to be $I_m{}^2/2$. The square root of the average i^2 is called the root mean i^2, and this must then be expressed as $I_m/\sqrt{2} = 0.707 I_m$.

The average of the squares of instantaneous voltage values is also important. Average power may be expressed in terms of this:

$$P = \text{average } p = \text{average } (v^2/R) = (\text{average } v^2)/R \tag{13.8}$$

Note that average power may be computed when the *equation* of the instantaneous voltage across a resistance is known. The average of the squares of the instantaneous voltage values may be obtained readily from that equation.

Returning to the idea of equivalence of an alternating current to a direct current, we should like to write average power $P = I^2 R$ in which I means amperes of alternating current. Remember that an alternating current varies continually in instantaneous magnitude, and it changes its direction of flow every half-cycle.

We may express average power this way if we make use of Equation 13.7 and let $I^2 = \text{average } i^2$. This would mean that

$$I = \sqrt{\text{average } i^2} \tag{13.9}$$

Thus I is equal to the square root of the average of the squares of the instantaneous values. Writers often say "*root*" instead of *square root* and "*mean*" instead of *average*. Thus we conclude that the *effective value* of an alternating current is the *root-mean-square (rms) value*. This is shown by the top dashed line in Fig. 13.7.

Summarizing, we shall say that P always represents average power unless specified otherwise, I and V represent effective values when they designate

[3]Shown in Appendix. A.8.

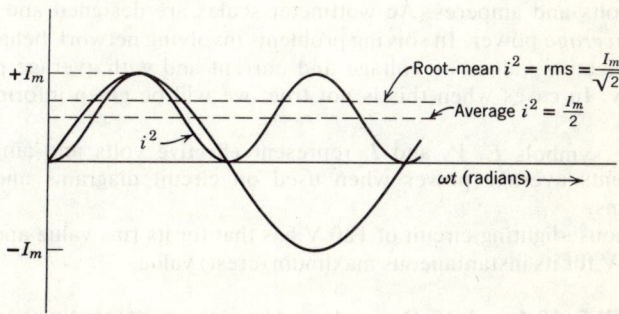

FIGURE 13.7   Curves of current and (current)^2 when $I_m = 1$ A.

alternating current and voltage. It becomes laborious and unnecessary to write I_{rms} and V_{rms}.

When a *periodic current* of any kind of wave form flows through a resistor, the average power delivered to the resistor is given by Equations (13.7), (13.8):

$$P = I^2R, \qquad \text{or } P_{avg} = I_{rms}^2 R \qquad (13.10)$$

$$P = V^2/R, \qquad \text{or } P_{avg} = V_{rms}^2/R \qquad (13.11)$$

$$V_{rms} = \sqrt{\text{average } v^2} \qquad (13.12)$$

Because Equations (13.10) and (13.11) are correct, we may write, for alternating current or periodic current of any wave form:

$$(P_{avg})^2 = (V_{rms}^2/R)(I_{rms}^2 R) = V_{rms}^2 I_{rms}^2$$

$$P_{avg} = V_{rms} I_{rms} \qquad \text{or } P = VI$$

The power expressions in Equations (13.10), (13.11), and (13.13) for power is a resistance are very useful. They hold for direct currents of constant value also.

Every student of electrical engineering should memorize the following forms and *should be sure* to remember that they apply to sine- and cosine-wave forms only:

$$I_{rms} = I_m/\sqrt{2} = 0.707 I_m \qquad (13.13)$$

$$V_{rms} = V_m/\sqrt{2} = 0.707 V_m \qquad (13.14)$$

It is incorrect to use these relations for a wave of any other shape.

13.6 Scales of AC Meters

Alternating current voltmeters and ammeters give readings of *rms* (that is, *effective*) values. Their scales are designed and calibrated to show rms (effective) volts and amperes. Ac wattmeter scales are designed and calibrated to show *average* power. In solving problems involving network behavior we work with effective values of voltage and current and with average power almost entirely. In cases when this is not true, we will be given information to that effect.

The symbols E, V, and I, represent **effective** volts and amperes, and P represents average power when used on circuit diagrams and in problem solutions.

A house-lighting circuit of 120 V has that for its rms value and $\sqrt{2} \times 120 = 169.68$ V for its instantaneous maximum (crest) value.

EXAMPLE 13.4 A 15-Ω resistance carries an alternating current of 10 A as measured with an ammeter. We are now prepared to calculate the following quantities:

(a) Voltage applied, $V = 10 \times 15 = 150$ V effective
(b) Assume sine-wave form. $V_m = 150 \sqrt{2} = 212.1$ V maximum
(c) $I_{max} = 10\sqrt{2} = 14.14$ A
(d) $P = I^2R = 10^2 \times 15 = 1500$ W $= 1.5$ kW

Because the current is in a pure resistance,

$$P = VI = 150 \times 10 = 1500 \text{ W} = 1.5 \text{ kW}$$

Also,

$$P = V^2/R = 150^2/15 = 1500 \text{ W}$$

Problems

9. A 1200-Ω resistor is found to have 30 V drop across it as measured with an ac voltmeter. What are the current and the power in it?
10. What should be the resistance of an electric heater that takes 2500 W on 120 V? What is the maximum instantaneous value of the alternating current taken by the heater?
11. What is the maximum instantaneous power in the resistor of Problem 9?
12. The equation for the instantaneous ac voltage that is applied to a pure resistance is $e = V_m \sin \omega t$. The current is, as we know, in phase with the voltage. Combine this equation with the one that represents the current and obtain the equation for instantaneous power in a resistance. What is the frequency of the pulsations of power if f hertz is the frequency of the current and voltage? See Fig. 13.6.

13.7   Power in an Inductance

We learned in Chapter 11 that when voltage is applied to the terminals of a coil of wire there is a delay in the flow of current and time is required for the current to reach maximum value. Opposing voltage produced by increasing flux linkages causes this. We shall now consider the case of an alternating voltage applied to a coil that has *zero resistance*. This impractical case of zero resistance need not bother us because we shall be dealing *separately* with resistance effect and inductive effect in a coil. Our analysis in this section is based on the assumption that only inductance exists in our circuit element, that is, we have a "pure" inductance, an "ideal" element for study.

Instantaneous values of current in an inductance *lag behind* corresponding instantaneous values of a varying applied voltage by $\pi/2$ rad (90°). The proof is given in the appendix. Thus the current curve i in Fig. 13.8a reaches its maximum at $\omega t = \pi/2$ *if we begin counting time at the instant the applied volt-*

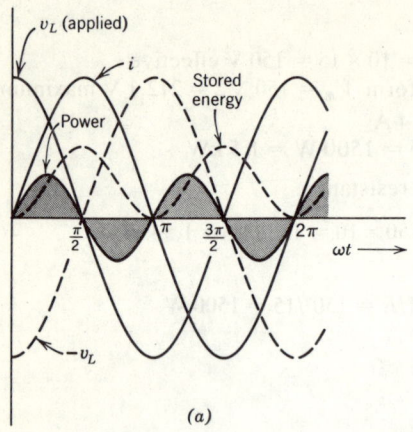

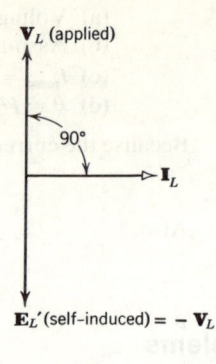

(a)   (b)

**FIGURE 13.8   (a) Curves showing instantaneous values of electrical quantities of a pure inductance:
$v_L = V_m \cos \omega t$, $i_L = I_m \cos (\omega t - \pi/2)$. (b) Phasors of voltages and current of a pure inductance. Current lags 90° behind applied voltage.**

age is maximum. When $v_L = V_m \cos \omega t$, $i_L = I_m \cos (\omega t - \pi/2)$. The student should substitute various values for ωt in these equations and satisfy himself that they are the equations of the *e* and *i* curves in Fig. 13.8.

The power curve, as expected, crosses the zero axis every time either the voltage or current is zero. Negative values of power exist when either voltage or current is negative, but when both are negative, power is positive. What does negative power mean? We found in Chapter 10 that energy stored in a magnetic field is given by $\frac{1}{2}Li^2$ joules. During the parts of the cycle when $p = vi$ is positive, energy is being put into the field (it is building up), and during the parts of the cycle when $p = vi$ is negative, the field energy is being returned to the source of supply. The positive-power values are canceled out by the negative-power values over a cycle so that *the average power in a pure inductance is zero.*

The phasors in Fig. 13.8b represent the applied voltage, the equal and opposite voltage of self-induction, and the current of a pure inductance. A simple calculus integration[4] results in the equation for the current. It not only shows the lag in phase by $\pi/2$ radians, but it gives the relation between I_m and V_m.

$$I_m = \frac{V_m}{\omega L} = \frac{V_m}{2\pi f L}$$

Instantaneous energy is represented by

$$w = \frac{1}{2}L(I_m \sin \omega t)^2 = \frac{1}{2}LI_m^2 \sin^2 \omega t$$

[4]Shown in Appendix A.9.

It is evident from this equation that the energy curve has no negative values. It has zero values, however, at even multiples of $\pi/2$ for ωt and maxima for odd multiples. Therefore a full cycle of the energy curve spans only π radians.

We must recognize that either the sine or the cosine may be used in expressing an ac voltage or current mathematically. Changing from one form to the other merely amounts to shifting the vertical axis, at which $t = 0$, $\pi/2$ radians.

The energy is always positive (when it is not zero) because it is equal to a positive constant ($\frac{1}{2}L$) times the square of the current. The energy curve is a double-frequency curve as is the power curve.

13.8   Power in a Capacitance

When an alternating voltage is applied to a capacitance, the *average power* supplied is zero. An examination of the curve of Fig. 13.9, in the light of our conclusions about the power supplied to a pure inductance, will reveal that there is as much positive power as negative power per cycle. Both parts of Fig. 13.9 show that *the current in a capacitor leads* the applied voltage by 90°, that is, the current reaches successive maximum and zero values, as time goes on, 90° before the voltage does.[5]

$$i_c = \omega C V_m \cos(\omega t + \pi/2)$$

[5]Shown in Appendix A.10.

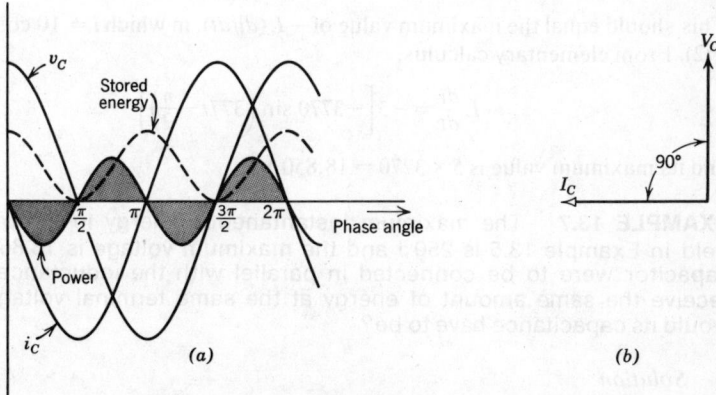

FIGURE 13.9   (a) Curves showing instantaneous values of electrical quantities of a capacitance. $v_c = V_m \cos \omega t$, $i_c = I_m \cos(\omega t + \pi/2)$. (b) Phasors of voltage and current of a pure capacitance. Current leads voltage 90° in phase. The power curve, with shaded areas, has equal positive and negative portions. The average power over any cycle is zero.

It is interesting and important to note that energy is at a maximum when the voltage is at a maximum. This is in contrast to the inductance case in which energy is a maximum when the *current* is at a maximum. We recall that capacitor energy is given by $\frac{1}{2}CE^2$ in the steady state. As in the inductance case, energy is given back to the source in alternate half-cycles of the voltage. The capacitance consumes no energy. In both cases the product *vi* is called volt-amperes instead of watts.

EXAMPLE 13.5 The instantaneous current in a pure inductance of 5 H is expressed as $10 \cos(377t - \pi/2)$. Determine (a) the rms value of the current, (b) the maximum instantaneous energy in the magnetic field produced, (c) the frequency.

Solution

(a) $I = 0.707 \times 10 = 7.07$ A rms
(b) $W = 0.5 \times 5 \times 10^2 = 250$ J
(c) $\omega = 2\pi f = 377$ rad/s
   $f = 377/2\pi = 60$ Hz

EXAMPLE 13.6 What is the maximum instantaneous voltage across the inductance of Example 13.5?

Solution. According to Section 13.7, the maximum instantaneous current $I_m = V_m/\omega L$

$$V_m = \omega L I_m = 377 \times 5 \times 10 = 18,850 \text{ V}$$

This should equal the maximum value of $-L(di/dt)$, in which $i = 10 \cos(377t - \pi/2)$. From elementary calculus,

$$-L\frac{di}{dt} = -5\left[-3770 \sin\left(377t - \frac{\pi}{2}\right)\right]$$

and its maximum value is $5 \times 3770 = 18,850$ V.

EXAMPLE 13.7 The maximum instantaneous energy in the magnetic field in Example 13.5 is 250 J and the maximum voltage is 18,850 V. If a capacitor were to be connected in parallel with the inductance and to receive the same amount of energy at the same terminal voltage, what would its capacitance have to be?

Solution

$$0.5C \times 18,850^2 = 250$$
$$C = 1.407 \ \mu\text{F}$$

EXAMPLE 13.8 The current in the capacitor of Example 13.7 is given by $i = \omega C V_m \cos(\omega t + \pi/2)$. Calculate its maximum instantaneous value. It should be 10 A because the current has the same value as that in the inductance which gave all of its energy to the capacitor.

Solution

$$I_m = \omega C V_m = 377 \times 1.407 \times 10^{-6} \times 18,850$$
$$I_m = 10 \text{ A}$$

13.9   Review Questions

1. Why does the voltage generated by a machine constructed like that shown in Fig. 13.1 pass through negative as well as positive values?

2. What is the difference between the voltages represented by 100 sin 377t and 100 cos 377t?

3. What is the name of the quantity whose value is 377 in the two expressions in Question 2?

4. Define radian. A circle has a central angle of one radian. The arc subtended by the angle is 10 cm long. What is the circumference of the circle?

5. What conditions exist when an alternating current or voltage is in a steady state?

6. Distinguish between mechanical and electrical degrees as encountered in a study of an ac generator.

7. What is meant by phase angle? Answer with reference to sinusoidal wave forms of voltage or current. Then comment on phase angle between the two.

8. What is a cycle of an alternating current or voltage?

9. Define frequency and give the name of the unit.

10. Define angular frequency. What is its symbol?

11. Define effective value of a sinusoidal current or voltage.

12. What is the phase relation between an alternating current in a resistor and the voltage drop across it?

13. Why does the curve of instantaneous power in a resistance have two positive crest values while the current is having one positive and one negative crest value?

14. Why is the average value of a sine-wave-form current or voltage zero?

15. A voltage across a coil is expressed as $e = 100 \sin 377t$, and a current as $i = 5 \sin (377t - \pi/4)$. (a) What is the frequency? (b) What are the effective values of voltage and current? (c) What is the phase of i with respect to e? (d) What is the maximum instantaneous power? (e) What is the average power?

16. Even though current in, and voltage across, a pure inductance are present during substantial parts of every half-cycle, explain how the average power can be zero. (b) Answer for a capacitance.

17. Are ac meters calibrated in average, maximum, or effective values? How can you determine an instantaneous maximum value of a sinusoidal quantity if given the rms value?

18. A capacitance C has an instantaneous maximum voltage V_m across it. How would you calculate the maximum energy, and where is the energy located?

19. A current $i = 2 \cos 754t$ flows in an inductor that has 500 mH inductance. What is the maximum instantaneous stored energy? What is the frequency of the alternating current?

20. What is the phase angle between the voltage and current in an inductance? In a capacitance?

21. How many rpm must an 8-pole generator make to generate the EMF $e = 170 \sin 377t$?

13.10   Problems

13. A voltage $v = 50 \sin 2513t$ exists across a 10-Ω resistor. (a) Write the equation of the instantaneous current. (b) What are the effective values of the voltage and current? (c) How much power is present in the resistor? Obtain the answer three different ways. (d) What is the maximum instantaneous power?

14. (a) How many *pairs* of poles has a 60-Hz hydroelectric generator which runs at 300 rpm? (b) A six-pole ac generator in a power station in Europe has a speed of 1000 rpm. What is its frequency? (c) How many electrical degrees correspond to one mechanical degree on this machine? Think this one out.

15. The current in a pure inductance for which $\omega L = 5\ \Omega$ is given by $i = 2 \cos 377t$. (a) Write the equation for the instantaneous voltage. (b) Sketch the current and voltage waves roughly to scale. (c) Draw the phasor diagram.

16. An EMF is given by $e = 340 \sin 314.16t$ volts. What are the (a) rms value, (b) frequency, (c) radians through which this phasor has moved in 0.01 s (d) number of degrees in part c, (e) EMF at the instant $t = 0.01$ s?

17. The voltage applied to a circuit element is $v = 100 \sin \omega t$ and the current is $i = 4 \sin (\omega t - \pi/6)$. (a) Draw the phasor representing rms values at $\omega t = \pi$ rad. (b) What are the instantaneous values of v and i in part a? (c) What would a voltmeter read if connected across the element?

18. An inductance $L = 1.8$ H carries a current of 0.3 A rms. (a) How much energy does it receive each cycle? (b) What is the average power each cycle?

19. The voltage on a circuit element is $v = 170 \sin \omega t$ volts and the current is $i = 5 \sin (\omega t + \pi/4)$ amperes. (a) Draw the phasors at $\omega t = \pi/2$ radians. (b) What are the instantaneous values of v and i at $\omega t = \pi/2$ radians?

20. What is the instantaneous power in Problem 19 when $\omega t = \pi/2$?

21. The sinusoidal wave form of a 50-Hz current, 10 A maximum value, is plotted so that a positive maximum occurs at $t = 0$. (a) Write the equation of this current. (b) Write the equation of the current squared. (c) Plot the curve represented by the current-squared equation. (d) Notice that the curve in (c) is symmetrical about a horizontal line at 50 A². What is the average value of the ordinate of the current-squared curve? (d) Obtain the square root of the average in part (d). What is the significance of your answer?

22. (a) Draw the phasor for $e = 5 \sin (\omega t + \pi/2)$. (b) Draw the phasor for $e = 5 \cos \omega t$. (c) Expand $5 \sin (\omega t + \pi/2)$, using a relation in trigonometry. (d) How are the phasors of parts a and b related?

23. Sketch the curves of Problem 22 on separate pairs of axes.

24. Show that the phasor for $e = 5 \cos (\omega t + \pi/2)$ and the phasor for $e = -5 \sin \omega t$ are one and the same.

CHAPTER FOURTEEN

Phasors in Complex-Number and Polar Forms

Voltage is not a vector quantity nor is current. Voltage, a convenient name for electrical potential difference, is classified as energy or work. The work that must be done to move a coulomb of charge from one point to another is the number of volts potential difference between the two points.

Current is not a vector quantity because it is simply the rate of transfer of charge past a point. When we insert an ammeter into a circuit to measure the current, or connect a voltmeter between two points to measure the potential difference (voltage), direction with reference to any set of axes is of no consequence. Therefore, neither voltage nor current is a vector quantity.

14.1 Complex Forms of E and I

Alternating current and voltage change in instantaneous value as time goes on. Phasor representation is very convenient in working with them. The arrow *representing* a phasor quantity may be represented as the *diagonal of a rectangle*. See Fig. 14.1a. The magnitude of the diagonal is equal to the square root of the sum of the squares of the magnitudes of the two adjacent sides. The

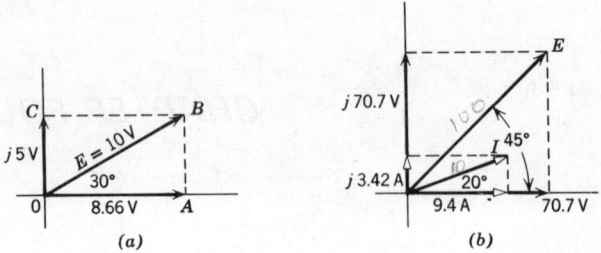

FIGURE 14.1 Representation of phasors in complex-number form. (a) E = 8.66 + j5V. (b) E = 70.7 + j70.7V. I = 9.4 + j3.42A.

sides are arrows representing two *rectangular components* of the given phasor. This component representation is very useful in ac circuit analysis.

In this chapter we shall consider phasors, which represent the effective values of voltages and currents, as the diagonals of *rectangles*. One adjacent side of the rectangle will be on the horizontal axis and the other will be on the vertical axis. It is common practice in electrical terminology to use the letter *j* with the quantity that expresses the "vertical component" of the phasor. In Fig. 14.1*a*, the voltage phasor **E** has advanced 30° beyond the $\omega t = 0$ angular position. It may be represented by its two *rectangular components*; *OA* and *OC*. The length of *OB* represents 10 V. Now let's see how much the lengths *OA* and *OC = AB* represent.

$$\cos 30° = \frac{OA}{OB} = \frac{OA}{10}$$

From tables,

$$\cos 30° = 0.866; \qquad OA = 10 \times 0.866 = 8.66 \text{ V}$$
$$\sin 30° = \frac{AB}{OB} = \frac{AB}{10}$$

From tables,

$$\sin 30° = 0.5; \qquad AB = OC = 10 \times 0.5 = 5 \text{ V}$$

To indicate the 90° *phase advance* (or *lead*) of *OC*, we precede the number 5 by the "*operator*" *j*. The operator *j* does not change the magnitude of the *OC* component of **E**.

We now represent **E** in complex-number form:

$$\mathbf{E} = 8.66 + j5 \text{ V}$$

Note that this form automatically includes the fact that **E** is at a 30° angle of lead. The *polar form* of **E** is quite simple, and easily understood:

$$\mathbf{E} = 10 \underline{/30°} \text{ V}$$

Figure 14.1*b* shows $\mathbf{E} = 100\underline{/45°}$ V and $I = 10\underline{/20°}$ A. Note that *we use the cosine* of the phase angle *to express the horizontal* (called *the real) component, and the sine* of the phase angle *to express the vertical* (called the *imaginary) component*:

$$\mathbf{E} = 100 \cos 45° + j\,100 \sin 45° \text{ volts}$$
$$\mathbf{E} = 70.7 + j70.7 \text{ volts}$$
$$\mathbf{I} = 10 \cos 20° + j\,10 \sin 20° \text{ volts}$$
$$\mathbf{I} = 9.40 + j3.42 \text{ amperes}$$

In polar form:

$$\mathbf{E} = 100 \underline{/45°} \qquad \mathbf{I} = 10 \underline{/20°}$$

14.2   Addition and Subtraction of Phasors

Only phasors of the same kind may be added together or subtracted from one another. Common sense tells us we should not try to add volts to amperes.

First we shall combine phasors that differ in phase by multiples of 90°. We can consider $\mathbf{E}_{OA} = 8.66$ V and $\mathbf{E}_{OC} = j5$ V in Fig. 13.1 as two separate phasors. Evidently, the magnitude of their sum, which must be their *phasor* (or graphical) sum, if $\sqrt{8.66^2 + 5^2} = 10$. But their sum is a phasor that makes an angle θ, with the horizontal, whose tangent is 5/8.66. We may express the sum $\mathbf{E}$ as

$$\mathbf{E} = \sqrt{8.66^2 + 5^2} \underline{/ \text{arc tan } (5/8.66)}$$
$$\mathbf{E} = 10 \underline{/30°} \text{ volts}$$
$$\mathbf{E} = 10 \cos 20° + j10 \sin 30° = 8.66 + j5 \text{ volts}$$

Using only complex number operations. we simply write

$$\mathbf{E} = \mathbf{E}_{OA} + \mathbf{E}_{OC} = 8.66 + j5 \text{ volts}$$

The heavy (bold-face) type is used to identify *phasor quantities*.

When two or more. phasors that are *not at right angles* to one another are added. it is much more convenient to use the complex-number forms. We can convert from polar form to complex-number form by using cosines and sines. as we did above where $\mathbf{E} = 10 \underline{/30°}$ volts.

EXAMPLE 14.1   Add the phasors $\mathbf{E}_1 = 25 \underline{/15°}$, $\mathbf{E}_2 = 15 \underline{/60°}$.

Solution

$$
\begin{array}{l|l}
\mathbf{E}_1 = 25(\cos 15° + j\sin 15°) & \mathbf{E}_2 = 15(\cos 60° + j\sin 60°) \\
\quad = 25 \times 0.966 + j25 \times 0.258 & \quad = 15 \times 0.5 + j15 \times 0.866 \\
\quad = 24.15 + j6.45 & \quad = 7.5 + j12.99
\end{array}
$$

$$\mathbf{E}_1 + \mathbf{E}_2 = 24.15 + 7.5 + j6.45 + j12.99$$
$$= 31.65 + j19.44$$
$$\mathbf{E}_1 + \mathbf{E}_2 = \sqrt{31.65^2 + 19.44^2} \,\underline{/\, \text{arc tan } 19.44/31.65}$$
$$\mathbf{E}_1 + \mathbf{E}_2 = \sqrt{1379} \,\underline{/\, \text{arc tan } 0.615} = 37.1 \,\underline{/\, 31.6°}$$

EXAMPLE 14.2   Subtract phasor $\mathbf{E}_2$ from phasor $\mathbf{E}_1$ of Example 14.1.

Solution

$$\mathbf{E}_1 - \mathbf{E}_2 = 24.15 + j6.45 - 7.5 - j12.99$$
$$= 24.15 - 7.5 + j6.45 - j12.99$$
$$= 16.65 - j6.54$$
$$\mathbf{E}_1 - \mathbf{E}_2 = \sqrt{16.65^2 + 6.54^2} \,\underline{/\, \text{arc tan } (-6.54/16.65)}$$
$$= \sqrt{277.2 + 42.77} \,\underline{/\, \text{arc tan } -0.393}$$
$$\mathbf{E}_1 - \mathbf{E}_2 = 17.88 \,\underline{/\, -21.45°}$$

GRAPHICAL ADDITION AND SUBTRACTION OF PHASORS

Phasors can be added or subtracted graphically. They can be drawn to scale and their sum and difference can be scaled off and thus evaluated. Figure 14.2 shows graphical addition and subtraction of the $\mathbf{E}_1$ and $\mathbf{E}_2$ phasors treated in Examples 14.1 and 14.2. The sum and difference can be computed without measuring the lengths of the arrows in terms of a scale.

EXAMPLE 14.3   Determine the sum and difference of $\mathbf{E}_1$ and $\mathbf{E}_2$ as defined in Examples 14.1 and 14.2, using graphical construction and trigonometry.

Solution. In Fig. 14.2*a*, the sum of $\mathbf{E}_1$ and $\mathbf{E}_2$ is represented by the diagonal of the parallelogram with the two phasors as sides. If $\mathbf{E}_1$ and $\mathbf{E}_2$ are drawn to scale and placed accurately at their respective angles, we can scale off this diagonal and obtain the value of their sum. It is more accurate however, to use trigonometry and calculate the sum.

The real component of the sum is

$$E_1 \cos 15° + E_2 \cos 60° = 24.15 + 7.5 = 31.65$$

as found in Example 1.

The component that is 90° ahead of the real component is

$$E_1 \sin 15° + E_2 \sin 60° = j6.45 + j12.99 = j19.44$$

as found in Example 14.1 also.

The sum, $\mathbf{E}_1 + \mathbf{E}_2$, has a magnitude and phase given by

$$\mathbf{E}_1 + \mathbf{E}_2 = \sqrt{31.65^2 + 19.44^2} \,\underline{/\, \text{arc tan } 19.44/31.65}$$
$$= 37.1 \,\underline{/\, 31.6°}$$

as found in Example 14.1.

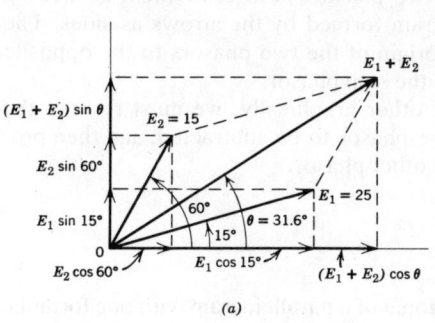
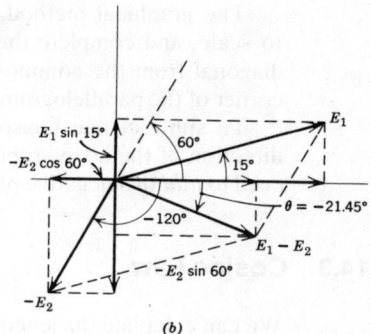

FIGURE 14.2 Graphical addition and subtraction of phasors. (a) Addition: $E_1 + E_2$. (b) Subtraction: $E_1 - E_2$.

The difference between $\mathbf{E}_1$ and $\mathbf{E}_2$, obtained as $\mathbf{E}_1 + (-\mathbf{E}_2)$, in Fig. 14.2, is shown as the diagonal of the parallelogram formed by $\mathbf{E}_1$ and $-\mathbf{E}_2$. Note that the diagonal is the one that starts at 0, the point common to the two phasors.

The real component of $-\mathbf{E}_2$ is toward the left on the horizontal axis and the 90° component is downward because of the direction of $-\mathbf{E}_2$. The *net horizontal component* (which is the horizontal component of $\mathbf{E}_1 - \mathbf{E}_2$) is obtained by combining the two horizontal components:

Horizontal component of $(\mathbf{E}_1 - \mathbf{E}_2) = 24.15 - 7.5 = 16.65$

Combining 90° (vertical) components:

Vertical component of $(\mathbf{E}_1 - \mathbf{E}_2) = j6.45 - j12.29 = -j6.54$

$$\mathbf{E}_1 - \mathbf{E}_2 = \sqrt{16.65^2 + 6.54^2} \,/\, \text{arc tan } 16.65$$
$$= \sqrt{319.5} \,/\, \text{arc tan}(-0.402) = 17.88 \,/\, -21.45° \text{ V}$$

In review, let's remember that to add or subtract phasors (and they must represent the same kind of quantity) we may first express each one as a complex number $(a + jb)$ of which a represents the real part, often called the horizontal component on the phasor diagram, and b with the operator j attached represents the imaginary part, often called the vertical component of the phasor. In addition of phasors, the real parts of the phasors are added together to get the horizontal component (on the phasor diagram) of the sum phasor, and the parts containing the j operator are added together to get the vertical component of the sum phasor.

If phasor $\mathbf{E}_2 = a_2 + jb_2$ is to be subtracted from phasor $\mathbf{E}_1 = a_1 + jb_1$, we simply change the signs of the terms of $\mathbf{E}_2$, making it $-\mathbf{E}_2 = -a_2 - jb_2$, then add $\mathbf{E}_1$ and $-\mathbf{E}_2$ together.

The graphical method of adding two phasors is to draw them as arrows, to scale, and complete the parallelogram formed by the arrows as sides. The diagonal from the common point of origin of the two phasors to the opposite corner of the parallelogram represents the sum phasor.

To subtract one phasor from the other graphically, we must reverse the direction of the arrow representing the phasor to be subtracted, and then proceed to *add* this negative phasor to the other phasor.

14.3   Cosine Law

We can calculate the length of the diagonal of a parallelogram with one formula, called the *cosine law*. The angle α in the cosine law formula is the angle between the phasors to be added. Let's apply the cosine law to the parallelogram in Fig. 14.2a, the two sides of which are represented by $|\mathbf{E}_1| = 25$ V, $|\mathbf{E}_2| = 15$ V.

EXAMPLE 14.4   What is the magnitude of the sum of E_1 and E_2 in Fig. 14.2a?

Solution

$$|\mathbf{E}_1 + \mathbf{E}_2| = \sqrt{E_1{}^2 + E_2{}^2 + 2E_1E_2 \cos \alpha} \qquad (14.1)$$
$$|\mathbf{E}_1 + \mathbf{E}_2| = \sqrt{25^2 + 15^2 + 2 \times 25 \times 15 \cos (60° - 15°)}$$
$$= \sqrt{625 + 225 + 530} = \sqrt{1380}$$
$$|\mathbf{E}_1 + \mathbf{E}_2| = 37.1 \text{ V}$$

This method does not give the phase angle of the sum phasor. It must be determined by means of other applications of trigonometry that we need not present here.

In applying the cosine law to obtain the sum of $\mathbf{E}_1$ and $-\mathbf{E}_2$ (Fig. 14.2b) we need only to be sure to use the minus sign on the cosine factor because the included angle is 135°. Cosines of angles between 90° and 270° are negative.

14.4   The Operator j

A few words should be said about the operator j. Although the characteristic most often used is the one in which *multiplying by j rotates the phasor through 90° counterclockwise*, it is also useful to know that j is the symbol for $\sqrt{-1}$ and that $j^2 = -1$.

The 90° rotation characteristic is made use of in complex-number representation of phasors as we have seen in examples thus far. Soon we shall be multiplying two complex numbers together, and this is where $j^2 = -1$ comes in.

EXAMPLE 5   Multiply $E = 5 + j2$ volts by $I = 2 + j1$ amperes.

Solution

$$
\begin{array}{l}
5+j2 \\
2+j1 \\
\hline
10+j4 \\
\quad +j5+j^23 \\
\hline
10+j9-3 \\
(5+j^2)\,(2+j1) = 7+j9
\end{array}
\qquad
\begin{array}{l}
\text{since} \qquad
\end{array}
\qquad
\begin{array}{l}
j^23 = -1 \times 3 = -3 \\
\\
j = \sqrt{-1} \\
\\
j^2 = -1
\end{array}
$$

DIVISION OF COMPLEX NUMBERS

Soon we shall be concerned with the division of a voltage **E** in complex form by a so-called impedance **Z**, also in complex form. The process, which is easy to perform, gives the answer in complex form, from which its magnitude and phase can be computed. An example will make all of this clear.

EXAMPLE 14.6   Divide **E** $= 100 + j20$ by **Z** $= 3 + j4$. We shall find, later, that the answer in a case of this kind is a current whose magnitude and phase angle, with respect to $\theta = 0$ on the horizontal to the right, can be determined.

Solution

$$\frac{100+j20}{3+j4} \times \frac{3-j4}{3-j4} = \frac{300+j60-j400+80}{9+j12-j12-j^216}$$

Multiplying both numerator and denominator by the denominator *after its sign has been reversed* is called rationalizing the denominator. The result has a denominator that is a *rational number*; it has no imaginary part (no j part). When a complex number is rewritten with its sign reversed, the new complex number is called the *complex conjugate* of the old one.

Our quotient is

$$\frac{380-j340}{25} = 15.2 - j13.6$$

The magnitude is $\sqrt{15.2^2 + 13.6^2} = 20.39$

The phase is $\theta = $ arc tan $(13.6/15.2) = $ arc tan $0.895 = 41.8°$

EXAMPLE 14.7   Divide $(-50 + j60)$ by $(8 - j12)$.

Solution

$$\frac{-50+j60}{8-j12} \times \frac{8+j12}{8+j12} = \frac{-1120-j120}{64+144}$$

$$= \frac{-1120-j120}{208} = -5.39 - j0.577$$

The answer in polar form is

$$\sqrt{5.39^2 + 0.577^2} \; \underline{/\text{arc tan}\,(-5.39/-0.577)}$$
$$= \sqrt{29.38} \; \underline{/\text{arc tan}\,9.34} = 5.42 \; \underline{/-173.9°}$$

Notice that $-5.39 - j0.577$ is in the *third quadrant* on the axes, just a little below the horizontal. Since arc tan $9.34 = 83.9°$, we must increase it by $90°$ to put it in the third quadrant and we must call this a negative angle. Using a positive phase angle, we could give the answer as $5.32 \; \underline{/186.1°}$.

14.5   Polar Forms

We have seen that a voltage $e = E_m \sin \omega t$ is represented by a phasor whose effective value is $E = E_m/\sqrt{2}$, and the position of the arrow used to represent E graphically depends on the value of ωt. We start with the arrow horizontally to the right at $t = 0$; at $\omega t = \pi/2$ the arrow had advanced to a vertically upward position because $\omega t = \pi/2$ means an advance, counter-clockwise, of 90 electrical degrees. At $\omega t = \pi/4$, the arrow makes a $45°$ angle with the horizontal. Polar-form representation of E at these three instants of time would be $E\underline{/0°}$, $E\underline{/90°}$, $E\underline{/45°}$.

ADDITION AND SUBTRACTION OF POLAR FORMS

Unless we resort to graphical representation, as we did in Fig. 14.2 with $E_1\underline{/15°}$ and $E_2\underline{/60°}$, and solve with the aid of sine, cosine, and tangent formulas, we usually convert the phasors into complex-number form if we want to add or subtract them.

MULTIPLICATION OF PHASORS IN POLAR FORM

Multiplication of phasors expressed in polar form is a simple procedure. All we need to do is multiply the magnitudes and add the angles.

EXAMPLE 14.6   Multiply $\mathbf{E}_1 = 120\underline{/25°}$ by $\mathbf{E}_2 = 10\underline{/-10°}$.

Solution

$$\mathbf{E}_1\mathbf{E}_2 = 120\underline{/25°} \times 10\underline{/-10°}$$
$$= 1200\underline{/25° + (-10°)} = 1200\underline{/15°}$$

EXAMPLE 14.7   Multiply $\mathbf{I}_1 = 10\underline{/-30°}$ by $\mathbf{Z}_1 = 25\underline{/70°}$.

Solution

$$\mathbf{I}_1\mathbf{Z}_1 = 10\underline{/-30°} \times 25\underline{/70°}$$
$$= 250\underline{/-30° + 70°} = 250\underline{/40°}$$

DIVISION OF PHASORS IN POLAR FORM

To divide a phasor in polar form by another phasor in polar form, we simply divide their magnitudes and subtract their angles.

EXAMPLE 14.8   Divide $\mathbf{E} = 120\underline{/0°}$ by $\mathbf{Z} = 50\underline{/40°}$

Solution

$$\frac{120\underline{/0°}}{50\underline{/40°}} = 2.4\underline{/0° - 40°} = 2.4\underline{/-40°}\ \text{A}$$

EXAMPLE 14.9   Divide $\mathbf{E} = 50\underline{/-30°}$ by $\mathbf{I} = 8\underline{/-90°}$.

Solution

$$\frac{50\underline{/-30°}}{8\underline{/-90°}} = 6.25\underline{/-30° - (-90°)} = 6.25\underline{/60°}$$

SQUARES AND SQUARE ROOTS OF PHASORS

To square a quantity, we simply multiply it by itself. We know how to do this in both polar and complex form. To extract the square root we use the polar form because it is easy to do. We simply get the square root of the magnitude and divide the angle by 2.

EXAMPLE 14.10   Obtain $\mathbf{I}^2$ when $\mathbf{I} = 12\underline{/30°}$.

Solution

$$\mathbf{I}^2 = 12^2\underline{/2 \times 30°} = 144\underline{/60°}$$

EXAMPLE 14.11   Obtain the square root of $324\underline{/70°}$.

Solution

$$(324\underline{/70°})^{1/2} = (324)^{1/2}\underline{/(70/2)}° = 18\underline{/35°}$$

14.6   Exponential Forms

We learn in advanced mathematics that when ϵ, the base of the system of natural logarithms, is raised to a power that is an imaginary number, like $j\theta$, it is equal to a special trigonometric form:

$$\epsilon^{j\theta} = \cos\theta + j\sin\theta \tag{14.2}$$

and any number that multiplies $\epsilon^{j\theta}$ becomes a complex number. If the number represents a phasor quantity, such as $E\underline{/\theta}$ volts or $I\underline{/\alpha}$ amperes, we get the expression for the phasor in complex form. That is,

$$E \epsilon^{j\theta} = E \cos \theta + jE \sin \theta \qquad (14.3)$$
$$I \epsilon^{j\alpha} = I \cos \alpha + jI \sin \alpha \qquad (14.4)$$

This means that the expressions to the left in the following:

$$25\epsilon^{j15°} = 25 \cos 15° + j25 \sin 15°$$

$$15 \epsilon^{j\pi/3} = 15 \cos \frac{\pi}{3} + j15 \sin \frac{\pi}{3}$$

are *exponential forms* that fully describe $\mathbf{E}_1$ and $\mathbf{E}_2$ in Fig. 14.2. If θ is negative, Equation (13.2) becomes

$$\epsilon^{j(-\theta)} = \cos(-\theta) + j \sin(-\theta)$$

or

$$\epsilon^{j(-\theta)} = \epsilon^{-j\theta} = \cos \theta - j \sin \theta$$

This is true because $\cos(-\theta) = \cos \theta$ and $\sin(-\theta) = -\sin \theta$.

If we examine Equation (14.2) closely, we shall find that $\epsilon^{j\theta}$ is just an operator that turns a numerical quantity into a complex number. The two parts on the right side of the equation are at right angles and if we square them and extract the square root we get unity:

$$\cos^2 \theta + \sin^2 \theta = 1$$

The factor $\epsilon^{j\theta}$ turns the quantity counterclockwise through θ degrees (or radians). It is interesting that $\epsilon^{j\pi} + 1 = 0$ is a relation that brings five important numbers together in a single equation: ϵ, the base of the system of natural logarithms, j the 90° rotator, π the ratio of the circumference of a circle to its diameter, 1 the first numeral in the universally used Arabic system of numbers, and zero the number most often used when we start to count.

Exponential forms are not used as much as are complex-number and polar forms in elementary work in electrical engineering, so we shall not expand on this topic.

14.7 Review Questions

1. What are the meanings of the following forms: (a) $E \underline{/30°}$, (b) $I \underline{/-20°}$, (c) $E = 6 + j8 = 10 \underline{/53.1°}$?

2. An applied voltage $E = 100 \underline{/10°}$ sends a current $I = 10 \underline{/-25°}$ through a branch of a circuit. Does the current lead or lag the voltage, and by how much?

3. The voltage drop between two points A and C is made up of two parts $V_{AB} = 5 + j3$ volts, $V_{BC} = 4 - j2$ volts. What is the total voltage drop between A and C?

4. Tell what you think would be the easiest way to add a voltage drop $V_1 = 10 \underline{/20°}$ to a voltage drop $V_2 = 5 \underline{/-30°}$ and get an answer that is accurate to three decimal places.

5. Consider $E_1 + E_2$ in Example 14.4 to be represented by the diagonal of a parallelogram that has adjacent sides with lengths that represent E_1 and E_2

to scale. The angle between these adjacent sides is alpha (α). State the cosine law in your own words as concisely as you can, and apply it to this case.

6. In Question 1(c) what does the operator j mean?

7. A current is expressed as $25\,\underline{/30^\circ}$ A. (a) What is the expression for the square of the current? What is the expression for the square root of the current?

8. Write $E = 120\,\underline{/45^\circ}$ in exponential form.

9. Write $I = 10\,\underline{/0^\circ}$ in exponential form.

10. Write E, then I, of Questions 8 and 9 in j notation, which is properly called complex-number form.

14.8 Problems

1. Write the complex number form for each of the following phasors: (a) 100 Ω at $+45^\circ$, (b) 30 V at 70°, (c) 50 V at $+120^\circ$, (d) 120 V at 210°, (e) 100 V at -50°.

2. Write the polar form for each of the following: (a) $\mathbf{I} = 10 + j5$A (b) $\mathbf{I} = -4 + 6.928$ A (c) $\mathbf{E} = 103.9 + j60$ (d) $\mathbf{E} = 40 - j30$ V (e) $\mathbf{E} = -50 - j75$ V.

3. (a) Obtain the sum of these two voltages:

$$\mathbf{E}_1 = 50 + j75, \ \mathbf{E}_2 = 4 - j30$$

(b) Express the sum in polar form.

4. Obtain the sum of $\mathbf{E}_1 = 100\,\underline{/45^\circ}$ and $\mathbf{E}_2 = 80\,\underline{/120^\circ}$ in complex form and in polar form.

5. Subtract $\mathbf{E}_3 = 40\,\underline{/-60^\circ}$ from the sum of $\mathbf{E}_1$ and $\mathbf{E}_2$ of Problem 4.

6. Apply the cosine law to solve Problem 4 and check the magnitude of the sum with the result obtained in that problem solution.

7. Multiply $\mathbf{E} = 50 + j75$ V by $\mathbf{I} = 3 + j5$ A.

8. If $\mathbf{I} = 10 + j20$, obtain the complex form of I^2.

9. Multiply (a) $15\,\underline{/30^\circ}$ by $40\,\underline{/75^\circ}$, (b) $20\,\underline{/-12^\circ}$ by $16\,\underline{/40^\circ}$.

10. Divide (a) $75\,\underline{/30^\circ}$ by $15\,\underline{/10^\circ}$, (b) $E = 100\,\underline{/45^\circ}$ by $Z = 20\,\underline{/60^\circ}$, (c) $\mathbf{E} = 120\,\underline{/0^\circ}$ by $\mathbf{I} = 40\,\underline{/-30^\circ}$.

11. Express $100\,\epsilon^{\,j30^\circ}$ in polar form and in complex form.

12. Express $100\,\epsilon^{-j\pi/3}$ in polar form and in complex form.

13. Add $5 + j12$ to $8 - j9$ and divide the sum by $3 + j4$.

14. Add $10\,\underline{/40^\circ}$ to $15\,\underline{/60^\circ}$ and divide the result by $(20\,\underline{/45^\circ} - 12\,\underline{/180^\circ})$.

15. Multiply $\epsilon^{\,j20^\circ}$ by $\epsilon^{\,j40^\circ}$. (Note x^3 times $x^2 = x^{(3+2)} = x^5$).

16. Divide $\epsilon^{\,j20^\circ}$ by $\epsilon^{\,j40^\circ}$. (Note $x^3 \div x^2 = x^3$ times $x^{-2} = x^{(3-2)} = x^1 = x$.)

17. Add: $\mathbf{E}_1 = 100\,\underline{/45^\circ}$ V, $\mathbf{E}_2 = 60 + j30$ volts, $\mathbf{E}_3 = 50\epsilon^{\,j\pi/2}$ volts.

18. Two adjacent sides of a parallelogram are $\mathbf{L}_1 = 25 + j0$m, $\mathbf{L}_2 = 5 + j12$m. Compute the length of the longer diagonal using the cosine law.

19. Compute the length of the short diagonal of the parallelogram of Problem 18, using the cosine law.

20. Express the following phasor quantities using smaller angles: (a) $20\,\underline{/315^\circ}$ V, (b) $5\,\underline{/260^\circ}$ A, (c) $120\,\underline{/280^\circ}$ V, (d) $12\,\underline{/355^\circ}$ A.

21. Obtain (a) square root of $256\,\underline{/110^\circ}$, (b) cube root of $64\,\underline{/90^\circ}$, (c) $(5 + j12)^2$.

CHAPTER FIFTEEN

Resistance, Reactance, and Impedance

When we began to study applications of Ohm's law we worked with dc only and saw that when voltage is applied to a resistance, or a combination of resistances, the ohms value determined the amount of current. That is, current was equal to volts divided by ohms resistance.

We now know that an increasing current in an inductance is delayed in reaching its maximum value by the EMF of self-induction $(-L(di/dt))$. We also know that if a potential difference (a voltage) is suddenly applied to a capacitor in series with a resistance, the current starts out $(t = 0)$ at maximum value and decreases to zero while the potential difference between the plates builds up to a maximum. Inductances (henrys) and capacitances (farads) have ohms values associated with them, and these ohms values are called *reactance* instead of resistance. The ohms *reactance* of a coil of wire cannot be added directly to the ohm *resistance* of the coil for reasons that we shall explain. It is also true that ohms reactance of a capacitor cannot be added directly to ohms resistance that may be in series with the capacitor. There are matters of 90° phase relations that must be taken into account.

351

15.1   Inductive Reactance

The footnote in Section 13.7 contains some information at least equal in importance to the fact it sets out to prove. It shows not only that current in a pure inductance lags the applied voltage by 90° (when the voltage is sinusoidal in wave form), but also that the alternating current crest value is obtained by dividing the crest value of the voltage by ωL.

$$I_m = \frac{V_m}{\omega L} = \frac{V_m}{2\pi f L} \tag{15.1}$$

Dividing by $\sqrt{2}$,

$$\frac{I_m}{\sqrt{2}} = \frac{V_m}{\sqrt{2}\omega L} = \frac{V_m}{\sqrt{2} \times 2\pi f L}$$

Expressing Equation (15.1) in rms values,

$$I = \frac{V}{\omega L} = \frac{V}{2\pi f L} \tag{15.2}$$

The quantity $\omega L = 2\pi f L$ must have *ohms for its units*. This quantity is called *inductive reactance*.

Inductive reactance = $2\pi \times$ frequency $\times$ inductance

The symbol for inductive reactance is X_L

$$X_L = 2\pi f L \text{ ohms} \tag{15.3}$$

in which f = hertz, L = henrys.

We can derive the expression for inductive reactance and phase angle for an inductance in a somewhat more straightforward way.
Let

$$i_L = I_m \sin \omega t \tag{15.4}$$

$$L \frac{di_L}{dt} = \omega L I_m \cos \omega t = v_L \tag{15.5}$$

The applied voltage, $V = L(di/dt)$ is properly equal and opposite to the EMF of self-induction $-L(di/dt)$.
The maximum values are $\omega L I_m = V_m$.

Rms values are

$$\omega L \frac{I_m}{\sqrt{2}} = \frac{V_m}{\sqrt{2}}$$

$$\omega L I_L = V$$

$$I_L = \frac{V}{\omega L} = \frac{V}{X_L}$$

An examination of Equations (15.4) and (15.5) reveals that i_L (represented by a sine curve) is 90° behind v_L. So, the phasor I_L is drawn 90° behind the phasor V.

15.2 Alternating Current in an Inductance

We shall first consider an inductance coil that has no resistance (Fig. 15.1a). We know from previous study (Fig. 13.8a) that an alternating current in a pure inductance lags 90° in phase behind the ac voltage applied. If we let the inductive reactance be represented by $j2\pi fL$, we can see how this lag is shown mathematically. Remember, the operator $-j$ rotates the phasor $-90°$.

$$\mathbf{I}_L = \frac{V + j0}{j2\pi fL} = \frac{V}{j2\pi fL}$$

Multiply the right side by j in both numerator and denominator:

$$\mathbf{I}_L = \frac{jV}{j^2 2\pi fL} = \frac{jV}{-2\pi fL} = \frac{-jV}{2\pi fL} \quad \text{rationalized}$$

The phasor diagram (Fig. 15.1b) shows this current and this voltage.

EXAMPLE 15.1   A voltage $\mathbf{V} = 100 \sin 377t$ is applied to a pure inductance of 0.1 H. Calculate the current. What is its phase with respect to the phase of V? At the instant $377t = \pi$ radians, what are the positions of the voltage and current phasors on the phasor diagram?

Solution
$$2\pi f = 377, \qquad f = 60 \text{ Hz}$$
$$\mathbf{X}_L = j2\pi fL = j377 \times 0.1 = j37.7 \ \Omega$$
$$\mathbf{I}_L = 100/j377 = -j0.265 \text{ A, } 90° \text{ behind } V$$

At $\omega t = \pi$, $\mathbf{V}$ is horizontally to the left and $\mathbf{I}_L$ is vertically upward.

EXAMPLE 15.2   A coil has an inductance of 200 mH. Calculate its inductive reactance at frequencies of (a) 1000 Hz, (b) 100 kHz, (c) 10 MHz.

Solution
(a) $\omega L = 2\pi \times 1000 \times 0.2 = 1256 \ \Omega$
(b) $\omega L = 2\pi \times 10^5 \times 0.2 = 1.256 \times 10^5 \ \Omega$
(c) $\omega L = 2\pi \times 10^7 \times 0.2 = 1.256 \times 10^7 \ \Omega$

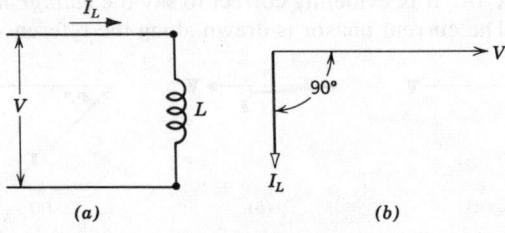

(a)   (b)

FIGURE 15.1   (a) Voltage *V* across pure inductance *L*. (b) Current in inductance lags voltage 90° in phase.

15.3 Resistance and Inductance in Series; Impedance

When an ac voltage is applied to a practical coil, the amount and phase angle of the current depend on the inductance and resistance of the coil. When the resistance is very small compared with the inductive reactance, the current lags the terminal voltage of the coil by nearly 90°. Conversely, when the resistance is very large compared with the inductive reactance, the current is almost in phase with the voltage. Then the angle of lag is nearly zero. See Fig. 15.2.

Because the effects of resistance and inductive reactance differ in influencing the *phase* of the current, we are not permitted to add their ohms values in the way we add resistances in series. We may represent R and X_L by two sides of a right triangle, as shown in Fig. 15.3a, and make the lengths proportional to the ohms values. The hypotenuse of the triangle also represents an ohms value called *impedance*. When properly used the impedance represents the combined effects of the resistance and inductance. Clearly,

$$Z^2 = R^2 + X_L{}^2$$

and

$$Z = \sqrt{R^2 + X_L{}^2} \text{ ohms}$$

(15.6)

Instead of adding the values of R and X_L directly, we must add them as if they were represented by the sides of a right triangle.

Note that an *impedance* Z is not fully described by its magnitude alone. Its angle θ with respect to a reference axis must be given also. We could draw an arc of a circle with the length representing Z as a radius, drop a perpendicular line from any point on the arc to the horizontal reference axis, and obtain different pairs of values for R and X_L which would fit Equation 15.4. So we are obliged to *define* Z *in magnitude and direction.* Accordingly, we may write impedance, inductive reactance, and resistance, in *polar form* as follows:

$$Z \underline{/\theta}, \qquad X_L \underline{/90°}, \qquad R \underline{/0°}$$

(15.7)

Figure 15.3b is the phasor diagram for a coil. V represents the applied voltage and I the current. The current lags behind the voltage by the angle θ, given by $\tan \theta = X_L/R$. It is evidently correct to say the *voltage leads the current* by that angle. The current phasor is drawn along the reference axis in the figure,

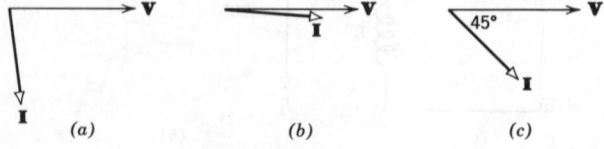

FIGURE 15.2 Phasor diagrams for three different coils. (a) $X_L \gg R$. (b) $R \gg X_L$ (c) $R = X_L$.

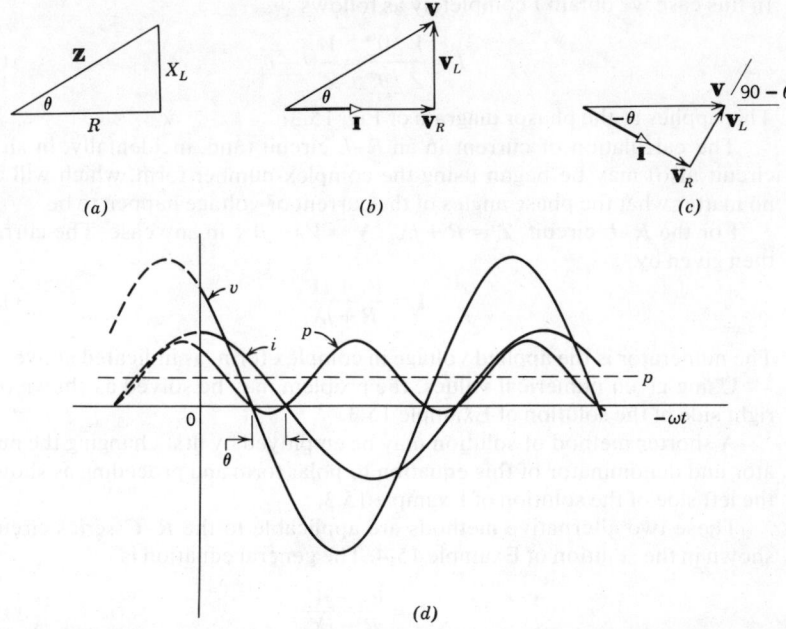

FIGURE 15.3 Diagrams for a series *RL* circuit. (a) Impedance diagram: $Z = R + jX$. (b) Phasor diagram: $I \angle 0°$, $V \angle \theta°$, $V_L \angle 90°$, $I = I + j0$, $V = V_R + jV_L$, $V_L = 0 + jV_L$. (c) Phasor diagram: $V \angle 0°$, $I \angle -\theta°$, $V_L \angle 90° - \theta$. $V = V + j0$, $I = I \cos \theta - jI \sin \theta$, $V_R = V_R \cos \theta - jV_R \sin \theta$, $E_L = V_L \cos (90 - \theta) + jV \sin (90 - \theta)$. (d) $i = I_m \cos \omega t$, $e = V_m \cos (\omega t + \theta)$.

and we could designate the current and voltage as $I \angle 0°$, $V \angle \tan^{-1} X_L / R$.

The applied voltage is said to have *a component* $V_R \angle 0°$ in phase with the current and *a component* $V_L \angle 90°$ which *the current lags* by 90°. We might suspect that the ampere value of the current may be computed by the relation $I = |V| / |Z|$, and this is correct. But this is not all the information we need for defining the current completely. As indicated before, we must know *its phase*. If the phase of **V** is specified as θ degrees in advance of the reference axis, θ being the impedance angle in Fig. 15.3*a*, the current phasor coincides with the reference axis (Fig. 15.3*b*).

$$\mathbf{I} = \frac{V \angle \theta}{Z \angle \theta} = \frac{V}{Z} \angle \theta - \theta = \frac{V}{Z} \angle 0° \qquad (15.8)$$

Quite often, *and this is important*, the *V* phasor is drawn on the reference axis and the *I* phasor at its proper angle. In the case of a coil we would have

$$V \angle 0°, \quad I \angle -\theta°$$

In this case we obtain I completely as follows

$$I = \frac{V \underline{/0°}}{Z \underline{/\theta°}} = \frac{V}{Z} \underline{/-\theta°} \tag{15.9}$$

This applies to the phasor diagram of Fig. 15.3c.

The calculation of current in an R–L circuit (and, incidentally, in an R–C circuit also) may be begun using the complex-number form, which will apply no matter what the phase angles of the current or voltage happen to be.

For the R–L circuit, $\mathbf{Z} = R + jX_L$, $\mathbf{V} = V_R + jV_X$ in any case. The current is then given by

$$\mathbf{I} = \frac{V_R + jV_X}{R + jX_L} \tag{15.10}$$

The numerator is the applied voltage in complex form, as indicated above.

Using given numerical values, the problem may be solved as shown on the right side of the solution of Example 15.3.

A shorter method of solution may be employed by first changing the numerator and denominator of this equation to polar form and proceding as shown on the left side of the solution of Example 15.3.

These two alternative methods are applicable to the R–C series circuit, as shown in the solution of Example 15.4. The general equation is

$$\mathbf{I} = \frac{V_R + jV_C}{R - jX_C} \tag{15.11}$$

EXAMPLE 15.3   A coil that has $R = 9.07\ \Omega$, $X_L = 4.21\ \Omega$, has $V = 50\ \text{V}$ rms impressed upon it. Determine the (a) impedance, (b) phase angle between the voltage and current, (c) current, (d) voltage drop due to resistance, (e) voltage drop due to inductive reactance. Draw the impedance diagram and the phasor diagram.

Solution. See Fig. 15.4. Both the polar and complex-number methods will be shown.

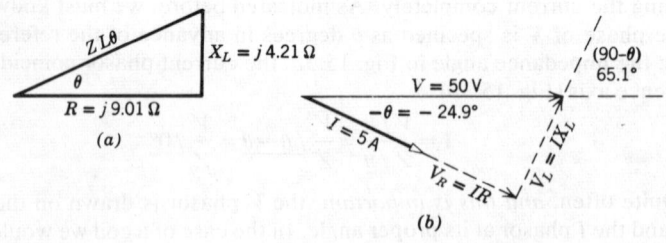

FIGURE 15.4   (a) Impedance diagram. (b) Phasor diagram.

(a) $Z = \sqrt{9.07^2 + 4.21} \underline{/\theta°}$
    $Z = \sqrt{100} = 10 \underline{/\theta°}$

(b) $I = 50 \underline{/0°}/10 \underline{/\theta°} = 5 \underline{/-\theta°}$ A

(c) $\tan \theta = 4.21/9.07 = 0.464$
    $\theta = 24.9°$

(d) $IR = 5 \times 9.07 = 45.35$ V

(e) $IX_L = 5 \times 4.21 = 21.05$ V

Check: $V = \sqrt{45.35^2 + 21.05^2}$
       $= \sqrt{2057 + 443}$
       $V = \sqrt{2500} = 50$ V

(a) $\mathbf{Z} = 9.07 + j4.21$
    $= 10 \underline{/\arc \tan (4.21/9.07)}$

(b) $\theta = \arc \tan 4.21/9.07$
    $\arc \tan \underline{/0.464} = 24.9°$

(c) $\mathbf{I} = \dfrac{50 + j0}{9.07 + j4.21} \times \dfrac{9.07 - j4.21}{9.07 - j4.21}$

    $= \dfrac{453.5 - j210.5}{82.26 + 17.72}$

    $\mathbf{I} = \dfrac{453.5 - j210.5}{99.98}$

    $= 4.54 - j2.11 = 5 \underline{/-24.9°}$ A

$\mathbf{I}R = 4.54 - j2.11)(9.07 + j0)$
    $= 41.17 - j19.13 =$
    $= 45.4 \underline{/-24.9°}$

$\mathbf{I}X_L = (4.54 - j2.11)(j4.21)[$
    $= 8.88 - j19.11$
    $= 21.05 \underline{/-65.1°}$

Note that the complex number method[1] gives us the angle of lag of the current in the expression for **I**. It also gives us the phase angles of **I**R and **I**X_L with respect to the reference angle for **V** which is assumed zero. In calculating the voltage drop due to inductive reactance by the complex number method, we must use X_L in its proper angular relation, which is at 90° advance with respect to R: $X_L \doteq j0.421$ ohms.

Problems

1. A coil has 25 Ω resistance and 80 mH inductance. What are its reactance and impedance when carrying current for which $f = 120$ Hz?

2. A voltage $V = 100$ V, 120 Hz is applied to the coil in Problem 1. Calculate the current and its phase with respect to that of V.

3. Given the coil, voltage, and other data of Problems 1 and 2, determine how much voltage drop occurs due to resistance and how much due to inductive reactance. Check your answers to see if their phasor sum equals the applied voltage.

4. A coil for which $L = 0.05$ H, $R = 56.5$ Ω must carry a 60 Hz current of 3 A. How much voltage must be applied to it? What will be the phase of the voltage with respect to the current?

[1]An extensive treatment of the algebra of complex notation and polar and exponential forms is given in the appendix. Also given are instructions in changing from complex-number to polar form *by means of the slide rule.*

5. An inductance coil has $L = 1$ H, $R = 20\ \Omega$. It is necessary that current in the coil lag the applied voltage by $45°$ when the frequency is 60 Hz. How much resistance must be added in series?

6. A coil in a radio receiver circuit has $L = 20\ \mu H$, $R = 80\ \Omega$. A voltage of 10 mV, 1.592 MHz is applied to it. How much current is in the coil and what is its phase angle with respect to the voltage?

7. In Problem 6, determine V_R and V_L. What is the phase of V_L with respect to the current, and also with respect to the applied voltage?

15.4   Capacitive Reactance

We have seen in Section 12.11 that the amount of current in a capacitor is given by the product of the capacitance and the rate of change of the capacitor voltage with respect to time:

$$i_c = C\,\frac{dv_c}{dt}$$

We can let $v_c = V_{cm} \cos \omega t$, then $dv_c/dt = -\omega V_{cm} \sin \omega t \sin \omega t$ so that

$$i_c = -\omega C V_{cm} \sin \omega t$$
$$i_c = \omega C V_{cm} \cos (\omega t + \pi/2)^* \qquad (15.12)$$

This equation shows that i_c is $\pi/2$ radians *ahead* of v_c. *The current in a capacitor leads the voltage on it by $\pi/2$ radians, or $90°$, in phase.*

Equation (15.4) tells us that the maximum instantaneous value of current is given by $\omega C V_{cm}$, so we have

$$I_m = \omega C V_{cm} \qquad (15.13)$$

Changing to rms values,

$$\frac{I_m}{\sqrt{2}} = \omega C\,\frac{V_{cm}}{\sqrt{2}}$$

$$I = \omega C V_c \qquad (15.14)$$

or

$$I = \frac{V_c}{1/\omega C} \qquad (15.15)$$

Evidently $1/\omega C$ is an *ohms* value. It is called **capacitive reactance**, and is represented by X_C.

$$X_C = \frac{1}{\omega C} \text{ ohms capacitive reactance} \qquad (15.16)$$

15.5   Resistance and Capacitance in Series

When an ac voltage is applied to a resistance and a capacitance in series, the amount and phase angle of the current depend on the values of the resistance

*$\cos (\omega t + \pi/2) = \cos \omega t \cos (\pi/2) - \sin \omega t \sin (\pi/2) = -\sin \omega t$.

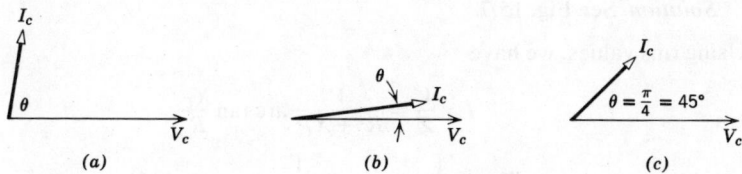

FIGURE 15.5 Phasor diagrams of capacitor current and voltage of an *R–C* series circuit. (a) $X_c \gg R$. (b) $X_c \ll R$. (c) $X_c = R$.

and capacitance. When C is relatively large, the capacitive reactance, X_C, is relatively small and the current is limited mostly by the resistance. Its angle of lead is small. See Fig. 15.5. When C is relatively small, X_C is relatively large and the current is limited mostly by the capacitive reactance. When C has a value that makes $1/\omega C = R$, the angle of lead is 45°. Waves of current and voltage in a series $R–C$ circuit are shown in Fig. 15.6. The current leads the voltage by the angle θ radians. The applied voltage is $v = V_m \cos \omega t$ and the current is $i = I_m \cos (\omega t + \theta)$.

EXAMPLE 15.4   A resistance $R = 30\ \Omega$ and a capacitor $C = 0.025\ \mu\text{F}$ are in series and a voltage $V = 141.4 \sin 10^6 t$ is applied to the a pair. Calculate the impedance, current, phase of current with respect to voltage, and voltages across R and C.

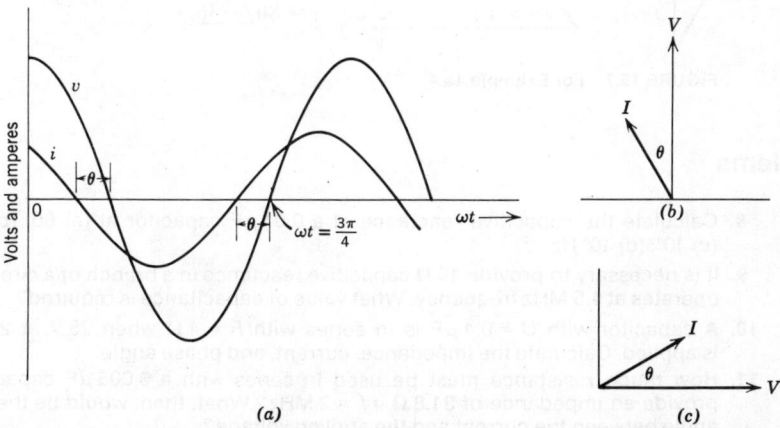

FIGURE 15.6 (a) Cosine-wave forms of *v*, the voltage applied to an *R–C* series circuit, and *i*, the current in the circuit. The current leads the voltage by $\theta = \tan^{-1} X_C/R$. (b) Phasors at $t = 0$. (c) Phasors at $\omega t = 3\pi/2$ radians.

Solution. See Fig. 15.7

Using rms values, we have

$$I = \frac{V}{Z} = \frac{V}{R^2 + X_C^2} \text{ arc tan } \frac{X_C}{R}$$

$$X_C = \frac{1}{\omega C} = \frac{1}{10^6 \times 0.025 \times 10^{-6}} = 40 \ \Omega$$

$Z = \sqrt{30^2 + 40^2} = \sqrt{2500} = 50 \ \Omega$

$\theta = \text{arc tan} -40/30 = -53.1°$

$E = 141.4/\sqrt{2} = 100/0° \text{ V}$

$I = 100/50 \underline{/-53.1°} = 2 \underline{/53.1°} \text{A}$

$V_R = 2 \underline{/53.1°} \times 30 \underline{/0°} = 60 \underline{/53.1°} \text{ V}$

$Z = 30 - j40 = 50 \underline{/- \text{arc tan } 1.333}$

$\theta = -\text{arc tan } 1.333 = -53.1°$

$E = 100 + j0$

$I = \frac{100}{30 - j40} \times \frac{30 + j40}{30 + j40} = \frac{3000 + j4000}{2500}$

$I = 1.2 + j1.6 = 2 \underline{/\text{arc tan } 1.6/1.2}$

$\quad = 2\underline{/53.1°}\text{A}$

$X_C = -1/\omega C$, and the phase angle is positive with respect to the angle ωt at which V is expressed, If $t = 0$, v has its maximum instantaneous value.

$V_C = 2\underline{/53.1°} \times 40 \underline{/-90°} = 80 \underline{/-36.9°}$

$R = 30 \Omega$

$X_c = j40 \Omega$

$\theta = 53.1°$

$I = 2\underline{/\text{arc tan } 1.333} = 2\underline{/53.1°}$

$V_R = (1.2 + j1.6)(30 + j0) = 36 + j48$

$V_R = \sqrt{36^2 + 48^2} \underline{/\text{arc tan } 48/36}$

$\quad = 60\underline{/53.1°}$

$V_C = (1.2 + j1.6)(-j48) = 64 - j48$

$V_C = \sqrt{64^2 + 48^2} \underline{/\text{arc tan } -48/64}$

$\quad = 80\underline{/-36.1°}$

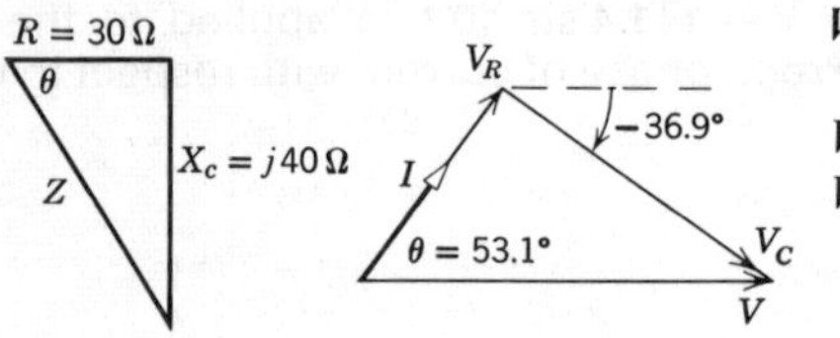

FIGURE 15.7   For Example 15.4.

Problems

8. Calculate the capacitive reactance of a 0.01 μF capacitor at (a) 60, (b) 1000, (c) 10^6, (d) 10^8 Hz.

9. It is necessary to provide 10 Ω capacitive reactance in a branch of a circuit that operates at 1.5 MHz frequency. What value of capacitance is required?

10. A capacitor with $C = 0.1 \mu$F is in series with $R = 4 \Omega$ when 25 V at 200 KHz is applied. Calculate the impedance, current, and phase angle.

11. How much resistance must be used in series with a 0.005 μF capacitor to provide an impedance of 31.8 Ω if $f = 2$ MHz? What, then, would be the phase angle between the current and the applied voltage?

12. It is necessary to provide a current of 64 mA in an *R-C* series circuit and have it lead the applied voltage, 120 V rms, 60 Hz by 45°. What size capacitor is needed to go with a 1.327 Ω resistor?

15.6   Resistance, Inductance, and Capacitance in Series

We are now ready to study an ac series circuit that has one or more of all three kinds of elements: resistance, inductance, and capacitance. The schematic diagram of the circuit may be drawn as shown in Fig. 15.8a in which R represent the sum of the resistances, L represents the sum of the inductances, and C represents the *equivalent-series* capacitance. It will be recalled that the equivalent capacitance of capacitors in series is the reciprocal of the sum of the reciprocal of the sum of the reciprocals of the separate capacitances.

We have learned that a reactance is expressed as the imaginary part of a complex number, an inductive reactance by $j\omega L$ and a capacitive reactance by $-j/\omega C$. The series impedance of this circuit is then

$$\mathbf{Z} = R + j\omega L - \frac{j}{\omega C} \qquad (15.17)$$

This may also be written

$$\mathbf{Z} = \sqrt{R^2 + (X_L - X_C)^2} \; /\text{arc tan } X_L - X_C/R \qquad (15.18)$$

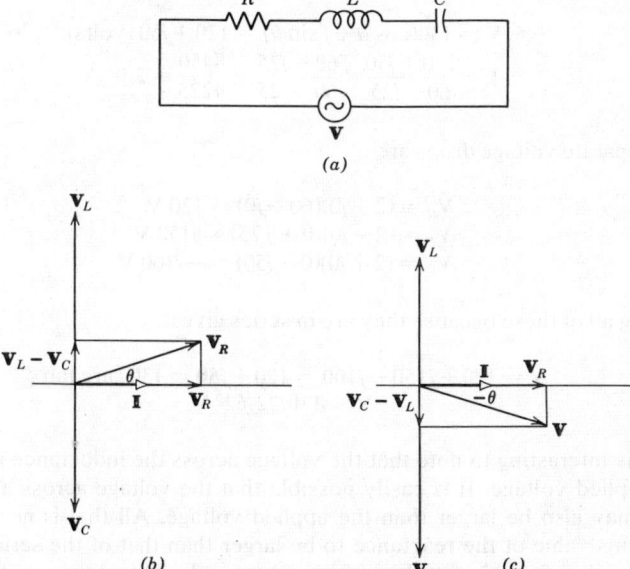

(a)

(b)   *(c)*

FIGURE 15.8   For a series *RLC* circuit. (a) Schematic. (b) Phasor diagram when $V_L > V_c$. (c) Phasor diagram when V > V . The phasor diagrams may be drawn with V on the horizontal. Then all phasors will be rotated an angle $-\theta$ about the origin in (b) and $+\theta$ in (c).

If we assume $X_L > X_C$, the current will lag behind the applied voltage. The phasor diagram for this condition is shown in Fig. 15.8*b*. If $X_C > X_L$ the current leads the applied voltage. This is shown in Fig. 15.8*c*.

Because the current has the same value in all three elements of the circuit, it is convenient to draw the current phasor on the reference axis. When $X_L > X_C$, $V_L > V_C$ and $[V_L + (-V_C)]$ is a phasor which coincides with $\mathbf{V}_L$ on the diagram. The applied voltage is the phasor sum of all three voltage drops: $\mathbf{V} = \mathbf{V}_R + \mathbf{V}_L - \mathbf{V}_C$.

EXAMPLE 15.5   We shall use numerical values to illustrate this situation. Let the ohms values be: $X_L = 75$, $X_C = 50$, $R = 60$; let $V = 130$ V rms.

Solution

$$\mathbf{Z} = 60 + j25 = 65 \,\underline{/\text{arc tan } \tfrac{5}{12}}$$

The angle θ between the applied voltage $\mathbf{V}$ and the current $\mathbf{I}$ is therefore $\theta = \text{arc tan } \tfrac{5}{12}$, and $\mathbf{V}$ leads $\mathbf{I}$ because $\mathbf{I}$ lags $\mathbf{V}$. The magnitude of the component of $\mathbf{V}$ which is in phase with $\mathbf{I}$ is $V \cos \theta$ and the magnitude of the quadrature component is $V \sin \theta$. In this case $\cos \theta = \tfrac{12}{13}$, $\sin \theta = \tfrac{5}{13}$.

$$\mathbf{V} = 130(\cos \theta + j \sin \theta) = 120 + j50 \text{ (volts)}$$
$$\mathbf{I} = \frac{120 + j50}{60 + j25} \times \frac{60 - j25}{60 - j25} = \frac{8450}{4225} = 2\,\underline{/0°}\text{ A}$$

The separate voltage drops are

$$\mathbf{V}_R = (2 + j0)(60 + j0) = 120 \text{ V}$$
$$\mathbf{V}_L = (2 + j0)(0 + j75) = j150 \text{ V}$$
$$\mathbf{V}_C = (2 + j0)(0 - j50) = -j100 \text{ V}$$

Adding all of these because they are in series gives

$$\mathbf{V} = 120 + j150 - j100 = 120 + j50 = 130 \,\underline{/\text{arc tan } \tfrac{5}{12}}$$
$$\mathbf{V} = 130\,\underline{/22.62°}\text{ V}$$

It is interesting to note that the voltage across the inductance is larger than the applied voltage. It is easily possible that the voltage across a series capacitor may also be larger than the applied voltage. All that is necessary is for the ohms value of the reactance to be larger than that of the series resistance R. It is also possible for the voltages across the inductance and capacitance to be equal. In such a case, $\mathbf{V}_L = -\mathbf{V}_C$ and their sum is zero. *The impedance is then a pure resistance* (see Equations (15.17), (15.18)) and a phenomenon called *series resonance* exists. Series resonance is discussed in Chapter 20.

Problems

13. A series circuit contains the following elements. $R_1 = 8\,\Omega$, $L_1 = 20$ mH, $C_1 = 50\,\mu$F, $R_2 = 7\,\Omega$, $L_2 = 5$ mH. The applied voltage is 100 V rms at $\omega = 10^3$ rad/s. What are the total (a) resistance, (b) inductive reactance, and (c) capacitive reactance? Express (b) and (c) in complex notation.

14. (a) Sketch the impedance diagram for the circuit of Problem 13. Evaluate the impedance and the current. Express both in polar and complex forms. Sketch the phasor diagram.

15. Calculate the voltage drop across each of the five elements of the circuit of Problem 13.

16. A network in a radio circuit is made up of $R = 20\,\Omega$, $L = 1\,\mu$H, $C = 5$ nano-farads in series. The applied voltage is 20 V rms at 1.592×10^6 Hz. Calculate the current in and the voltage across each element. What is the angle between the current and the voltage? Draw the phasor diagram.

17. (a) Express each of the three voltages found in Problem 16 in complex form. (b) Add the voltages and check their sum to equal the applied voltage.

15.7   Review Questions

1. We learned in an earlier chapter that when alternating current is in an inductance, an induced EMF accounts for voltage across the inductance. We can now express this as the product of the current times an ohms value. What is this ohms value called, and how is it calculated?

2. What equations in Section 15.1 show that the current in an inductance lags 90° behind the voltage impressed on it?

3. Define the impedance of a series circuit that contains resistance and inductance.

4. Write the equation for the impedance in complex form.

5. How is the voltage applied to a coil related to the voltage drops due to resistance and inductive reactance?

6. State the general form of "Ohm's law for an ac circuit."

7. Define capacitive reactance.

8. Consider two capacitors C_1 and C_2, C_2 the larger. (a) Which has the larger capacitive reactance at a given frequency? (b) Consider two capacitors with values of C that are equal. How would you connect them for minimum total reactance, in series or in parallel?

9. What is the equation for calculating the current in an R-C series circuit? Does the current lag or lead the applied voltage? Repeat for an R-L series current.

10. Which equation in this chapter shows the phase angle between a capacitor current and the voltage on the capacitor?

11. What effect does increasing R in a series R-C circuit have on the phase angle between the voltage applied and the current in the circuit?

12. A series R-C circuit has a resistance R, a capacitance C, and an applied voltage with ω angular frequency. Write the equation for the impedance in complex form.

13. A series RLC circuit has $v = V_m \sin \omega t$ applied. Write the equation for the impedance in (a) polar form, (b) complex-number form.

14. Draw the impedance diagram for a series RLC circuit that has $X_L > X_C$.

15. Repeat Problem 14 for the case when $X_C > X_L$.

16. Assume the separate voltages across R, L, and C in a series RLC circuit are, respectively, $10+j4$, $20+j6$, $5-j8$. Can you obtain from these the complex form of the voltage applied to the circuit?

15.8 Problems

18. Change to complex form: (a) $20\underline{/30°}$, (b) $40\underline{/-60°}$, (c) $25\underline{/125°}$, (d) $100\underline{/-270°}$.

19. Change to polar form: (a) $3+j4$; (b) $5-j12$; (c) $-7.07+j7.07$; (d) $-7.07-j7.07$.

20. Given $\epsilon^{j\theta} = \cos\theta + j\sin\theta$, convert the following to complex form: (a) $\epsilon^{j30°}$; (b) $10\epsilon^{j45°}$; (c) $\epsilon^{-j67.4°}$; (d) $100\,\epsilon^{j(\pi/2)}$; (e) $10\epsilon^{-j(\pi/4)}$.

21. Evaluate the sum, difference, and product of the phasors $\mathbf{A} = 2+j3$, $\mathbf{B} = 5-j4$. Show the results graphically.

22. Convert $1/(3-j4)$ into a complex number by rationalizing the denominator, and sketch the phasor representing it.

23. Convert the following into a single complex quantity:
$$[(4+j3)(1-j2)] \div [(6-j8)(-1+j2)]$$

24. Obtain the square of each of the following, first using the polar form and then the complex form of the same quantity in each case: (a) $10\underline{/30°}$, (b) $5\underline{/45°}$, (c) $10\underline{/120°}$.

25. Obtain the square root of each of the following: (a) $81\underline{/60°}$, (b) $144\underline{/130°}$, (c) $25\underline{/-45°}$. Watch this last one because the angle is not only $-22.5°$. Check your answer by squaring it. Is $5\underline{/157.5°}$ an answer also?

26. A "reactor" coil has $L = 2$ H, $R = 25\ \Omega$. When used on a 120-V 60-Hz power line, what is the current and its phase angle?

27. Two impedances, Z_1 and Z_2, are in series. The voltages across them are: $V_1 = 10 \sin(\omega t + 60°)$, $V_2 = 5 \sin(\omega t + 30°)$. The applied voltage to the circuit is $V = V_1 + V_2$. Sketch the curves of V_1 and V_2 freehand, with reasonable accuracy, and also sketch the curve of V by adding ordinates.

28. An inductive reactance of $10\ \Omega$, a capacitive reactance of $8\ \Omega$, and a resistance of $4\ \Omega$ are in series. A voltage $V = 10+j0$ volts, $\omega = 10^6$ rps, is applied. (a) What is the impedance in ohms and in complex form? (b) Determine the complex expression for the current, also the polar form and the ampere value. (c) How much voltage drop is due to R, to L, and to C? (d) Calculate the values of L and C. (e) Draw the impedance diagram and the phasor diagram.

29. A series circuit with $R = 15\ \Omega$, $X_L = 20\ \Omega$ is supplied with current when $V = 150$ V rms at 60 Hz is applied. (a) What value of capacitance must be connected in series in order that the current will be in phase with the voltage? (b) Calculate the voltage across the inductance, the capacitance, and the resistance. tance.

CHAPTER SIXTEEN

Parallel and Series-Parallel AC Circuits

In our study of circuits containing resistance only, we found it desirable, in some cases, to represent a group of two or more resistors that were in parallel by a single equivalent resistance. Then we could add the equivalent resistance to one or more that were *in series* with the parallel group.

We shall, in some cases, follow the same procedure when we have parallel branches that contain combinations of R, L, and C. We shall obtain the value of the equivalent impedance and combine it properly with other impedances in the circuit.

The formula for conductance will involve reactance as well as resistance. New quantities called susceptance and admittance will be defined and used in ac circuit analysis in this chapter.

16.1 Resistance and Inductance in Parallel

In the simple circuit of Fig. 16.1*a*, containing R and L in parallel, the total current $\mathbf{I}_T$ is the phasor sum of $\mathbf{I}_R$ and $\mathbf{I}_L$. The phasor diagram in (*b*) shows $\mathbf{I}_L$ lagging $\mathbf{V}$ by 90°. The applied voltage is the same on all branches of a parallel

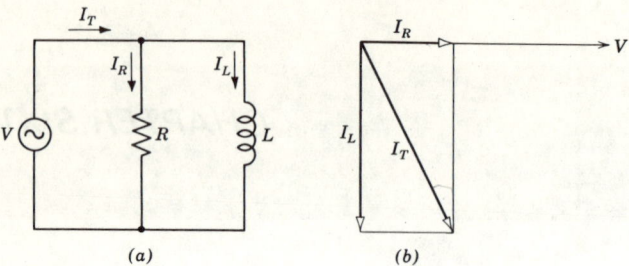

FIGURE 16.1 **(a) Parallel *RL* circuit. (a) Circuit diagram. (b) Phasor diagram. *V* = 100 V, *R* = 20 Ω,**
X_L = 10 Ω, *θ* = − 63.4°.

circuit, so we can calculate the separate currents directly:

$$I_R = \frac{100}{20} = 5 \text{ A}, \qquad I_L = \frac{100}{10} = 10 \text{ A}$$

Using complex notation,

$\mathbf{I}_R = (100 + j0)/(20 + j0) = 5 + j0 \text{ A}; \mathbf{I}_L = (100 + j0)/j10 = 10/j = -j10 \text{ A}$

$\mathbf{I}_T = \sqrt{5^2 + 10^2} \; \underline{/\theta°} = 11.18 \; \underline{/\theta} \text{ A}; \; |\mathbf{I}_T = 5 + j0 - j10 = 5 - j10 \text{ A in } j \text{ notation}$

$\theta = \text{arc tan} (-10/5) = -63.4° \qquad |\mathbf{I}_T = \sqrt{125} \; \underline{/-\theta} = 11.18 \; \underline{/-63.4°}$

When only two impedances are in parallel we can calculate their equivalent
impedance by getting their product divided by their sum:

$$\mathbf{Z} = \frac{20 \, (j10)}{20 + j10} \times \frac{20 - j10}{20 - j10} =$$

$$\mathbf{Z} = \frac{2000 + j4000}{500} = 4 + j8 = 8.94 \; \underline{/63.4°}$$

Let's get $\mathbf{I}_T$ from **V** and **Z**,

$$\mathbf{I}_T = 100 \; \underline{/0°} \div 8.94 \; \underline{/63.4°} = 11.18 \; \underline{/-63.4°}$$

16.2   Equivalent Series Circuit

On occasion we may wish to represent a parallel *R-L* circuit by an *equivalent
series circuit*. This would be a circuit that would take the same current and
power from the source and have the same phase angle between its current and
voltage.

Because the equivalent series circuit must have the same impedance, we
find that the real and "reactive" components of **Z** will be the resistance R_e

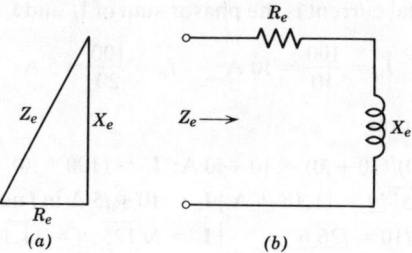

FIGURE 16.2   Series-circuit equivalent of a parallel _RL_ circuit. (_a_) Impedance diagram. (_b_) Circuit diagram.

and reactance X_e. The impedance and schematic diagram of the equivalent circuit for Fig. 16.1a àre in Fig. 16.2.

$$R_{se} = Z \cos \theta = 8.94 \cos 63.4° = 4\ \Omega$$
$$X_{se} = Z \sin \theta = 8.94 \sin 63.4° = 8\ \Omega$$

Note that these are the values we got when we took the product over the sum. The values are different from the original R and X_L, as we would expect.

16.3   Resistance and Capacitance in Parallel

In the simple circuit of Fig. 16.3a, containing R and C in parallel, we shall find that the total current I leads the voltage V by an angle that depends on the relative values of R and C. The phasor diagram in (b) shows the capacitor current leading the applied voltage V by 90°.

EXAMPLE 16.1   Calculate the total current and impedance of the parallel circuit in Fig. 16.3a. What is the impedance of the equivalent series circuit?

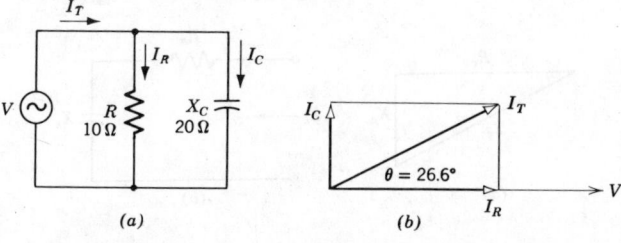

FIGURE 16.3   For Example 16.1. Parallel _RC_ circuit. (_a_) Circuit diagram. (_b_) Phasor diagram.

Solution: The total current is the phasor sum of $\mathbf{I}_R$ and $\mathbf{I}_c$.

$$I_R = \frac{100}{10} = 10 \text{ A} \qquad I_c = \frac{100}{20} = 5 \text{ A}$$

Using complex notation

$$\mathbf{I}_R = (100 + j0)/(10 + j0) = 10 + j0 \text{ A}; \ \mathbf{I}_C = (100 + j0)/-j20 = j5 \text{ A}$$

$$\mathbf{I}_T = \sqrt{10^2 + 5^2} \ \underline{/\theta} = 11.18 \ \underline{/\theta} \text{ A} \ \Big| \ \mathbf{I}_T = 10 + j5 \text{ A in } j \text{ notation}$$

$$\theta = \text{arc tan } 5/10 = \underline{/26.6°} \qquad \Big| \ \mathbf{I}_T = \sqrt{125} \ \underline{/\theta} = 11.18 \ \underline{/26.6°} \text{ A}$$

The equivalent impedance is

$$\mathbf{Z} = \frac{10\,(-j20)}{10 - j20} \times \frac{10 + j20}{10 + j20}$$

$$\mathbf{Z} = \frac{4000 - j2000}{500} = 8 - j4 = 8.94 \ \underline{/-26.6°} \ \Omega$$

The *equivalent series circuit* has this identical impedance. That is $R_e = 8 \ \Omega$, $X_e = -j4 \ \Omega$ as shown in Fig. 16.4.

16.4 *R, L,* and *C* in Parallel

Let's first discuss the treatment of the three-branch parallel circuit of Fig. 16.5a in general terms. Assume X_L is smaller than X_C. The ampere value of $\mathbf{I}_L$ will then be larger than that of $\mathbf{I}_C$, as shown in (*b*). The total current, $\mathbf{I}_T$, will lag $\mathbf{V}$ by an angle θ that we can easily evaluate. The separate currents are

$$\mathbf{I}_R = \frac{V \ \underline{/0°}}{R \ \underline{/0°}} = \frac{V}{R} \ \underline{/0°}$$

$$\mathbf{I}_L = \frac{V \ \underline{/0°}}{X_L \ \underline{/90°}} = \frac{V}{X_L} \ \underline{/-90°}$$

$$\mathbf{I}_C = \frac{V \ \underline{/0}}{X_C \ \underline{/-90°}} = \frac{V}{X_C} \ \underline{/90°}$$

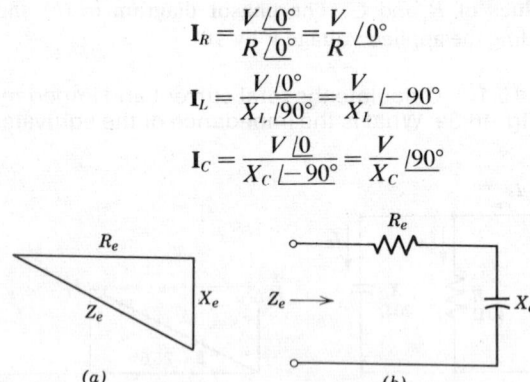

(a)　　　　　(b)

FIGURE 16.4 Series circuit equivalent of a parallel *RC* circuit. (a) Impedance diagram. (b) Circuit diagram.

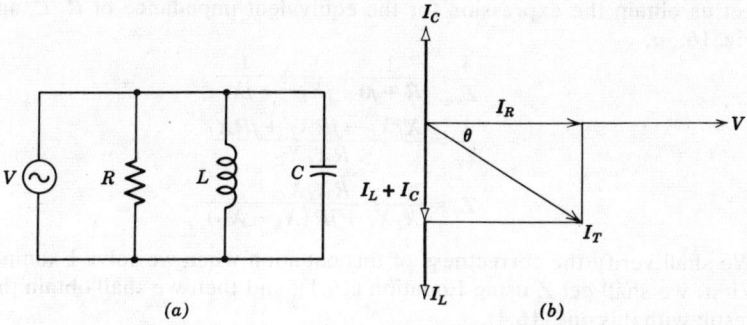

FIGURE 16.5   Parallel *RLC* circuit. (*a*) Circuit diagram. (*b*) Phasor diagram.

The *component of total current* that must account for $\mathbf{I}_L$ and $\mathbf{I}_C$ is their sum. Call this "reactive component" $\mathbf{I}_X$.

$$\mathbf{I}_X = -j\mathbf{I}_L + j\mathbf{I}_C = -j(\mathbf{I}_L - \mathbf{I}_C)$$

The total current is

$$\mathbf{I}_T = \sqrt{I_R^2 + I_X^2}\,\underline{/\text{arc tan}\,(-I_X/I_R)}$$

in which $I_X = I_L + I_C$, the net reactive component.

We can get the total impedance of the circuit by dividing the applied voltage by the total current:

$$\mathbf{Z} = \frac{V\underline{/0°}}{I_T\underline{/\theta}} \tag{16.1}$$

in which θ is the phase angle of $\mathbf{I}_T$.

16.5   Equivalent Impedance of Impedances in Parallel

In our study of parallel resistances we learned that the equivalent resistance of any number of resistances in parallel is obtained by the reciprocal relation. See Equation (5.7).

When impedances are in parallel we use the same kind of relation:

$$\frac{1}{\mathbf{Z}_e} = \frac{1}{\mathbf{Z}_1} + \frac{1}{\mathbf{Z}_2} + \frac{1}{\mathbf{Z}_3} + \cdots + \frac{1}{\mathbf{Z}_n} \tag{16.2}$$

When there are only two in parallel, the "product over the sum" relation holds,

$$\mathbf{Z}_e = \frac{\mathbf{Z}_1 \mathbf{Z}_2}{\mathbf{Z}_1 + \mathbf{Z}_2} \tag{16.3}$$

Let us obtain the expression for the equivalent impedance of R, L, and C in Fig. 16.5a,

$$\frac{1}{\mathbf{Z}_e} = \frac{1}{R+j0} + \frac{1}{jX_L} + \frac{1}{-jX_C}$$

$$\frac{1}{\mathbf{Z}_e} = \frac{X_LX_C - jRX_C + jRX_L}{RX_LX_C}$$

$$\mathbf{Z}_e = \frac{RX_LX_C}{X_LX_C + jR(X_L - X_C)} \tag{16.4}$$

We shall verify the correctness of this equation when we solve Example 16.2. First, we shall get $\mathbf{Z}$ using Equation (16.1), and then we shall obtain the same result with this one (16.4).

EXAMPLE 16.2 In Fig. 16.5, assume $R = 15\ \Omega$, $X_L = 7.5\ \Omega$, $X_C = 12\ \Omega$, $V = 120$ V 60 Hz. Determine the (a) current in each branch, (b) total current, (c) phase angle of the total current, (d) impedance, in two ways.

Solution

$$\mathbf{I}_R = 120/0°/15/0° = 8/0°\ \mathrm{A}$$
$$\mathbf{I}_L = 120\underline{/0°}/7.5\underline{/90°} = 16\underline{/-90°}\ \mathrm{A}$$
$$\mathbf{I}_C = 120\underline{/0°}/12\underline{/-90°} = 10\underline{/90°}\ \mathrm{A}$$
$$\mathbf{I}_T = \sqrt{8^2 + (16-10)^2}/\theta$$
$$\mathbf{I}_T = 10\underline{/-36.9°};\ \tan\theta = -6/8 = -0.750$$
$$\mathbf{Z} = 120/10\underline{/-36.9°} = 12\underline{/36.9°}\ \Omega$$

Using Equation (16.4),

$$\mathbf{Z} = \frac{15 \times 7.5 \times 12}{7.5 \times 12 + j15(7.5 - 12)}$$

$$\mathbf{Z} = \frac{1350}{90 - j67.5} = \frac{1350\underline{/0°}}{112.5\underline{/-36.9°}}$$

$$\mathbf{Z} = 12\underline{/36.9°}\ \Omega$$

Problems

1. A resistance $R = 25\ \Omega$, and a capacitor for which $X_C = 10\ \Omega$ are in parallel and supplied by a source for which $V = 50$ V rms. Calculate all currents, phase angles, and the total impedance.
2. An inductance L for which $X_L = 20\ \Omega$ is in parallel with a resistance $R = 25\ \Omega$ and $V = 50$ V rms is applied to the pair. Calculate all currents, phase angles, and the total impedance.
3. A three-element parallel circuit in which $R = 25\ \Omega$, $X_L = 20\ \Omega$, $X_C = 10\ \Omega$ has 50 V rms applied to it. What are the (a) total current and its phase angle, (b) imped-

ance? Check the impedance by calculating its value in two ways. Draw the phasor diagram.

4. The expression for the total impedance of the R-L parallel circuit of Fig. 16.1a is obtained by writing the product of the separate branch impedances over the sum:

$$\mathbf{Z}_r = \frac{R(jX_L)}{R + jX_L}$$

Start with this, rationalize the denominator, and obtain the impedance of the *equivalent series* circuit:

$$\mathbf{Z}_{se} = R_{se} + jX_{se}$$

$$\mathbf{Z}_{se} = \frac{RX_L{}^2}{R^2 + X_L{}^2} + j\frac{R^2 X_L}{R^2 + X_L{}^2} \tag{16.5}$$

5. Obtain the expression for the impedance of the R-C parallel circuit of Fig. 16.3a:

$$\mathbf{Z}_{se} = \frac{RX_C{}^2}{R^2 + X_C{}^2} - j\frac{R^2 X_C}{R^2 + X_C{}^2} \tag{16.6}$$

16.6   Admittance, Conductance, Susceptance

In the section following this one we shall study how to analyze series-parallel circuits, which are the ones that have a series group of elements connected to a parallel group so that the two groups are in series. But first we must become familiar with two new quantities, admittance and susceptance, and with a general form for conductance that is something more than merely the reciprocal of resistance. You will recall that in a network that contains only resistance units and no inductances or capacitances, $G = 1/R$.

ADMITTANCE

Admittance is defined as the reciprocal of impedance. Its unit is the *mho*. In general, it has a real part and an imaginary part, as we might suspect, because rationalizing a denominator gives a complex number as a result. Let's express a series impedance in a general form:

$$\mathbf{Z}_s = R_s + jX_s \tag{16.7}$$

The admittance of this series circuit is

$$\mathbf{Y} = \frac{1}{\mathbf{Z}_s} = \frac{1}{R_s + jX_s} \tag{16.8}$$

$$\mathbf{Y} = \frac{1}{R_s + jX_s} \times \frac{R_s - jX_s}{R_s - jX_s} = \frac{R_s - jX_s}{R_s{}^2 + X_s{}^2} \tag{16.9}$$

Important: This is also the expression for the admittance of the parallel circuit that is equivalent to the assumed series circuit.

CONDUCTANCE

Note that the admittance of a series circuit has, in general form, two parts. Equation (16.9) is

$$Y = \frac{R_s}{R_s^2 + X_s^2} - j\frac{X_s}{R_s^2 + X_s^2} \qquad (16.10)$$

The *real part* is the mathematical expression for *conductance*, whose symbol is G, as we have learned.

$$G = \frac{R_s}{R_s^2 + X_s^2} \text{ mhos} \qquad (16.11)$$

Observe that if there is no reactance in the series circuit, X_s is zero and $G = 1/R_s$, as we learned when studying circuits containing only resistances.

SUSCEPTANCE

Susceptance, expressed in mhos, is the *imaginary part* of Y, the admittance. Its symbol is B.

$$B = \frac{X_s}{R_s^2 + X_s^2} \text{ mhos} \qquad (16.12)$$

The admittance diagram is shown in Fig. 16.6.

$$Y = \sqrt{G^2 + B^2} \text{ arc tan } B/G$$
$$B = B_L + B_C = \text{net susceptance}$$
$$B_L \text{ is negative, } B_C \text{ is positive}$$

We must explain why the susceptance is shown vertically upward on the admittance diagram when it is for a capacitance. If a branch of a network

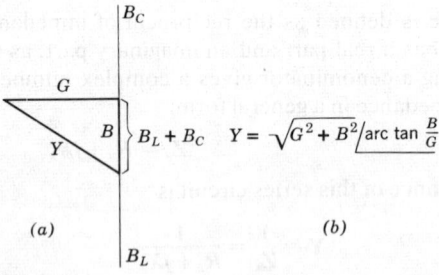

(a)   (b)

FIGURE 16.6 (a) Admittance diagram. (b) Mathematical relations. Net susceptance $B = B_L + B_C$; B_L is negative, B_C positive.

contains only a capacitance, its susceptance is

$$\mathbf{B}_C = \frac{1}{-jX_C} = \frac{j}{X_C} = \frac{1}{X_C} \underline{/90°}$$

Also, the susceptance of a branch that contains only an inductance is

$$\mathbf{B}_L = \frac{1}{jX_L} = \frac{j}{-X_L} = \frac{1}{X_L}(-j) = \frac{1}{X_L} \underline{/-90°}$$

Now that we have complicated things by introducing two new quantities (**Y** and **G**) and have changed the definition of conductance, what is to be gained by this? We shall soon see that we can add admittances (in their $G + jB$ forms) when they are in parallel. (Recall that we add impedances in their $R + jX$ forms when they are in series.) An example will show how we must handle these forms.

You might question why we keep insisting on putting the words *series circuit* into this discussion of G and B which have *parallel circuit applications* as their reasons for existence. Soon we shall be working with parallel circuits in which each branch will have either R and X_L or R and X_C. In order to write the admittances of such branches we will resort to Equation (1.8) to get the Y for one branch at a time. Now, *this is important*: If only R is present in one parallel branch, Equation 16.11 gives us $G = 1/R$ which takes the place of R_s/R_s^2 in the equation. We are still allowed to call *this branch* a series circuit which has no reactance.

If a branch of a parallel circuit has only X_L, then $G = 0$ and $B = -1/X_L$. This is plotted vertically downward on the **Y** diagram. Therefore, a parallel circuit with only R in one branch and only X_L in the other has

$$\mathbf{Y} = \frac{1}{R} + j\frac{-1}{X_L} = \frac{1}{R} - j\frac{1}{X_L}$$

If you go back to Section 16.1 where the impedance of R and L in parallel is found to be $\mathbf{Z} = 4 + j8$, you can see that the admittance $\mathbf{Y} = 1/\mathbf{Z}$ is $0.05 - j0.10$.

A parallel circuit with only R in one branch and only X_C in the other has $Y = (1/R) + j(1/X_C)$. The impedance is of the form $\mathbf{Z} = R_s - jX_s$. This is seen by reference to Section 16.3 where **Z** turned out to be $8 - j4$ ohms.

EXAMPLE 16.3   Given the parallel circuit of Fig. 16.7a in which $R = 5\,\Omega$ $X_L = 0.2\,\Omega$, $X_C = 0.25\,\Omega$. Obtain R_s and X_s, the elements of the equivalent series circuit.

Solution. For the branch containing R only, from Equations (16.9) and (16.10),

$$Y = G = \frac{1}{R} = \frac{1}{5} = 0.2 \text{ mho}, \qquad B = 0$$

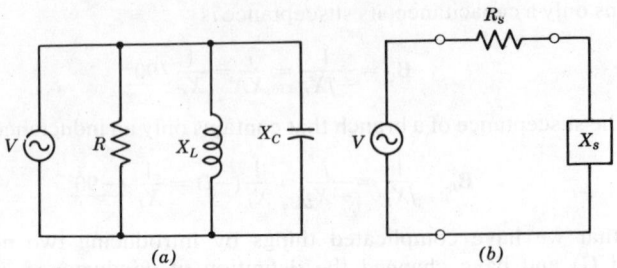

FIGURE 16.7   For Example 16.3. (a) *RLC* parallel circuit. (b) Series-circuit equivalent of (a).

For the branch containing X_L only,

$$Y = B_L = 0 - j\frac{0.2}{(0.02)^2} = -j5 \text{ mhos}$$

For the branch containing X_C only,

$$Y = B_C = 0 - j\frac{(-0.25)}{(-0.25)^2} = j4 \text{ mhos}$$

The total admittance of the parallel circuit (and also of the equivalent series circuit) is

$$Y = 0.2 + j4 - j5 = 0.2 - j1 \text{ mhos}$$

The impedance of the equivalent series circuit is

$$Z = \frac{1}{Y} = \frac{1}{0.2 - j1} = 0.192 + j0.961$$

From this, we see that X_s is inductive, so the square representing X_s in Fig. 16.7b must have an inductance in it.

Note. B in Equation (16.10) will be *negative* if X_C exceeds X_L in a series branch. This will make the second term (imaginary part) of Y [Equation (16.8)] *positive*. Conversely, if $X_L > X_C$ in a series branch, X_s will be positive and so will B in Equation (16.10). This will make the imaginary part of Y [Equation (16.8)] negative. We must draw Y at $-\theta$ when Z is at $+\theta$.

We shall now analyze a parallel circuit that has two different kinds of component in each branch.

EXAMPLE 16.4   In the circuit of Fig. 16.8, $R_1 = 10 \, \Omega$, $X_1 = 5 \, \Omega$, $R_2 = 25 \, \Omega$, $X_2 = 10 \, \Omega$, and $V = 100$ V. (a) Obtain the admittances Y_1 and Y_2, (b) Calculate I_T using the total admittance. (c) What is the impedance of the equivalent series circuit?

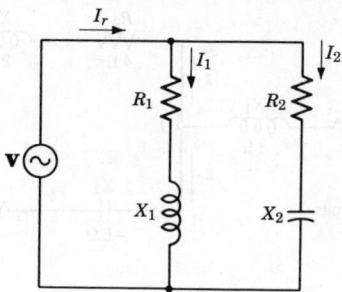

FIGURE 16.8   For Example 16.4.

Solution

(a) $Y_1 = G_1 + jB_1 = \dfrac{10}{10^2 + 5^2} - j\dfrac{5}{10^2 + 5^2} = 0.08 - j0.04$

$Y_2 = G_2 + jB_2 = \dfrac{25}{25^2 + 10^2} + j\dfrac{10}{25^2 + 10^2} = 0.0345 + j0.0138$

(b) $Y_T = Y_1 + Y_2 = 0.1145 - j0.0262 = 0.1174\underline{/-12.9°}$ mho

$I_T = VY_T = 100(0.1145 - j0.0262)$

$I_T = 11.45 - j2.62 = 11.74\underline{/-12.9°}$ A

(c) $Z_e = \dfrac{1}{Y_T} = \dfrac{1}{0.1174\underline{/-12.9°}} = 8.51\underline{/12.9°}\ \Omega$

16.7   Series-Parallel Circuits

In our study of series-parallel circuits that contained only resistance, we made use of the "product-over-the-sum" procedure to replace a parallel pair with an equivalent series pair. This enabled us to simplify the circuit so that we could calculate the total resistance.

Now that we have parallel branches containing R, X_L, and X_C, we make use of admittances of the branches which can be added. The equivalent impedance of a parallel group is obtained as the reciprocal of its admittance, as we have seen. We shall now analyze a series-parallel ac circuit and again illustrate the use of admittances.

EXAMPLE 16.5   Analyze the circuit of Fig. 16.9. Obtain the values of all currents, voltages, and phase angles. Construct the phasor diagram.

Solution. Let Y_3 be the admittance of the upper branch of the parallel group in which I_3 flows, and let Y_2 be the admittance of the lower branch in which I_2 flows.

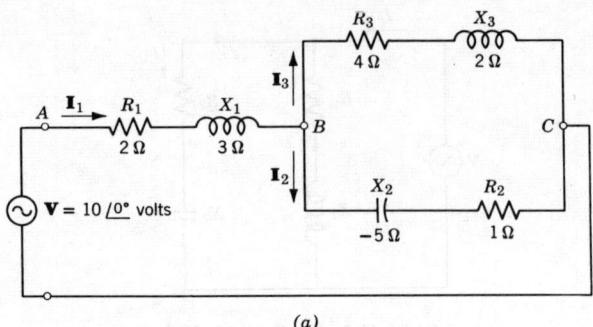

(a)

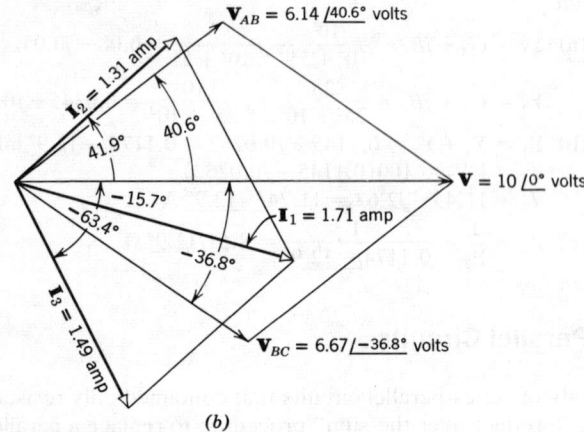

(b)

FIGURE 16.9 (a) A series-parallel circuit. (b) Phasor diagram for the circuit of (a).

$$Y_3 = \frac{1}{4+j2} \times \frac{4-j2}{4-j2} = 0.2 - j0.1 \text{ mho}$$

$$Y_2 = \frac{1}{1-j5} \times \frac{1+j5}{1+j5} = 0.0384 + j0.192 \text{ mho}$$

The admittance between points B and C is

$$Y_{BC} = 0.2 - j0.1 + 0.0384 + j0.192 = 0.2384 + j0.092 \text{ mho}$$

The impedance of the parallel branch is

$$Z_{BC} = \frac{1}{0.2384 + j0.092} \times \frac{0.2384 - j0.092}{0.2384 - j0.092} = 3.65 - j1.41 \text{ ohms}$$

The total impedance $\mathbf{Z}_T$ of the circuit is the sum of $\mathbf{Z}_{BC}$ and $\mathbf{Z}_{AB}$, the impedance of the series branch:

$$\mathbf{Z}_T = 3.65 - j1.41 + 2 + j3 = 5.65 + j1.59 = 5.87 \underline{/15.7°}$$

The input current is $I_1 = E/Z_T$

$$I_1 = \frac{10/0°}{5.87 \underline{/15.7°}} = 1.71 \underline{/-15.7°} \text{ A}$$

Using complex notation,

$$I_1 = \frac{10 + j0}{5.65 + j1.59} \times \frac{5.65 - j1.59}{5.65 - j1.59} = \frac{56.5 - j15.9}{34.45}$$
$$= 1.64 - j0.462 = 1.71 \underline{/-15.7°} \text{ A}$$

The branch voltages $\mathbf{V}_{AB}$ and $\mathbf{V}_{BC}$ will now be computed.

$$\mathbf{V}_{AB} = \mathbf{I}_1\mathbf{Z}_{AB} = (1.64 - j0.462)(2 + j3)$$
$$= 4.66 + j4 = 6.15 \underline{/40.6°} \text{ V}$$
$$\mathbf{V}_{BC} = \mathbf{I}_1\mathbf{Z}_{BC} = (1.65 - j0.462)(3.65 - j1.41)$$
$$= 5.34 - j4 = 6.67 \underline{/-36.8°} \text{ V}$$

These are seen to add to give $10 + j0$ volts, the applied voltage $\mathbf{V}$. The currents in the parallel branches are:

$$\mathbf{I}_2 = \mathbf{V}_{BC}\mathbf{Y}_Z = (5.34 - j4)(0.0384 + j0.192) = 0.975 + j0.872 = 1.31 \underline{/41.9°} \text{ A}$$
$$\mathbf{I}_3 = \mathbf{V}_{BC}\mathbf{Y}_3 = (5.34 - j4)(0.2 - j0.1) = 0.668 - j1.334 = 1.49 \underline{/-63.4°} \text{ A}$$
$$\mathbf{I}_1 = \mathbf{I}_2 + \mathbf{I}_3 = 1.643 - j0.462 = 1.71 \underline{/-15.7°} \text{ A as found above}$$

Problems

6. A resistance of 50 Ω is in parallel with a capacitor for which $X_C = 20 \, \Omega$. What are the conductance and susceptance of the circuit? What is the admittance? Draw the admittance diagram.

7. Obtain the impedance of the series-circuit equivalent of the parallel circuit of Problem 6. First get 1/$\mathbf{Y}$, then get $\mathbf{Z}$ by product over sum.

8. An inductance of 5 Ω reactance is in series with a resistance of 12 Ω resistance. What are the conductance and susceptance of the circuit? What is the admittance? Draw the admittance diagram.

9. Obtain the impedance of the series circuit equivalent of the parallel circuit of Problem 8. First get 1/$\mathbf{Y}$ and then get $\mathbf{Z}$ by product over sum.

10. An inductive reactance of 40 Ω is connected in parallel with the resistance and capacitor of Problem 6. Determine the total admittance. What is the total current if the applied voltage is 50 V rms? Draw the admittance and phasor diagrams.

11. A capacitive reactance of 20 Ω is connected in parallel with the inductance and resistance of Problem 8. Determine the total admittance. What is the total

current if the applied voltage is 50 V rms? Draw the admittance and phasor diagrams.

12. Calculate the admittance Y_{BCD} and the total impedance of the circuit of Fig. 16.10.

13. What are the branch currents and the total current in the circuit of Fig. 16.10?

16.8   Review Questions

1. Define "parallel circuit." The source is included when we say "circuit." The branches that are in parallel constitute a *network* or part of a network.

2. What is the conductance of a 100-Ω resistor? Express it in polar and complex-number form.

3. A pure inductance L is connected between two points. Express its reactance in polar and in complex-number forms. Also express its admittance in the two forms. What are the units of admittance?

4. What special name is given to an admittance that has only the j term, such as $j0.2$?

5. Express the reactance and the susceptance of a capacitor that has 2 Ω reactance. Give polar and complex-number forms.

6. The impedance of a coil is $Z = 0.1 \underline{/30°}$ Ω. Express its admittance in polar and in complex number form. How do the angles of Z and the admittance compare?

7. A capacitor and resistance are in series. Their impedance is $0.125 \underline{/-45°}$. Express the admittance of the branch in polar and in complex-number form.

8. What are the conductance and susceptance of the admittance in Question 6? What are their units?

9. An inductor has $Z = 12 + j5$ ohms $= 13 \underline{/22.5°}$ Ω. Express the admittance, giving the magnitude as a common fraction.

10. Two parallel circuits have admittances $Y_1 = 4 + j15$, $Y_2 = 6 - j5$. What is their combined admittance? What is it in polar form?

11. How would you determine the impedance of a series network that is equivalent to the parallel network of Question 9?

12. Give the general expression for conductance in mathematical form.

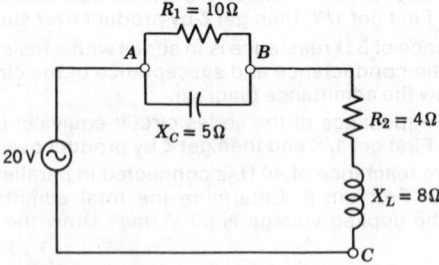

FIGURE 16.10   For Problem 23.

13. Give the general expression for susceptance in mathematical form.
14. State briefly the procedure you would follow in calculating the total current in a circuit that has two impedances Z_1 and Z_2 as a parallel pair that is in series with a third impedance Z_3.
15. Given $Y = G + jB$, under what circumstances will the jB term be negative? Your answer concerns the relative values of inductive and capacitive reactance.

16.9 Problems

Group A

14. A 25 μH inductance and 50 kΩ resistance are in parallel across a 10-V 200 mHz source. What is (a) the current in each unit? (b) Obtain the total current by adding the answer to (a). (c) What is the total admittance? Obtain this as the sum of the separate admittances. Multiply the total admittance by the impressed voltage and check your answer for the total current.
15. What is the magnitude of the total impedance when a 150 μH inductance is in parallel with an 8000-Ω resistance at 15 MHz?
16. A resistor $R = 398$ Ω is in parallel with a capacitor $C = 1$ μF, and the network is supplied with 100 V at 400 Hz. What are the total admittance and total current?
17. In the network of Problem 16, what are the branch currents?
18. (a) How much total current flows in a parallel network of $R = 20$ Ω, $X_C = 50$ Ω when 20 V, 10^7 Hz is applied? (b) What is the phase angle between the voltage and current? (c) What is the capacitance of the capacitor?
19. What is the equivalent series impedance of the parallel network of Problem 18?
20. Assume an inductive reactance of 40 Ω is connected in parallel with R and C of Problem 18. Determine (a) the total current, (b) the equivalent series impedance using voltage and current values.
21. Verify your answer for the equivalent impedance of the network of Problem 20 by calculating it directly.
22. $R = 6250$ Ω, $L = 4$ H, $C = 0.01$ μF are connected in parallel on a 10 V, 1 kHz line. What is their total current? What phase angle exists between the total current and the applied voltage?
23. Calculate the total admittance of the network of Fig. 16.10. Multiply this by the applied voltage and evaluate the total current.

Group B

24. Calculate separately, the voltages V_{AB} and V_{BC} in Fig. 16.10. Check their sum against the applied voltage.
25. In Fig. 16.11, $V = 100$ V, $R_1 = 3\Omega$, $R_2 = 6\Omega$, $X_L = 8\Omega$, $X_C = 5\Omega$, $R_3 = 12\Omega$. Determine the two branch currents that flow between nodes B and C. What are the voltages across (a) R_2; (b) X_L? What would a voltmeter read if connected to points D and E? Assume the voltmeter takes a negligible amount of current.

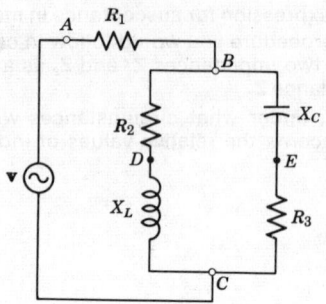

FIGURE 16.11   For Problem 25.

26. It is desired to replace the $R_3 = 4$, $X_3 = j2$ branch of the circuit in Fig. 16.9 with the impedance the branch should have in order that it may receive maximum power. What should the replacement impedance be?

27. Calculate the total impedance of the new circuit after the replacement of the impedance has been made as specified in Problem 26. What will the total input current be?

28. By current division, calculate the current in the new impedance, using the answer to Problem 27.

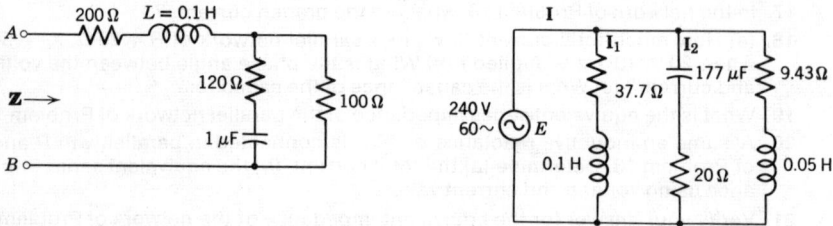

FIGURE 16.12   For Problems 29 and 30.   **FIGURE 16.13   For Problem 31.**

29. The network in Fig. 16.12 is supplied with an applied voltage that has a frequency of 796 Hz. What is its input impedance?

30. Assume the current in the 100-Ω resistance in Fig. 16.12 is 15 mA at the given frequency. Calculate (a) the total input current, (b) the input voltage V_{AB}.

31. Figure 16.13 shows three impedances in parallel across a 240-V, 60 Hz line. Determine the (a) branch currents, (b) power factor of each branch impedance, (c) power (I^2R) in each branch, (d) total power in the whole circuit, (e) total current, (f) phase angle between the input voltage and total current, (g) total power, using $VI \cos \theta$. Check this with your answer for part d.

CHAPTER SEVENTEEN

Power and Energy

in AC Circuits

We have learned that when voltage is suddenly applied to the terminals of a coil of wire, there is a delay in the flow of current and time is required for the current to reach its maximum value. Opposing EMF, induced by increasing flux linkages, causes this.

When an alternating voltage is applied to a coil, the effect of the inductance is to cause the current to *lag* in time-phase behind the applied voltage. This happens not only while the applied voltage is increasing, but also while it is decreasing. The alternating voltage applied to the coil becomes zero before the current does, reaches its maximum negative value, comes back up to zero, and starts going positive before the current does. The curves of Fig. 13.4 show this.

When an alternating voltage is applied to a capacitor and a resistance in series, the capacitor resists change in its terminal voltage and the result is that the capacitor current *leads the capacitor terminal voltage* by 90°. The capacitor current is the same as the current in the resistance, and this current leads the voltage applied to the pair, which are in series, by an amount that depends on the values of capacitance, resistance, and frequency of the applied voltage.

17.1 AC Power in Resistive Networks

When we developed the expressions for effective (rms) values of an ac voltage and current, we were obliged to discuss power dissipated in a resistance. Curves of instantaneous power and the meaning of average power are shown in Fig. 13.6. Reviewing a bit, we see that *instantaneous power* is the product of instantaneous values of the voltage across and the current in a network unit: $p = ei$. Average power is a much more useful and practical quantity. The average *ac power in a resistance* is given by each of the following expressions if V is the effective voltage drop across the resistance and I is the effective value of the current through it.

$$P = VI = I^2R = \frac{V^2}{R} \text{ watts} \qquad (17.1)$$

If we have a network that contains resistance units only (no inductance or capacitance) these formulas will give the power in any one resistance unit or in a single *equivalent resistance unit* if it carries I amperes, has V volts drop across it, and has R ohms resistance. This means that if a complicated network of pure resistance elements can be reduced to a *single equivalent resistance* unit, its power can be calculated as we do with any other single element whose values of R, I, and V we know.

17.2 AC Power and Energy in an Inductance

Although we are ready to discuss power in a network containing inductance and resistance, we think it desirable at first to refer you to Section 13.7 in which we discussed power and energy in an *inductance only*. There we found that when a sinusoidal voltage is applied to a pure inductance, the average power is zero because all of the energy received from the source in a half cycle is returned to the source the next half cycle. The instantaneous value of the energy stored in the magnetic field varies from zero to maximum as the instantaneous current varies from zero to maximum values. The instantaneous energy is given by $W = \frac{1}{2}Li^2$, and its maximum value is

$$W_m = \frac{1}{2}L(\sqrt{2}I)^2 \qquad (17.2)$$

in which I is the rms value of the current. L is in henrys, and I is effective amperes.

The instantaneous power in a pure inductance is the product of instantaneous voltage times instantaneous current. When the instantaneous power values are plotted, a double-frequency curve results whose average value is zero. This is shown in Fig. 13.8a, to which we have already referred.

17.3 AC Power and Energy in a Capacitance

Section 13.8 discusses power in a capacitance. The current leads the applied voltage 90° and the average power is zero. The energy is stored in the electric field between the plates; it attains maximum value and then zero value twice each cycle of the applied voltage. Its instantaneous value is given by $\frac{1}{2}Cv_C^2$, and its maximum value is

$$W_m = \tfrac{1}{2}C(\sqrt{2}V_C)^2 \tag{17.3}$$

in which C is in farads and V_C is effective volts on the capacitor.

The curve of instantaneous *power* is a double-frequency curve whose average value is zero, as was the case with a pure inductance. Power and energy curves are shown in Fig. 13.9a.

17.4 Power and Energy in a Coil

In Example 15.3 we had a coil with 9.07 Ω resistance and 4.21 Ω inductive reactance to which 50 V rms was applied. The current was calculated to be 5 A at an angle of 24.9° lagging.

We learned that the average power (and that's what a wattmeter reads) in the inductive part is zero. But we also know that I amperes in a resistance of R ohms represents an average power of I^2R watts.

The phasor diagram in Fig. 17.1a shows 50 V applied, the 5 A lagging the voltage in phase by 24.9°.

The power supplied to the coil is

$$P = I^2R = 25 \times 9.07 = 226.75 \text{ W}$$

The phasor labeled $I \cos \theta$ represents the *in-phase component* of the current. That amount of current is the *power-producing* (*also energy-producing*) part of the total current. We may calculate the power by multiplying the applied voltage by this amount of current.

$$P = VI \cos \theta$$
$$P = 50 \times 5 \cos 24.9° = 226.75 \text{ W} \tag{17.4}$$

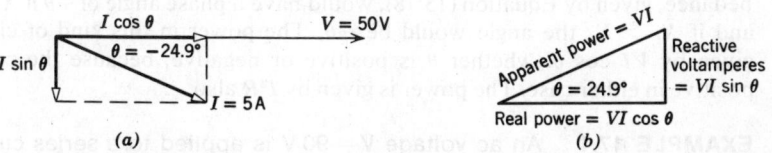

FIGURE 17.1 For a coil that has 10 Ω impedance. (a) In-phase component of current: $I \cos \theta =$ 4.535 A. Quadrature component: $I \sin \theta = 2.105$ A. (b) Real power $= VI \cos \theta = 226.75$ W. Apparent power $= VI = 250$ VA. Reactive VA $= VI \sin \theta = 105$ VA.

It would be logical for us to inquire about the other component of current, namely, $I \sin \theta$. This is called the *quadrature component*. It contributes no power, although its combination with the voltage gives a *reactive volt-ampere* value.

$$VI \sin \theta = \text{reactive voltamperes} \qquad (17.5)$$

The product VI is *total voltamperes*. It is represented by the hypotenuse of the triangle in Fig. 17.1*b* and is called the apparent power.

The base represents real power: power-producing component of current times voltage. The altitude represents reactive voltamperes. The angle 0 is the same as that between the current and applied voltage.

Equation (17.4) ($P = VI \cos \theta$) is the most useful form for calculating real power. It holds for a single component of a network, for any branch, combination of branches, or for the whole network. If it is used on a branch of a network, V must be the rms voltage applied to the branch and I must be the total current in the branch.

The energy expended in a coil that has R ohms resistance is equal to the average power multiplied by the time during which the average power is constant. We have known for some time now that energy is power multiplied by time: joules = watts times seconds. For example, if a coil that has 5 Ω resistance (no matter what its inductance is) carries 2 A for one minute, the energy expended is

$$W = I^2 R t = 2^2 \times 5 \times 60 = 1200 \, \text{J}$$

In this expression, I is the effective (rms) value of the current in R ohms and flows for t seconds.

17.5   Power in an RLC Series Circuit

We learned in Chapter 15 how to calculate the current in a series circuit containing resistance, inductance, and capacitance. We found that the total impedance, given by Equation (15.18), would have a phase angle of $+\theta$ if $X_L > X_C$, and if $X_C > X_L$ the angle would be $-\theta$. The power in this kind of circuit is given by $VI \cos \theta$, whether θ is positive or negative, because the cosine is positive in either case. The power is given by $I^2 R$ also.

EXAMPLE 17.1   An ac voltage $V = 90$ V is applied to a series circuit of $R = 18 \, \Omega$, $X_L = 12 \, \Omega$, and $X_C = 36 \, \Omega$. Calculate the current, power, and the phase angle of the current with respect to the voltage.

Solution

$$\mathbf{Z} = \sqrt{18^2 + (12-36)^2}\,\underline{/\theta}$$
$$\mathbf{Z} = \sqrt{324+576}\,\underline{/\text{arc tan }(-24/18)} = 30\,\underline{/-53.1^\circ}\,\Omega$$
$$\mathbf{I} = 90\,\underline{/0^\circ} \div 30\,\underline{/-53.1^\circ} = 3\,\underline{/53.1^\circ}\,\text{A}$$

$$P = 90 \times 3 \cos 53.1^\circ = 270 \times 0.6 = 162\,\text{W}$$
$$P = I^2R = 3^2 \times 18 = 162\,\text{W}$$

Note that $X_C > X_L$, so the current leads the applied voltage by the phase angle 53.1°. All of the power creates lost energy in the resistance; none in the inductance or capacitance.

17.6   Power Factor

Notice that the only difference between the equation giving the amount of apparent power and that giving the amount of real power in Fig. 17.1 is the factor $\cos\theta$. This is given the special name *power factor*. Therefore,

$$\text{Real power} = \text{apparent power} \times \text{power factor}$$

$$\text{Power factor} = \frac{\text{real power}}{\text{apparent power}} = \frac{\text{watts}}{\text{total voltamperes}}$$

$$\text{Power factor} = \cos\theta \tag{17.6}$$

The angle θ is the phase angle between the voltage and the current in an element, a network branch, or a whole network. This angle is the same as θ in the impedance diagram, and in the power triangle as shown in Fig. 17.1.

The power factor in Example 17.1 is $\cos 53.1^\circ = 0.6$.

EXAMPLE 17.2   The parallel circuit of Fig. 17.2, with its phasor diagram, was analyzed and the following results obtained:

$$\mathbf{I}_1 = 3.55 - j8.52 = 9.23\,\underline{/-67.4^\circ}\,\text{A}$$
$$\mathbf{I}_2 = 7.20 + j9.60 = 12\,\underline{/53.1^\circ}\,\text{A}$$
$$\mathbf{I} = 10.75 + j1.08 = 10.8\,\underline{/5.7^\circ}\,\text{A}$$

What is the power expended in each branch, and the total power in the whole circuit?

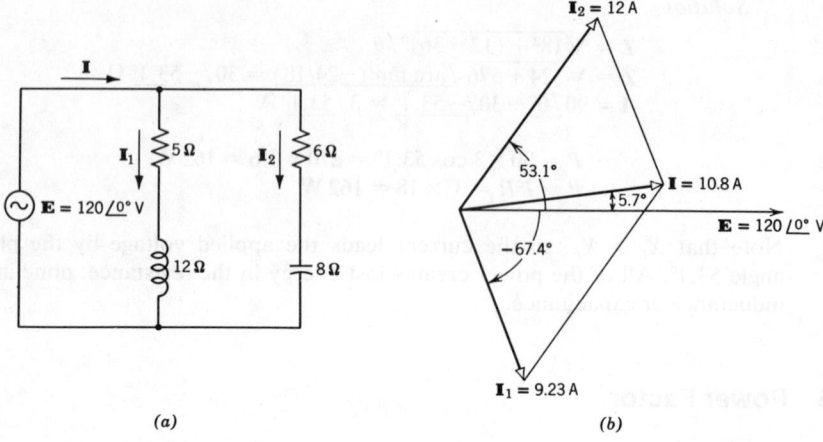

(a)　　　　　　　　　　　　　　　(b)

FIGURE 17.2 For Example 17.2. (*a*) A parallel ac circuit. (*b*) Phasor diagram of circuit in (*a*).

Solution. The power in the first branch is obtained using the voltage across the branch and the current through it. Don't forget the power factor.

$$P_1 = VI_1 \cos \theta_1 = 120 \times 9.23 \ \cos (-67.4°)$$
$$P_1 = 426 \text{ W}$$

The power in the second branch is calculated in the same manner.

$$P_2 = VI_2 \cos \theta_2 = 120 \times 12 \cos 53.1°$$
$$P_2 = 864 \text{ W}$$

The total power is

$$P_T = 426 + 864 = 1290 \text{ W}$$

Let's check these values, using I^2R.

$$P_1 = 9.23^2 \times 5 = 426 \text{ W}; \qquad P_2 = 12^2 \times 6 = 864 \text{ W}$$

EXAMPLE 17.3 Determine the following for the circuit of Fig. 17.2. (a) Power factor in each branch, (b) power factor of the whole circuit, (c) total power, using total current, (d) total voltamperes, (e) reactive voltamperes, (f) ratio of true power to apparent power.

Solution. (a) In the first branch the power factor is

$$\text{P.F.} = \cos \theta_1 = \cos (-67.4°) = 0.384$$

In the second branch,

$$P.F. = \cos \theta_2 = \cos 53.1° = 0.600$$

(b) P.F. of the whole circuit is

$$P.F. = \cos 5.7° = 0.995$$

(c) Total power is

$$P = VI \cos \theta = 120 \times 10.8 \cos 5.7°$$
$$P = 1290 \text{ W}$$

(d) Total voltamperes

$$VI = 120 \times 10.8 = 1296 \text{ VA}$$

(e) Reactive voltamperes:

$$VI \sin \theta = 120 \times 10.8 \sin 5.7° = 1296 \times 0.0995 = 128.9 \text{ RVA}$$

(f) True power ÷ apparent power:

$$1290 ÷ 1296 = 0.995, \text{ the power factor of the whole circuit}$$

17.7 Power Calculation Using Complex Notation

Real power can be calculated using complex forms of voltage and current. It is obtained by adding together two products, one of which is the product of the *in-phase* components of V and I and the other is the product of the *quadrature components*. If the quadrature components are both positive or both negative (both up or both down on the phasor diagram) their product is positive. If the quadrature components are opposite in nature (one up, the other down) their product is considered to be negative power.

EXAMPLE 17.3   Figure 17.3 shows two pairs of current and voltage that have the same values but one pair is for a circuit whose power factor is

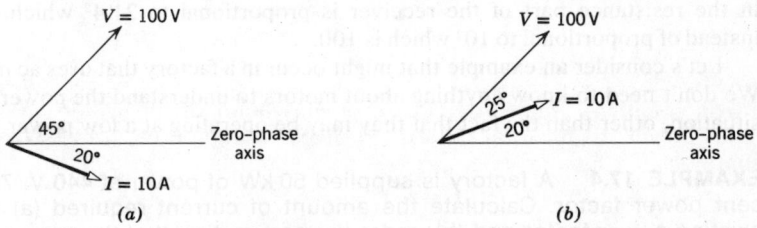

FIGURE 17.3   For Example 17.3.

cos 65° = 0.422, the other is for a circuit that has a power factor cos 25° = 0.906.

Solution. In (a)

$$P = (100 \cos 45°)(10 \cos 20°) - (100 \sin 45°)(10 \sin 20°)$$
$$= 70.7 \times 9.4 - 70.7 \times 3.42 = 423 \text{ W}$$

Checking: $P = 100 \times 10 \cos (45° + 20°) = 1000 \cos 65° = 423$ W

In (b) $P = (100 \cos 45°)(10 \cos 20°) + (100 \sin 45°)(10 \sin 20°)$ (17.9)
$$= 70.7 \times 9.4 + 70.7 \times 3.42 = 70.7 \times 12.82 = 906 \text{ W}$$

Check: $VI \cos \theta = 100 \times 10 \cos 25° = 906$ W

17.8   Significance of Power Factor

Example 17.3 shows that we can supply 906 W to a network, or some other kind of receiver of electrical power (even an ac motor), by delivering to it 10 A at 100 V if the phase angle is 25°. The power factor, cos 25°, is 0.906. Suppose the phase angle is 65°, as in Fig. 17.3a. We shall find that much more than 10 A of current is needed to supply 906 W to the receiver. Using our general power equation,

$$EI \cos \theta = 100I \cos 65° = 906$$

$$I = \frac{906}{100 \times 0.423} = 21.4 \text{ A}$$

Let's think about this for a minute. In order to deliver 906 W when the power factor is only 0.423 (which is poor), more than twice as much *current* is needed than when the power factor is good (0.904). Wires need to be larger to carry 21.4 A than to carry 10 A. Furthermore, the power lost in the supply lines and in the resistance part of the receiver is proportional to 21.4^2 which is 458 instead of proportional to 10^2 which is 100.

Let's consider an example that might occur in a factory that uses ac motors. We don't need to know anything about motors to understand the power factor situation, other than the fact that they may be operating at a low power factor.

EXAMPLE 17.4   A factory is supplied 50 kW of power at 440 V, 70 percent power factor. Calculate the amount of current required (a) at the existing power factor and (b) under the assumption that the power factor has been improved and brought up to 85 percent.

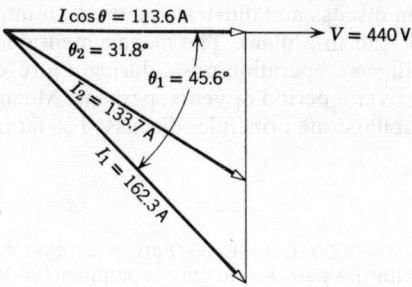

FIGURE 17.4   For Example 17.4. A 50 kW load at 440 V takes 162.3 A at 70 percent power factor, but only 133.7 A at 85 percent power factor.

Solution. The voltage and load currents are represented in Fig. 17.4.

$$\text{(a)} \quad 440 I_1 \times 0.7 = 50,000$$
$$I_1 = 162.3 \text{ A}$$
$$\text{(b)} \quad 440 I_2 \times 0.85 = 50,000$$
$$I_2 = 133.7 \text{ A}$$

The in-phase component of the current, if the power is to be 50,000 W at 440 V, must be calculated. We can get its value from $I \cos \theta$, using either $I_1 \cos \theta_1$ or $I_2 \cos \theta_2$.

$$I_1 \cos \theta_1 = 162.3 \times 0.7 = 113.6 \text{ A}$$
$$I_2 \cos \theta_2 = 133.7 \times 0.85 = 113.6 \text{ A}$$

Or, we can get it from

$$440 I_R = 50,000$$
$$I_R = 113.6 \text{ A}$$

in which I_R is the magnitude of the in-phase component of the total current.

This shows that if operation is at 85 percent power factor, 133.7 A are required to deliver 50 kW to the load, whereas 162.3 A are required at 70 percent power factor.

We shall soon see how to improve the power factor of this *electrical load* in the plant. The first important point for you to get is that the power company must run larger cables into a plant that will operate at low power factor than would be needed if the power factor were high. It must also install larger-capacity transformers near the plant's building to handle the larger amounts of current that are needed when the power factor is low. The power company is justified in charging more in their monthly billing (higher rates) than they would charge if the plant management had assured them that the power factor would be as high as 85 percent, so that they could have installed smaller transformers and smaller cables in their original construction.

We shall soon discuss and illustrate methods of improving the power factor of the electrical load in a plant. The management does this when an analysis shows that it will save operation costs during a foreseeable interval of operation – extending over a period of years, perhaps. Meanwhile, we are presenting ten problems that illustrate principles discussed so far in this chapter.

Problems

1. Two resistors $A = 3000\ \Omega$, $B = 6000\ \Omega$ are in parallel. A third resistor $C = 8000\ \Omega$ is in series with the pair. An ac source applies 100 V rms to this network. (a) What is the power in each resistor? (b) Calculate the total power using the applied voltage and current supplied by the source. (c) What is the power factor of the whole network?

2. An amplifier supplies power to an $R–L$ series circuit at 10 V, 1 kHz. $R = 8\ \Omega$, $L = 1.592$ mH. Calculate the power output of the amplifier in two ways. What is the power factor of the load?

3. A timing circuit has $R = 1000\ \Omega$ in series with $C = 1.75\ \mu$F. The circuit has 120 V at $\omega = 377$ rad/s impressed upon it. What is the power dissipation while full voltage is applied? What is the power factor?

4. If the circuit of Problem 3 is actuated for a period of 1 min before the applied voltage is removed, how much energy is supplied to it during that time interval?

5. A branch of a network carries $I = 5 + j3$ mA. Its impedance is $10 + j4$ ohms. Calculate the power dissipated in it using the complex-number method of solution.

6. Check your answer to Problem 5 using the polar forms of I and V.

7. An ac motor takes 20 A at 230 V while receiving 3.68 kW of power. At what power factor does it operate? Calculate the in-phase component of the current. Calculate the reactive voltamperes and the total voltamperes. Check the power factor using power and total voltamperes. Draw the voltampere triangle.

8. A factory is supplied with 100 kW at 440 V. The total current is 303 A. What is the power factor? Calculate the reactive voltamperes. If it were possible to make an arrangement in the factory so that the 100 kW could be received at 90 percent power factor, how much would the reactive voltamperes amount to? How much reduction would occur in reactive voltamperes?

9. A bank of capacitors must draw a current of 101.14 A from a 440-V, 60 Hz line. How many reactive voltamperes is this? What is the angle between the current and the voltage?

10. How much capacitance would be required to draw the current specified in Problem 9?

17.9 Power Factor Improvement

The R, X_L series branch in Fig. 17.5a represents the factory load to which $I_1 = 162.3$ A is supplied when the power factor is 0.7 as assumed in Example

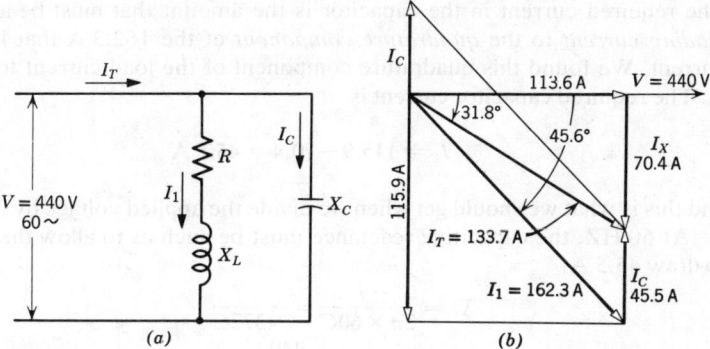

FIGURE 17.5 (a) A capacitive reactance, X_C, draws a leading current to improve the power factor of a factory load. (b) Factory load current I_1 has 113.6 A in-phase component and 115.9 A quadrature component, lagging. Capacitor current, I_C, raises power factor from 70 to 85 percent.

17.4. This factory load current lags the applied voltage by an angle of 45.6°. This is shown in the phasor diagram Fig. 17.5b.

If we can draw a *leading current*, I_C in Fig. 17.5a, which will flow through a capacitor C that we can connect to the supply cables as indicated, we can make that leading current sufficiently large so that when it is added to the 162.3 A their sum will be 113.6 A, which will lag the applied voltage the required angle $\theta_2 = 31.8°$ shown in Fig. 17.4. Let's proceed to calculate the required amount of capacitance.

EXAMPLE 17.5   Calculate the amount of capacitance that must be connected across the supply cables, at the factory load in Example 17.4, in order to raise the power factor from 70 to 85 percent.

Solution: The *quadrature component* of I_L, the factory load current, is

$$I_L \sin 45.6° = 162.3 \times 0.714 = 115.9 \text{ A lagging } 90°$$

The quadrature component of the total current after power-factor improvement can be determined without knowing the new total current value. Observe in Fig. 17.5b that the *in-phase component must be* 113.6 A in order that the real power may be 50,000 W. The quadrature component of the total current *after improvement* is represented by the short vertical side of the right triangle whose base represents 113.6 A and whose adjacent angle is 31.8°. The *reactive component*, I_X, of the total current is obtained from the relation

$$\frac{I_X}{113.6} = \tan 31.8°$$
$$I_X = 113.6 \times 0.62 = 70.4 \text{ A}$$

The required current in the capacitor is the amount that must be added *as a leading current* to the *quadrature component* of the 162.3 A that is the load current. We found this quadrature component of the load current to be 115.9 A. The required capacitor current is

$$I_C = 115.9 - 70.4 = 45.5 \text{ A}$$

and this is what we should get when we divide the applied voltage by X_C.

At 60 HZ, the capacitive reactance must be such as to allow the capacitor to draw 45.5 A:

$$X_C = \frac{-j}{2\pi \times 60C} = -\frac{j}{377C}$$

$$I_C = j45.5 = \frac{440}{-j/377C}$$

$$C = 45.5/(440 \times 377)$$

$$C = 274.3 \times 10^{-6} \text{ F} = 274.3 \ \mu\text{F}$$

Large capacitors, probably two of $100 \ \mu\text{F}$ each and one of $75 \ \mu\text{F}$, would be connected in parallel across the two main supply cables just inside the plant. They would draw 45.5 A of current that would lead the applied voltage by 90° in phase, thus neutralizing 45.5 A of the quadrature component of the lagging current taken by the plant's motors. This would make the *net quadrature component* of current equal to 70.4 A lagging.

The total current supplied by the power company is then

$$I_T = \sqrt{113.6^2 + 70.4^2} = 133.7 \text{ A}$$

Checking the power factor,

$$\cos \theta = \frac{113.6}{133.7} = 0.85$$

We found the required capacitor current to be 45.5 A. We can determine it in a different and perhaps shorter way. The method involves use of the reactive voltamperes taken by the plant before improvement and the amount of *VAR* reduction necessary. A little serious thinking will reveal the logic of the procedure.

Figure 17.6a shows the triangle of voltamperes and real power for the unimproved case, with its phase angle of 45.6°. The reactive voltamperes (VAR) is evaluated by use of the tangent of the angle.

$$\text{VAR} = 50,000 \tan 45.6°$$
$$= 50,000 \times 1.0212 = 51,060 \text{ VA}$$

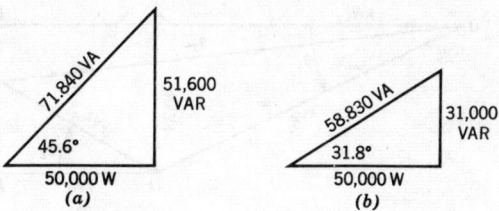

FIGURE 17.6 **Reactive voltamperes, and total voltamperes when 50 kW is supplied at (a) 70 percent power factor, and (b) 85 percent power factor.**

After improvement the phase angle will be $\cos^{-1} 0.85 = 31.8°$, and the reactive voltamperes will be

$$VAR = 50,000 \tan 31.8°$$
$$= 50,000 \times 0.6200 = 31,000 \text{ VA}$$

The reduction in voltamperes to be achieved by use of capacitor current is $51,060 - 31,000 = 20,060$. The voltage on the capacitors is to be 440 V, so

$$440 I_C = 20,060$$
$$I_C = 45.5 \text{ A}$$

From this we calculate the value of capacitance required, as was done before.

17.10 Use of Synchronous Motor for Power Factor Improvement

When we study alternating-current motors we shall learn a little bit about a so-called synchronous motor that can be made to draw a *leading current* from the supply line while driving a load. This leading current has a quadrature component that has the same effect as the current in a capacitor. At the same time there is an in-phase component of the synchronous motor current that will add to the in-phase component of the current supplied to the other factory motors. The result is a decrease in the lagging phase angle of the total current and an improvement in power factor.

Figure 17.7 shows how the addition of a synchronous motor load can increase power factor while doing useful work in driving a load. The length OA, called kW_L, represents the factory power without the synchronous motor in operation; kW_S represents the power taken by the synchronous motor; OB represents the total true power supplied to the factory; BC represents the net reactive kilovolt-amperes; OC represents the net kVA; and O_T is the power factor angle for the whole load.

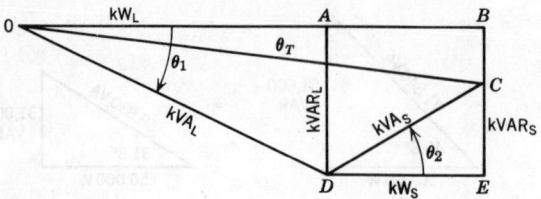

FIGURE 17.7   A synchronous motor taking kVAR$_s$ improves power factor of an inductive load, decreasing the power factor angle from θ_1 to θ_T.

Solution

$$\text{Useful power} = 100 \times 0.8 + 60 \times 0.86 = 131.6 \text{ kW}$$
$$\text{Reactive kilovolt amperes} = 100 \sin \theta_1 - 60 \sin \theta_2$$
$$= 100 \times 0.6 - 60 \times 0.51 = 29.4 \text{ kVAR}$$
$$\text{Tan } \theta_T = 29.4/131.6 = 0.223, \theta_T = 12.6°$$
$$\text{P.F.} = \cos \theta_T = 0.976 = 97.6 \text{ percent}$$

17.11   Effective Resistance

When alternating current flows in a wire there is a larger number of flux linkages affecting the portion near the center of the wire than near the surface. The result is a larger counter EMF of self-induction opposing current flow near the wire's axis than near its surface. Because of this, and especially at higher frequencies, the current tends to crowd more into the parts of the wire near its surface. This phenomenon is called *skin effect*. The resistance of the wire is therefore larger than it would be with direct current.

The skin effect is very pronounced in radio and television circuits. Most of the current may flow in only the outer few percent of the wire's cross-section area. Because of this, a special kind of conductor, called Litz wire, is used in winding some of the coils. It is made of many strands of fine wire, each separately insulated from the others. You can see that this would be more suitable than a solid piece of wire of the same gauge, and that even a smaller total cross section of copper in a Litz conductor could be used.

At microwave frequencies of 10,000 MHz or higher, a plating of gold or silver on the surface of a hollow conductor that has a very thin wall is sometimes employed. The current is essentially surface current. In the transmission of 60-Hz power, an aluminum cable with a steel core (for strength) is used. The current in the steel core is negligible compared with that in the aluminum owing to skin effect and to the higher conductivity of aluminum.

The resistance offered by a wire to the flow of alternating current is greater than the resistance to the flow of direct current. The increase is larger the higher the frequency.

$$R_{ac} > R_{dc}$$

EFFECT OF A METALLIC CORE

If we measure the ac power taken by a coil that has an air core and then insert a nonmagnetic core made of aluminum or brass (a nonferrous metal) we shall find that even though we are careful to make the currents alike in magnitude, the power input with the metallic core is the larger in amount. This is because eddy currents induced in the metal cause an energy loss (Section 6.19) that is not present with a nonmetallic core. Furthermore, an iron core with the same dimensions as those of the aluminum or brass core will use up additional energy because of hysteresis loss (Section 6.18).

When the core is metallic the input power $VI \cos \theta$ is, therefore, not all I^2R loss, where R is the dc resistance. However, we say that the input power is equal to I^2R_e in which R_e is the value of R_{ac} that will be obtained when we divide the input power by I^2.

EXAMPLE 17.8 A high-frequency current flows in a coil with an iron core. The dc resistance of the coil is 25 Ω and the coil takes 10 mA and 3.5 mW of power. What is the equivalent resistance at the existing frequency?

Solution

$$I^2R_e = 3.5 \times 10^{-3}$$

$$R_e = \frac{3.5 \times 10^{-3}}{(10^{-2})^2} = 35 \ \Omega$$

17.12   Review Questions

1. What is the power factor of a network that contains resistance units only? Can we calculate power delivered to such a network by use of each of the three equations that apply to dc power calculation? What are the equations? What precautions must be observed in using $V1$ and V^2/R?

2. A series RL circuit has an ac voltage applied. How do we calculate the current? How do we calculate the impedance?

3. Answer Question 2 for a series RC circuit.

4. A series circuit has R, X_L and X_C and carries a current I rms amperes. What is the expression for the impedance?

5. When will the current I in Question 4 lag the applied voltage, and when will it lead?

6. Give two equations by which the total power in the circuit of Question 4 can be evaluated.

7. Define power factor. First, relate it to voltage and current; second, to impedance.

8. Tell how to calculate the power in a circuit in which $I = a - jb$ amperes flows as a result of an applied voltage $V = c + jd$ volts. Is j^2 important here, or are the signs of the j terms important?

9. (a) $I_1 = 3 - j2$, $V_1 = 10 + j10$, (b) $I_1 = 3 + j2$, $V_1 = 10 + j10$. Calculate the power in each case. Why is the power in (b) the larger amount?

10. The voltage $V = 100 \underline{/0°}$ V rms is applied to an impedance which carries current $I = 10\underline{/-30°}$ A. How much power is involved and what is the power factor?

11. Why is the average power produced by a sine-wave current in a resistor one half the maximum instantaneous power? See Fig. 13.6.

12. Why is the average power in a pure inductance zero? What about the average power in a pure capacitance?

13. What is meant by reactive power? This has been called the "wattless component" of the total volt amperes. Why?

14. A series circuit contains R, L, and C components. If a voltage V is applied, how could you calculate the net reactive voltamperes?

15. Why is the angle θ in the voltampere-power triangle the same as θ between the total voltage applied to, and the total current in, a network?

16. Why do you suppose generators and transformers are rated in kilovolt amperes rather than in kilowatts? Would a 50,000 kVA generator be required to deliver more current to a 50,000 kW load at 80 percent power factor than at 100 percent power factor? Under which load would it be operating at a higher temperature, and why?

17. Why should the power factor of a factory load be as high as possible?

18. Explain how connecting capacitors across the supply lines to a factory load improves the power factor. First emphasize components of current in your explanation, then discuss reactive kilovolt amperes.

19. What is the special characteristic of a synchronous motor that makes possible its use in improving the power factor of a factory load?

20. Referring to Question 19, sometimes a synchronous motor is run *at no load* merely to obtain power factor improvement. How would this change the diagram in Fig. 17.7? Keep in mind the useful output of the motor would be zero.

21. What is skin effect in an electrical conductor, and what causes it?

22. Why is the resistance of a conductor increased by skin effect? How is this increase related to frequency?

23. Explain the effects of a nonmagnetic metallic core (like aluminum) and a magnetic core on the effective resistance of a coil of wire.

24. Two identical coils of wire carry the same amount of alternating current. Coil A has a wooden core, Coil B has an iron core. One of them has a larger in-phase component of current than the other has. Which is it, and why is its in-phase component larger?

25. Assume the difference in the amounts of in-phase components of current in Question 24 is 0.2 A and the applied voltage is 10 V. What does the 2 W increase in power represent?

17.13 Problems

11. A coil for which $R = 113\,\Omega$, $X_L = 0.1$ H takes 226 W of power from a 60-Hz line. What is the applied voltage? What is the power factor?

12. Calculate the reactive voltamperes of Problem 11.

13. Two appliances are connected in parallel to a 120-V 60-Hz source. One carries 8 A at 0.85 P.F. lagging the other 5 A at 0.8 P.F. leading. Determine the total current, power, and overall power factor.

14. A coil has 80 Ω resistance and 159 mH inductance. At what frequency does it take 1.2 A on a 120-V line? How much power does it receive and what is the power factor?

15. A 12-Ω resistance added between terminals A and B in Fig. 17.8 must receive 48 W of power. What must be the applied voltage V_{AB}? What is the total power? What is the overall power factor?

16. A current $I = 10 - j6$ A flows in a load that has $V = 100 + j30$ V applied to its terminals. Calculate the power using these complex forms. What is the power factor?

17. Voltage $V = 100 + j30$ V sends a current $I = 10 + j6$ A through an impedance. Calculate the power using these complex forms. What is the power factor?

18. A number of motors with a total output of 15 hp operate on a 208-V 60-Hz line at a lagging power factor of 80 per cent. The efficiency is 80 percent. What is the kVA rating of a capacitor required to bring the overall power factor to unity? What is the amount of capacitance required?

19. Solve Problem 18, assuming improvement to 90 percent power factor is adequate.

20. Assume a load carries $I = 12 - j30$ A and $V = 120 \underline{/0°}$ V is across its terminals. How much capacitance is required to increase the power factor to 90 percent? The frequency is 60 Hz.

21. $V = 110 + j40$ V, 60 Hz is measured across a load that carries $I = 12 - j5$ A. How much capacitance must be connected in parallel with the load to bring the power factor up to unity?

22. A 208-V, 60-Hz motor load of 100 hp, 83 percent P.F., 88 percent efficiency, is to be brought up to unity power factor by means of adding the leading current

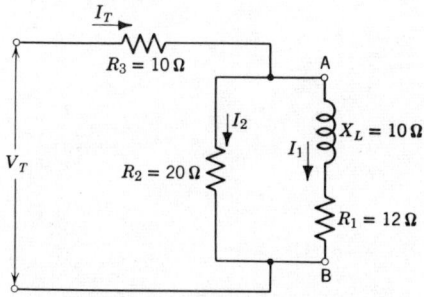

FIGURE 17.8   For Problem 15.

taken by a synchronous motor. Assume the motor is to "float on the line" unloaded. The only power it takes is that to supply its losses which can be considered negligible. What must be its kVA rating? How much current will the motor take?

23. The motors in Problem 18 are operating at full load. Calculate the kVA rating of a synchronous motor load that will bring the power factor up to unity. What must be its kVA rating if its input is 11 kW? At what power factor must it operate?

24. A coil of wire is supplied with 5 A at 120 V. Its dc resistance is 24 Ω. It has an iron core. The power input is measured and found to be 625 W. How much more power is supplied to it than the amount required to supply loss caused by the resistance of the wire. Why is this additional power required? What is the equivalent resistance?

25. A load made up of small motors on a 120-V 60-Hz line totals 5 kW at 70 percent power factor. A lighting load on the same line totals 10 kW at unity power factor. Determine the total current and the total power factor.

26. Refer to Problem 23 and determine the kVA capacity of a synchronous motor that will raise the total power factor to 0.9 while taking 1.5 kW from the line.

CHAPTER EIGHTEEN

Alternating-Current Meters and Measurements

When we studied meters used in the measurement of direct current and voltage, we learned that the D'Arsonval movement is most popular. Owing to the fact that it employs a permanent magnet, it is not adaptable to alternating-current use. If its moving coil carried alternating current, the net torque on it would be zero and the pointer would remain at zero. This is because the force tending to turn the coil would reverse direction every half cycle. In order to produce torque to move the coil and its pointer up-scale, the magnetic field in which the coil lies would have to reverse direction each time the current reverses direction. And we know this happens every half cycle. We shall discuss a type of meter movement in which this happens, but it will not employ a permanent magnetic field.

18.1 Iron-Vane Movement

A popular type of commercial ac voltmeter employs thin, curved strips of soft iron, one stationary and one movable, to produce deflection of a meter pointer. Figure 18.1 is a cut-away view of the iron-vane movement developed by the Weston Instrument Division of Daystrom Corporation.

399

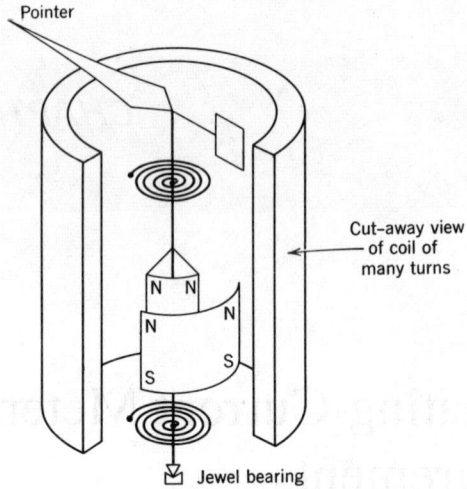

FIGURE 18.1   Iron-vane meter movement. Courtesy of Weston Instrument Division of Daystrom Corporation.

One of the two curved metal strips is mounted in a nonferrous cylinder attached to the inside wall of the coil; the other is supported by the movable shaft. During the half-cycle of coil current when its flux is upward inside the coil, the soft-iron vanes are magnetized with north polarity on top edges and south polarity on bottom, as shown. Forces of repulsion cause the inner vane to rotate the shaft and pointer clockwise. During negative half-cycles the polarities of the edges are south at the top and north at the bottom. The forces are still repulsive and clockwise torque continues. The average torque is proportional to the flux density which, in turn, is proportional to the current, although not exactly so. The reluctance of the magnetic path will vary somewhat with the relative positions of the iron vanes. Nevertheless, precision (about 0.5 percent) is achieved with careful calibration. Friction is minimized by the use of very light-weight moving parts and jeweled bearings. Air damping, to prevent oscillations of the pointer, is achieved by the use of a small aluminum vane attached to the rear end of the pointer and moving in a restricted enclosure. This vane helps balance the pointer also.

Another form of iron-vane instrument has the field coil in an inclined position with its axis about 45° from the vertical. A thin iron vane pierced at its center and supported by the rotating shaft is positioned at the axis of the coil. When acted upon by the magnetic field the vane receives a torque that causes it and the shaft to turn to allow the vane to come more nearly in alignment with the axis of the coil. That is, the vane becomes a magnet and, quite naturally, tries to line up as completely as possible with the direction of the coil flux along the axis of the coil.

There are iron laminations around the coil to prevent its field from being influenced by stray magnetic fields or the earth's field. A thin, rectangular aluminum vane is mounted in a vertical position at the back end of the pointer structure. Air damping of the pointer is produced by the air in a chamber in which the vane moves laterally. The scale of the iron-vane meter is practically linear in the upper two-thirds of its range.

The torque is proportional to the *average of the squares of the meter currents*. This is because (*a*) the magnetic field produced by the current induces, in the soft-iron vanes, magnetic poles the strength of which is proportional to the intensity of the field. (*b*) The intensity of the field is proportional to the amount of current. The force on the iron vane which turns is, therefore, proportional to the square of the field intensity, and therefore to the square of the current in the meter. Thus the meter scale can be laid out in terms of rms values.

18.2   Meter Damping

When an ac meter is first energized there is a tendency for the moving coil and its pointer to oscillate around the position where they soon come to rest. This is objectionable to the user. To *damp out* these oscillations a small flat rectangular vane made of thin aluminum strips is attached to the shaft just below the pointer. This moves, with small clearance, in a cylindrical air chamber where air friction opposes the tendency to oscillate.

Air damping is employed, in the same way, in dc meters. Some of them have the aluminum damping vane mounted so it moves in the permanent magnetic field. In this case, induced eddy currents absorb the energy contributed by forces causing oscillation and quickly reduce it to zero.

Multiscale ac voltmeters employ additional series resistors in the same manner as that described in Chapter 9, for the multiscale dc voltmeters.

18.3   Iron-Vane AC Ammeters

It is obvious that the small currents in the stationary coil of an iron-vane voltmeter may be obtained from a shunt, as in the case of a dc ammeter. This is done when the iron-vane movement is used to measure ac amperes.

When large values of alternating current are to be measured, a *current transformer* is employed to provide a current that is a small fraction of the current to be measured and send it through the meter coil. This small amount of current is not part of the current being measured (which is the case when a shunt is used power.) By transformer action the smaller current is *induced* in a separate coil which is connected to the meter coil. Instrument current transformers are employed with other types of meter movement, including one used in wattmeters.

18.4   Electrodynamometer Principle

Meters used on alternating current have alternating magnetic fields that must be produced by the current being measured. Iron cannot be used in the magnetic circuit and voltage-drop values in an ac meter are large compared with those of the D'Arsonval-type meter, and therefore more power is required by the electrodynamometer type. Inductive effects in coils and in metal near the coils, and variations caused by changes in frequency, affect the accuracy of ac meters. The design of such meters is therefore more involved than that of dc meters.

Figure 18.2 illustrates the electrodynamometer principle. Two coils C and C' are firmly mounted in the case and connected in series. Their magnetic fields aid each other. The movable coil M, made of very fine wire, is mounted on a vertical shaft that turns in jeweled bearings and supports the meter pointer. Two spiral springs are fastened to the shaft, one above the coil and one below. They carry current to the coil M and offer mechanical opposition to its turning.

The direction of the magnetic flux through a coil depends on the direction of current flow in the turns of the coil. In the electrodynamometer *voltmeter*, the movable coil M is connected in series with the stationary coils and with a large amount of resistance to keep the current value small — a few milliamperes. The coil connections are arranged so that the two fluxes ϕ_M and $\phi_{CC'}$ are directed as shown during half a cycle. From what we learned about the force on a conductor carrying current in a magnetic field, we deduce that there is an upward force on the lower-left side of the movable coil and a downward

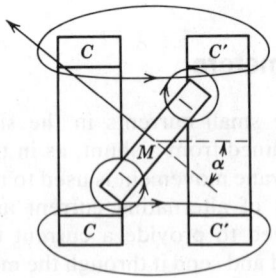

FIGURE 18.2   Electrodynamometer movement, sectional view.

force on the upper-right side. This produces clockwise motion of the coil until the torque is balanced by that of the retarding springs. It is characteristic in a system like this for the moving coil to try to move to a position where its *flux linkages will be a maximum*. If the springs did not stop the motion, the turning coil would stop of its own accord only when the pointer is directed horizontally to the right and the axes of the coils are coincident. The flux through the moving coil would then be a maximum.

The torque produced is proportional to the product of the fluxes times sin α. With no magnetic material in the system, the fluxes are proportional to the currents in the coils. Therefore, the torque is proportional to the product of the currents times the sine of the angle α. The deflection is directly proportional to the torque, owing to the physical property of the retarding springs.

In the *electrodynamometer voltmeter*, the deflection is very nearly proportional to the *square of the voltage* since the current is the same in the stationary and movable coils. As a result, the scale markings on the meter dial are not uniformly spaced. They are crowded in the lower part of the dial, but near the middle and upper portions they are farther apart and more uniformly spaced so that readings may be taken with precision. This type of voltmeter takes much more power than the D'Arsonval voltmeter (about 5 times more) and it is adversely affected by stray magnetic fields because it operates with comparatively weak fields so that its power consumption may be a minimum. An important reason for keeping voltmeter current as small as possible is to prevent disturbing voltage drops to occur in circuits containing large series resistances or impedances. Vacuum-tube voltmeters, which take current in the microampere range (or no current at all), are preferred for use in high-impedance circuits.

The electrodynamometer instrument may be used for either alternating or direct current. On direct current, pairs of readings should be taken by reversing the current direction to eliminate the effect of the earth's magnetic field and any other stray fields that may be present. The presence of iron near the meter will cause a stray field to exist, one which may be produced by the meter itself. Because the deflections of the pointer depend on the square of the voltage, the meter scale is marked in rms values.

18.5   Meters for Use at High Frequencies

If a moving-coil meter element were used to measure alternating current or voltage at high frequencies, the effect of inductive reactance of the coil would cause appreciable error. Because measurements must generally be made at frequencies covering an appreciable range, anywhere between 60 cycles and 1 megacycle, for example, moving-coil instruments are impractical unless the alternating current in the meter is first rectified into direct current.

Ac vacuum-tube voltmeters provide unidirectional current for the moving element that carries the pointer. Vacuum tubes and transistors respond to voltages that are proportional to the voltage being measured, so that the current through the indicating meter element is caused to be a measure of the voltage applied to the whole instrument.

A convenient way to measure alternating *current* at any frequency is to measure the voltage drop across a small noninductive resistance by means of a high-impedance voltmeter which is accurate at the existing frequency. Ohm's law is then applied. A precision resistor of low ohms value is preferred in this method of current measurement.

Thermocouple instruments are used to measure rms (effective) values of alternating currents and voltages. They are capable of accurate measurement when the wave forms are nonsinusoidal and also when the frequency is high. Conventional meters are inaccurate and unreliable under these conditions.

Thermocouple current meters have excellent accuracy (only about 1 per cent error) up to 50 MHz, even though they are calibrated at 60 Hz. Some milliammeters measure with fair accuracy (5 to 10 percent) up to 300 MHz.

The thermocouple instrument has a short, straight, heater wire supported at its ends by two metal blocks. A thermocouple junction is bonded to the center of the heater wire. Each of the wires of the thermocouple extends from the hot junction to a metal block, which serves as a cold junction. Copper wires connect the ends of the thermocouple leads to the moving coil of the instrument. The thermocouple produces a *direct EMF* by what is known as the *Seebeck effect*.

The metal blocks provide *cold-junction* compensation by preventing temperature changes, due to ambient conditions, from affecting the cold ends to a degree different from their effect on the hot junction, and thus introducing error. The heater wire has almost zero temperature coefficient of resistance. The thermal EMF is proportional to the temperature difference between the hot junction and the points on the metal blocks where the thermocouple leads are connected to the copper wires that go to the meter coil. The instrument scale can be calibrated in terms of the current in the heater wire.

Because the temperature of the heater increases as the square of the current, the deflection of the conventional permanent-magnet, moving-coil, dc millivoltmeter would have to be provided with a *square-law* scale and this is decidedly nonuniform. A nearly uniform scale is made possible by increasing the length of the air gap in the meter and flattening the pole faces somewhat. This decreases the meter sensitivity which accounts for the more nearly uniform scale.

By using a heater wire of very small cross section and a high resistance in series, the instrument becomes a voltmeter.

ELECTRONIC VOLTMETER

When measurements of voltage are made in circuits that would be disturbed in their operation if the voltmeter took even very small amounts of current (of the order of 1 mA), an electronic voltmeter is used. Before the advent of the transistor they were called *vacuum-tube voltmeters*. Some manufacturers continue using that name, even though they make some types that are all transistorized and contain no vacuum tubes. Other types contain both tubes (for high impedance at the input end of their circuit) and transistors, the latter in amplifier circuits.

Input impedances of electronic voltmeters are in the million-ohm range, some having input impedances in excess of 100 million ohms. Indeed, at least one type of input circuit has *infinite ohms impedance*. This means the meter takes no current at all from the network to which it is connected. Internal batteries or a voltage supply from the laboratory lighting circuit supplies the current needed to deflect the meter pointer.

18.6   Wattmeter

A wattmeter measures power, which is given by volts times amperes times power factor. Because volts and amperes direct current can be measured more accurately than watts, and since power factor is unity in circuits containing steady direct current, the use of a voltmeter and an ammeter is preferred for dc power measurement.

The wattmeter operates on the electrodynamometer principle (Section 18.4). Coils and connections of a wattmeter are illustrated in Fig. 18.3. The current through the fixed coils A, A' is load current and it produces a flux density that is proportional to the current. The very small current through the moving coil C and the voltage-dropping resistor R is proportional to the voltage. The flux density produced by C is, therefore, proportional to the load

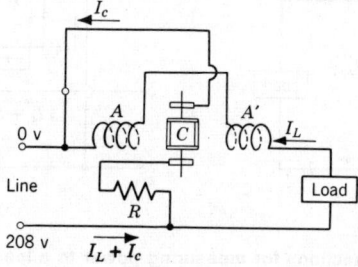

FIGURE 18.3   Wattmeter coils and internal connections.

voltage. Since the torque on the turning coil at any instant is proportional to the product of the two currents, it is also proportional to the product of the current and voltage or to the instantaneous power.

If the power factor is not unity, there are negative values on the power curve in each cycle. During these negative-power intervals the directions of current flow in the coils are such that negative torque results. The meter movement, however, does not have time to reverse the reading, owing to the inertia of the moving parts. The moving-coil system assumes a position where the deflection is proportional to the average torque or *average power*. The torque produced depends on the angle through which the moving coil has turned, but the scale calibration (nonlinear) takes care of this variation. Notice that the coils are all at the same potential. The upper wire is marked 0 volts and may be considered grounded to earth, which is actually the case in commercial electrical systems. Of course, with alternating current the upper wire is at 120 V rms above the lower wire during half of every cycle. But the important point with the connections as shown is that there is no potential difference between the series coils A, A' (often called field coils) and the moving coil C. This removes insulation hazards and inaccuracies caused by electrostatic forces between the coils.

WATTMETER CONNECTIONS IN A CIRCUIT

In Fig. 18.4 are shown two ways in which to connect the voltage-coil circuit of a wattmeter. The two terminals on the right lead to the series-connected current (field) coils, which must be in series with the device receiving the power. In part *a* the voltage on the moving-coil circuit (small terminals on the left) includes the voltage drop across the current coils. The total line voltage, not just that across the load, is applied to the meter. This means that the power loss in the current coils is included in the meter reading. To compute this loss, for purposes of correction, would involve substantial error because the

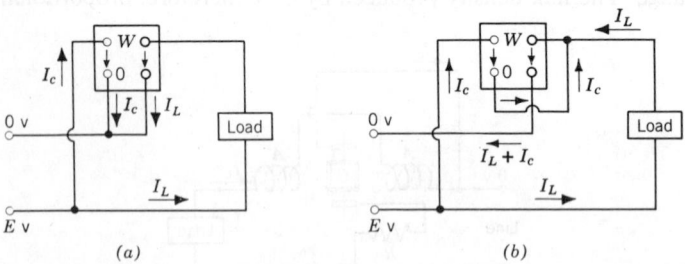

FIGURE 18.4   Wattmeter connections for measuring power to a load. (*a*) Reading includes power loss in current coils (connected to terminals to the right). (*b*) Reading includes power loss in voltage coil. I_c arrows show path of current in voltage coil.

resistance of the series coil is very low and is usually neither given nor deter-minable with high accuracy. Under these conditions the true power P is equal to the wattmeter reading minus I^2R of the series coils. This I^2R is only a few watts, at rated current of the wattmeter, and it may often be neglected.

In part b the current in the fixed coils is larger than the load current by the amount taken by the moving coil and its series resistance. Observe that the moving-coil system is in parallel with the load and connected on the load side of the wattmeter. The power in the potential circuit should be deducted from the meter reading. This may be determined easily because the resistance R_p of the potential circuit is specified on a chart in the meter lid. The true power P to the load is then

$$P = W_m - \frac{V^2}{R_p}$$

where W_m is the meter reading, V is the load voltage (which needs to be measured), and R_p is the resistance of the moving coil and its series resistor. The reactance of the moving (potential) coil is small enough to be ignored.

The connections in part a are preferred when the load current is small. Since the resistance of the current coils is very small, the I^2R loss in them is so small that the error is usually negligible, and correction need not be applied unless the instrument must be a high-precision type because high accuracy is required.

Large values of load current mean large values of power inasmuch as the voltage is usually appreciable. In this case the connections in part a may be used, of course, and those in part b also. The I^2R of the voltage coil, although not entirely negligible, will be very small compared with the power measured and may usually be ignored. Note that this loss is constant at all values of load current if the voltage V is constant.

EXAMPLE 1   A wattmeter connected as shown in Fig. 18-4b reads 192 W. The line voltage is 208 and the resistance of the potential circuit is 3825 Ω. What is the load power?

Solution

$$P = 192 - \frac{(208)^2}{3825} = 180.7 \text{ W}$$

If the power taken by the potential circuit were neglected, there would be a 6.25 percent error in the measurement of power to the load.

Automatic compensation for the loss in the potential circuit of a wattmeter is accomplished by interwinding with the series coils a like number of turns of fine wire connected in series with the potential coil and its resistor. The field of this compensating coil *bucks the field* of the current coils and produces a small countertorque that reduces the needle deflection an amount necessary to compensate for the loss in the potential circuit.

WATTMETER RATINGS

Wattmeters have been burned up while their readings stood at much less than full-scale value. Suppose a meter connected for use on a 120-V circuit has a full-scale capacity of 100 W and the current coils have a maximum safe current capacity of 15 A. This meter, connected properly to measure power to a load of 840 W at 120 V and 25 *percent power factor* reads 840 W, but has current in its series coils given by

$$840 = 120I(0.25)$$

from which $I = 28$ A. This amount of current will burn out the current coils of the meter.

This meter is capable of handling an 1800-W load at *unity power factor* without burning out, but in this application the series-coil current is only 15 A.

Because a wattmeter reading gives no indication of current and voltage values, an ammeter and a voltmeter should be used in conjunction with a wattmeter when low power factor is suspected, or whenever the power factor is not known. When the ammeter and voltmeter are connected between the wattmeter and the load, the I^2R_m of the ammeter and the V^2/R_v of the voltmeter must be deducted from the wattmeter reading to get the true power to the load. If the voltmeter is connected directly across the load the ammeter will read too high by the amount of current taken by the voltmeter.

18.7   Watthour Meters

A watthour meter measures electrical energy supplied to each residence or commercial establishment that is supplied from a commercial electric power station, unless an agreement has been made to determine the charge for the service on a basis other than the actual amount of energy used. As most of us know, readings of the meter are taken at regular intervals, usually monthly but in some cases every two months, and the amount used is determined by subtracting the last previous reading from the latest reading.

There are five small dials on most modern meters for domestic use. Each has ten digits with zero at the top and the numeral 5 at the bottom. Because adjacent dial pointers rotate in opposite directions, the numerals increase in value first in clockwise and then counterclockwise direction. Look at the meter at your own residence sometime.

To read the meter it is necessary merely to observe the numeral on each dial that its "hand" has *just passed* and record them from left to right as they are observed. A reading 05324 is obtained on a meter whose first hand (the one on the extreme left) has passed zero, whose second hand has passed 5, third has passed 3, and so on. Note that the first hand must be about half-way between 0 and 1 because the gear that has the second hand on its shaft is half-way through one revolution as indicated when its hand is on 5. Experience shows that when reading a meter of this kind, especially when learning, it is better to read and record the digits from right to left. The distance past zero moved by each hand gives a good clue to how far the one on its left could have gone.

Because electrical energy consumed by a load adds up as time goes on (watthours = watts × hours), it is evident that a watthour meter is a *summing-up* or *integrating instrument*. A register that records kilowatthours is driven by a rotating shaft through a gear train. A thin, light-weight, aluminum disk, mounted on the turning shaft, receives torque through an electromagnetic induction process.

Figure 18.5 is a sketch of a single-phase, induction-type watthour meter as seen after it has been removed from its iron case. The shaft, register, and damping magnets are not shown. The disk turns in the magnetic fields of two powerful permanent magnets which provide opposing torque such that the speed of rotation of the disk will be kept constant and at a practical value. The speed is directly proportional to the *actual power* passing through the meter.

The current coils *A* correspond to the current coils of a wattmeter; they carry the line current and are wound on the lugs so that they produce unlike poles. The voltage coil *V* corresponds to the potential coil of the wattmeter. Since it is stationary, it can be made with sufficient turns of small wire with

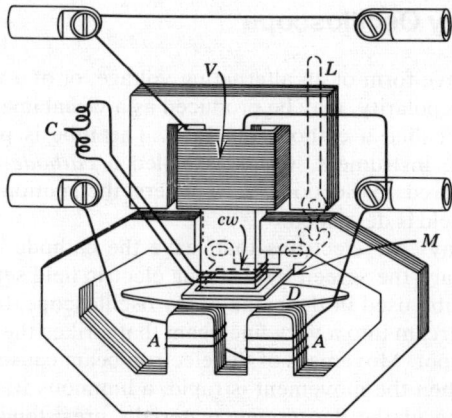

FIGURE 18.5   Assembly of a single-phase induction watthour meter. *A, A* are the current coils, *V* the potential coil; *C*, a variable resistance for adjusting impedance of *cw*, the compensating winding which provides for proper phase relations between fluxes of *V* and *A, A*. Metal stamping *M* moved horizontally by the lever *L*, provides the small torque necessary to compensate for the friction of moving parts. The rotating disc *D* drives its shaft (not shown), which moves the gear train and its pointers to register kilowatthours.

enough *impedance* to permit connection across the line voltage without a series resistor. The current in the voltage coil produces flux that reaches the air gaps above and below the disk by means of the laminated core extending downward through the coil. This flux is made to lead, by 90°, the flux produced by the series coils *A*.

Problem

1. A wattmeter is connected as in Fig. 18.4*a* to measure power to a load. A voltmeter with an internal resistance of 15,000 Ω is connected across the load and reads 120 V. An ammeter with negligible internal resistance is connected between the voltmeter and the wattmeter and it reads 12.5 A. The wattmeter current coils have negligible resistance; its voltage-coil and series resistor have 10,000-ohms resistance. The watt-meter reads 1200 W.

    (*a*) Draw the circuit.
    (*b*) Compute the power taken by the voltmeter.
    (*c*) Compute the true values of load current and power.
    (*d*) Compute true load power factor.

2. In the circuit of Problem 1, what would the wattmeter read if its potential circuit had been connected as in Fig. 18.4*b*, assuming the power factor of the potential circuit is unity?

18.8   Cathode-Ray Oscilloscope

The actual wave form of an alternating voltage, or of a voltage pulse that may have only one polarity, may be produced as a visual image on the screen of an electron tube called a cathode-ray tube. The tube is powered by electronic circuits of the instrument, which is called a *cathode-ray oscilloscope*. The reader is referred to Section 12.15 where the motion of an electron in an electrostatic field is described.

Cathode rays are electrons that leave the cathode in a stream. They are propelled toward the screen by a strong electric field set up in the neck of the tube. In the tube used in the cathode-ray oscilloscope, focusing devices shape the electron stream into a very fine beam that strikes the screen of the tube in a bright, small spot. Movement of the electron beam causes the luminous spot to move, and, when the movement is rapid, a luminous trace is produced on the screen because of the fluorescent material's presistence in emitting light for a small fraction of a second at each point. Rapid retracing of the beam over the same path, and the presistence of vision of the human eye, cause the moving spot to appear as a continuous line. Variations in strength of horizontal and vertical electric fields through which the electron beam passes cause it to have motions simultaneously in horizontal and vertical directions.

The strength of the electric field which produces horizontal motion of the beam can be made to increase directly with time. A constant field strength would produce a constant force on the electron beam, which would cause a sidewise deflection of constant magnitude. The result would be a stationary spot on the cathode-ray tube screen. It can be shown that, when the intensity (strength) of the deflecting electrostatic field *increases at a constant rate with respect to time*, the beam and its spot will move horizontally at constant speed.[1] That is, if the beam spot starts near the left-hand end of the horizontal diameter of the screen and moves horizontally to the right at constant speed, it will move equal distances (Δx) in equal time intervals (Δt). Keeping this fact in mind, we may readily understand that a sine-wave trace will show on the screen if, while the beam spot has a constant horizontal-velocity component, it is also given a vertical-velocity component that varies as a sine function of time.

To explain this further, let us consider the total time required for the spot to go from left to right on the screen, and divide that time into quarters. At the beginning of the first quarter, the vertical component of spot velocity is a maximum, but it decreases to zero by the time the first quarter of horizontal travel is completed. At that instant the spot is at the tip of the wave (point *a*) shown in Figure 18.6). During the second quarter of horizontal travel, the vertical component of spot velocity has reversed and increases to a maximum value by the time the spot is halfway across the screen (point *b*). It continues downward with decreasing vertical-component velocity until that component is again zero at the end of the third quarter (point *c*). During the fourth quarter the vertical component is again directed upward and is increasing. It reaches a maximum upward velocity at the end of the fourth quarter of horizontal travel (point *d*). At that instant the electric field giving horizontal velocity to the electron beam suddenly reverses, building up very rapidly in the reverse direction. This causes the luminous spot to fly back to the point where the

---

[1] The electron beam in the tube of a cathode-ray oscilloscope has a deflection (Y) which is directly proportional to the voltage (V_d) between the deflection plates. That is, the *steady deflection* produced by a *constant V_d* is given by

$$Y = KV_d$$

in which K is a constant determined by linear dimensions in the tube and by the constant *accelerating* voltage V_a used to produce the electron beam. The horizontal velocity of the spot on the screen is the rate of change of Y with respect to time:

$$\frac{dY}{dt} = K\frac{dV_d}{dt}$$

This shows that when the voltage V_d between the deflecting plates *increases at a constant rate with respect to time*, the spot velocity (dY/dt) *is constant*. A constant rate of change of V_d means that the electric-field intensity $\mathscr{E}_d$ between the plates must also be changing at a constant rate. That is, if dV_d/dt is constant, so must $d\mathscr{E}_d/dt$ be constant, because $\mathscr{E}_d = V_d/d$, when d is the distance between the plates.

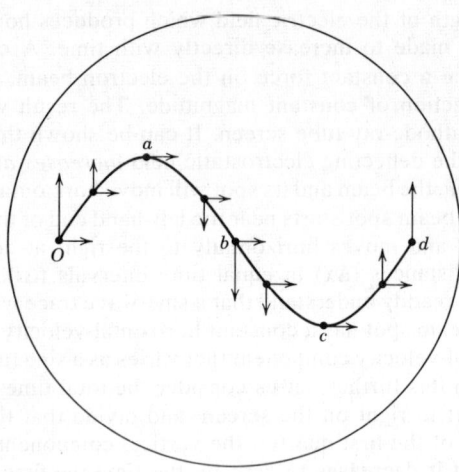

FIGURE 18.6   Sine-wave pattern on CRO screen. Arrows show that the spot's horizontal-velocity component is constant and its vertical component varies from zero to maximum in both directions.

screen trace started (point O). The field immediately reverses again and produces a horizontal component toward the right identical with the one before. The vertical-velocity component goes through the same cycle of values as just described. The result is a retracing of the same sinusoidal path by the spot.

The electric field is established between parallel-deflection plates. The strength of the field varies directly with the potential difference established on the plates. If the potential difference varies with time according to a sine-wave variation, the field will vary in that manner and produce a component of electron-beam velocity that varies in the same manner. In the example just described, the plates which produce the field that contributes the vertical component of beam velocity were supplied with a voltage having sine-wave form. The luminous trace on the screen had exactly the same wave form as that voltage. It is a common practice to say that we looked at the voltage, instead of being more precise, that is, using more words and time, and saying that we looked at a pattern which has exactly the same wave form as the voltage we applied to the vertical-deflection plates of the cathode-ray tube.

A detailed description of the construction of the tube is not given here; it may be found in booklets obtainable from the manufacturers. In the neck of the tube are mounted the cathode, a control grid, and an accelerating anode. Other electrodes, mounted between the grid and the accelerating anode, are used to focus the beam so that it produces a spot of desired size on the screen. The action of the grid is to control the strength of the electron beam and thus the brightness of the spot.

The cathode-ray oscilloscope principally is used to observe and examine wave forms of voltages and currents in both steady-state and transient conditions. Single pulses of extremely short duration occur in the circuits of digital computers, television, radio, and other communications and control apparatus. Observations and measurements of amplitude and width (in terms of nanoseconds, for example) of such pulses must be made. This is done conveniently and quickly by means of the scope. Other applications include measurements of phase angles, peak-to-peak values of voltages, the calibration of oscillators, frequency determination, and the detection of small variations in dc voltages and currents. The wave form of a current is observed by connecting the scope leads across a noninductive resistance, of small ohms value, through which the current flows. In some cases a connecting wire which has small cross-section area will serve this purpose. The amplifiers in the scope can change extremely small ac input voltages into the hundred or more volts required on the deflection plates for screen patterns of one or two inches of wave amplitude.

Cathode-ray oscilloscopes that will accept dc voltages are in common use. The dc voltage causes constant deflection of the cathode-ray beam. This is shown as a steady displacement of the spot trace from the center (no-voltage) position. An ac voltage of sine-wave form superimposed on a steady dc voltage will show on the screen as a sine wave on an invisible horizontal axis displaced vertically from the normal central position.

18.9   Measurement of Impedance

We have learned that impedance is an electrical phenomenon that expresses, quantitatively, the opposition offered to the flow of alternating current. Its value is influenced by the frequency and by inductance or capacitance or both. Accordingly, we can understand the necessity of using an *alternating-current bridge* to measure impedance accurately.

The general form of the alternating-current bridge is that of the Wheatstone bridge shown in Fig. 18.7. Impedances take the place of resistances in the arms. The detector is usually a pair of telephone receivers when the frequency at which the impedance is to be determined is in the "audio" band. This means that a current through the headphones will produce an audible sound. At frequencies above the limit of human hearing the detector may be a vacuum-tube voltmeter or an oscilloscope. The detector may be fed through an amplifier to make it more sensitive. At balance, the current through the detector is zero, as is the case of the Wheatstone bridge.

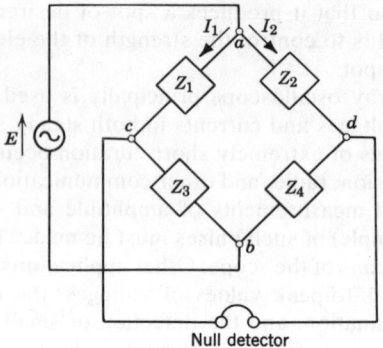

Null detector

FIGURE 18.7   AC impedance bridge. Z_3 is the unknown in the analysis in sections 18.9, 18.10, 18.11.

When the bridge is balanced, the junctions c and d are at the same potential and no current flows in the detector no matter what its impedance. The following relations then hold:

$$I_1 Z_1 = I_2 Z_2 \tag{18.1}$$

$$I_1 Z_3 = I_2 Z_4 \tag{18.2}$$

So, at balance

$$\frac{Z_1}{Z_3} = \frac{Z_2}{Z_4} \quad \text{or} \quad Z_1 Z_4 = Z_2 Z_3 \tag{18.3}$$

Equation (18.3) is the *equation of balance* of the ac Wheatstone bridge.

It is convenient and advantageous to make Z_1 and Z_2 pure resistances. Then Equation (18.3) becomes

$$\frac{R_1 + j0}{R_3 + jX_3} = \frac{R_2 + j0}{R_4 + jX_4} \tag{18.4}$$

When two quantities are equal their reciprocals are equal;

$$\frac{R_3}{R_1} + j\frac{X_3}{R_1} = \frac{R_4}{R_2} + j\frac{X_4}{R_2} \tag{18.5}$$

In order for this equation to be satisfied, the real parts of both sides must be equal and the imaginary parts must be equal. Let us assume that Z_3 is the unknown impedance. Its components may be determined as follows: Equating reals,

$$\frac{R_3}{R_1} = \frac{R_4}{R_2}; \qquad R_3 = \frac{R_1}{R_2} R_4 \tag{18.6}$$

Equating imaginaries,

$$\frac{X_3}{R_1} = \frac{X_4}{R_2}; \qquad X_3 = \frac{R_1}{R_2} X_4 \tag{18.7}$$

R_1 and R_2 are noninductive resistances arranged so that their *ratio* is variable in decades, that is, $0.001, 0.01, 0.1, 1, 10, 100, 1000$.

We can easily show that when the bridge is balanced, the phase angles of Z_3 and Z_4 are equal. The first form of Equation (18.3) may be written.

$$\frac{Z_1\underline{/\theta_1}}{Z_3\underline{/\theta_3}} = \frac{Z_2\underline{/\theta_2}}{Z_4\underline{/\theta_4}} \qquad (18.8)$$

From which

$$\left|\frac{Z_1}{Z_3}\right| = \left|\frac{Z_2}{Z_4}\right| \qquad \text{and} \qquad \theta_1 - \theta_3 = \theta_2 - \theta_4 \qquad (18.9)$$

The vertical lines mean absolute values, phase angles excluded.
From this,

$$\left|\frac{Z_1}{Z_2}\right| = \left|\frac{Z_3}{Z_4}\right| = \frac{R_1}{R_2} \text{ since } |Z_1| \text{ is replaced by } R_1 \text{ and } \|Z_2\| \text{ by } R_2$$

From the middle and right-hand members,

$$|Z_3| = \frac{R_1}{R_2}|Z_4| \qquad (17.10)$$

and

$$\theta_3 = \theta_4 \text{ because } \theta_1 = 0 = \theta_2$$

18.10 Capacitance Bridge

A bridge that can measure the capacitance and "loss resistance" of a capacitor quite accurately compares its values C_3 and R_3 with those of a *standard capacitor*, C_4, and a known resistance, R_4, connected in series with the standard. The circuit is shown in Fig. 18.8. Again we use R_1 and R_2 for $|Z_1|$ and

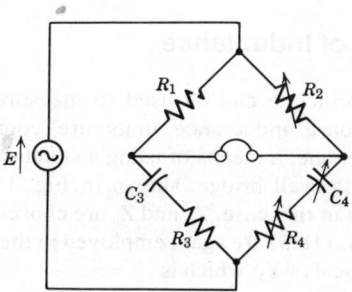

FIGURE 18.8 Capacitance bridge. Comparison of unknowns C_3 and R_3 with knowns C_4 and R_4 are made.

$|\mathbf{Z}_2|$ and we make them variable. $\mathbf{Z}_3$ and $\mathbf{Z}_4$ are expressed as indicated:

$$\mathbf{Z}_3 = R_3 - j\frac{1}{\omega C_3}; \ \mathbf{Z}_4 = R_4 - j\frac{1}{\omega C_4}; \ |\mathbf{Z}_1| = R_1; \ |\mathbf{Z}_2| = R_2$$

Substituting in Equation (18.3),

$$R_2 R_3 - j\frac{R_2}{\omega C_3} = R_1 R_4 - j\frac{R_1}{\omega C_4} \qquad (18.11)$$

Equating reals,

$$R_2 R_3 = R_1 R_4$$
$$R_3 = \frac{R_1}{R_2} R_4 \qquad (18.12)$$

Equating imaginaries,

$$\frac{R_2}{\omega C_3} = \frac{R_1}{\omega C_4} \qquad (18.13)$$

$$C_3 = \frac{R_2}{R_1} C_4 \qquad (18.14)$$

We are, as before, considering R_3 and C_3 to be the unknowns. Notice that when we equated the imaginaries the omegas dropped out. This means that the accuracy of the measurements is independent of variations in frequency and also of the value of the frequency. This required, however, that the impedance values of all other elements of the bridge be independent of frequency. The magnitude of the applied voltage used in the measurement need not be known, although it is good practice to use as small a voltage as possible consistent with good response of the detector and the current-carrying capacity of the elements in the bridge arms.

18.11   Measurement of Inductance

The comparison principle can be used to measure an unknown inductance, but because standard inductance units are comparatively expensive and somewhat cumbersome, a means of using a standard capacitance was found in the form of the Maxwell bridge, shown in Fig. 18.9. It is convenient to let $\mathbf{Z}_4$ be the unknown in this case. $\mathbf{Z}_2$ and $\mathbf{Z}_3$ are chosen to be the pure resistances. The basic equations (18.3) are again employed in the derivation.

$\mathbf{Z}_1$ is the reciprocal of $\mathbf{Y}_1$, which is

$$\mathbf{Y}_1 = \frac{1}{R_1} + \frac{1}{1/j\omega C_1} = \frac{1}{R_1} + j\omega C_1 = \frac{1}{\mathbf{Z}_1}$$
$$\mathbf{Z}_2 = R_2, \qquad \mathbf{Z}_3 = R_3, \qquad \mathbf{Z}_4 = R_4 + j\omega L_4$$

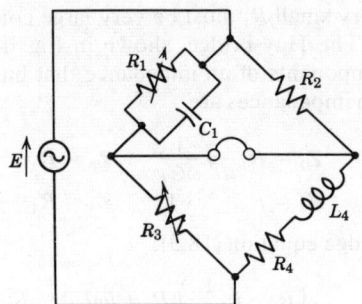

FIGURE 18.9   Maxwell bridge.

From Equation (18.3),

$$\mathbf{Z}_4 = \frac{\mathbf{Z}_2 \mathbf{Z}_3}{\mathbf{Z}_1} = \mathbf{Y}_1 \mathbf{Z}_2 \mathbf{Z}_3$$

$$\mathbf{Z}_4 = \left(\frac{1}{R_1} + j\omega C_1\right) R_2 R_3$$

$$R_4 + j\omega L_4 = \frac{R_2}{R_1} R_3 + j\omega R_2 R_3 C_1 \tag{18.15}$$

Equating reals,

$$R_4 = \frac{R_3}{R_1} R_2 \tag{18.16}$$

Equating imaginaries,

$$L_4 = R_2 R_3 C_1 \tag{18.17}$$

Again we see that the results are independent of frequency so long as the resistance and capacitance values are not affected by changes in frequency.

The Maxwell bridge is especially adaptable to the measurement of large inductances, because the impedances of the four arms can easily be made large, that is, in the same order of magnitude. Air-cored standard inductances are not ordinarily available in sizes larger than 1 H.

In radio circuit applications it is common practice to measure the incremental inductance of an audio-frequency transformer, such as one that would couple a loudspeaker to the output of a power amplifier. The inductance that exists while a direct current is flowing through the transformer winding can be measured. The direct current may be introduced by a dc supply in series with the ac supply.

18.12   Hay Bridge

In practical application, the Maxwell bridge is not as well adapted to the measurement of inductive impedances for which L_4 is very large compared with

R_4. When R_4 is very small R_1 must be very large compared with R_2 in order to obtain a balance. The Hay bridge, shown in Fig. 18.10, is more suitable for measuring the components of an impedance that has a very low power factor ($X_L \gg R$). The arm impedances are

$$\mathbf{Z}_1 = R_1 - j\frac{1}{\omega C_1} \qquad \mathbf{Z}_2 = R_2$$
$$\mathbf{Z}_3 = R_3 \qquad \mathbf{Z}_4 = R_4 + j\omega L_4$$

Using the basic bridge equation (18.3),

$$\left(R_1 - j\frac{1}{\omega C_1}\right)(R_4 + j\omega L_4) = R_2 R_3$$
$$R_1 R_4 + \frac{L_4}{C_1} + j\omega L_4 R_1 = R_2 R_3 + j\frac{R_4}{\omega C_1}$$

Equating reals and imaginaries,

$$R_1 R_4 + \frac{L_4}{C_1} = R_2 R_3 \tag{18.18}$$

$$\omega L_4 R_1 = \frac{R_4}{\omega C_1} \tag{18.19}$$

Simultaneous solution of these last two equations gives

$$L_4 = \frac{C_1 R_2 R_3}{1 + \omega^2 C_1{}^2 R_1{}^2} \tag{18.20}$$

$$R_4 = \frac{\omega^2 C_1{}^2 R_1 R_2 R_3}{1 + \omega^2 C_1{}^2 R_1{}^2} \tag{18.21}$$

Evidently about twice as much computation is required to evaluate the unknown impedance components as when the Maxwell bridge is used. This is a small price to pay for the much better accuracy obtained.

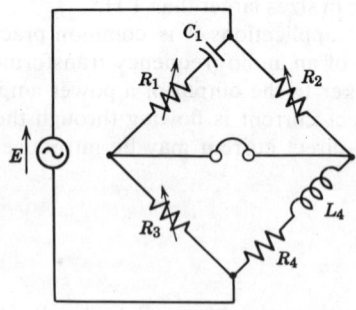

FIGURE 18.10   Hay Bridge.

Correct manipulation of ac bridges in the making of measurements is somewhat of an art. The instrument should be handled with care and some precautions should be observed.

The applied voltage, usually called the signal voltage, should not be larger than that which gives adequate response in the null detector. Overdriving the detector may not only be harmful, but it may contribute to loss of accuracy. If the detector responds to voltage only (when it is an electronic voltmeter or an oscilloscope) and not to power (when it is a pair of headphones) its impedance should be as high as possible.

When either of the variable resistances is very small, from 1 to 20 Ω, for example, inaccuracies may be far greater than they would be if that resistance were made 10 times as large and the other one increased correspondingly.

When audio frequency or radio frequency measurements are made, *stray capacitances* (which can exist between connecting wires, and between circuit and operator) in the bridge circuit are likely to be present. Their effects can be neutralized by the connection of what is called a "Wagner Earth," or "Wagner Ground."[2] A temporary connection is made to one side of the null detector and the other end of the connecting wire is fastened to a ground wire leading into the earth.

Although numerical examples of the use of impedance bridges in a textbook on theory amount to little more than substitution into formulas, some will be given here.

EXAMPLE 18.2 A manufacturer of small inductances uses the bridge circuit of Fig. 18.7 in a test laboratory to measure its products. The signal generator supplies E at 1000 Hz. The arm Z_4 has a standard inductance of 0.100 H and a resistance of 500 Ω in series. At balance, $R_1/R_2 = 2$. What is the value of the unknown inductance and how much resistance is in series with it in the Z_3 arm?

Solution. From Equations (18.6) and (18.7)

$$R_3 = \frac{R_1}{R_2} R_4 = 2 \times 500 = 1000$$

$$X_3 = \omega L_3 = 2 \times \omega \times 0.100 = \frac{R_1}{R_2} \omega L_4$$
$$L_3 = 0.200 \text{ H}$$

EXAMPLE 18.3 When an unknown capacitance was measured with the bridge of Fig. 18.8, the following readings were recorded: $R_1 = 195$ Ω, $R_2 = 10^5$ Ω, $R_4 = 1000$ Ω, $C_4 = 0.00002 \mu$F. What were the capacitance and loss resistance of the capacitor found to be?

[2]Everett and Anner, *Communication Engineering*, Third Edition, p. 194, New York, McGraw-Hill, 1956.

Solution

$$R_3 = \frac{R_1}{R_2} R_4 = \frac{195}{10^5} \times 1000 = 1.95 \ \Omega$$

$$C_3 = \frac{R_2}{R_1} C_4 = \frac{10^5}{195} \times 2 \times 10^{-5} = 0.1025 \ \mu F$$

Note. R_3 is the leakage resistance expressed as a series value. The power loss would be computed using $I_c^2 R$. If I_c were 100 μA, for example, the power loss would be $1.95 \times 10^{-2} \ \mu W$. Leakage resistance is usually shown as a resistance in parallel with the capacitance in the equivalent circuit. If the capacitive reactance were such that the voltage across the capacitor were 100 V while the power lost was $1.95 \times 10^{-2} \ \mu W$, the parallel leakage resistance would be such that $V^2/R = 100^2/R = 1.95 \times 10^{-2} \ \mu W$.

Solving for R, we get

$$R = \left(\frac{100^2}{1.95 \times 10^{-8}} \right) = 513,000 \ M\Omega$$

Remember that C_4 is a standard capacitor, high in quality, with zero leakage current. R_4 is used to serve as a balancing effect for R_3. If the "unknown" capacitance had practically zero leakage, the ratio R_1/R_2 would have been much smaller than that recorded.

EXAMPLE 18.4 The Maxwell bridge was used to measure an inductance suspected of having a value between 2 H and 3 H. Readings obtained were

$$R_1 = 1 \ M\Omega, R_2 = 11,250 \ \Omega, R_3 = 4160 \ \Omega, C_1 = 0.05 \ \mu F$$

What values were obtained for L_4 and R_4?

Solution. From Equation (18.16) and (18.17),

$$L_4 = R_2 R_3 C_1 = 11.250 \times 4160 \times 5 \times 10^{-8} = 2.34 \ H$$
$$R_4 = R_2 R_3/R_1 = 11,250 \times 4160/10^6 = 46.8 \ \Omega$$

18.13 Review Questions

1. Why will the D'Arsonval meter movement not work on alternating current?
2. Explain the principle of operation of the iron-vane meter movement.
3. Describe the electrodynamometer meter movement. What are its advantages and disadvantages?
4. When large amounts of alternating current are to be measured, what arrangement is made in connecting an ammeter for the purpose?
5. Tell how an alternating current can be measured using a noninductive resistance of small ohms value.
6. Why is it not practical to measure a high-frequency alternating current using a

moving-coil instrument when the moving coil carries all or part of the alternating current?

7. Describe the thermocouple instrument for measuring alternating current. What about its accuracy?

8. What is the most important advantage in the use of an electronic voltmeter to measure voltages in a radio or television receiver?

9. How does the circuit inside a wattmeter differ from ammeter and voltmeter circuits?

10. Describe two possible ways to connect a wattmeter. Why do the readings differ in the two cases?

11. Tell how it is possible that a wattmeter with 1 kW full-scale capacity may be burned out while carrying only 0.5 kW.

12. How does the moving part of a watthour meter differ from the moving part of a wattmeter?

13. Tell how to read a residence-type watthour meter.

14. What are cathode rays?

15. What happens to the electrons after they are focused to a thin stream while in the neck of a cathode ray oscilloscope tube?

16. What is required of the voltage between the horizontal deflecting plates of a cathode ray oscilloscope tube in order that the spot will move at *constant velocity* across the screen of the tube?

17. Tell how a sine wave is formed on the screen of a CRO tube.

18. Sketch the general form of the impedance bridge, label the impedances, and write the basic equation.

19. Name three devices that can serve as a null detector in an ac bridge circuit, and tell how you know when balance is achieved when using each.

20. When is it advisable to use a Wagner Earth connection, and what is accomplished by its use?

18.14 Problems

3. An unknown impedance, Z_3, was measured by means of the impedance bridge (Fig. 18.10) at a frequency of 1000 Hz. Data obtained were: $R_1/R_2 = 1/10$, $R_4 = 50\ \Omega$, $L_4 = 0.150$ H (inductance component of Z_4). What were (a) the two components of the unknown impedance, and (b) its inductance?

4. The bridge of Fig. 18.7 was used to measure an unknown inductive reactance at 10 KHz. It was found to be $8 + j\,110\ \Omega$. Determine the settings of the control impedance, Z_4, if $R_1/R_2 = 1/100$.

5. The capacitance comparison bridge (Fig. 18.8) was balanced at $f = 10$ kHz when $R_1/R_2 = 0.1$, $R_4 = 10$ kΩ, $C_4 = 50$ pF. Evaluate the unknown impedance that was measured. Express it in complex-number form.

6. A capacitor used in the starting winding of a single-phase motor was measured on a capacitance comparison bridge. The data taken at balance were: $R_1/R_2 = 0.01$, $R_4 = 100\ \Omega$, $C_4 = 0.041\ \mu$F. What was the measured capacitance value? Why is the value of R_4 not needed in the determination but necessary in the measuring process?

7. The components of an unknown inductive impedance were measured by means of the Maxwell bridge. The values of the resistances were found, at balance, to be $R_1 = k\Omega$, $R_2 = 20\,\Omega$, $R_3 = 377\,\Omega$, and C_1 was 318.5 nF. The frequency was 1 kHz. Determine the inductance and the unknown impedance.

8. The inductance and resistance of a radio coil were measured on a Hay bridge at 100 kHz. The readings were $R_1 = 10\,\Omega$, $R_2 = 50\,\Omega$, $R_3 = 38\,\Omega$, $C_1 = 60$ pF. What values were found by calculations from these data?

CHAPTER NINETEEN

Network Theorems Applied to AC Circuits

The most important network theorems used in elementary circuit analysis were introduced and described, with numerical examples, during our study of direct-current circuits. This is desirable because the objective was to have you learn the meaning of each theorem and its usefulness, as illustrated by an application, in the easiest possible manner. We were not obliged to include reactances, so this reduced the amount of mathematical manipulation to a minimum.

The theorems apply equally well to ac networks. We shall review the meaning of each in an informal way and illustrate its application by an example.

19.1 Thévenin's Theorem

Suppose we have a complicated network containing one or more ac sources and we want to know how much current flows in a particular one of its branches. We call this branch the "load," and its impedance is the load impedance. Thévenin's theorem allows us to obtain a simple series circuit that has a *new source* and a *single series impedance* to which we can connect our original load

423

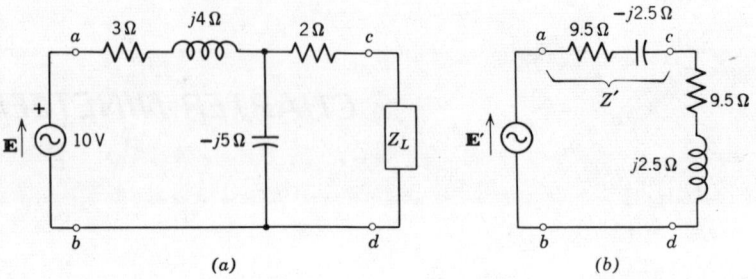

FIGURE 19.1   Circuits illustrating Thévenin's theorem. (a) Original circuit. (b) Thévenin equivalent.

impedance. Thus, we have a single current, supplied by the new source, that flows through the two impedances connected in series. The amount and phase of this current will be the same as the amount and phase of the "load current" in the original circuit.

The circuit of Fig. 19.1*a* is our original circuit. We want to know how much current flows in $\mathbf{Z}_L$, the load impedance. The circuit in (*b*) is the Thévenin equivalent circuit. We shall now show how we determine its source voltage $\mathbf{E}'$ and its series impedance $\mathbf{Z}'$.

The source voltage $\mathbf{E}'$ of the equivalent circuit is the voltage at the load terminals c, d *after the load impedance $\mathbf{Z}_L$ has been removed*. Note that then there will be no current in the 2-Ω resistor and the open-circuit voltage will be the capacitor voltage.

We can get the open-circuit voltage by simple voltage division. We have

$$\mathbf{V}_{c,d} = \mathbf{V}_{oc} = \frac{-j5}{3+j4-j5} \times 10$$

$$\mathbf{V}_{oc} = \frac{-j50}{3-j} \times \frac{3+j}{3+j} = \frac{50-j150}{10} = 5-j15 \text{ V}$$

The internal impedance of the source is zero, so the impedance looking back at terminals c, d is

$$\mathbf{Z}' = 2+j0+\frac{-j5(3+j4)}{3+j4-j5} = 2+\frac{20-j15}{3-j}$$

$$\mathbf{Z}' = 2+\frac{75-j25}{10} = 9.5-j2.5$$

The Thévenin equivalent circuit, therefore, has a source voltage $\mathbf{E}'$ of $5-j15$ V and a series impedance of $9.5-j2.5$ Ω. Let us assume the load impedance

Z_L to be $9.5 + j2.5\ \Omega$. The only reason these values were chosen is to shorten the computation. The load current is then

$$I_L = \frac{E'}{Z' + Z_L} = \frac{5 - j15}{9.5 - j2.5 + 9.5 + j2.5} = \frac{5 - j15}{19 + j0} = 0.263 - j0.789\ A$$

The current will now be determined using series-parallel circuit theory as a check. The impedance of the parallel section is

$$Z_p = \frac{-j5(11.5 + j2.5)}{11.5 + j2.5 - j5} = \frac{-j5(11.5 + j2.5)(11.5 + j2.5)}{(11.5 - j2.5)(11.5 + j2.5)} = 2.076 - j4.55$$

$$V_p = \frac{2.076 - j4.55}{3 + j4 + 2.076 - j4.55} \times 10 = \frac{20.76 - j45.5}{5.076 - j0.55} = 5 - j8.43$$

$$I_L = \frac{5 - j8.43}{11.5 + j2.5} \times \frac{11.5 - j2.5}{11.5 - j2.5} = \frac{36.43 - j109.4}{132.25 + 6.25} = 0.263 - j0.789\ A$$

19.2   Norton's Theorem

We shall use the same original network (Fig. 19.1a) to illustrate Norton's theorem. You may recall that the impedance "looking back from the open-circuited load terminals," which we call Z', is used in Norton's theorem also, but it is connected in parallel with the load instead of in series. Also, we use the "short-circuit current" at the load terminals instead of the open-circuit voltage. That is, we imagine an ammeter that has zero internal impedance connected in place of the load impedance and then calculate the current it would carry. Replacing the load impedance in Fig. 19.1a with an ammeter and calculating the equivalent impedance of the remaining parallel section, we have

$$Z_p = \frac{-j5(2)}{2 - j5} = \frac{-j10(2 + j5)}{29} = 1.724 - j0.69\ \Omega$$

The total current is

$$I = \frac{10 + j0}{3 + j4 + 1.724 - j0.69} = \frac{10}{4.724 + j3.31}$$

$$= \frac{47.24 - j33.1}{33.26} = 1.421 - j0.995$$

The short-circuit current is found by current division.

$$I_{SC} = \frac{(1.421 - j0.995)(-j5)}{2 - j5} = \frac{-4.975 - j7.105}{2 - j5} = 0.881 - j1.35\ A$$

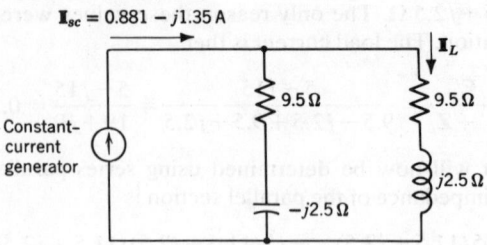

FIGURE 19.2   Norton equivalent circuit for circuit of Fig. 19.1a.

The Norton equivalent circuit shown in Fig. 19.2 has a constant-current generator supplying $0.881 - j1.35$ A to $9.5 - j2.5\,\Omega$ ($\mathbf{Z}'$, which is the impedance looking back at the output terminals and found for the Thévenin circuit) in parallel with the load impedance, $\mathbf{Z}_L = 9.5 + j2.5\,\Omega$ (Fig. 19.2). The load current is computed using the current-division principle:

$$\mathbf{I}_L = \frac{(0.881 - j1.35)(9.5 - j2.5)}{9.5 - j2.5 + 9.5 + j2.5} = \frac{5 - j15}{19} = 0.263 - j0.789 \text{ A}$$

Problems

1. In the circuit of Fig. 19.3 assume $R_1 = 5\,\Omega$, $R_2 = 10\,\Omega$, $X_2 = j10\,\Omega$, $Z_L = 20 + j0\,\Omega$, $\mathbf{E} = 100$ V. Convert the circuit into its Thévenin equivalent circuit and label all elements with correct numerical values.
2. Calculate the load current using the Thévenin equivalent circuit in Problem 1 and verify your result by applying a series-parallel method of solution.
3. Repeat Problem 1 using the following constants: $R_1 = 8\,\Omega$, $R_2 = 12\,\Omega$, $X_2 = -j5\,\Omega$, $\mathbf{Z}_L = 15 + j0\,\Omega$, $V = 120$ V.
4. Calculate the load current in the circuit of Problem 3 using the Thévenin equivalent circuit and verify your answer.
5. In the circuit of Fig. 19.3 assume the following values: $R_1 = 2{,}000\,\Omega$, $R_2 = 50{,}000\,\Omega$, $X_2 = -j12{,}000\,\Omega$, $\mathbf{Z}_L = 10{,}000 - j20{,}000\,\Omega$, $\mathbf{E} = 10$ V. Convert the circuit

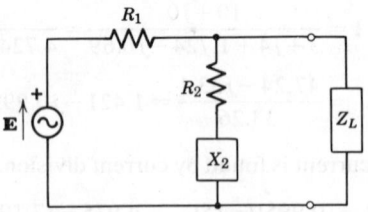

FIGURE 19.3   For Problems 1–8.

into its Norton equivalent circuit and label all elements with correct numerical values. Calculate the load current and the voltage drop across the load.

6. Assume the original circuit described in Problem 1 and calculate the current that would flow in a zero-resistance ammeter that is connected in place of $\mathbf{Z}_L$.

7. Set up the Norton equivalent circuit for the original circuit of Problem 3 and calculate the load current.

8. Set up the Thévenin equivalent circuit for the original circuit of Problem 5 and calculate the load current.

19.3   Reciprocity Theorem

Suppose we have an ac network of elements connected in series-parallel and they are supplied with current from an ac generator connected to terminals a, b as in Fig. 19.4. The reciprocity theorem says that if the generator and ammeter are interchanged (that is, the generator connected to terminals c, d and the ammeter to terminals a, b) and the generator voltage is unchanged, the reading of the ammeter will be the same as before. We shall now show that this theorem holds true.

The impedance of the parallel pair is the product over the sum:

$$\mathbf{Z}_p = \frac{-j25}{5-j5} \times \frac{5+j5}{5+j5} = 2.5 - j2.5$$

The total impedance is

$$\mathbf{Z}_T = 2.5 - j2.5 + j = 2.5 - j1.5$$

The voltage drop across the parallel pair can be found by voltage division:

$$\mathbf{V}_p = \frac{2.5 - j2.5}{2.5 - j1.5} + 20 = \frac{50 - j50}{2.5 - j1.5}$$

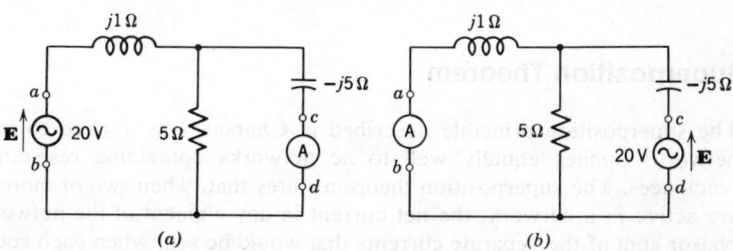

FIGURE 19.4   **Illustrating the reciprocity theorem. (a) Original circuit. (b) Generator and ammeter interchanged.**

Rationalizing the denominator gives

$$V_p = \frac{200 - j50}{8.5} = 23.5 - j5.88$$

$$I_A = \frac{23.5 - j5.88}{-j5} = 1.176 + j4.7 = 4.84\underline{/76°}\ \text{A}$$

Now, when we interchange the 20-V source and the ammeter, the meter should read the same. In Fig. 19.4b, the parallel impedance is

$$Z_p = \frac{j5}{5+j} \times \frac{5-j}{5-j} = \frac{5+j25}{26} = 0.192 + j0.962$$

The total impedance is

$$Z_T = 0.192 + j0.962 - j5 = 0.192 - j4.038$$
$$Z_T = 4.04\underline{/-87.3°} \text{ and } Z_p = 0.98\underline{/78.6°}$$

The voltage drop across the parallel pair is

$$V_p = \frac{0.98\underline{/78.6°}}{4.04\underline{/-87.3°}} \times 20 = 4.85\underline{/165.9°}$$

The ammeter reading is

$$I_A = \frac{4.85\underline{/165.9°}}{1\underline{/90°}} = 4.85\underline{/76.9°}\ \text{A}$$

So, we can see that the reciprocity theorem holds for ac as well as for dc. Observe that we could have assumed the components of the circuit to be 1000 Ω inductive reactance, 5000 Ω capacitor reactance and 5000 Ω resistance. We could have solved exactly as was done, calling the units kilohms and our answers would have been in milliamperes.

19.4 Superposition Theorem

The superposition principle described in Chapter 7 as applied to resistance networks applies equally well to ac networks containing resistances and reactances. The superposition theorem states that, when two or more sources are active in a network, the net current in any element of the network is the phasor sum of the separate currents that would be sent when each source acts alone. The sources that have been temporarily removed must be replaced by their internal impedances.

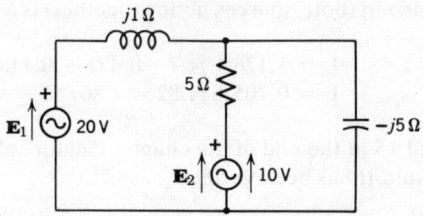

FIGURE 19.5   Illustrating the superposition theorem.

Figure 19.5 will be recognized as Fig. 19.4 with a 10-V source added in series with the 5-Ω resistance. The capacitive reactance, $-j5$ Ω, will be called *the load* and we shall find the amount of load current by adding together the amount of current produced by each source separately. The sources produce voltages that are in phase with each other, as indicated by the $+$ signs.

(a) $\mathbf{E}_1$ operating alone; $\mathbf{E}_2$ replaced by a zero impedance. The impedance of the parallel branch is

$$\mathbf{Z}_p = \frac{-j25}{5-j5} \times \frac{5+j5}{5+j5} = \frac{125-j125}{50} = 2.5 - j2.5$$

By voltage division, the voltage drop across the parallel pair is

$$\mathbf{V}_p = \frac{2.5-j2.5}{2.5-j1.5} \times 20 = \frac{200-j50}{8.5}$$

The load current is given by this voltage divided by the load impedance:

$$\mathbf{I}_{L1} = \frac{200-j50}{8.5 \times (-j5)} = 1.176 + j4.70 = 4.85 \underline{/76°}$$

(b) $\mathbf{E}_2$ operating alone; $\mathbf{E}_1$ replaced by a zero impedance. The parallel branch, this time, has $j1$ and $-j5$. The impedance of the parallel branch is

$$\mathbf{Z}_p = \frac{-j^2 5}{j(1-5)} = \frac{5}{-j4} = \frac{j5}{4} = j1.25$$

The total impedance is

$$\mathbf{Z}_T = 5 + j1.25$$

The voltage drop across the parallel pair is

$$\mathbf{V}_p = \frac{j1.25 \times 10}{5+j1.25} = \frac{j50}{20+j5}$$

The load current is given by this voltage divided by $-j5$:

$$\mathbf{I}_{L2} = \frac{j50}{(20+j5)(-j5)} = \frac{j50}{25-j100} = -0.471 + j0.118$$

The true load current (both sources acting together) is

$$I_L = 1.176 + j4.7 - 0.471 + j0.118$$
$$I_L = 0.705 + j4.82 = 4.86 \underline{/80.8°}$$

Problems 14 and 15 at the end of the chapter call for solution by other methods to verify this result. It has been verified.

19.5   Maximum-Power-Transfer Theorem

We have already applied this theorem in working with the Thévenin-equivalent circuit in Section 18.10. The series impedance for the Thévenin circuit was found to be $9.5 - j2.5 \, \Omega$. The load impedance Z_L was chosen to be the conjugate of this: $9.5 + j2.5 \, \Omega$. The purpose of this choice was to simplify the computations for load current.

It is not as easy as it is with direct-current circuits to show that any other choice of load impedance results in less power to the load. One could keep the real part of Z_L constant, vary the quadrature part both above and below the $j2.5$ value, and compute load power. Then one could keep the quadrature part constant and change the real part to a few values greater or less than 9. One would find the power in the load smaller than the value 6.56 W that is maximum.

Consider an ac source with internal impedance $Z_G = R_G + jX_G$ applying a voltage E_G to a receiver with impedance $Z_R = R_R + jX_R$. The current in the receiver (the load impedance) is

$$I_R = \frac{E_G}{\sqrt{(R_G + R_R)^2 + (X_G + X_R)^2}} \underline{/\theta}$$

The power to the receiver (the load) is

$$P_R = I_R^2 R_R = \frac{E_G^2 R_R}{(R_G + R_R)^2 + (X_G + X_R)^2}$$

With R_G and R_R already fixed and equal, the power to the load will be a maximum when the reactance term in the denominator is zero. This means the reactances are equal and opposite in sign. This requires the complex load impedance to be the *conjugate of the generator impedance*. If $Z_G = R_G + jX_G$ (i.e., inductive), then Z_R must be $R_R - jX_R$ and $R_R = R_G$, $|X_R| = |X_G|$. If $Z_G = R_G - jX_G$, (capacitive), then $Z_R = R_R + jX_R$ and $R_R = R_G$, $|X_R| = |X_G|$. This means, of course, that if Z_G is inductive Z_R must be capacitive, and vice versa.

19.6   Problems

9. An ac generator with an EMF $\mathbf{E}_g = 10\,\text{V}$ and internal impedance $\mathbf{Z}_G = 2 + j5\,\Omega$ supplies power to a variable load impedance. At what value should the load impedance be set so that it will receive maximum power, and how much power will that be?

10. Assume a 50-Ω resistance is connected to the generator terminals of Problem 9. What should be the impedance of a load connected to the terminals of the generator in order that the load may receive maximum power?

11. In Fig. 19.6, $\mathbf{E}_g = 100\,\text{V}$; $R_g = 5\,\Omega$; $R = 95\,\Omega$; $\mathbf{X} = 200\,\Omega$ inductive. What are the components of the load impedance, $\mathbf{Z}_L$, that will receive maximum power?

12. Change the reactance $\mathbf{X}$ of Problem 11 to 200 Ω capacitive and then solve the problem.

13. In Fig. 19.5, interchange the $-j5$ reactance and the 5 Ω resistance and solve for the current in the resistance, using the superposition theorem.

14. Consider the $-j5$ reactance the load, and set up the Thévenin equivalent circuit for the network of Fig. 19.5 and solve for the load current.

15. Set up the Norton equivalent circuit for Fig. 19.5 and determine the load current.

16. The answer to Problem 26 in Chapter 16 says that changing R_3 from 4 Ω to 5 Ω, and X_3 from 2 Ω to 1 Ω will cause maximum power to be delivered to that branch. Show that this impedance must be as indicated, that is $Z = 5 + j1\,\Omega$.

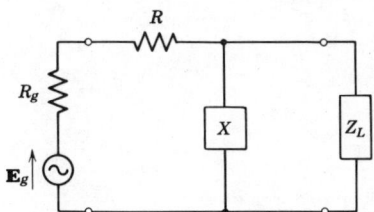

FIGURE 19.6   For Problem 11.

Resonance

Many people have observed that the back-and-forth motion (oscillation) of a child on a playground swing can be greatly increased by the application of small, short pushes that are properly timed. This is an example of *mechanical resonance*.

If we have an electrical circuit containing a resistance R, an inductance L, and a capacitance C connected *in series* and apply a small voltage E from a source that can keep the magnitude of E constant but can vary its frequency, we shall find that the amount of current will change as the *frequency* of E is changed. There will be *one value of frequency* at which *the current is a maximum. Electrical resonance* exists when this condition is reached.

In this chapter we shall discuss this phenomenon (more precisely termed series resonance) and also the situation when a constant voltage E of variable frequency is applied to a *parallel circuit*. The simplest parallel circuit encountered in practice has a coil with an inductance, L, and a resistance, R, connected in parallel with a capacitance C. The resonant condition in this case is called *parallel resonance*, and sometimes *antiresonance*. The latter name is prompted by the fact that *at resonance the input current to the parallel circuit is a minimum.*

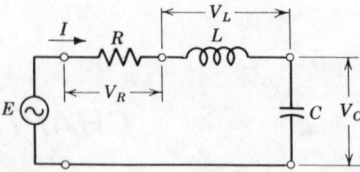

FIGURE 20.1   A series-resonant circuit. The frequency of the applied voltage E is variable; R, L, and C are constant.

20.1   Series-Resonant Circuit

The simple series circuit, containing R, L, and C (Fig. 20.1) is very important in electrical-engineering practice, and, fortunately, the study of its characteristics and response to applied ac voltage of variable frequency is quite interesting. Its most important and useful characteristic is that it offers low impedance to the flow of current of a particular (resonant) frequency, which is determined by the values of L and C, and it offers high impedance to current flow when the frequency of the driving voltage E differs significantly from that particular frequency. The circuit is said to be *resonant* when the *frequency of E is adjusted to produce maximum current while the magnitude of E is kept constant.* The frequency of E, and of this maximum current, is called the *resonant frequency*.

In our study of resonance phenomena we shall keep the voltage applied to the circuit constant and allow the frequency to vary. Then X_L and X_C will vary, X_L increasing with frequency and X_C decreasing with increase in frequency. We already know that when X_L is in series with X_C their *difference* is the net reactance. The curves of Fig. 20.2 show how the reactances change with frequency and that resonant frequency is the value on the frequency scale where $2\pi fL + (-1/2\pi fC) =$ zero.

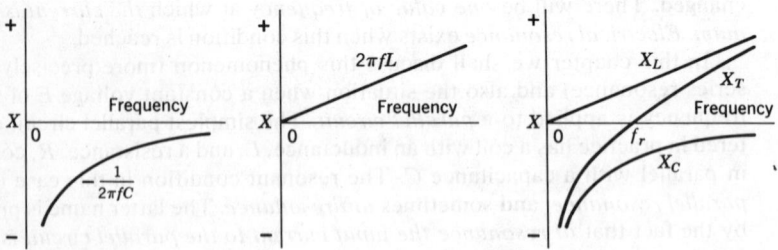

FIGURE 20.2   Variations of reactances with frequency in a series R, L, C circuit. At f_r, the positive and negative reactances add to zero.

The impedance expression for the *RLC*-series circuit is written in the usual manner:

$$\mathbf{Z} = R + j\omega L + \frac{1}{j\omega C} = R + j\left(\omega L - \frac{1}{\omega C}\right) \qquad (20.1)$$

It is evident that for the current to be a maximum, $\mathbf{Z}$ must be a minimum. So, in order to have resonance, the quantity in parenthesis must be zero. We shall let the subscript r designate values at resonance: f_r (resonant frequency), $2\pi f_r = \omega_r$ (resonant *angular frequency*). At *resonance,*

$$\omega_r L - \frac{1}{\omega_r C} = 0, \qquad \text{or } \omega_r L = \frac{1}{\omega_r C} \qquad (20.2)$$

$$\omega_r{}^2 = \frac{1}{LC}, \qquad \text{or } \omega_r = \frac{1}{\sqrt{LC}} \qquad (20.3)$$

The resonant frequency is

$$f_r = \frac{1}{2\pi\sqrt{LC}} \qquad (20.4)$$

where f_r is in Hertz, L in henrys, and C in farads.

Evidently $Z_r = R$ of the series circuit at resonance, and Z is a minimum at resonant frequency. When Z is a minimum, I is a maximum. Let's assign some values to R, L, and C and calculate the current at several different frequencies. We'll choose $E = 1$ V rms; $R = 50\ \Omega$; $L = 159\ \mu$H; $C = 159$ pF. Computations at seven values of frequency give us data for curve ① in Fig. 20.3a. When R is changed to 25 Ω we get the results with which curve ② is plotted. The resonant frequency [Equation (20.4)] comes out to be 1 MHz.

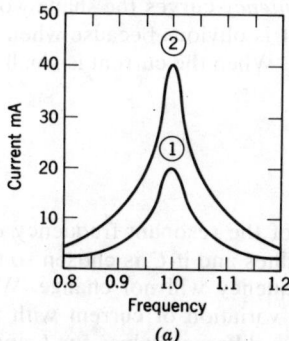

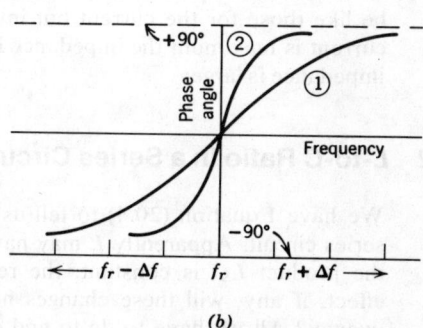

FIGURE 20.3 Series circuit performance as frequency of an applied constant voltage varies through the resonant value. (a) Current when ① $R = 50\ \Omega$, ② $R = 25\ \Omega$. (b) Phase angle between I and E when ① $R = 50\ \Omega$, ② $R = 25\ \Omega$.

TABLE 20.1   **Performance Data for Series Circuit at Frequencies Near Resonance**

f MHz	R ohms	X_L ohms	X_C ohms	$X_L + X_C$ ohms	Z $\sqrt{R^2 + X^2}$	$0°$ $\tan^{-1}(X/R)$	$I = E/Z$ mA
0.80	50	800	−1250	−450	453	−83.7	2.20
0.90	50	900	−1111	−211	217	−76.7	4.61
0.95	50	950	−1053	−103	114	−64.1	8.77
1.00	50	1000	−1000	0	50	0	20.00
1.05	50	1050	− 953	+ 97	108	+62.7	9.26
1.10	50	1100	− 909	+191	197	+75.3	5.07
1.20	50	1200	− 833	+367	370	+82.2	2.70

The results in Table 20.1 show that **at series resonance:**

1. The current is maximum.
2. The impedance is minimum and equal to the circuit resistance.
3. The phase angle between the current and voltage is zero and, therefore, the power factor is unity.
4. The net reactance is capacitive when the frequency is below resonance and inductive when it is above resonance.

Figure 20.3a is a plot ① of current versus frequency for this series RLC circuit, and Fig. 20.3b shows how the phase angle ① between the current and voltage varies above and below resonance. If R is decreased to 25 Ω, the current at resonance becomes twice as large (40 mA) and the sides of the curve are steeper at a given frequency. The phase angle approaches zero more rapidly as the frequency approaches resonance.

If we should plot the *impedance versus frequency* curves the shape would be like those for the current but inverted. That is obvious because when the current is maximum the impedance is minimum. When the current is small the impedance is large.

20.2   L-to-C Ratio in a Series Circuit

We have Equation (20.4) to tell us the value of the resonant frequency of a series circuit. Apparently L may have many values and if C is chosen so that the product LC is constant, the resonant frequency will not change. What effect, if any, will these changes have on the variation of current with frequency? All we have to do to find out is choose different values for L and C and observe what happens to the impedance at various frequencies.

At $f = 1.05$ MHz the impedance is 108 Ω and the current is 9.26 mA. Let's choose L four times its value (636 μH) and C one-fourth its value (39.75 pF).

Calculations give the following results:

$$X_L = 4200 \ \Omega; X_C = -3812 \ \Omega; \quad X_L + X_C = 388 \ \Omega; \quad Z = 392 \ \Omega; I = 2.55 \ \text{mA}$$

The impedance, at the same frequency, was increased when the L/C ratio was increased, and therefore the current was decreased. This happens at all other frequencies with the result that the current curve is narrowed and steepened. We shall soon find that this is desirable in a radio circuit.

Increasing the L/C ratio of a series resonant circuit decreases its input near resonant frequency and increases the quality factor Q.

20.3 Quality Factor Q

The equations related to resonance become more meaningful with the introduction of the quality factor Q of a resonant circuit. One speaks of the Q of a coil. A coil of good quality has a high Q value. What do we mean by a good quality coil? It is one that can store a required amount of energy in its magnetic field with a minimum of energy loss.

Basically, the Q of a coil is a measure of its efficiency of energy storage when it carries alternating current. The mathematical definition is

$$Q = 2\pi \ \frac{\text{maximum energy stored}}{\text{energy dissipated per cycle}} \tag{20.5}$$

Multiplying the numerator and denominator by the frequency f,

$$Q = \omega \ \frac{\text{maximum energy stored}}{\text{average power dissipated}} \tag{20.6}$$

Remember that energy per cycle × cycles per second equals energy per second = power. Another popular name for quality factor is *figure of merit*.

Later we shall talk about the quality factor of a capacitor, so we shall use Q_L or a coil (or inductor) and Q_C for a capacitor.

Q may be expressed in terms of R_L, L, and f. Let I (rms) at f(Hz) flow through the inductor represented in Fig. 20.4a.

The maximum energy stored *per cycle* in the field of a coil is

$$W_M = \tfrac{1}{2}LI_m{}^2$$
$$W_m = \tfrac{1}{2}L(\sqrt{2}I)^2 = LI^2 \tag{20.7}$$

So that the average power, P, dissipated is

$$P = I^2R \tag{20.8}$$

Substituting these last two quantities into Equation (20.6) gives the quality

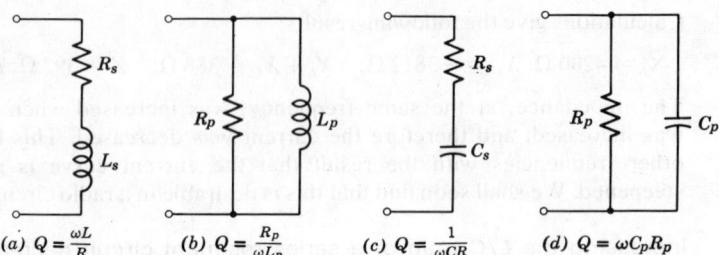

(a) $Q = \dfrac{\omega L}{R}$ (b) $Q = \dfrac{R_p}{\omega L_p}$ (c) $Q = \dfrac{1}{\omega CR}$ (d) $Q = \omega C_p R_p$

FIGURE 20.4 A coil represented by (a) a series network, (b) an equivalent parallel network. A capacitor represented by (c) a series network, (d) a parallel network.

factor of an inductor in terms of L and R:

$$Q_L = \frac{\omega(LI^2)}{I^2 R}$$
$$Q_L = \frac{\omega L}{R}$$

(20.9)

Observe that Q_L depends on frequency, and hence the value of the quality factor of an inductor has meaning only when the frequency is known. At high frequencies, L and R are not necessarily constant, so Q_L does not necessarily vary linearly with frequency.

The quality factor of a series resonant circuit can also be expressed in terms of L and C. When in resonance at frequency f_r the numerator of Equation (20.9) is $\omega_r L$ and from the frequency equation [Equation (20.4)]

$$2\pi f_r = \frac{1}{\sqrt{LC}} = \omega_r$$

Substituting into Equation (20.9) we have

$$Q_r = \frac{\omega_r L}{R} = \frac{L}{R\sqrt{LC}} = \frac{1}{R}\sqrt{\frac{L}{C}}$$

(20.10)

EXAMPLE 20.1 A series circuit has $R = 100\ \Omega$; $L = 100\ \mu H$; $C = 100\ pF$, and has $E = 10\ V$ rms applied to it. What are the (a) resonant frequency, (b) current at resonance, (c) voltages across R, L, and C, (d) the Q of the circuit at resonance?

Solution

(a) $f_r = \dfrac{1}{2\pi(100 \times 10^{-6} \times 100 \times 10^{-12})^{1/2}} = 1.59 \times 10^6\ \text{Hz} = 1.59\ \text{MHz}$

(b) $I = 10/100 = 0.1\ \text{A} = 100\ \text{mA}$

(c) $V_R = 10\ \text{V};\ V_L = (0.1)(2\pi \times 1.59 \times 10^6 \times 100 \times 10^{-6}) = 100\ \text{V}$

    $V_C = (0.1)(10^{12}/(2\pi \times 1.59 \times 10^6 \times 100)) = 100\ \text{V}$

$$\text{(d) } Q = \frac{\omega L}{R} = \frac{2\pi \times 1.59 \times 10^6 \times 100 \times 10^{-6}}{100} = 10$$

Also

$$Q = \frac{1}{R}\sqrt{\frac{L}{C}} = \frac{1}{100}\left(\frac{100 \times 10^{-6}}{100 \times 10^{-12}}\right)^{1/2} = 10$$

EXAMPLE 20.2   What is the maximum instantaneous energy in the magnetic field of the inductor of the circuit of Example 20.1? Calculate the maximum instantaneous energy in the capacitor. Also calculate the average power dissipated in the circuit and, using Equation (20.6), calculate the Q of the circuit.

Solution. Maximum energy in the magnetic field [from Equation (20.7)]:

$$\text{Max. } W_L = LI^2 = 100 \times 10^{-6} \times 0.1^2 = 10^{-6}\,\text{J}$$

Maximum energy in the electric field [from Equation (20.13)] which will next be derived:

$$\text{Max. } W_C = \frac{I^2}{\omega^2 C} = \frac{0.1^2}{(2\pi \times 1.59 \times 10^6)^2 (100 \times 10^{-12})} = 10^{-6}\,\text{J}$$

Average power =

$$P = I^2 R = 0.1^2 \times 100 = 1\,\text{W}$$

$$Q = 2\pi \times 1.59 \times 10^6 \frac{10^{-6}}{1} = 10$$

An inductor that has resistance R and inductance L may be represented by an equivalent parallel circuit, represented in Fig. 20.4b, in which R_p is in parallel with L_p. These may be evaluated in terms of the resistance and inductance of the inductor which may be represented by R_s and L_s, with the subscript indicating that they are in series. The expression[1] for the quality factor in this case is

$$Q = \frac{R_p}{\omega L_p} \tag{20.11}$$

The concept of quality factor may be applied to a capacitor. The efficiency of energy storage is the ratio of the maximum energy stored in the capacitor field to the average power dissipated.

The maximum energy stored is [Equation (12.43)]

$$W_{Cm} = \tfrac{1}{2}CV_m{}^2 = \tfrac{1}{2}C(\sqrt{2}V)^2$$
$$W_{Cm} = CV^2$$
$$V = IX_C = I/\omega C$$
$$W_{Cm} = \frac{CI^2}{\omega^2 C^2} = \frac{I^2}{\omega^2 C} \tag{20.13}$$

[1]The derivation of this is begun with Equation (20.15).

The average power dissipated is I^2R. Substituting this and Equation (20.13) into Equation (20.6),

$$Q = \frac{\omega C I^2}{\omega^2 C^2} \times \frac{1}{I^2 R}$$

$$Q = \frac{1}{\omega CR} \qquad (20.14)$$

We should discuss the relationship between R_p, L_p of Fig. 20.4b to R_s, L_s in Fig. 20.4a. At a given frequency an inductor may also be specified in terms of its *effective shunt* resistive and inductive components, which are represented in Fig. 20.4b. The input impedance, Z_p, of the parallel pair may be separated into two components of its equivalent series pair. This impedance is

$$Z_p = \frac{(R_p)jX_p}{R_p+jX_p} \times \frac{R_p-jX_p}{R_p-jX_p} = \frac{R_p X_p^2}{R_p^2+X_p^2} + j\frac{R_p^2 X_p}{R_p^2+X_p^2} \qquad (20.15)$$

The first term is the resistive component of the equivalent series circuit and the j term is the reactive component, i.e., $Z_p = R_s + jX_s$:

$$Q_L = \frac{\omega L}{R} = \frac{X_s}{R} = \frac{R_p^2 X_p}{R_p^2+X_p^2} \times \frac{R_p^2 X_p^2}{R_p X_p^2}$$

This becomes

$$Q_L = \frac{R_p}{X_p} = \frac{R_p}{\omega L_p} \qquad (20.16)$$

in Fig. 20.4b.

In a similar manner it may be shown that $Q_C = \omega C_p R_p$ as shown in Fig. 19.4d.

Equation (20.15) is an important relationship. It specifies the components of a *series RL* branch that is equivalent to a given parallel *RL* branch.

EXAMPLE 20.3   An inductance of 200 mH is in parallel with a resistance of 1000 Ω. What are the components of an equivalent series branch at 1590 Hz?

Solution

$$X_p = 2\pi \times 1590 \times 0.2 = 2000 \ \Omega$$

$$R_s = \frac{R_p X_p^2}{R_p^2+X_p^2} = \frac{1000 \times 2000^2}{1000^2+2000^2} = 800 \ \Omega$$

$$X_s = j\frac{R_p^2 X_p}{R_p^2+X_p^2} = j\frac{1000^2 \times 2000}{1000^2+2000^2} = j400 \ \Omega$$

Let's see if the Q's of the two equivalent networks are the same. For the original parallel pair of components,

$$Q_p = \frac{R_p}{\omega L_p} = \frac{1000}{2000} = 0.5$$

And for the equivalent series pair,

$$Q_s = \frac{\omega L_s}{R_s} = \frac{400}{800} = 0.5$$

Evidently the resistances are much too large to give values of Q that would be acceptable in a radio circuit. In most such applications the Q should be at least 10 and preferably much larger.

20.4   Resonant Rise in Voltage in a Series Circuit

The data in Table 20.1 are for a series R, L, C circuit to which 1 V, at variable frequency, is applied. At resonance the voltage across the inductance is equal in magnitude and opposite in phase to the voltage across the capacitance. Each is calculated as the product of current and reactance.

The voltage across the inductance is

$$V_L = -j0.020\,(j1000) = j20 \text{ V}$$

and the voltage on the capacitance is

$$V_C = 0.020\,(-j1000) = -j20 \text{ V}$$

The Q_r of the circuit on which Table 20.1 is based is

$$Q_r = \frac{\omega L}{R} = \frac{1000}{50} = 20$$

Note that V_L is 20 times the input voltage. So is V_C. The ratio of these voltages to the input voltage can easily be shown equal to Q.

$$V_L = I\omega L$$

At resonance, $I = V/R$, therefore,

$$V_L = \frac{\omega L V}{R} = QV \tag{20.17}$$

Likewise

$$V_C = I/\omega C$$

At resonance, $I = V/R$, therefore,

$$V_C = \frac{V}{\omega CR} = QV \tag{20.18}$$

These voltages (V_L and V_C), being equal and opposite in phase, cancel each other as shown in Fig. 20.5 but their values are 20 times the applied voltage. Had the inductance been twice the value chosen ($2 \times 159 \times 10^{-6}$) and the capacitance one-half the value chosen ($0.5 \times 159 \times 10^{-12}$) the voltages would have been 40 V each instead of 20. This amount of rise is quite realizable.

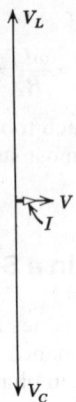

FIGURE 20.5   Voltages and current in a series resonant circuit. R   assumed zero, I ≃ 0.

20.5   Tuning a Series Resonant Circuit

Series resonant circuits in radio equipment are tuned by means of a variable capacitor. In order to bring in a station it is necessary to vary the capacitance until its reactance at that frequency becomes equal in magnitude to the inductive reactance of the coil that is in series with it. The voltage of the desired signal will then send current at its own frequency through the circuit and it will be considerably larger than currents at any other frequency.

Signals from broadcast stations operating at frequencies in adjacent channels will also appear as voltages applied to the same series circuit. If the circuit is adequately *selective*, that is, if its Q_r is large enough, the current versus frequency curve will be steep enough on both sides of the peak value so that the unwanted currents from the other stations will be too small to cause interference with reception of the resonant-frequency signal.

When a radio signal voltage is induced in the coil of a series circuit tuned to the frequency of the signal voltage, the voltage across the variable capacitor is "picked off" and applied to the input of a voltage amplifier as indicated in Fig. 20.6. This capacitor voltage, which is Q times the induced voltage in L, has the same frequency. Hence, a few microvolts of original signal is converted to perhaps 100 or more times its value. At resonance, the induced voltage, E, in L minus the current times R is equal to V_C, the voltage across the capacitor, and $V_C = QE$. The amplifier receives the greatly increased signal voltage and increases it many times more.

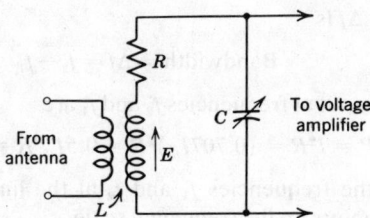

FIGURE 20.6   Tuned series resonant circuit. *E* represents voltage induced in secondary coil *L*.

20.6   Bandwidth

Consider the current-versus-frequency curve in Fig. 20.7, the shape of which is justified by the discussion related to Fig. 20.3*a*. Currents at frequencies just above and just below resonant frequency will flow in the circuit and produce voltages that have effects of their own. Therefore a certain *band of frequencies* will be represented in the amplified signal. This is desirable because sound and speech is made up of a group of frequencies, so the radio receiver must respond to signals within a justifiable *bandwidth*.

There are upper and lower boundaries of the bandwidth; their frequency values are denoted by f_1 and f_2. A reasonable ratio of the current I at each boundary frequency to the maximum current has been chosen so that

$$I = \frac{I_m}{\sqrt{2}} = 0.707 I_m$$

so that

$$I_{f_1} = I_{f_2} = 0.707 I_m$$

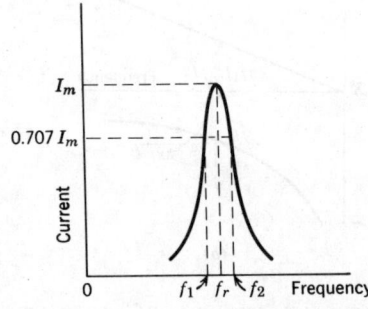

FIGURE 20.7   Current-frequency curve of series resonant circuit. Half-power points on curve are at f_1 and f_2, where instantaneous current = 0.707 *I* max and power is 0.5 *P* max. Bandwidth = $f_2 - f_1$.

and the bandwidth Δf is

$$\text{Bandwidth} = \Delta f = f_2 - f_1 \tag{20.19}$$

The actual power input at frequencies f_2 and f_1 are

$$P = I^2R = (0.707I_m)^2R = 0.5I_m^2R = 0.5\,P_m$$

This means that the frequencies f_1 and f_2 at the limits of the bandwidth are called *half power points* on the frequency scale.

The impedance of the tuned circuit must be $\sqrt{2}$ times its impedance at resonance in order that the current can be $I_m/\sqrt{2} = 0.707I_m$. But the impedance at resonance is $Z = R$, so *at the half power points the impedance is 1.414R.* Since $Z = \sqrt{R^2 + X^2}$, if we let $X = R$ we have $Z = \sqrt{2R^2} = \sqrt{2}R$ which is what we want. This is shown in Fig. 20.8a.

Now let us consider the changes in X_L and X_C while the frequency changes from f_1 to f_2, that is, over the bandwidth. At f_r their absolute values are equal. Figure 20.8b shows that over a narrow band of frequencies the magnitude of X_C at f_2 is, for all practical purposes, equal to the magnitude of X_L at f_1 owing to the fact that over the short frequency range, X_C varies almost linearly as does X_L. At f_2, $X_L = 2\pi f_2 L$ and $X_C = 1/2\pi f_2 C$. Their difference is the net reactance at f_2. But we justified taking the magnitude of X_C at f_2 equal to the magnitude of X_L at f_1. Thus, the vertical leg of the impedance diagram in Fig. 20.8c which should represent a magnitude $2\pi f_2 L_2 - 1/2\pi f_2 C$ can be expressed as $2\pi f_2 L - 2\pi f_1 L$ and this must equal R at the half-power point.

$$2\pi f_2 L - 2\pi f_1 L = R, \qquad \text{or,} \quad 2\pi L(f_2 - f_1) = R$$

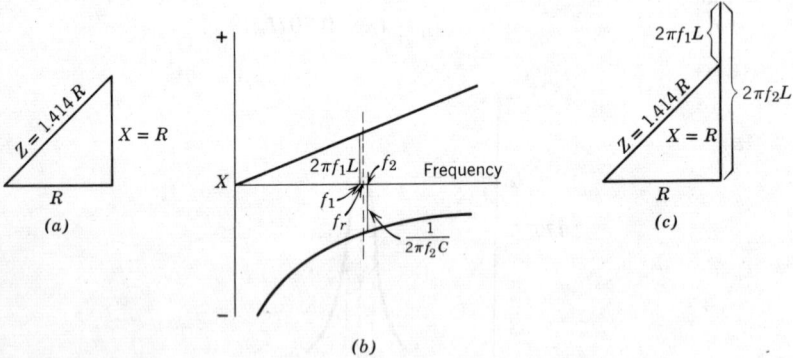

FIGURE 20.8 Conditions at frequencies that determine bandwidth. (a) Magnitude of $Z = \sqrt{2}R$ at half power points. (b) $f_1 = f_r - \Delta f/2$, $f_2 = f_r + \Delta f/2$, and the net reactance X equals the resistance R. $X_1 = 2\pi f_1 L = 1/2\pi f_2 c = X_2$. This is used in (c).

so that

$$\Delta f = \frac{R}{2\pi L}$$

$$\frac{\Delta f}{f_r} = \frac{R}{2\pi f_r L} = \frac{R}{X_L} = \frac{1}{Q}$$

$$Q = \frac{f_r}{\Delta f} \qquad (20.20)$$

This means that increasing the Q of a resonant circuit narrows the bandwidth and thus improves the selectivity, i.e., sharpens the tuning. We recall from Equation (20.18) that it also increases the resonant rise in voltage.

EXAMPLE 20.4 Calculate the bandwidth of the circuit of Example 20.1 and the voltage across the tuning capacitor if the applied voltage at resonance is 50 μV.

Solution

$$\Delta f = \frac{f_r}{Q} = \frac{1.59}{10} = 0.159 \text{ MHz} = 159,000 \text{ Hz}$$

$$V_C = QV = 10 \times 50 \ \mu\text{V} = 0.5 \text{ mV}$$

Problems

1. A series resonant circuit has $R = 10\,\Omega$, $L = 50\,\mu$H, and a variable capacitance C. When C is set at 2.45 nF, what is the resonant frequency? Calculate the Q of the circuit at resonance and the voltage across the capacitor when 1 V is applied to the circuit. Draw the phasor diagram.

2. A series resonant circuit has $R = 100\,\Omega$, $L = 50\,\mu$H, and C is set at 4.9 nF. Calculate the resonant frequency, the voltage across the capacitor, and the circuit Q.

3. What should be the setting of the capacitance in the circuit of Problem 1 to make it resonate at 750 KHz? Calculate Q and V_C.

4. What should be the setting of the capacitor in Problem 2 to provide resonance at 1 mHz? Calculate Q and V_C.

5. A variable capacitor has $C_{min} = 50$ pF, $C_{max} = 300$ pF, and is in series with an inductor that has $12\,\Omega$ resistance. Assume the capacitor is set at its middle value: 175 pF. What value of L will produce resonance at 1 MHz?

6. Calculate the bandwidth of the circuit of Problem 5 when it is resonant at the frequency specified. What is the voltage across the capacitor?

7. Using the calculated value of L from Problem 5, calculate C required to produce resonance at 550 KHz and at 1550 KHz. Also calculate the Q at each frequency. Commercial AM radio stations broadcast in this range of frequencies.

8. Determine the bandwidths in Problem 7, and the voltage across the capacitor at each frequency.

9. A leaky capacitor is represented by the parallel circuit in Fig. 20.4*d*. $R_p = 250\ k\Omega$, $C_p = 0.0005\ \mu F$. What are the R_s and C_s components of the equivalent series circuit at 3184 kHz?

20.7   Parallel Resonant Circuits

The most common form of parallel resonant circuit in practical use is represented in Fig. 20.9. As indicated by the absence of leakage resistance in the capacitor (no R_C is present), we assume the capacitor is of good quality and its leakage current, if any, is negligible at the operating voltages.

20.8   Input Impedance

The input impedance, in the general case, is obtained by means of the familiar product-over-sum method.

$$Z = \frac{(R + j\omega L)(-j1/\omega C)}{R + j\omega L - j1/\omega C} = \frac{(1 + R/j\omega L)\ L/C}{R + j(\omega L - 1/\omega C)} \qquad (20.21)$$

Parallel-resonant circuits are designed so that they have as little loss as possible, which means the highest Q possible is desired. At frequencies at and near resonant value, R is very small compared with ωL, so that the numerator becomes L/C, and we have, to a very good approximation,

$$Z = \frac{L}{C} \cdot \frac{1}{R + j(\omega L - 1/\omega C)} \qquad (20.21)$$

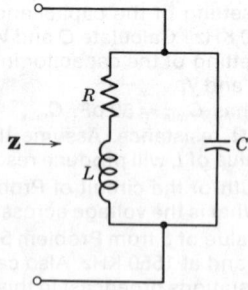

FIGURE 20.9   Representing a coil in parallel with a capacitor.

When resonance exists the *power factor is unity* and the inductive reactance effect cancels the capacitive reactance effect, so we get the following expression for the **input impedance at resonance**:

$$Z_r = \frac{L}{CR}$$

in which Z_r is the ohms *input impedance at resonance* of a coil and high-quality capacitor connected in parallel. L is the coil inductance in henrys, and C is the capacitance in farads. Obviously this impedance is a pure resistance because there is no frequency term present. This relation is very useful.

Let's first multiply the right side of Equation (20.23) by ω_r in both numerator and denominator:

$$Z_r = \frac{\omega_r L}{\omega_r CR}$$

Since $\omega_r L = 1/\omega_r C$ at resonance,

$$Z_r = \frac{\omega_r^2 L^2}{R} = \frac{1}{\omega_r^2 C^2 R} \qquad (20.23)$$

The Q of the two-branch circuit should now be considered. From the standpoint of the definition of Q in terms of the ratio of energy stored to energy expended, what difference should a change of terminals make when in one instance the R, L, and C form a series circuit and in the other a parallel circuit? This suggests the same expression for Q as we had for the series circuit, and this is true.

The Q of a parallel resonant circuit made up of an inductance coil, with resistance r, and a capacitor, at resonant frequency, is

$$Q_r = \frac{\omega_r L}{R} \qquad (20.24)$$

Using this relation in Equation (20.23), we get some convenient and useful expressions for the input impedance of a parallel-resonant circuit,

$$Z_r = (\omega_r L)Q_r = \frac{1}{\omega_r C}Q_r = RQ_r^2 \qquad (20.25)$$

These relations, which hold very well when Q_r is greater than 10 (preferably 20 or larger), are worth memorizing. That is, *the input impedance of a two-branch, parallel-resonant circuit is equal to Q_r times the reactance of either branch, or it is also equal to the Q_r^2 times the resistance in the loop.*[2] That this is applicable to high-Q circuits is an important observation here.

---

[2]This assumes there is no "load" connected across the parallel-resonant circuit. If a voltmeter or any other kind of load, such as a resistor at the input stage of an amplifier, is so connected, the Q could be lowered appreciably. The load resistance should be at least ten times Z_r to prevent this.

20.9 Frequency of Resonance of Parallel Circuit

We have already made use of the fact that the resonance requirement for the coil-capacitor parallel circuit is that X_L and X_C be equal in magnitude. This brings us to the familiar equation for resonant frequency of a series circuit [Equation 20.4]. This holds also for the high-Q parallel circuit (Fig. 20.9),

$$f_r = \frac{1}{2\pi\sqrt{LC}} \tag{20.26}$$

in which L is the inductance of a coil connected in parallel with a capacitance C.

EXAMPLE 20.5 A coil for which $L = 0.2\,\mathrm{H}$, $R = 4\,\Omega$ is in parallel with a capacitor that has $C = 0.8\,\mu\mathrm{F}$ and no leakage. Determine the (a) resonant frequency, (b) Q at resonance, (c) input impedance at resonance.

Solution

$$f_r = \frac{1}{2\pi\sqrt{LC}} = \frac{1}{2\pi\sqrt{0.2 \times 0.8 \times 10^{-6}}} = 398\,\mathrm{Hz}$$

$$Q_r = \frac{\omega_r L}{R} = \frac{2\pi \times 398 \times 0.2}{4} = 125$$

$$Z_r = \frac{L}{CR} = \frac{0.2}{0.8 \times 10^{-6} \times 4} = 62{,}500\,\Omega$$

Note how very much higher this impedance is than the coil resistance. At parallel resonance the input impedance is a maximum. This value of Z_r is obtainable from all other forms of Equation (20.26).

$$\omega_r L Q_r = 2500 \times 0.2 \times 125 = 62{,}500$$
$$Q_r/\omega_r C = 125/(2500 \times 0.8 \times 10^{-6}) = 62{,}500$$
$$Z_r = R Q_r{}^2 = 4 \times 125^2 = 62{,}500\,\Omega$$

20.10 Resonant Rise of Current

The current in the capacitor of the simple parallel circuit of Fig. 20.9 is related to Q_r. Recall that we found the *capacitor voltage* of a series circuit at resonance given as $V_C = Q_r V$ in which V was the total voltage applied to the circuit. We shall find the amount of *capacitor current* given by Q_r times the input current to the parallel-resonant network.

Let V be the voltage applied to the capacitor and coil connected in parallel. The total current at resonance is

$$I = \frac{V}{Z_r} \tag{20.27}$$

The capacitor current is

$$I_C = \frac{V}{1/\omega C} = \omega_r C V$$

$$\frac{I_C}{I} = \frac{\omega_R C V}{V/Z_r} = \omega_r C Z_r$$

From Equation (20.25), $\omega_r C Z_r = Q_r$, therefore,

$$I_C = QI \tag{20.28}$$

In high-Q parallel circuits I is much smaller than I_C, so that I_C is practically equal to I_L, the coil current.

From Equations (20.27) and (20.28),

$$\frac{V}{X_C} = Q\frac{V}{Z_r}$$
$$\therefore\ Z_r = QX_C \tag{20.29}$$

20.11 A Coil as a Resonant Circuit

It is possible for a coil to act as a resonant circuit all by itself. This is because *distributed capacitance* is present in high-frequency operation. Parallel, insulated conductors (the turns and layers of a coil) have capacitance. This capacitance value can be measured, at operating frequency, by means of a capacitance bridge. It is represented by the schematic of Fig. 20.10a. As a circuit element it is represented by part b, in which the measured value of distributed capacitance is shown "lumped" in parallel with the coil inductance and resistance.

At low frequency a coil acts as an inductance of practically constant value. Both resistance and inductance increase with frequency as self-resonance is approached. At self-resonant frequency the coil appears as a pure resistance and at above-resonance frequencies it appears as a capacitance in the circuit with resistance also present.

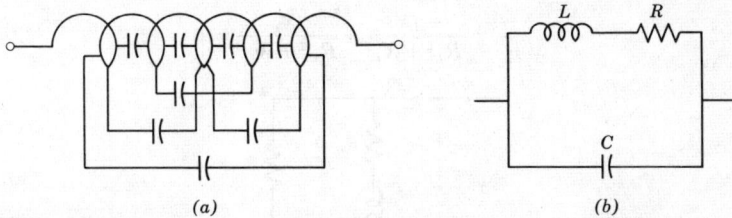

(a)   (b)

FIGURE 20.10 (a) Small capacitances between turns of a coil combine to make up "distributed capacitance." (b) Schematic diagram representing a coil in an ac circuit operating at high frequency. The total distributed capacitance is represented as a "lumped" capacitance C.

Problems

10. An inductance $L = 25.3$ H and a capacitance $C = 10^{-9}$ F are in parallel. Resistance is assumed zero in each branch. A voltage $E = 1$ V at resonant frequency is applied. Determine the (a) resonant frequency, (b) input impedance, (c) Q, (d) current in the capacitance, (e) current in the inductance, (f) energy loss per cycle.

11. Assume a 10-Ω resistor is in series with L in Problem 10. Solve for the quantities specified. Also determine the input current and the power loss.

12. In Problem 11, the capacitor current at resonance should equal Q times the input current. See if it does.

13. A coil that has an inductance L and a resistance of $8\,\Omega$ is in parallel with a standard variable capacitance. At $\omega = 1.5 \times 10^6$ rad/s, resonance is achieved when $C = 0.0037\,\mu$f. What is the inductance of the coil?

14. An inductor for which $L = 40$ mH is to be used in parallel with a 1 μF capacitor in a circuit that must have Q at least as large as 25 when the frequency is 1 kHz. What is the minimum amount of resistance the coil may have? What will the input impedance be when the minimum R is used?

20.12 Resistance in Both Branches of a Parallel Circuit

The circuit shown in Fig. 20.11 represents the general case for parallel resonant circuits. We shall develop the equation for resonant frequency. Both R_L and R_C will appear in it. The equation should be used when either, or both, are appreciable with respect to the L-to-C ratio and, therefore, should not be considered negligible. If one or the other resistance values happens to be zero, or is small enough to be neglected, it may be ignored. If *both R_L and R_C satisfy* this condition, the equation for resonant frequency reduces to $f_r = \frac{1}{2}\pi\sqrt{LC}$, which we have been using up to this point.

An exact equation for use in computing resonant frequency when there is resistance in both branches of a parallel-resonant circuit is readily obtained. Consider Fig. 20.11. The susceptances of the branches will be equal, but of opposite sign. Their sum will be zero:

$$\frac{-jX_L}{R_L{}^2 + X_L{}^2} + \frac{jX_C}{R_C{}^2 + X_C{}^2} = 0$$

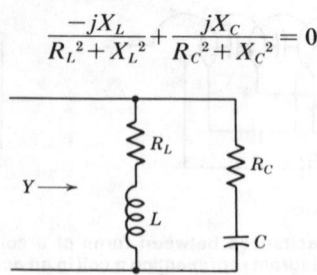

FIGURE 20.11   A parallel-resonant circuit with resistance in both branches.

Canceling j and inserting resonant angular frequency ω_r

$$\frac{\omega_r L}{R_L{}^2 + \omega_r{}^2 L^2} = \frac{1/\omega_r C}{R_C{}^2 + 1/\omega_r{}^2 C^2} = \frac{\omega_r C}{\omega_r{}^2 R_C{}^2 C^2 + 1}$$

From the first and third forms

$$\omega_r{}^2 R_C{}^2 L C^2 + L = R_L{}^2 C + \omega_r{}^2 L^2 C$$

$$\omega_r{}^2 (R_C{}^2 L C^2 - L^2 C) = R_L{}^2 C - L$$

$$\omega_r{}^2 = \frac{R_L{}^2 C - L}{R_C{}^2 L C^2 - L^2 C} = \frac{R_L{}^2 - L/C}{R_C{}^2 L C - L^2}$$

$$\omega_r{}^2 = \frac{1}{LC} \left(\frac{R_L{}^2 - L/C}{R_C{}^2 - L/C} \right)$$

$$f_r = \frac{1}{2\pi \sqrt{LC}} \sqrt{\frac{R_L{}^2 - L/C}{R_C{}^2 - L/C}} \tag{20.30}$$

It is evident that if $R_L{}^2$ and $R_C{}^2$ are very small compared with L/C, this equation reduces, practically, to Equation (20.4).

Let us consider practical values of L and C for a circuit that would resonate at 10 kHz: $L = 80$ mH, $C = 0.00316\,\mu$F. With these values $L/C = 25.3 \times 10^6$. Apparently R_L or R_C can be quite large and the square of either may be quite large and still be small compared with L/C. The resistance can be as much as 1000 Ω, and R^2 will still be less than $\frac{1}{25}$ of L/C. Using $R_L = 1000\ \Omega$, $R_C = 0$, the resonant frequency computed with Equation (20.30) turns out to be $f_r = 9600$ Hz. Assuming this to be the exact value, the approximate value of 10,000 Hz, given by Equation (20.4), is 4.16 percent high.

By rather involved analytical procedures the bandwidth of the parallel resonant circuit can be shown to be

$$\text{Bandwidth} = f_2 - f_1 = \frac{f_r}{Q}\ \text{Hz} \tag{20.31}$$

Q_r of this circuit (reactance of either branch divided by the total resistance) is $(0.08 \times 2\pi \times 10^4)/10^3 = 5$. Obviously this would be increased to 200 if R_L were a more practical 25 Ω. In that case, Equation (20.30) would give a value for f_r which would be less than one one-hundredth of 1 percent larger than the 10,000 cps value obtained from the approximate equation.

An interesting case occurs when $R_C = R_L = \sqrt{L/C}$ in Equation (20.30). This produces 0/0 under the radical, and the value of f_r, the resonant frequency, is indeterminate. It turns out, however, that *the input current at any frequency is in phase with the applied voltage, so that the circuit is resonant at all frequencies*. The in-phase nature of the input current can be shown by proving that the susceptance components of the admittances of the two branches are equal and opposite in sign. These are:

$$B_1 = \frac{-\omega L}{R^2 + \omega^2 L^2} \quad \text{and} \quad B_2 = \frac{\omega C}{\omega^2 R^2 C^2 + 1} \tag{20.31}$$

in which R represents both R_L and R_C. Substituting $C = L/R^2$ in the B_2 expression converts it into the expression for B_1, but with positive sign.

EXAMPLE 20.6   Assume the following constants for the circuit of Fig. 20.11: $R_C = 100\ \Omega$, $R_L = 200\ \Omega$, $L = 0.2$ H, $C = 0.5\ \mu$F. Calculate the resonant frequency by the exact equation and compare the result with that obtained with the approximate equation for this four-element parallel network.

Solution

$$f_r = \frac{1}{2\pi\sqrt{0.2 \times 0.5 \times 10^{-6}}} \sqrt{\frac{200^2 - 0.2/(0.5 \times 10^{-6})}{100^2 - 0.2/(0.5 \times 10^{-6})}}$$

$$= \frac{1000}{2\pi\sqrt{0.10}} \sqrt{\frac{40{,}000 - 400{,}000}{10{,}000 - 400{,}000}}$$

$$= 503 \sqrt{\frac{-360{,}000}{-390{,}000}} = 483\ \text{Hz}$$

The approximate equation gives $f_r = 503$ Hz. This is only 4.14 percent high.

20.13   Q of a Parallel Circuit

The Q of a parallel resonant circuit is calculated by the reactance of either arm divided by the sum of the resistances in both arms. In the circuit for Example 20.6, the sum of R_L and R_C is 300 Ω and $Q = 2\pi \times 483 \times 0.2/300 = 2$, which is too low for many applications. In a problem at the end of the chapter we shall encounter a circuit with a suitable Q for communication-circuit applications.

Parallel-resonant circuits are generally used to obtain relatively large voltages at a particular frequency, or relatively large currents. The voltage may be "picked off" the capacitor terminals by means of a "coupling capacitor" as in electronic amplifiers, or the high energy in the field of the inductance coil may be "magnetically coupled" by mutual induction into an external circuit, as in inductance heating. In either, and in similar cases, the resonant circuit is said to be loaded. Figure 20.12b properly represents Fig. 20.12a with a load resistor R_1 not connected, R_r is the input impedance of the original circuit at resonance. The impedance of the L and C pair is infinite, so the total impedance presented to the generator is only R_r. However, there will be voltage across L and C.

Let us now imagine that an amplifier stage, or some other *load*, is coupled to the original circuit of Fig. 20.12a. Its input impedance, represented by R_1, would be shown connected in parallel with R_r, L and C of Fig. 20.12b. It would then be desirable to combine R_r and R_1 into a single branch using the "product over the sum" rule to obtain R_e, their equivalent. An example will now be

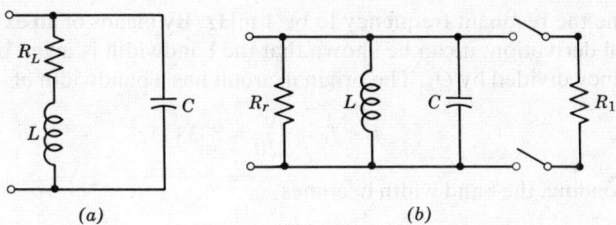

(a)   *(b)*

FIGURE 20.12   *(a)* A parallel-resonant circuit. *(b)* A circuit equivalent to part *a* with a load resistance R_1 to be added.

given to illustrate this, and to demonstrate the effect of such loading on band width and on Q.

EXAMPLE 20.7   The two-branch circuit of Fig. 20.13 has a 60,000-Ω load connected to it. The Q of the two branches is $Q_r = \frac{1200}{10} = 120$ and their input impedance is

Solution

$$Z_r = Q_r X_L = Q_r X_C = 120 \times 1200 = 144,000$$

When the 60,000-Ω resistance load is added, the circuit is represented as shown Fig. 20.12*b*. R' is given by

$$R' = \frac{144,000 \times 60,000}{144,000 + 60,000} = 42,350 \ \Omega$$

This reduces the Q to

$$Q' = \frac{42,350}{1200} = 35.3$$

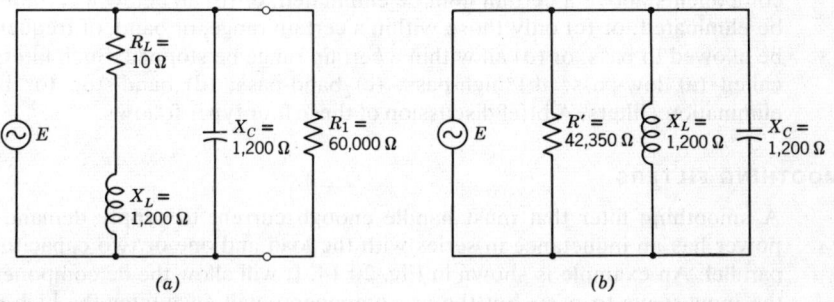

(a)   *(b)*

FIGURE 20.13   *(a)* A parallel-resonant circuit with loading. *(b)* Circuit equivalent.

Assume the resonant frequency to be 1 mHz. By means of an extensive mathematical derivation, it can be shown that the bandwidth is given by the resonant frequency divided by Q_r. The original circuit has a bandwidth of

$$f_2 - f_1 = \frac{10^6}{120} = 8333 \text{ Hz}$$

With loading, the band width becomes

$$f_2' - f_1' = \frac{10^6}{35.3} = 2.83 \times 10^4 \text{ Hz}$$

The bandwidth is nearly three and one-half times the no-load value. The sharpness of tuning is, evidently, very materially reduced. In order to maintain a desired bandwidth *with loading*, the load impedance must be as high as practicable. Of course, it will always be larger than the bandwidth without loading unless the impedance of the load is infinite, which means there is no output current.

20.14   Simple Filter Networks

When a dc voltage is needed, which is the case in many applications such as electronic circuits using transistors and/or vacuum tubes, for example, it is usually obtained by *rectifying* an ac voltage. This means that a system called a *rectifier* accepts ac power and delivers dc power. The dc output voltage of the rectifier is not steady because it contains halves of sine waves following one another. This wave form consists of a dc component and many ac components that give the wave form its half-sine-wave shape. A "smoothing filter" takes out the ac components and lets the dc component pass on. We shall learn about the action of such a filter in this section.

It is common in communications networks to require that (a) all frequency components above a certain limit be eliminated; or (b) all below a certain limit be eliminated; or (c) only those within a certain range, or band, of frequencies be allowed to pass; or (d) all within a certain range be stopped. Such filters are called (a) low-pass; (b) high-pass; (c) band-pass; (d) band-stop (or band-elimination) filters. A brief discussion of these four types follows.

SMOOTHING FILTERS

A smoothing filter that must handle enough current to supply demands for power has an inductance in series with the load and one or two capacitors in parallel. An example is shown in Fig. 20.14. It will allow the dc component of the input wave to pass, but the ac components will encounter the high reactance of L. Those that get through will be greatly reduced in amplitude. These

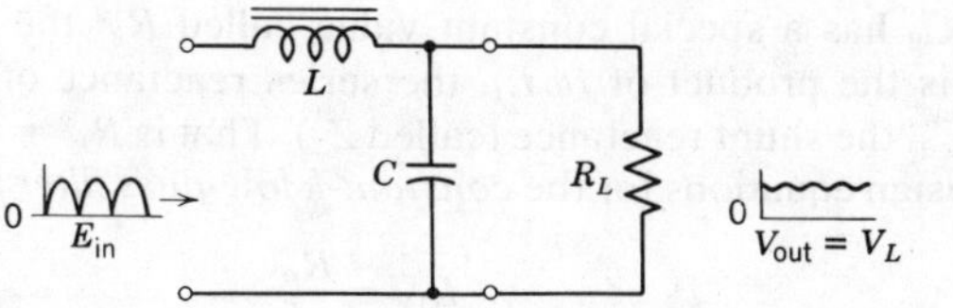

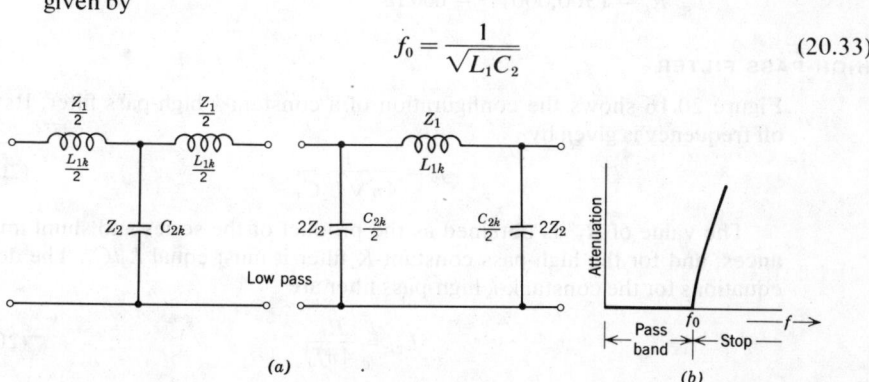

FIGURE 20.14   Smoothing filter section. V_m is a rectified sine wave. V_{out} has very small ripple.

will be bypassed by the relatively low reactance of C. In practice, a voltage at that frequency exists across C, but it is very much smaller than the amount that exists at the filter input terminals. The efficiency of the filter is increased when a second capacitor is connected across the lines ahead of L. The smoothing filter is of the *low-pass* type. That is, it passes low-frequency currents and greatly attenuates (reduces) high-frequency currents.

Design equations for a smoothing filter contain symbols representing ripple factor, smoothing factor, and load resistance. The power supply of a radio receiver may have an inductance of 10 H and a capacitance of 8 μF or more.

When very little power is required, as is the case when a high dc voltage with only a few microamperes is needed, a resistor is used in place of an inductance. This is called an *RC* filter.

LOW-PASS FILTER IN COMMUNICATION CIRCUITS

The currents and voltages needed in telephone circuits are very small. Frequencies are, basically, in the *audio range* (approximately 30–15,000 Hz). A low-pass filter section may be of the Tee or Pi type as shown in Fig. 20.15. The upper limit of the *pass band* is called the *cut-off frequency*, f_0, and it is given by

$$f_0 = \frac{1}{\sqrt{L_1 C_2}}$$
(20.33)

FIGURE 20.15   Low-pass constant k filter. (a) Schematics of T and π prototypes. (b) Performance curve.

When L_1/C_2 has a special constant value called $R_k{}^2$ the filter is a *constant-k* type. $R_k{}^2$ is the product of $j\omega L_1$, the series reactance of the filter (called Z_1) and $-j/\omega C_2$, the shunt reactance (called Z_2). That is $R_k{}^2 = Z_1 Z_2 = L_1/C_2$.

The design equations for the *constant-k low-pass filter* are

$$L_{1k} = \frac{R_k}{\pi f_0} \tag{20.34}$$

$$C_{2k} = \frac{1}{\pi f_0 R_k} \tag{20.35}$$

$$f_0 = \frac{1}{\pi\sqrt{L_{1k}C_{2k}}} \tag{20.36}$$

$$R_k = \sqrt{\frac{L_{ik}}{C_{2k}}} \tag{20.37}$$

EXAMPLE 20.7   A low-pass filter in the form of a T section has $L_{1k}/2 = 0.0955$ H, $C_{2k} = 0.531\ \mu$F. What are the cut-off frequency and the product of its total series and shunt impedance?

$$f_0 = \frac{1}{\pi\sqrt{2 \times 0.0955 \times 0.0531 \times 10^{-6}}} = 1000 \text{ Hz}$$

$$R_k = \sqrt{(0.191/0.531) \times 10^6} = 600\ \Omega$$

Note that

$$\begin{aligned}
Z_1 Z_2 &= (j2\pi f L_{1k})(-j/2\pi f C_{2k}) = R_k{}^2 \\
&= (j2\pi \times 1000 \times 0.191)[-j/(2\pi + 1000 \times 0.531 \times 10^{-6})] \\
&= 0.191/(0.531 \times 10^{-6}) = 360{,}000 = R_k{}^2 = L_{1k}/C_{2k} \\
R_k &= (360{,}000)^{1/2} = 600\ \Omega
\end{aligned}$$

HIGH-PASS FILTER

Figure 20.16 shows the configuration of a constant-*k* high-pass filter. Its cut-off frequency is given by

$$f_0 = \frac{1}{4\pi\sqrt{L_2 C_1}} \tag{20.38}$$

The value of $R_k{}^2$ is obtained as the product of the series and shunt impedances, and for the high-pass constant-*K* filter it must equal L_2/C_1. The design equations for the constank-*k* high-pass filter are

$$L_{2k} = \frac{R_k}{4\pi f_0} \tag{20.39}$$

$$C_{1k} = \frac{1}{4\pi f_0 R_k}$$

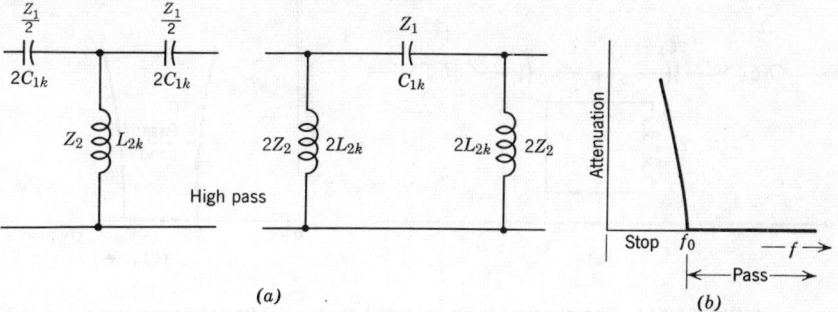

FIGURE 20.16   High-pass constant-k filter. (a) Schematics of T and π prototypes. (b) Performance curves.

EXAMPLE 20.8   A high-pass, constant-k filter is to have $R_k = 500\ \Omega$ and a cut-off frequency of 20 kHz. What must be the capacitance of each arm of a T-section and the inductance of the shunt arm? Also specify the arms of the filter when it has the π-section configuration.

Solution

$$C_{1k} = \frac{1}{4\pi f_0 R_k} = \frac{1}{4 \times 20{,}000 \times 500\pi} = 0.796 \times 10^{-8}\ \text{F}$$

$$L_{2k} = \frac{R_k}{4\pi f_0} = \frac{500}{4 \times 20{,}000\pi} = 1.99 \times 10^{-3}\ \text{H}$$

T-section: Series arms have $2C_{1k} = 1.592 \times 10^{-8}$ F
             Shunt arm has $L_{2k} = 1.99$ mH
π-section: Shunt arms have $2L_{2k} = 3.98$ mH
             Series arm has $C_{1k} = 0.796 \times 10^{-8}$ F

20.15   Band-Pass and Band-Stop Filters

A band-pass filter is a network that attenuates (decreases the magnitudes of) currents of all frequencies outside a certain specified band width. One way to accomplish this is to connect a low-pass filter (cut-off frequency $= f_{ol}$) in cascade with a high-pass filter (cut-off frequency $= f_{oh}$, which is lower than f_{ol}). The combination theoretically will stop all currents with frequencies below f_{oh} and all frequencies above f_{ol}, giving a pass-band width of $f_{ol} - f_{oh}$. The zero attenuation will, lie, naturally, between f_{oh} and f_{ol}. Be sure to note, here, that $f_{ol} > f_{oh}$ in this case.

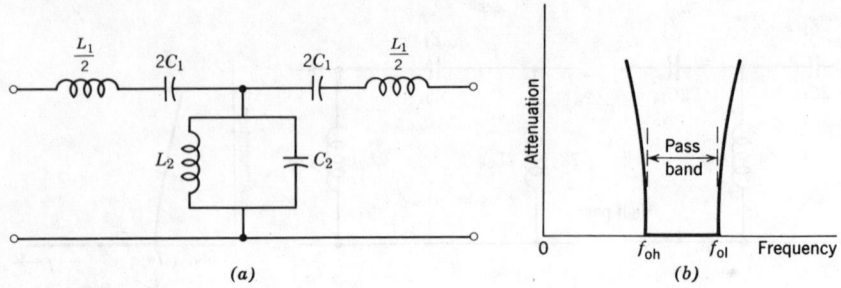

FIGURE 20.17   The band-pass filter. (a) Schematic. (b) Performance curve.

A single section may be designed to have band-pass properties. See Fig. 20.17. This network is classified as a *constant-k, band-pass filter* if $L_2C_2 = L_1C_1$, because it is then true that

$$Z_1Z_2 = \frac{L_2}{C_1} = \frac{L_1}{C_2} = \text{a constant}$$

There are resonant circuits and an antiresonant circuit in Fig. 20.17. Currents having frequencies just above and just below the series-resonant frequency of $L_1/2$ and $2C_1$ will pass with little attenuation, whereas those outside this frequency range will be greatly attenuated. The parallel-resonant (antiresonant) circuit formed by L_2 and C_2 will offer high impedance to currents in the pass band because the resonant frequency for L_2 and C_2 is near the middle of the pass band. Currents with frequencies outside this band will find a low-impedance through this shunt arm *and will not reach the load.*

Figure 20.18 shows a *band-stop* filter section, which is also a *constant-k* type when $L_1C_1 = L_2C_2$. The student may use reasoning similar to that given above to satisfy himself that resonance effects make possible high-impedance values in a band-stop range of frequencies.

T-sections have *one half* of the full series impedance in each series arm; hence the symbol $Z_1/2$. If it is an inductance, the value is $L_{1k}/2$. If it is a capacitance, its value can be obtained only by putting two of the C_{1k} capacitances in *parallel*, therefore we use $2C_{1k}$.

The shunt arm of a T-section has impedance Z_2. If it is an inductance, it is L_{2k}; if it is a capacitance, it is C_{2k}.

The series arm of a π-section has impedance Z_1. If it is an inductance, it is L_{1k}; if it is a capacitance, it is C_{1k}.

The shunt arm of a π-section has impedance $2Z_2$. If it is an inductance it is L_{2k}; if it is a capacitance it is $C_{2k}/2$. Half the capacitance is needed to provide $2Z_2$.

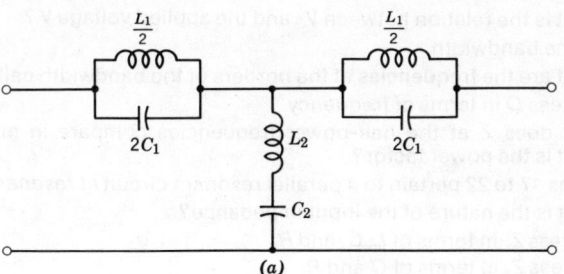

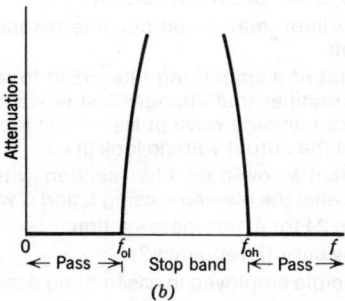

FIGURE 20.18   Band-stop filter. (a) Schematic. (b) Performance curves.

20.16 Review Questions

The first 16 questions apply to a series circuit *at resonance*.

1. What are the elements of a series resonant circuit?
2. Explain the comparison of V_L with V_C.
3. How do we calculate the current value?
4. How do we calculate the resonant frequency?
5. What is the power factor?
6. State the nature of the imput impedance at frequencies (a) below and (b) above resonance.
7. What effect does increasing the L/C ratio have?
8. How is the Q calculated?
9. What is the formula for the maximum energy stored per cycle in the inductance in terms of (a) instantaneous value of current, (b) effective value of current?
10. Express Q in terms of L and C.
11. What is the most practical way to tune the circuit?

12. What is the relation between V_C and the applied voltage V?

13. Define bandwidth.

14. What are the frequencies at the borders of the bandwidth called?

15. Express Q in terms of frequency.

16. How does Z at the half-power frequencies compare in magnitude with R? What is the power factor?

Questions 17 to 22 pertain to a parallel resonant circuit *at resonance.*

17. What is the nature of the input impedance?

18. Express Z_r in terms of L, C, and R.

19. Express Z_r in terms of Q and R.

20. What is the simple equation for calculating the resonant frequency, and what conditions limit its usefulness?

21. How is I_c related to the total input current?

22. Under what condition may a coil become resonant? Explain and sketch the equivalent circuit.

23. Sketch the circuit of a smoothing filter used to remove the ac components in the output of a rectifier that changes a sine-wave input voltage into a voltage that is a series of half-sine wave pulses. What jobs do L and C do? What does the wave form of the output voltage look like?

24. Sketch a constant-k low-pass filter section with (a) T configuration, (b) π configuration. Label the elements using L and C with proper subscripts.

25. Repeat Question 24 for a high-pass section.

26. What is meant by cut-off frequency?

27. Explain the principle employed in assembling a band-pass filter.

28. Explain the principle employed in assembling a band-stop filter.

20.17   Problems

15. A coil with inductance L and resistance R is in parallel with a capacitor C. The capacitor has minimum and maximum values of 25 pF and 250 pF. What value of inductance is required for resonance to occur at 550 kHz when C is maximum? Assume $R = 40 \, \Omega$. What will be the value of Q?

16. Assume L of the coil in Problem 15 is 335 μH. At what frequency will resonance occur when C is minimum? What is the Q at that frequency?

17. What are the input impedances of the circuit of Problems 16 and 17 at the minimum and maximum resonant frequencies? What are the bandwidths?

18. Calculate the input current and capacitor current in the two cases of Problem 17.

19. The parallel circuit of Fig. 20.11 has $L = 100 \, \mu$H; $R_L = 200 \, \Omega$; $C = 40$ pF; $R_C = 500 \, \Omega$. Calculate the resonant frequency using the exact and approximate equations and express the difference in percent.

20. A coil carries current that operates at a frequency high enough to cause the coil to become self-resonant. The values of the lumped constants in the equivalent parallel circuit are $L = 0.1 \, \mu$H; $R = 60 \, \Omega$; $C = 10$ pF. Determine the frequency by means of the exact expression.

21. The fundamental ac component of the input voltage to the smoothing filter in Fig. 20.12 is $e = 399 \sin 754t$. The filter constants are $L = 10$ H, $C = 16 \mu F$. What are the crest and rms values of this component of the capacitor voltage when the load resistor R is disconnected? What percent voltage reduction is achieved?

22. The series arms of a T-section of a low-pass filter have an inductance of 0.06 H and zero resistance. The shunt arm has a capacitor for which $C = 0.4 \mu F$. What is the cut-off frequency? Evaluate R_k^2, the product of the total series and shunt impedances.

23. A π-section high-pass filter has a series arm of $C = 160$ pF and two shunt arms with inductances of 100 μH each. Calculate the cut-off frequency and determine R_k.

24. A parallel-resonant circuit has a coil with $R_L = 100 \, \Omega$, $L = 0.2$ H in parallel with a capacitor that resonates with it at 50 kHz. (a) What is the bandwidth? (b) If an amplifier stage with 0.1 $M\Omega$ input resistance is added as a load, what will be the bandwidth?

25. It is desired to limit the bandwidth of the circuit of Problem 24 to 1 kHz. What is the minimum input resistance the amplifier may have?

CHAPTER TWENTY-ONE

Coupled Circuits

When a generator and a load are connected by means of a network instead of by direct connections, they are said to be *coupled* through the *coupling* network. Some familiar examples are the *LC* smoothing filter, discussed in Chapter 19, that connects a rectifier to a load, a transformer that inductively couples a load to a transmission line or a generator, and the load resistor, R_1, that is a common parallel impedance between the parallel resonant circuit of Fig. 20.13a and an amplifier that follows it.

We shall encounter some new terminology in our study of coupled circuits, such as mutual impedance designated as Z_{12} to represent the *ratio* of a voltage induced into circuit 1 to the current in circuit 2, and Z_{21} to represent the *ratio* of voltage induced in circuit 2 to the current in circuit 1. This is a *mutual impedance* which can be, and often is, designated as Z_m.

Let's try to prevent some confusion "right-off-the-bat" (as not-so-modern slang would put it). It soon becomes evident that keeping the meanings of Z_{12} and Z_{21} separate and understandable could become a chore. It will be helpful to observe that Z_{12} appears in the Kirchhoff voltage equation for the mesh, or loop, in which I_1 flows. It is part of the term that represents the voltage induced in the mutual impedance due to I_2 and is therefore multiplied by I_2. The mutual impedance is common to mesh 1 and mesh 2. Looking ahead in Fig. 21.1 we see that the shunt arm of the T is the mutual impedance.

463

You may wonder why we go into all this new stuff when we have already learned ways of solving for unknown currents, etc, by methods that are simpler to understand. A few of the reasons are (1) the methods make some solutions easier and more applicable in general; (2) the techniques are adaptable to the "black box" case where the network inside is sealed up and unknown, and there are only two input and two output terminals available. The network may be represented by the equivalent T of Fig. 21.1; (3) the parameters of circuits of transistor amplifiers, called *hybrid parameters*, involve mutual impedances like Z_{12} and Z_{21}. There is little use for hybrid parameters in an introductory nonelectronic textbook like this, so we shall not burden you with even an introductory study of them.

21.2   Coefficient of Coupling

Since $Z_{12} = Z_{21}$, let's call them both Z_m, the mutual impedance, and let's use Z_p for Z_{11} the total impedance of the primary and Z_s for Z_{22} the total impedance of the secondary.

In order to let the current equations have only voltage in the numerator and impedances in the denominator, we divide by the impedance factor in the numerators of Equations (21.3) and (21.4), and get

$$I_1 = \frac{E_1}{Z_p - Z_m^{\,2}/Z_s} \tag{21.3a}$$

$$I_2 = \frac{E_1}{Z_p Z_s/Z_m - Z_m} \tag{21.4a}$$

We shall simply give the expression for coefficient of coupling between two circuits in terms of their mutual and total impedances. Later, when we study inductively coupled circuits we shall express the coefficient of coupling in terms of inductances instead of impedances. In that case, merely multiplying each inductance by $\omega = 2\pi f$ will give the impedances that are involved in determining the coupling coefficient.

The coefficient of coupling, k, is given in terms of the circuit impedances as

$$k = \frac{Z_m}{\sqrt{Z_p Z_s}} \tag{21.5}$$

21.3   Transfer Impedance

The denominator of Equation (21.4a) represents an impedance value by which the *input voltage* can be divided to get the *output current*. This is called the *transfer impedance*, Z_{tr}. That is

$$I_2 = \frac{E_1}{Z_{tr}} \tag{21.6}$$

in which

$$Z_{tr} = \frac{Z_p Z_s}{Z_m} - Z_m \tag{21.7}$$

Now that $Z_m = k\sqrt{Z_p Z_s}$, we can substitute this and get

$$Z_{tr} = \frac{Z_p Z_s - k^2 Z_p Z_s}{k\sqrt{Z_p Z_s}}$$

$$Z_{tr} = \sqrt{Z_p Z_s} \left(\frac{1-k^2}{k} \right) \tag{21.8}$$

IMPORTANT OBSERVATION

The circuit in Fig. 21.1 has no load impedance in series with Z_2. Normally there would be some kind of load called Z_L. It is necessary to include the load impedance in Z_s which is also called Z_{22}.

EXAMPLE 21.2   Various impedances, the coefficient of coupling, and the currents in Fig. 21.2 will be evaluated.

Solution

$$Z_{11} = 40 + 60 = 100 \ \Omega = Z_p$$
$$Z_{22} = 60 + 20 + 10 = 90 \ \Omega = Z_s$$
$$Z_m = 60 \ \Omega$$

$$I_1 = \frac{E_1}{Z_p - Z_m^2/Z_s} = \frac{120}{100 - 3600/90} = 2 \ A$$
$$I_2 = \frac{E_1}{Z_p Z_s/Z_m - Z_m} = \frac{120}{9000/60 - 60} = 1\tfrac{1}{3} \ A$$

$$k = \frac{Z_m}{\sqrt{Z_p Z_s}} = \frac{60}{\sqrt{100 \times 90}} = 0.632$$

$$Z_{tr} = \frac{Z_p Z_s}{Z_m} - Z_m = \frac{9000}{60} - 60 = 90 \ \Omega$$

This is the denominator of Equation (20.6).

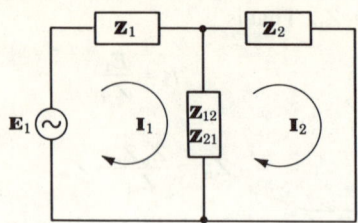

FIGURE 21.1   The branch containing I_2 is conductively coupled to the branch containing I_1.

Z_{21} appears in the Kirchhoff voltage equation for mesh 2 and is used with I_1 to represent the voltage induced in the mutual impedance by I_1. This voltage, $Z_{21}I_2$, must be assigned correct polarity, as must $Z_{12}I_1$ also. We shall see that this introduces no problem in writing the equations.

21.1   Conductively Coupled Circuits

A circuit branch is *conductively coupled* to another branch when it is directly connected so that the current at their node may divide so that part of it enters the second branch. An example is illustrated in Fig. 21.1 where we could say that part of the current in Z_1 may continue on through Z_2 at the top node. This network may be regarded as a combination of two circuits, the one in which I_1 flows may be called the primary circuit, and the one containing I_2, the secondary circuit. The impedance common to both circuits is called Z_{12} when used in the Kirchhoff voltage equation for circuit 1 and Z_{21} when used in the equation for circuit 2. In our treatment of coupled circuits these two quantities will be equal because the circuit parameters (R, L, C, M) will be constant.

When the driving voltage $\mathbf{E}_1$ and circuit parameters in Fig. 21.1 are given, the branch currents, voltages, and powers can be evaluated by relatively simple procedures. Employing mesh-current notation, with positive directions of currents arbitrarily chosen as in Fig. 21.1, we see that the current in $\mathbf{Z}_{12}$ in the $\mathbf{I}_1$ direction is designated as $\mathbf{I}_1 - \mathbf{I}_2$.

The "self impedance" of circuit 1, that is, the impedance of circuit 1 to $\mathbf{I}_1$, is, by definition,

$$\mathbf{Z}_{11} = \mathbf{Z}_1 + \mathbf{Z}_{12}$$

The self-impedance of circuit 2 (i.e., to $\mathbf{I}_2$) is

$$\mathbf{Z}_{22} = \mathbf{Z}_2 + \mathbf{Z}_{21}$$

And, as stated above, when the circuit parameters are constant,

$$\mathbf{Z}_{12} = \mathbf{Z}_{21}$$

Employing Kirchhoff's law:

$$\mathbf{Z}_{11}\mathbf{I}_1 - \mathbf{Z}_{12}\mathbf{I}_2 = \mathbf{E}_1 \tag{21.1}$$
$$-\mathbf{Z}_{21}\mathbf{I}_1 + \mathbf{Z}_{22}\mathbf{I}_2 = 0 \tag{21.2}$$

Solving for $\mathbf{I}_1$ by determinants, we replace the $\mathbf{I}_1$ terms by $\mathbf{E}_1$ and 0 in the numerator, and write the coefficients of $\mathbf{I}_1$ and $\mathbf{I}_2$ just as they are in the denominator.

$$\mathbf{I}_1 = \frac{\begin{vmatrix} \mathbf{E}_1 & -\mathbf{Z}_{12} \\ 0 & \mathbf{Z}_{22} \end{vmatrix}}{\begin{vmatrix} \mathbf{Z}_{11} & -\mathbf{Z}_{12} \\ -\mathbf{Z}_{21} & \mathbf{Z}_{22} \end{vmatrix}} = \frac{\mathbf{E}_1\mathbf{Z}_{22}}{\mathbf{Z}_{11}\mathbf{Z}_{22} - \mathbf{Z}_{12}^2} \tag{21.3}$$

and similarly for $\mathbf{I}_2$

$$\mathbf{I}_2 = \frac{\begin{vmatrix} \mathbf{Z}_{11} & \mathbf{E}_1 \\ -\mathbf{Z}_{21} & 0 \end{vmatrix}}{\begin{vmatrix} \mathbf{Z}_{11} & -\mathbf{Z}_{12} \\ -\mathbf{Z}_{21} & \mathbf{Z}_{22} \end{vmatrix}} = \frac{\mathbf{E}_1\mathbf{Z}_{21}}{\mathbf{Z}_{11}\mathbf{Z}_{22} - \mathbf{Z}_{12}^2}$$

EXAMPLE 21.1   Assume the following values for Fig. 21.1: $\mathbf{E} = 120\underline{/0°}$ V rms, $\mathbf{Z}_1 = 5+j12\ \Omega$, $\mathbf{Z}_{12} = 5+j0\ \Omega$, $\mathbf{Z}_2 = 7-j6\ \Omega$. The generator impedance is included in the value for $\mathbf{Z}_1$. Determine all current and power values in the system.

Solution

$$\mathbf{Z}_{11} = 5+j12+5+j0 = 10+j12 = 15.62\underline{/50.2°}$$
$$\mathbf{Z}_{22} = 7-j6+5+j0 = 12-j6 = 13.42\underline{/-26.6°}$$
$$\mathbf{Z}_{11}\mathbf{Z}_{22} = 209\underline{/23.6°} = 192+j83.6\ \Omega$$
$$\mathbf{Z}_{11}\mathbf{Z}_{22} - \mathbf{Z}_{12}^2 = 167+j83.6 = 186.5\ \underline{/26.6°}$$
$$\mathbf{I}_1 = \frac{(120\underline{/0°})\,(13.42\underline{/-26.6°})}{186.5\underline{/26.6°}} = 8.63\underline{/-53.2°}\ \text{A}$$
$$\mathbf{I}_1 = 5.18-j6.90\ \text{A}$$
$$\mathbf{I}_2 = \frac{(120\underline{/0°})\,(5\underline{/0°})}{186.5\underline{/26.6°}} = 3.22\underline{/-26.6°}\ \text{A}$$
$$\mathbf{I}_2 = 2.88-j1.44\ \text{A}$$

Current in $\mathbf{Z}_{12}$ is $\mathbf{I}_{12} = \mathbf{I}_1 - \mathbf{I}_2$

$$\mathbf{I}_1 - \mathbf{I}_2 = (5.18-j6.90) - (2.88-j1.44)$$
$$= 2.3 - j5.46 = 5.93\underline{/-67.2°}\ \text{A}$$

Total power $= E_1 I_1 \cos\theta = 120 \times 8.63 \cos 53.2° = 621$ W.
Total power absorbed by the network is

$$I_1^2 R_1 + R_2^2 R_2 + I_{12}^2 R_{12} = 8.63^2 \times 5 + 3.22^2 \times 7 + 5.93^2 \times 5 = 621\ \text{W}$$

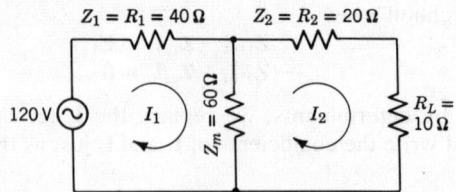

Figure 21.2   A source conductively coupled to a load.

Check

$$I_1 = \frac{120}{40 + 60(20 + 10)/90} = 2 \text{ A}$$

$$I_2 = 2 \times \frac{60}{90} = 1\tfrac{1}{3} \text{ A}$$

21.4   Reflected Impedance

In the study of transformer coupling and in applications of this theory to air-core and iron-core transformer operation, we shall learn how to represent the whole transformer and its load impedance as an *equivalent circuit* in which all secondary quantities (secondary current, self-impedance, load impedance) are converted into primary circuit symbols that represent them. The secondary impedance is then said to be *reflected into the primary*.

Equation (21.3a) says

$$I_1 = \frac{E_1}{Z_p - Z_m^2/Z_s} \tag{21.3a}$$

When we must include a load impedance Z_L,

$$Z_s = Z_{22} + Z_L = Z_2 + Z_m + Z_L$$

The term $-Z_m^2/Z_s$ is called the reflected impedance.

In the case of transformer coupling with only E_1 as a source, the mutual impedance will be found to be simply $2\pi f M = \omega M$ in which M is the mutual inductance. When added to other impedances, such as Z_1, this reactance should be expressed as $j\omega M$. Then $Z_m^2 = (j\omega M)^2 = -\omega^2 M^2$, and Z_{tr}, the denominator of Equation (21.3a), becomes $Z_p + (\omega^2 M^2/Z_s)$. The sign of Z_s may be such as to cause Z_{tr} to be smaller than Z_p rather than larger. It is larger in the case of a power transformer with normal loading.

The ratio of the output current to the input current of the coupling network can be expressed in terms of the mutual and secondary-circuit impedances.

Let's go back to Equations (21.3) and (21.4). From this we get, using the alternate symbols,

$$\frac{I_2}{I_1} = \frac{E_1 Z_m}{Z_p Z_s - Z_m^2} \times \frac{Z_p Z_s - Z_m^2}{E_1 Z_s}$$

$$\frac{I_2}{I_1} = \frac{Z_m}{Z_s} \tag{21.9}$$

This is called the *forward-current-transfer ratio*. Remember that Z_s is the total impedance, including the load impedance, through which I_2 flows. A little thought will reveal that this is simply the equation for current division. I_2 is obtained by multiplying it by the impedance of "the other branch connected to the junction (the node)" and dividing by the sum of the impedances of the two branches.

21.5   *R*- and *RC*-Coupling Networks

Figure 21.3 shows a pi network of resistances coupling two circuits connected to two pairs of terminals 1, 1′ and 2, 2′. The mutual impedance between circuits 1 and 2 is

$$\mathbf{Z}_m = \frac{\mathbf{V}_{22'}}{\mathbf{I}_1} \qquad \text{(when the switch } S \text{ is open)}$$

The product of R_C and its current gives $\mathbf{V}_{22'}$:

$$\mathbf{V}_{22'} = \frac{\mathbf{V}_{11'}}{R_b + R_c} R_c \tag{21.10}$$

$$\mathbf{I}_1 = \frac{\mathbf{V}_{11'}}{R_a} + \frac{\mathbf{V}_{11'}}{R_b + R_c} = \frac{\mathbf{V}_{11'}(R_a + R_b + R_c)}{R_a(R_b + R_c)} \tag{21.11}$$

Dividing Equation (21.10) by Equation (21.11) gives

$$\mathbf{Z}_{21} = \frac{R_a R_c}{R_a + R_b + R_c} \tag{26.12}$$

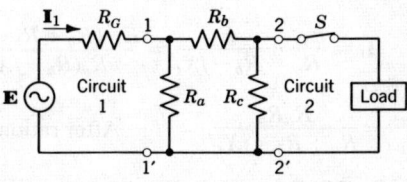

FIGURE 21.3   A network of resistances couples circuits 1 and 2.

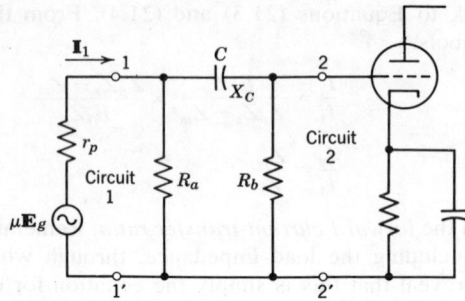

Figure 21.4   An *RC*-coupled amplifier stage.

Equation (21.12) is the same as Equation (6.7) in Section 6.4, which gives the value of R_3 in Fig. 6.6b. This means that $\mathbf{Z}_{21}$, derived here, is only one branch of a T network that must be used to join the generator and the load. The load current, represented as $\mathbf{I}_2$ in Fig. 20.1, will have the same value as that in the load when this T network replaces the π group of Fig. 20.3. To obtain the other two branches of the T, one may use Equations (6.5) and (6.6) which apply to Fig. 6.6b.

An *RC* network couples circuits 1 and 2 in Fig. 21.4. This kind of coupling is common in electronic circuits, particularly at the input of an amplifier stage where direct current in circuit 1 must be kept from entering circuit 2. Alternating-current energy is supplied to circuit 2 by means of the varying electric field in capacitor *C*.

In this system, circuit 2 is open-circuited because the capacitance between the grid and cathode of the tube is zero (and X_C is therefore infinite) for all practical purposes. The current in R_b is given by

$$\frac{\mathbf{V}_{11'}}{R_b - jX_C}$$

and the voltage developed in circuit 2 is

$$\mathbf{V}_{22'} = \frac{\mathbf{V}_{11'}R_b}{R_b - jX_C} \tag{21.13}$$

$$\mathbf{I}_1 = \frac{E_{11'}}{R_a} + \frac{E_{11'}}{R_b - jX_C} = \frac{E_{11'}(R_a + R_b - jX_C)}{R_a(R_b - jX_C)} \tag{21.14}$$

$$\mathbf{Z}_{21} = \frac{\mathbf{V}_{22'}}{\mathbf{I}_1} = \frac{R_a R_b}{R_a + R_b - jX_C}. \qquad \text{After rationalizing the denominator,}$$

$$\mathbf{Z}_{21} = \frac{R_a{}^2 R_b + R_a R_b{}^2 + jR_a R_b X_C}{(R_a + R_b)^2 + X_C{}^2} \tag{21.15}$$

Problems

1. Apply the equations of coupled-circuit analysis to Fig. 21.5 and determine the input current, load current, coefficient of coupling, transfer impedance, and forward-current-transfer ratio. Let $R_1 = 40\,\Omega$, $R_2 = 30\,\Omega$, $R_3 = 180\,\Omega$, $R_L = 60$, $E = 100$ V. Check the value of I_l using series-parallel circuit solution.

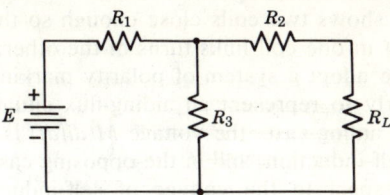

FIGURE 21.5   For Problems 1 and 2.

2. Solve for the quantities requested in Problem 1 using $E = 200$ V; $R_1 = 520\,\Omega$; $R_2 = 200\,\Omega$; $R_3 = 800\,\Omega$; $R_L = 1000\,\Omega$ in Fig. 21.5.

3. Use coupled-circuit analysis to solve for the following in Fig. 21.6: input current, transfer impedance, forward-transfer-current ratio, load current, and coupling coefficient. Let $E = 100$ V rms at 400 Hz, $L_1 = 2$ H, $L_2 = 4$ H, $L_3 = 6$ H, $Z_L = 15000 + j0$. (This is a long problem.)

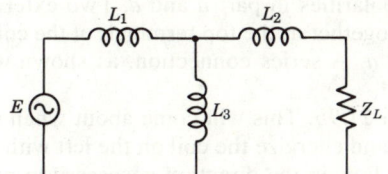

FIGURE 21.6   For Problem 3.

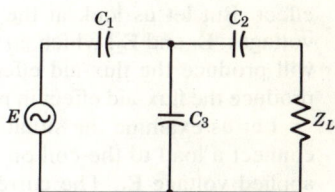

FIGURE 21.7   For Problem 4.

4. Solve for the quantities asked for in Problem 3, using coupled circuit analysis on the circuit of Fig. 21.7. Let $E = 100$ V rms at 60 Hz, $C_1 = 1.0\,\mu$F, $C_2 = 0.5\,\mu$F, $C_3 = 0.25\,\mu$F, $Z_L = 15,000\,\Omega$. (This is too!)

5. Draw a circuit equivalent to that of Fig. 21.8 after changing the π coupling network to its equivalent T. Apply coupled circuit theory and determine the values of all quantities requested in Problem 3. $E = 10\,\underline{/0°}$ V.

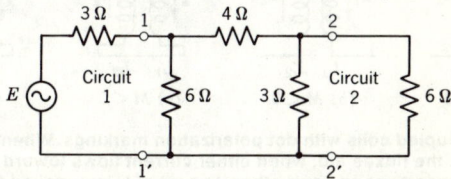

FIGURE 21.8   All elements are pure resistances. For Problem 5.

21.6   Mutual Inductance in a Series Circuit

Two coils may be connected in series in either of two ways, one in which their separate fluxes aid each other in producing mutual flux and the other in which their separate fluxes oppose each other and thus produce a mutual flux that is less in amount than in the aiding case.

Figure 21.9a shows two coils close enough so that some of the flux produced by current in one coil links turns of the other. In a sketch of this kind it is necessary to adopt a system of polarity marking so that one may write equations properly to represent an aiding-flux situation or an opposing-flux situation. In the aiding case, the voltage $M(di/dt)$ is positive with respect to the voltage of self-induction, and in the opposing case the voltage $M(di/dt)$ is negative with respect to the voltage of self-induction. The meaning and significance of this will be recalled if the reader refers back to Section 10.6

We shall agree to place the plus sign on the coil symbol at the tail of the current arrow. This will enable us to use plus signs with the voltage and current terms when we write Kirchhoff voltage equations around closed paths and progress in the direction of the current arrow.

We shall choose to represent a situation where the *fluxes* produced by the two currents *aid each other* by placing a dot at the tail end of each current arrow (Fig. 21.9a). The series connection shown in part *d* produces this effect. But let us look at the voltage polarities in part *a* and *d*. Two external voltages, $\mathbf{E}_1$ and $\mathbf{E}_2$, which go positive together at the top terminals of the coils, will produce the flux-aid effect in part *a*. A series connection, as shown will produce the flux-aid effect in part *d*.

Let us examine the situation in Fig. 21.9b. This will come about when we connect a load to the coil on the right and energize the coil on the left with an applied voltage $\mathbf{E}_1$. The current $\mathbf{I}_2$ will flow in the direction necessary to produce *opposing flux* (Lenz's law) so the arrow points toward the dot. The terms of voltage produced by mutual induction will be negative in both Kirchhoff equations for this case.

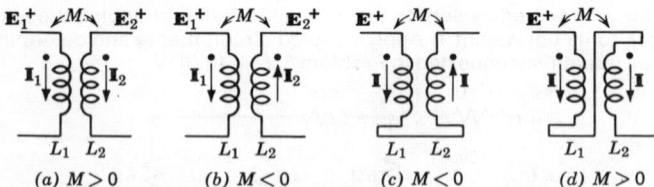

FIGURE 21.9   Mutually coupled coils with dot polarization markings. When both currents flow either toward or away from dots, the fluxes aid; when either current flows toward a dot and the other away from the other dot, the fluxes oppose each other. In part *a*, $\mathbf{I}_2$ is supplied from a second source; in part *b*, $\mathbf{I}_2$ is an induced current that flows through a load connected to L_2.

In Fig. 21.9c, a series connection has forced current $\mathbf{I_2}$ to flow toward the dot, so we again have opposing flux.

We conclude that if both current arrows *drawn beside the coils* point either away from or toward the dot, the case is one of *aiding flux* and M is positive, but if one arrow points toward a dot and the other away from a dot then the fluxes oppose each other and M is negative. This kind of labeling system removes the necessity of making a three-dimensional sketch to show the direction in which current flows around (through) the turns of a coil.

The coils in Fig. 21.10 are connected so that their fluxes aid each other in producing mutual flux at any instantaneous value, or direction, of current. In writing the voltage equation around the circuit we will therefore expect the voltage of mutual induction to be in phase with the voltage of self-induction and to have the same sign. The applied voltage is equal to the sum of the separate voltages in the circuit, so we have

$$e = ir_1 + L_1\frac{di}{dt} + M\frac{di}{dt} + ir_2 + L_2\frac{di}{dt} + M\frac{di}{dt} \qquad (21.16)$$

With a flux path of constant permeability, this equation becomes

$$e = (r_1 + r_2)i + (L_1 + L_2 + 2M)\frac{di}{dt} \qquad (21.17)$$

Using rms values of voltage and current when the circuit parameters are constant and a sinusoidal voltage is applied:

$$(r_1 + r_2)\mathbf{I} + j\omega(L_1 + L_2 + 2M)\mathbf{I} = \mathbf{E} \qquad (26.18)$$

Recognizing that ωM is a *mutual reactance*, the equivalent impedance of this series circuit may be found from $\mathbf{E}/\mathbf{I}$ using this equation:

$$\mathbf{Z_e} = \frac{\mathbf{E}}{\mathbf{I}} = [(r_1 + r_2)^2 + \omega^2(L_1 + L_2 + 2M)^2]^{1/2}\underline{/\theta} \qquad (21.19)$$

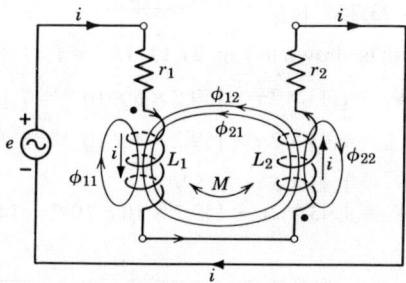

FIGURE 21.10   Coupled coils in series with fluxes aiding. *M* is positive.

where

$$\tan \theta = \frac{\omega(L_1 + L_2 + 2M)}{r_1 + r_2}$$

Dividing through Equation (21.18) by **I** gives the complex form of the equivalent impedance:

$$\mathbf{Z}_e = r_1 + r_2 + j\omega(L_1 + L_2 + 2M) = \mathbf{E}/\mathbf{I}$$

from which we get the self- and mutual impedances:

$$\mathbf{Z}_1 = r_1 + j\omega L_1, \qquad \mathbf{Z}_2 = r_2 + j\omega L_2, \qquad \mathbf{Z}_M = \pm j\omega M \qquad (21.20)$$

Recall that these equations are for a series circuit in which the coil fluxes aid each other. If the connections to one of the coils are reversed, causing the coil fluxes to oppose each other, the sign of every M term must be reversed.

It is recommended that the student review the material in Section 10.6 where the value of M is established from the self-inductances of the coils.

EXAMPLE 21.3   Two coils in series with their fluxes aiding, as represented in Fig. 21.10, are supplied with 34 V rms at 159.2 Hz. The circuit constants are: $r_1 = 0.7\ \Omega$, $L_1 = 5$ mH, $r_2 = 5\ \Omega$, $L_2 = 10$ mH, $M = 4$ mH. Find the (a) coefficient of coupling, (b) equivalent-series impedance, (c) current. Draw the phasor diagram.

Solution.

$$(a) \quad k = \frac{M}{\sqrt{L_1 L_2}} = \frac{4}{\sqrt{5 \times 10}} = 0.566$$

$$(b) \quad \mathbf{Z}_e = (r_1 + r_2) + j\omega(L_1 + L_2 + 2M)$$
$$= (0.7 + 5) + j2\pi \times 159.2(5 + 10 + 8)10^{-3}$$
$$= 5.7 + j23 = 23.75 \underline{/76.1^\circ}\ \Omega$$

$$(c) \quad \mathbf{I} = \frac{34}{23.75 \underline{/76.1^\circ}} = 1.43 \underline{/-76.1^\circ}\ A$$

The phasor diagram is shown in Fig. 21.11. $IR_1 = 1.43 \times 0.7 = 1$ V.

$$IX_1 = 1.43 \times 2\pi \times 159.2 \times 5 \times 10^{-3} = 7.15\ V$$

$$IX_M = 1.43 \times 2\pi \times 159.2 \times 4 \times 10^{-3} = 5.72\ V$$

$$IR_2 = 1.43 \times 5 = 7.15\ V$$
$$IX_2 = 1.43 \times 2\pi \times 159.2 \times 10 \times 10^{-3} = 14.3\ \text{volts}$$

Check:

$$E = \sqrt{(1 + 7.15)^2 + (7.15 + 14.3 + 2 \times 5.72)^2}$$
$$= \sqrt{8.15^2 + 32.89^2} = 34\ V$$

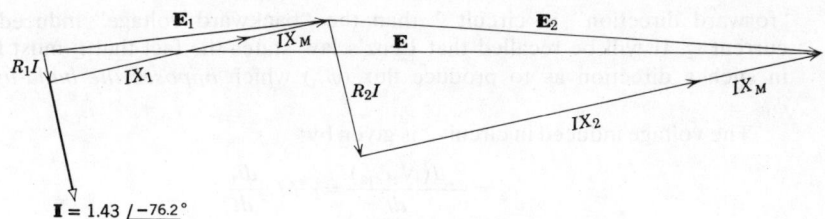

FIGURE 21.11   For Example 21.3. Phasor diagram.

21.7   The Air-Core Transformer

The primary and secondary windings of the conventional transformer are not connected physically, that is, there is no conductive connection between them. They are *magnetically coupled*, that is, magnetic flux produced by current in each winding links turns of the other winding. The primary circuit (circuit 1), which contains the primary winding, receives alternating current I_1 from a source of alternating voltage. The secondary winding, which is part of circuit 2, has an alternating voltage induced in it. This voltage causes an alternating current I_2 to flow in circuit 2.

Figure 21.12 is a schematic diagram of a conventional air-core transformer supplying current to a load consisting of R, L, and C elements in series. This type of load may be expected in practical applications of the air-core transformer, many of which are in high-frequency circuits used in radio and television receivers.

Instantaneous currents are shown in their proper directions which agree with the fact that their fluxes, ϕ_{12} and ϕ_{21}, are in opposition in the closed flux path that links both coils. The net common flux is in the direction of ϕ_{12}, produced by i_1, as expected, because this flux induces a larger voltage in the

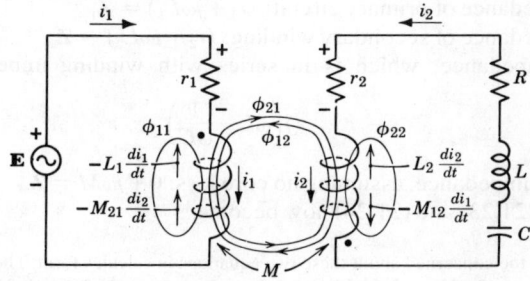

FIGURE 21.12   Schematic of an air-core transformer with an *RLC* load. Fluxes oppose each other.

"forward direction" in circuit 2 than the "backward voltage" induced by current i_2. It will be recalled that Lenz's law states the fact that i_2 must flow in such a direction as to produce flux (ϕ_{21}) which *opposes the build-up* of ϕ_{12}.

The voltage induced in circuit 2 is given by

$$e_2 = -\frac{d(N_2\phi_{12})}{dt} = -M_{12}\frac{di_1}{dt}$$

in which $N_2\phi_{12}$ represents the flux linkages in circuit 2 made by the part of ϕ_{11} which "reaches over" and treads through turns in the secondary, and M is the mutual inductance between the primary and secondary coils. The instantaneous applied voltage is equal to the sum of the i_1R_1 drop in the primary resistance and the voltages caused by self-induction and mutual induction. The Kirchoff equations[1] for the primary and secondary loops, using instantaneous values, are:

$$\mathbf{E}_1 = r_1i_1 + L_1\frac{di_1}{dt} - M_{21}\frac{di_2}{dt} \tag{21.21}$$

$$0 = (r_2 + R)i_2 + (L_2 + L)\frac{di_2}{dt} + \frac{1}{C_2}\int i_2\,dt - M_{12}\frac{di_1}{dt} \tag{21.22}$$

The $M_{21}(di_2/dt)$ term has a negative sign because the flux ϕ_{21} due to i_2 opposes the flux ϕ_{11} produced by i_1. The induced voltage in the secondary, $M_{12}(di_1/dt)$, is equal to the sum of all of the other voltages in circuit 2. Also $\int i_2\,dt = q_2$.

In the steady state when sinusoidal voltage is applied, these equations may be written in terms of effective values of voltage and current as follows:

$$(r_1 + j\omega L_1)\mathbf{I}_1 - j\omega M\mathbf{I}_2 = \mathbf{E}_1 \tag{21.23}$$

$$(r_2 + j\omega L_2)\mathbf{I}_2 + R\mathbf{I}_2 + j\left(\omega L - \frac{1}{\omega C}\right)\mathbf{I}_2 - j\omega M\mathbf{I}_1 = 0 \tag{21.24}$$

The following symbols will be used for simplification:
Self-impedance of primary circuit: $(r_1 + j\omega L_1) = \mathbf{Z}_1$
Self-impedance of secondary winding: $(r_2 + j\omega L_2) = \mathbf{Z}_2$
Load impedance, which is in series with winding impedance, $\mathbf{Z}_L = R +$

$$j\left(\omega L - \frac{1}{\omega C}\right)$$

Mutual impedance, assuming no core loss: $0 + j\omega M = \mathbf{Z}_M$
Equations (21.23) and (21.24) now become

[1] Do not be too concerned about these two equations in calculus form. They are merely to show where Equations (21.23) and (21.24) come from. Equations (21.25) and (21.26) are the principal forms that should concern us here.

$$\mathbf{Z}_1\mathbf{I}_1 - \mathbf{Z}_M\mathbf{I}_2 = \mathbf{E}_1 \tag{21.25}$$

$$-\mathbf{Z}_M\mathbf{I}_1 + (\mathbf{Z}_2 + \mathbf{Z}_L)\mathbf{I}_2 = 0 \tag{21.26}$$

Using determinants to solve for the currents,

$$\mathbf{I}_1 = \frac{\begin{vmatrix} \mathbf{E}_1 & -\mathbf{Z}_M \\ 0 & (\mathbf{Z}_2+\mathbf{Z}_L) \end{vmatrix}}{\begin{vmatrix} \mathbf{Z}_1 & -\mathbf{Z}_M \\ -\mathbf{Z}_M & (\mathbf{Z}_2+\mathbf{Z}_L) \end{vmatrix}} = \frac{\mathbf{E}_1(\mathbf{Z}_2+\mathbf{Z}_L)}{\mathbf{Z}_1(\mathbf{Z}_2+\mathbf{Z}_L) - \mathbf{Z}_M{}^2} \tag{21.27}$$

$$\mathbf{I}_2 = \frac{\begin{vmatrix} \mathbf{Z}_1 & \mathbf{E}_1 \\ -\mathbf{Z}_M & 0 \end{vmatrix}}{\begin{vmatrix} \mathbf{Z}_1 & -\mathbf{Z}_M \\ -\mathbf{Z}_M & (\mathbf{Z}_2+\mathbf{Z}_L) \end{vmatrix}} = \frac{\mathbf{E}_1\mathbf{Z}_M}{\mathbf{Z}_1(\mathbf{Z}_2+\mathbf{Z}_L) - \mathbf{Z}_M{}^2} \tag{21.28}$$

It is often desirable to determine the voltage $\mathbf{I}_2\mathbf{Z}_L$ across the load impedance. Since the voltage induced in the secondary circuit is given by $j\omega M\mathbf{I}_1$, the voltage across the load is this voltage minus the impedance drop in the secondary winding:

$$\mathbf{I}_2\mathbf{Z}_L = \mathbf{E}_2 = j\omega M\mathbf{I}_1 - \mathbf{I}_2\mathbf{Z}_2 \tag{21.29}$$

21.8   Equivalent Impedance Referred to Primary Terminals

One may obtain an expression for a single impedance $\mathbf{Z}_1'$ that will represent the the transformer and its load as seen when looking into the primary terminals. This will be the ratio of the applied voltage to the primary current. Thus from Equation (21.27) we obtain

$$\mathbf{Z}_1' = \frac{\mathbf{E}_1}{\mathbf{I}_1} = \frac{\mathbf{Z}_1(\mathbf{Z}_2+\mathbf{Z}_L) - \mathbf{Z}_M{}^2}{\mathbf{Z}_2+\mathbf{Z}_L} \tag{21.30}$$

This may be expressed in a form that brings out an additional meaning:

$$\mathbf{Z}_1' = \mathbf{Z}_1 - \frac{\mathbf{Z}_M{}^2}{\mathbf{Z}_2+\mathbf{Z}_L} = \mathbf{Z}_1 + \frac{\omega^2 M^2}{\mathbf{Z}_2+\mathbf{Z}_L} \tag{21.31}$$

Note that $\mathbf{Z}_M = j\omega M$, $\mathbf{Z}_M{}^2 = -\omega^2 M^2$, and the second term becomes positive. This term *represents the secondary circuit* in this simple series-circuit equation. The secondary-circuit impedance is said to be *reflected into the primary*. To reflect a secondary impedance into the primary one must *multiply* its reciprocal by $\omega^2 M^2$. This means that if $\mathbf{Z}_s$ represents the total secondary impedance of a loaded transformer, it may be represented by the single impedance $\omega^2 M^2/\mathbf{Z}_s$ in series with the primary winding when the equivalent circuit of the system is expressed in primary-circuit quantities only.

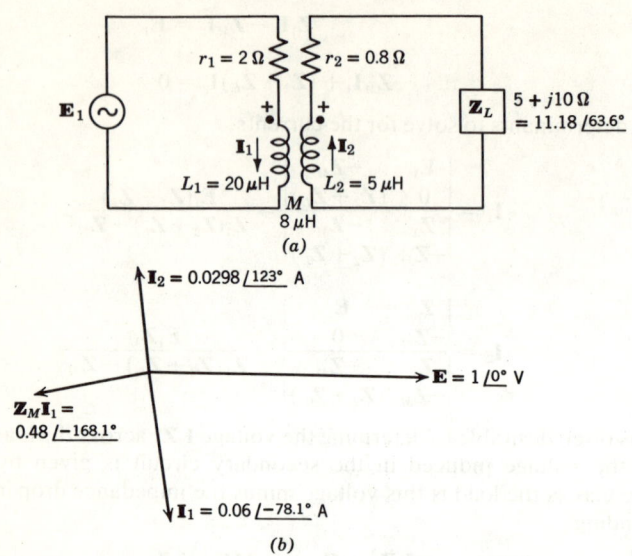

FIGURE 21.13   For Example 21.4. (a) Schematic diagram. (b) Phasor diagram.

EXAMPLE 21.4   Solve for currents, voltages, and powers in the circuit of Fig. 21.13, and draw the phasor diagram.

Solution. Using the equations developed above,

$$\mathbf{Z}_1 = 2 + j10^6 \times 20 \times 10^{-6} = 2 + j20 = 2.01 \underline{/84.3°}$$

$$\mathbf{Z}_2 = 0.8 + j10^6 \times 5 \times 10^{-6} = 0.8 + j5 = 5$$

$$\mathbf{Z}_M = 0 - j10^6 \times 8 \times 10^{-6} = -j8 = 8\underline{/-90°}$$

$$\mathbf{Z}_1' = \mathbf{Z}_1 - \frac{\mathbf{Z}_M{}^2}{\mathbf{Z}_2 + \mathbf{Z}_L} = 2 + j20 - \frac{64\underline{/-180°}}{0.8 + j5 + 5 + j10}$$

$$= 2 + j20 + \frac{64\underline{/0°}}{5.8 + j15}$$

$$\mathbf{Z}' = 2 + j20 + \frac{(64 + j0)(5.8 - j15)}{33.6 + 225} = 2 + j20 + 1.435 - j3.71$$

$$\mathbf{Z}' = 3.435 + j16.29 = 16.65 \underline{/78.1°}$$

$$\mathbf{I}_1 = \frac{\mathbf{E}_1}{\mathbf{Z}_1} = \frac{1\underline{/0°}}{16.65\underline{/78.1°}} = 0.060\underline{/-78.1°} \text{ A}$$

From Equation (21.26),

$$\mathbf{I}_2 = \frac{\mathbf{Z}_M \mathbf{I}_1}{\mathbf{Z}_2 + \mathbf{Z}_L} = \frac{(8\underline{/-90°})\,(0.060\underline{/-78.1°})}{5.8 + j15} = 0.0298\underline{/123°}$$

Voltage at the load terminals is $\mathbf{E}_2 = \mathbf{I}_2 \mathbf{Z}_L$

$$\mathbf{E}_2 = (0.0298\underline{/123°})\,(11.18\underline{/63.6°}) = 0.333\underline{/186.6°}\text{ V}$$

Power supplied by the source:

$$E_1 I_1 \cos\theta_1 = 1 \times 0.060 \cos 78.1° = 0.0124\text{ W} = 12.4\text{ mW}$$

Power received by the load:

$$E_2 I_2 \cos\theta_2 = 0.333 \times 0.0298 \cos(186.6° - 123°) = 0.00442\text{ W}$$
$$= 4.42\text{ mW}$$

Check:

$$I_2{}^2 R_L = (0.0298)^2 \times 5 = 0.00442\text{ W} = 4.42\text{ mW}$$

21.9   Equivalent Circuits

To simplify the procedure of analysis of an inductively coupled circuit and to assist in visualizing the problem, an equivalent circuit may be drawn. For example, the coupled circuits and the network in Fig. 21.14 are equivalent in the sense that the same loop equations hold for both.

$$(r_1 + j\omega L_1')\mathbf{I}_1 - j\omega M\mathbf{I}_2 = \mathbf{E}_1 \tag{21.32}$$

$$(r_2 + j\omega L_2')\mathbf{I}_2 - j\omega M\mathbf{I}_1 = -\mathbf{E}_2 \tag{21.33}$$

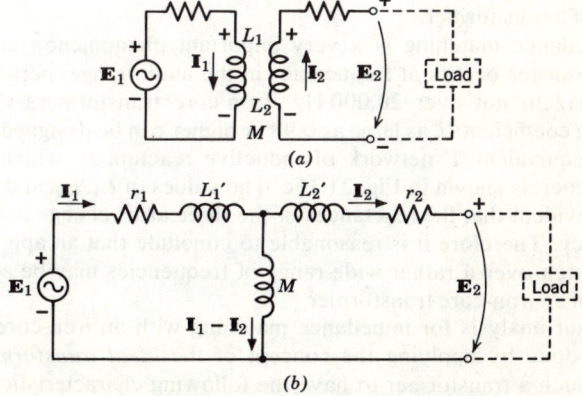

FIGURE 21.14   (a) Coupled circuit. (b) Equivalent circuit.

$L_1' = L_1 - M$ and $L_2' = L_2 - M$ in the equivalent network. M is usually less than L_1 and L_2 in an air-core-transformer application because the coupling is relatively loose.

It appears that the equivalent circuit may well represent a *step-down* transformer, inasmuch as the output voltage is less than the input voltage. But how could it represent a step-up transformer? In this case L_1' will be *negative* and a voltage *rise* will occur across L_1'.

This form of equivalent circuit is not applicable to iron-core transformers in which the core losses are not negligible. An additional resistance must be applied in parallel with M to take such losses into account. In Chapter 22 the equivalent circuit of the iron-core transformer is discussed in some detail, with an illustrative example.

21.10   Impedance Matching; The Ideal Transformer

Several examples of four-terminal circuits have been discussed in this chapter. Figures 21.2, 3, 4, 8, and others, show circuits of this type. Load voltage, current, and power are computed by means of equations developed from the actual circuit or from the equivalent circuit.

We have seen that the necessary condition for maximum power to be delivered to a load is that the load impedance must be the conjugate of the generator impedance (*Chapter 19*). It is not possible to fulfill this condition completely in many practical situations. Often we can vary only the magnitude of the load impedance and not the phase angle independently. In this case, the maximum power that can be delivered to the load will occur at only one frequency, that is, when $\mathbf{Z}_L = \mathbf{Z}_S$ in Fig. 21.15. If both of these impedances are fixed, that is, if neither can be varied, we can effect a matching condition by means of a transformer.

Impedance matching is a very important phenomenon in circuits where power transfer occurs at frequencies in the audio range, perhaps in the range of 100 Hz to not over 20,000 Hz. Iron-core transformers which may have coupling coefficients k as large as 0.98 or higher can be designed.

An equivalent T network of inductive reactances, which represents the transformer, is shown in Fig. 21.15c. The values of L_1, L_2, and M are constant, so it is evident that the reactances of the three arms change in like manner with frequency. Therefore it is reasonable to conclude that an approximate impedance match over a rather wide range of frequencies may be accomplished by means of an iron-core transformer.

Circuit analysis for impedance matching with an iron-core transformer is usually done by applying the concept of the *ideal transformer*. One would expect such a transformer to have the following characteristics: (*a*) no losses, (*b*) reactances of primary and secondary windings very much larger than the

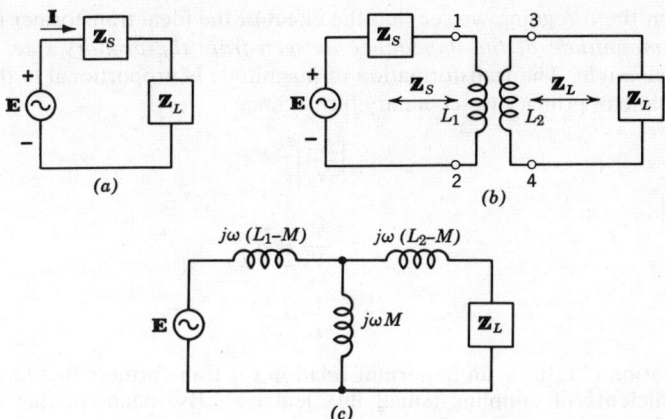

FIGURE 21.15   Impedance matching with a transformer. (*a*) Original circuit. (*b*) Circuit with matching transformer. (*c*) Equivalent circuit with "ideal" transformer.

impedance of an actual connected load, and (*c*) no leakage flux, which means unity coupling coefficient, that is, $k = 1$.

Iron-core transformers have efficiencies exceeding 90 percent, so characteristic *a* is satisfied from a practical point of view. The construction of a good quality iron-core transformer makes characteristic *c* possible, and characteristic *b* can be achieved in design and application.

With no leakage flux and no losses, we have

$$\frac{V_1}{V_2} = \frac{N_1}{N_2} \quad \text{and} \quad N_1 I_1 = N_2 I_2$$

from which,

$$V_2 = \frac{N_2}{N_1} V_1 \quad \text{and} \quad I_2 = \frac{N_1}{N_2} I_1$$

Since the load impedance, $\mathbf{Z}_2 = \mathbf{V}_2/\mathbf{I}_2$,

$$\mathbf{Z}_2 = \frac{\mathbf{V}_1 (N_2/N_1)}{\mathbf{I}_1 (N_1/N_2)} = \left[\frac{N_2}{N_1} \right]^2 \cdot \frac{\mathbf{V}_1}{\mathbf{V}_2} \tag{21.34}$$

Calling $\mathbf{Z}_1'$ the equivalent impedance on the primary side of the transformer,

$$\mathbf{Z}_1' = \frac{\mathbf{V}_1}{\mathbf{I}_1} = \left[\frac{N_1}{N_2} \right]^2 \cdot \mathbf{Z}_2 \tag{21.35}$$

It is customary, in the literature, to let $N_1/N_2 = a$, so that

$$\mathbf{Z}_1' = a^2 \mathbf{Z}_2 \tag{21.36}$$

From the foregoing we see that the effect of the ideal transformer is to *change the magnitude of the impedance as seen from the primary side, but not the phase angle*. The transformation of magnitude is proportional to the square of the ratio of primary to secondary turns. Since

$$\left[\frac{N_1}{N_2}\right]^2 = a^2$$

And

$$\frac{V_1}{V_2} = a \tag{21.37}$$

$$\frac{I_1}{I_2} = \frac{1}{a} \tag{21.38}$$

Equation (21.36) is an important relation for transformers that have very high coefficients of coupling (small flux leakage). By means of this relation any impedance on the secondary side can be represented by an impedance on the primary side.

In Fig. 21.15b we see a matching transformer which, if it has the correct turns ratio, will cause the impedance seen toward the right at terminals 1, 2 to be equal to Z_S, and the impedance seen toward the left at terminals 3, 4 to be equal to Z_L. In our terminology, Z_1' in Equation (21.36) equals the source impedance Z_S here. This quantity is easily determined using Thévenin's theorem.

EXAMPLE 21.5   A source with an impedance of 2500 Ω pure resistance is to supply a load of 25 Ω pure resistance and deliver maximum possible power. If the two were directly connected, what fraction of the total power would the load get?

Solution. Let I represent the current. Total power = $2525I^2$. Load power = $25I^2$. $P_L/P_T = 25/2525 = 0.009$. What should be the turns ratio of a matching transformer (Fig. 21.15b) to be used

$$Z_1' = a^2 Z_2, \qquad 2500 = 25\left(\frac{N_1}{N_2}\right)^2$$

$$\frac{N_1}{N_2} = \sqrt{\frac{2500}{25}} = 10$$

At the terminals to which the source is connected, the impedance looking toward the load is

$$Z_{1,2} = a^2 Z_2 = 100 \times 25 = 2500 \ \Omega$$

Therefore the power delivered to terminals 1, 2 is equal to the power dissipated in the source. Since the "ideal" transformer consumes no power, the load gets all the power delivered to terminals 1, 2. The actual load current is

$$I_2 = \frac{N_1}{N_2} I_1 = 10 I_1$$

The load power is $I_2^2 R_L = 100 I_1^2 \times 25 = 2500 I_1^2$, which is also the power dissipated in the source.

21.11   The Black Box

If we know that a "black box" contains only passive elements (no energy source) in its concealed network, we can set up a T-network that will transmit the same output current to a given load impedance that the black box would transmit if both are supplied with the same input voltage. In order to determine the values of Z_1 and Z_2, the series arms of the T (Fig. 21.16) and Z_3, the shunt arm, we must first measure the following at the terminals of the black box:

Z_{01} = input impedance at end 1 with the terminals at end 2 open-circuited.
Z_{S1} = input impedance at end 1 with the terminals at end 2 short-circuited.
Z_{02} = input impedance at end 2 with the terminals at end 1 open-circuited.
Z_{S2} = input impedance at end 2 with the terminals at end 1 short-circuited.

A little thought will reveal that these four measurements give the following symbolic results:

$$Z_{01} = Z_1 + Z_3 \tag{21.39}$$

$$Z_{S1} = Z_1 + \frac{Z_2 Z_3}{Z_2 + Z_3} \tag{21.40}$$

$$Z_{02} = Z_2 + Z_3 \tag{21.41}$$

$$Z_{S2} = Z_2 + \frac{Z_1 Z_3}{Z_1 + Z_3} \tag{21.42}$$

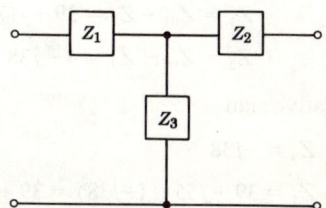

FIGURE 21.16   T network equivalent of unknown passive network in a black box.

We know we can solve these equations simultaneously for the three arm impedances, so we shall merely show the results:

$$Z_1 = Z_{01} - Z_3 \tag{21.43}$$

$$Z_2 = Z_{02} - Z_3 \tag{21.44}$$

$$Z_3 = \pm \sqrt{Z_{02}(Z_{01} - Z_{S1})} \tag{21.45}$$

There is an alternative equation for Z_3 because only three of the four impedances $Z_{01}, Z_{S1}, Z_{02}, Z_{S2}$ are unique. The alternative equation is

$$Z_3 = \pm \sqrt{Z_{01}(Z_{02} - Z_{S2})} \tag{21.46}$$

We can calculate Z_1 and Z_2 first using the plus sign for Z_3 and then the minus sign. With a resistive load on the network, one result for Z_2 will cause the load current to lag the input current, the other will cause it to lead the input current. We may choose the one that gives us the desired phase relation. The following example will illustrate this.

Example 6 Three of the four impedances were measured at the terminals of a black box containing only passive elements and found to be: $Z_{01} = 39 + j55 \ \Omega$; $Z_{S1} = 400 + j55 \ \Omega$; $Z_{02} = 4 + j0 \ \Omega$. When the output current I_2, through a resistive load was measured, it was found to lag I_1. Evaluate $Z_1, Z_2,$ and Z_3 of an equivalent T-network.

Solution

$$Z_3 = \pm \sqrt{(4\underline{/0°})(361\underline{/-180°})} = \pm \sqrt{1444\underline{/-180°}} \ \Omega$$

When we take the square root of a phasor quantity we may first express its angle as a *positive* number before we divide it by two.

$$Z_3 = \pm 38\underline{/90°} = \pm j338 \ \Omega$$

(a) Using the positive sign,

$$Z_3 = j38$$

$$Z_1 = Z_{01} - Z_3 = 39 + j17$$

$$Z_2 = Z_{02} - Z_3 = 4 - j38$$

(b) Using the negative sign

$$Z_3 = -j38$$

$$Z_1 = 39 + j55 - (-j38) = 39 + j93$$

$$Z_2 = 4 + j0 - (-j38) = 4 + j38$$

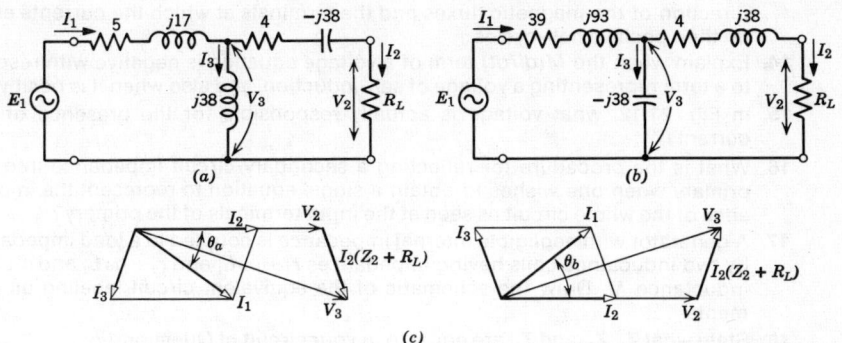

FIGURE 21.17   Elements and phasor diagrams for equivalent T coupling networks for positive and negative values of Z_3. (a) Equivalent of T with $Z_3 = j38\ \Omega$. (b) Equivalent T with $Z_3 = -j38\ \Omega$.

The phasor diagrams of Fig. 21.17 for the two cases show that the load current I_2 lags the input current in (a) but leads in (b). Therefore we use the negative sign for Z_3 and the equivalent T is that in (b).

21.12   Review Questions

1. When two circuits are said to be *coupled*, what does this mean?
2. How does conductive coupling differ from inductive coupling?
3. Sketch the general form of the schematic diagram of a conductively coupled circuit (using squares or boxes to represent impedances) and identify the mutual impedance.
4. Sketch the schematic of a generator with some internal resistance and an inductive load coupled by a π-network of pure resistance. How could the coupling network be replaced by a T-type resistance network that would produce the same effect as the π-network?
5. Explain a case in which a capacitor and resistor are used to couple voltage from a source to a load.
6. What is necessary in a coupled circuit to make $Z_{12} = Z_{21}$?
7. Define transfer impedance.
8. Define forward current transfer ratio.
9. Define reflected impedance as related to coupled circuits.
10. What is the general formula for calculating coefficient of coupling?
11. Why are the fluxes produced by I_1 and I_2 in opposition in a transformer with only one voltage source?
12. If the secondary and load impedances of a transformer are Z_2 and Z_L, how would you calculate the impedance reflected into the primary?
13. Describe fully the conventional method of indicating the polarity of a coil of transformer windings. Tell how one can determine the relation between

direction of the magnetic fluxes and the terminals at which the currents enter on their positive half-cycles.

14. Explain when the $M(di/dt)$ term of a voltage equation is negative with respect to a term representing a voltage of self-induction, and also when it is positive.

15. In Fig. 21.12, what voltage is actually responsible for the presence of the current i_2?

16. What is the procedure for reflecting a secondary-circuit impedance into the primary when one wishes to obtain a single equation to represent the impedance of the whole circuit as seen at the input terminals of the primary?

17. A generator with negligible internal impedance is coupled to a load impedance by two inductance coils having impedances $r_1 + j\omega L_1$ and $r_2 + j\omega L_2$ and mutual inductance M. Draw the schematic of the equivalent circuit labeling all elements.

18. State what $\mathbf{Z}_1$, $\mathbf{Z}_2$, and $\mathbf{Z}_M$ are equal to in your circuit of Question 17.

19. What are the terminal impedances called that one can measure on a "black box" network, and what measurement does each represent?

20. In calculating the arms of an equivalent T coupling network to represent the black box network, how should one choose between the positive and negative signs he gets for Z_3?

21.13 Problems

6. What are the self impedances (a) of circuit 1 in Fig. 21.18, (b) of circuit 2? What is the mutual impedance of the network?

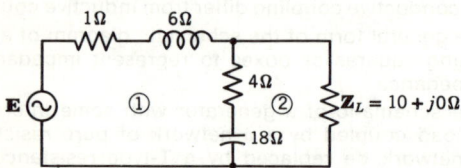

FIGURE 21.18   For Problem 6.

7. Calculate the generator and load currents in Fig. 21.18 when $E = 10\underline{/0°}\,\text{V}$, using self and mutual impedances.

8. Calculate the power in each branch in Fig. 21.18, and compare the total power with what you get using $EI \cos\theta$.

9. A π-network couples a generator to a load, as in Fig. 21.3. The series arm is $5\,\Omega$, the first shunt arm is $10\,\Omega$, and the second shunt arm is $15\,\Omega$. The induced EMF of the generator is 5 V and its internal resistance is $2\,\Omega$. The load is $5\,\Omega$ resistive. Determine the following: Z_{12}, Z_{21}, Z_{11}, Z_{22}, I_G, I_L.

10. What is the transfer impedance of the network in Problem 9?

11. Two coupled circuits arranged like those in Fig. 21.1. have $\mathbf{E} = 100\underline{/0°}\,\text{V rms}$, $Z_1 = 8 + j6\,\Omega$, $Z_2 = 5 + j12\,\Omega$, $Z_{12} = 0 + j10\,\Omega = Z_{21}$. (a) Calculate the generator

current, the current in Z_2, and the current in the mutual impedance, (b) Calculate the total power in two ways, (c) Draw the phasor diagram.

12. Determine I_1, I_2 and the current in R_3 in Fig. 21.19. Use Equations (21.1) and (21.2), using E_2 in place of zero in Equation (21.2). Assume both currents flow downward in R_3.

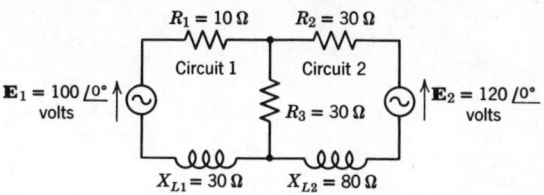

FIGURE 21.19   For Problem 12.

13. An air-core transformer couples a generator, with negligible internal impedance, to a 5 Ω resistive load. The winding inductances are $L_1 = 1$ mH, $L_2 = 1.2$ mH, $M = 0.85$ mH and their resistances are $r_1 = 2$ V, $r_2 = 4\,\Omega$. The terminal voltage of the generator is $10\underline{/0°}$ V at 398 Hz. What is the total equivalent impedance connected to the generator?

14. What are the values of the generator and load currents of Problem 13? How much power does the generator put out and how much of it does the load receive?

15. Two coils in series have their fluxes aiding and are supplied with 100 V rms, 1592 Hz. The circuit constants are: $r_1 = 10\,\Omega$, $L_1 = 10$ mH, $r_2 = 16\,\Omega$, $L_2 = 12$ mH, $M = 9$ mH. Find the (a) coefficient of coupling, (b) equivalent series impedance, (c) current. Draw the phasor diagram.

16. Assume the fluxes of the coils of Problem 15 are in opposition, $M = -9$ mH. Calculate the quantities requested. Why should the current be so much larger?

17. A transformer has a primary impedance $Z_1 = 2 + j10\,\Omega$, a secondary impedance $Z_2 = 1 + j4\,\Omega$, and a load impedance $Z_L = 5 + j10\,\Omega$. The mutual impedance is $-j10\,\Omega$. What would be the impedance, reflected into the primary, of the total secondary-circuit impedance? What, then, would be the total equivalent primary impedance?

18. An air-core transformer has $L_1 = 50$ mH, $r_1 = 10\,\Omega$, $L_2 = 40$ mH, $r_2 = 4\,\Omega$, $M = 30$ mH. Draw the equivalent circuit, showing a load $R_L = 50\,\Omega$ and an input voltage $E_1 = 1$ V at $\omega = 10^3$ rad/s. Calculate the input current and the load current. What is the transfer impedance?

19. A generator has an induced EMF of 10 V rms and an internal resistance of 1600 Ω. (a) How much power would it deliver to a 100-Ω load if it were directly connected? (b) Calculate the turns ratio of an ideal matching transformer that would make possible the delivery of maximum power to the load and also the maximum power.

CHAPTER TWENTY-TWO

Power Transformers

Transformers designed to change the voltage of electric power from one value to another are in everyday use throughout the world. Large transformers step up the voltage at which power is received from generators so that it can be transmitted long distances. As the voltage is being multiplied to hundreds of times its initial value, the current is stepped down in inverse proportion. A power transformer that receives 800 A at 6600 V may deliver 20 A at 264,000 V.

At the receiving end of a transmission line, transformers are employed to step down the voltage and, at the same time, to step up the current. Distribution lines serving a residential area may supply power to local "distribution" transformers at 7200 V. The secondaries of the transformers deliver power to a three-wire system which supplies residences with power at 120 V and 240 V at the same time.

A power transformer in a television receiver that may receive 7.5 mA at 120 V may deliver 50 μA at 18,000 V.

Power transformers have steel cores on which their coils are placed. Although it is customary to use the term *iron core*, the metal is really high-quality sheet steel alloy containing some silicon or other elements to improve its magnetic properties. The core is built up of layers of thin sheet-steel *laminations* that are covered with a varnish that insulates them electrically from one another. This is to minimize energy loss produced by eddy currents which we discussed in Section 8.19.

22.1   Basic Circuit

Figure 22.1 shows a "step-down" power transformer that is ready for connection to a load. In actual design the core has three legs, such as a letter E would have. The windings are slipped over the center leg and then a vertical piece of built-up core is added to complete the magnetic circuit. The windings are thus surrounded by a ferromagnetic path so that the magnetic flux can be a maximum. Each layer of the E lamination is reversed in direction so that the core will stay together and the very short air gaps where the sheets butt together will be distributed around the core instead of being all on one side.

With the switch open, the transformer is *unloaded*, but with E_1 applied the primary will take a small amount of current. The power input is all lost, since there is no power output, and the loss occurs in the iron (hysteresis and eddy current losses) and in the I^2R of the primary winding.

We should say a word here about using E and V. In this illustration E_2 is an induced voltage. We shall continue to use V for voltage drop, but if we use E occasionally for applied voltage it should not cause undue concern. It will be identified properly.

The basic theory of operation of inductively coupled coils (Chapter 10) applies here, but the coupling coefficient is very near unity. Energy is readily transferred from one coil (or set of coils) called the *primary* to the other coil (or set of coils) called the *secondary* by means of the magnetic flux which continuously varies in amount and periodically reverses its direction from clockwise to counterclockwise in the iron magnetic circuit.

A distribution system supplies the primary winding with an alternating current. This produces an alternating flux ϕ in the core, which, *while it is changing*, induces an alternating EMF in the turns of the secondary. This EMF has the same frequency as that of the primary current, and its presence means that if a "load," such as a resistance or an ac motor, is connected to the secondary terminals it will receive *current* and *energy* from the secondary

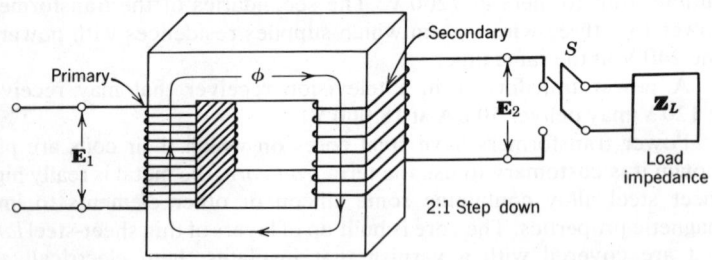

FIGURE 22.1   **Sketch of iron-core transformer with open secondary E_2.**

winding. In a transformer either winding may serve as the primary since *by definition* the primary is the winding that receives energy and the secondary is the winding that delivers energy. When a transformer is used to deliver energy at a *higher voltage* than that at which it is received, the primary winding is the one with the small number of turns and the larger wire in its coils.

An important point must be made clear here. The *amount of current* delivered by the secondary coil of a transformer *depends on the impedance connected to its secondary terminals.* If the impedance has a large value, the secondary current value will, of course, be relatively small, and vice versa. Except at no load and at very light loads (i.e., a few percent of full load), the secondary and primary currents are related inversely as the secondary and primary terminal voltages. This statement needs qualification and further explanation which will be given later, with the conclusion that if losses (which produce heat in the windings and in the iron) are neglected, and if unity power factor is assumed,

$$E_2 I_2 = E_1 I_1, \qquad \frac{I_2}{I_1} = \frac{E_1}{E_2}$$

The losses in the transformer are the $I^2 R$ losses in the windings and the losses due to hysteresis and eddy currents in the iron of the core.

22.2 Induced EMF's

The mutual flux ϕ links both primary and secondary windings and, therefore, induces voltages in them. The induced *volts per turn* is the same in the windings, so the total *induced EMF* in each winding is proportional to the number of turns in that winding. The ratio of the induced voltages must then be equal to the ratio of the numbers of turns:

$$\frac{E_1}{E_2} = \frac{N_1}{N_2} \qquad\qquad (22.1)$$

E_1 and N_1 are the primary induced voltage and turns, E_2 and N_2 are the secondary induced voltage and turns. In the practical power transformer the terminal voltage differs from the induced EMF by only a very small percentage. In most cases in practice it may be said that the primary and secondary *terminal voltages* are directly proportional to their number of turns.

Because induced voltage in a coil is proportional to the *time rate of change of flux*, one would expect the voltage induced in a transformer coil to depend on frequency, maximum value of flux, and number of turns. The rms value of EMF induced in a winding of N turns is given by Equation (22.2), and is derived as follows. In Fig. 22.2 the flux is seen to vary sinusoidally through the value $2\phi_m$ webers in $\frac{1}{2}T$ seconds, which is one-half the time of one complete cycle and this time is $1/2f$ where f is the frequency.

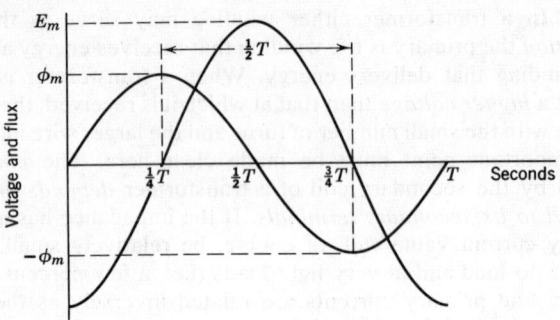

FIGURE 22.2   Instantaneous values of flux varying sinusoidally with time and of induced voltage.

The *average* induced EMF is the total change in flux divided by the time:

$$e_{avg} = -N\frac{2\phi_m}{1/2f}\text{volts}$$

$$= -4fN\phi_m \text{ volts}$$

With a sine wave, $e_{rms} = 0.707\,e_{max}$ and $e_{av} = 0.636\,e_{max}$. Therefore, $e_{rms} = 1.11\,e_{av}$. Let E' represent the rms value:

$$E' = -4.44fN\phi_m \text{ volts} \tag{22.2}$$

in which E' is the rms value of voltage induced in a coil of N turns when a flux of ϕ_m webers maximum value varies sinusoidally at a frequency of f cycles per second.

A more rigorous derivation of Equation (22.2) is

$$\phi = \phi_m \sin \omega t \qquad \text{instantaneous flux value}$$

$$e' = -N\frac{d\phi}{dt} = -N\phi_m\omega \cos \omega t \text{ volts}$$

The maximum instantaneous EMF is given by the coefficient,

$$E'_m = N\phi_m\omega = 2\pi fN\phi_m \text{ volts}$$

Rms volts:

$$E' = \frac{2\pi}{\sqrt{2}}fN\phi_m = 4.44fN\phi_m \text{ volts}$$

Notice that the induced voltage e' is expressed as a negative cosine function. This shows that it lags the sinusoidal flux inducing it by 90° (Figure 22.2).

EXAMPLE 22.1   A 60 Hz transformer has a core cross section of 20 in.², and a maximum flux density in the core of 6.1×10^{-4} W/in.². There are 720 turns in the primary winding and 72 turns on the secondary. What are the values of the induced EMF's?

In the secondary: $E_2' = 4.44 \times 60 \times 72 \times 6.1 \times 10^{-4} \times 20 = 233.5$ V

In the primary: $E_1' = 4.44 \times 60 \times 720 \times 6.1 \times 10^{-4} \times 20$
$$= 2335 \text{ V}$$

22.3   An Ideal Transformer at No Load

Equation (22.1) is exact for an ideal transformer (winding resistances assumed zero, iron losses zero, and no leakage flux) and this equation is accurate for an actual (practical) transformer at no load, because when I_2 is zero there can be no difference between the induced voltage in the secondary winding and its terminal voltage.

Assume the voltage $\mathbf{E}_1$, Fig. 22.3, is applied to the primary of a 2:1 step-down ideal transformer that has no load, that is, $\mathbf{I}_2 = 0$. The voltage $\mathbf{E}_2'$ induced in the secondary winding is one-half the voltage $\mathbf{E}_1'$ induced in the primary winding. As stated in Section 22.2, $\mathbf{E}_1'$ lags ϕ by 90°. $\mathbf{E}_2'$ is in phase with $\mathbf{E}_1'$ because it is induced by the same flux. $\mathbf{E}_1'$ opposes $\mathbf{E}_1$, the voltage applied to the primary, and is equal to it in magnitude in this "ideal" case, since there are no losses.

The input current $\mathbf{I}_0$ in the primary winding is called the *exciting current*. Note that, in this ideal case, the input power $E_1 I_0 \cos 90°$ is zero. $\mathbf{I}_0$ is very small compared with the full-load current. Equation (21.1) is in error by 2 or 3 percent in an actual transformer operating at good efficiency (i.e., at near full load and at high power factor).

22.4   Actual Transformer at No Load

The iron core of a practical transformer heats up as a result of energy losses caused by hysteresis and eddy currents in the iron. The windings have resis-

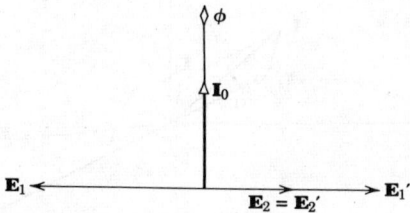

FIGURE 22.3   Phasor diagram for an ideal transformer with open secondary terminals.

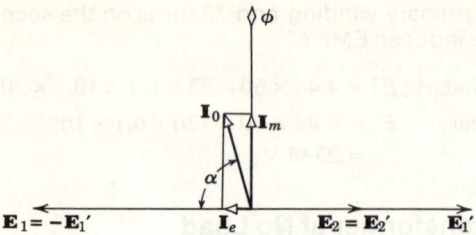

FIGURE 22.4 Phasor diagram for an actual power transformer at no load.

tance, and therefore I^2R losses are present in them. When rated voltage is applied to the primary terminals of the transformer, the exciting current $\mathbf{I}_0$ must not only produce flux but it must supply these *no-load* losses. This means there is an *energy component* $\mathbf{I}_e$ of the exciting current which is in phase with the applied voltage (Fig. 22.4). $E_1 I_e \cos 0° = E_1 I_e =$ no-load power loss. $\mathbf{I}_0$ lags $\mathbf{E}_1$ by the angle α, so that $I_e = I_0 \cos \alpha$, $I_m = I_0 \sin \alpha$. The iron losses are then given by $E_1 I_0 \cos \alpha$. Actually the primary copper loss *at no load* is included in the expression $E_1 I_e$, but it is negligibly small compared with the iron loss and, therefore, is usually ignored.

22.5 Power Transformer under Load

Assume that the switch in Fig. 22.1 is now closed and that rated voltage $\mathbf{E}_1$ is applied to the power transformer which has a 2 : 1 step-down ratio. The nature of the load impedance Z will determine the phase angle α_2 (Fig. 22.5) between $\mathbf{E}_2$ and $\mathbf{I}_2$, since $\mathbf{I}_2 = \mathbf{E}_2/\mathbf{Z}_2$. Obviously, $\mathbf{I}_2$ will lead $\mathbf{E}_2$ if $\mathbf{Z}_2$ has capacitive reactance. The following paragraph explains the rapid buildup of $\mathbf{I}_2$ and $\mathbf{I}_1'$.

$\mathbf{I}_2$ exerts a magnetomotive force of $N_2 I_2$ ampere-turns on the same magnetic circuit as the primary ampere-turns $N_1 I_0$ is acting. Now I_0 is very small and

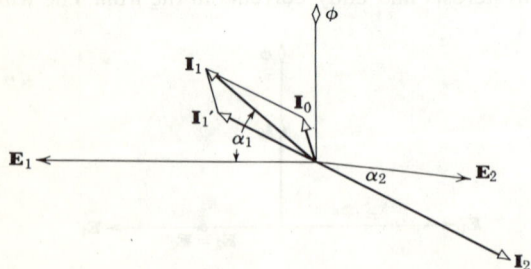

FIGURE 22.5 Phasor diagram for a 2 : 1 power transformer carrying a load.

it remains practically constant, so that N_2I_2 is about twenty-five times as large (at full load) as N_1I_0. Now, flux produced by a current *is in phase with the current*. Since $\mathbf{I}_2$ has a component of substantial magnitude which is opposite in direction to $\mathbf{I}_0$ (we should say $\mathbf{I}_m$, but there is negligible difference in the phase of these two), the effect of $\mathbf{I}_2$ is to oppose the flux that induces $\mathbf{E}_1'$. This reduces $\mathbf{E}_1'$, which is really responsible for limiting the input current $\mathbf{I}_1$ to a practical working value, and the result is that $\mathbf{I}_1$ increases to build up the flux that $\mathbf{I}_0$ has started. The net result is that current $\mathbf{I}_1'$ builds up in the primary winding until its ampere-turns oppose and neutralize the ampere-turns of $\mathbf{I}_2$, thus

$$N_1I_1' = N_2I_2 \tag{22.3}$$

$\mathbf{I}_1'$ is, of necessity, 180° out of phase with $\mathbf{I}_2$, as shown in Fig. 22.5. The total primary current is then $\mathbf{I}_1$, the phasor sum of $\mathbf{I}_1'$ and $\mathbf{I}_0$.

At and near full load, $\mathbf{I}_0$ is so small compared with $\mathbf{I}_1'$ that $\mathbf{I}_1$ may be substituted for $\mathbf{I}_1'$ in equation (22.3). Then, practically, $N_1I_1 = N_2I_2$ and

$$\frac{I_1}{I_2} = \frac{N_2}{N_1} \tag{22.4}$$

This equation is not applicable when the transformer is operated at light loads.

EXAMPLE 22.2   A power transformer at no load was supplied with 240 V. The current was found to be 1.52 A, and a wattmeter in the line read 43.8 W. The primary winding resistance was 0.8 Ω. The following important information can be calculated from these data.

1. Power factor = cos α
   = 43.8/(240 × 1.52)
   P.F. = 0.12

2. Magnetizing component of current:

$$I_m = I_0 \sin \alpha = 1.52 \sin (\text{arc } \cos 0.12)$$
$$I_m = 1.52 \sin 83.1° = 1.509 \text{ A}$$

3. Energy component of current:

$$I_e = I_0 \cos \alpha = 1.52 \cos 83.1°$$
$$I_e = 0.1824 \text{ A}$$

4. Copper loss:

$$I_0^2 R = 1.52^2 \times 0.8 = 1.85 \text{ W}$$

5. Core loss:

$$43.8 - 1.85 = 41.95 \text{ W}$$

Problems

1. The maximum instantaneous value of the core flux in a power transformer is 0.01 Wb. There are 2000 turns on the secondary winding and the frequency is 60 Hz. (a) Compute the voltage induced in the secondary at no load. (b) Compute the voltage induced in the primary, if the transformer is used as a step-up device with a ratio of 1 : 20.

2. A power transformer has 440 V at 60 Hz applied to its high-voltage terminals while its low-voltage terminals are open. The voltage at these terminals is determined to be 100 V by means of a high-impedance voltmeter. The input power at 440 V is 70.8 W and the input current is 1.4 A. (a) Draw a phasor diagram similar to Fig. 22.4 for this no-load situation; (b) What are the ampere values of the magnetizing current and the core-loss component of current?

3. Assume that the transformer of Problem 1 is carrying full-load current, 100 A, at 0.9205 power factor lagging. The secondary terminal voltage is 110 V. Assume the exciting current to be the same as that in Problem 1. Determine the values of the phasors in Fig. 22.5, and construct the diagram.

22.6 Leakage Reactance

We learned about leakage flux in our study of mutual inductance in Chapter 10. The coefficient of coupling between the primary and secondary windings of a practical power transformer is not unity, which would be perfect. There is some leakage flux.

The part of the primary flux that does not link the secondary turns produces a *primary-leakage reactance* X_1. The part of the flux produced by $\mathbf{I}_2$, that does not link the primary turns produces a *secondary-leakage reactance* X_2. The effects of these reactances and of the corresponding winding resistances are represented in Fig. 22.6 as causing voltage drops *external to a transformer that has no leakage flux and no resistance in its windings*. The whole illustration, however, represents a *practical power* transformer carrying a load, which means it has current $\mathbf{I}_2$ in its secondary winding.

We may use the same kind of phasor diagram for the portion of Fig. 22.6 between points *a, b* and *c, d* as shown in Fig. 21.5 for the no-load case where

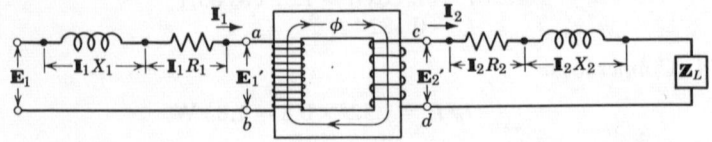

FIGURE 22.6 Winding resistances and leakage reactances cause voltage drops in a power transformer.

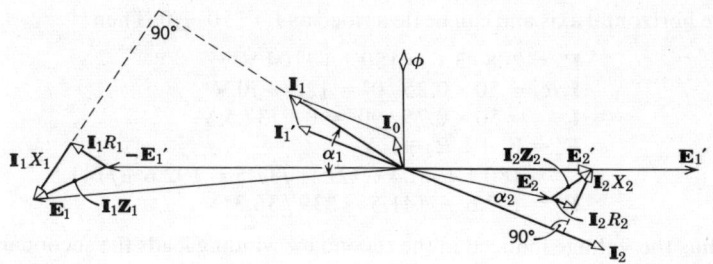

FIGURE 22.7   Phasor diagram for a loaded power transformer. Cos α_2 is the power factor of the load, cos α_1 is the power factor at the primary terminals of the transformer.

leakage fluxes were ignored. Figure 22.7 was started, therefore, by drawing $\mathbf{E}_2'$, $\mathbf{E}_1'$, $\mathbf{I}_2$, ϕ, $\mathbf{I}_0$, $\mathbf{I}_1'$, $\mathbf{I}_1$. Now $-\mathbf{E}_1'$ is the part of the applied primary voltage ($\mathbf{E}_1$) which is overcome by the "counter EMF" induced in the primary winding by the changing flux, ϕ. The applied voltage must be large enough, and properly phased, to supply the $\mathbf{I}_1R_1$ and $\mathbf{I}_1X_1$ voltage drops in the primary winding. Therefore, we must add $\mathbf{I}_1R_1$ in phase with $\mathbf{I}_1$, and $\mathbf{I}_1X_1$ 90° *ahead* of $\mathbf{I}_1$. Thus we obtain the phase position and magnitude of the phasor $\mathbf{E}_1$ representing the voltage applied to the primary winding.

The phasor $\mathbf{E}_2'$ represents the voltage induced in the secondary winding by the changing flux ϕ. The secondary terminal voltage will be what is left after the $\mathbf{I}_2R_2$ and $\mathbf{I}_2X_2$ drops are subtracted from $\mathbf{E}_2'$. $\mathbf{I}_2X_2$ leads $\mathbf{I}_2$ by 90°, so $-\mathbf{I}_2X_2$ is drawn equal and opposite to $\mathbf{I}_2X_2$. $\mathbf{I}_2R_2$ is in phase with $\mathbf{I}_2$, so $-\mathbf{I}_2R_2$ is drawn opposite and equal to $\mathbf{I}_2R_2$. The phasor $\mathbf{E}_2$ then represents the terminal voltage of the transformer secondary. This is, of course, the load voltage.

The angle α_2 is the power-factor angle of the load; the angle α_1 is the power-factor angle of the transformer and its load. Cos α_1 is the power factor that would be used in the equation giving the input power to the transformer. That is, P_1, the power input to the transformer, is given by

$$P_1 = \mathbf{E}_1\mathbf{I}_1 \cos \alpha_1$$

EXAMPLE 22.3   A power transformer is found, by measurement, to deliver 50 A full load at 208 V at a power factor of 0.866. Its secondary winding has 0.25 Ω resistance and the secondary leakage reactance is 0.75 Ω at 60 Hz. What is the voltage induced in the secondary winding?

Solution. We are to determine the value of $\mathbf{E}_2'$ in Fig. 22.7:

$$\cos \alpha_2 = 0.866, \qquad \alpha_2 = 30°$$

Assume that the phasor diagram is rotated counterclockwise until $\mathbf{I}_2$ is on the horizontal axis and can be described as $I_2 = 50 + j0$. Then

$$E_2 = 208 \underline{/30^\circ} = 180.1 + j104 \text{ V}$$
$$I_2 R_2 = 50 \times 0.25 \underline{/0^\circ} = 12.5 + j0 \text{ V}$$
$$I_2 X_2 = 50 \times 0.75 \underline{/90^\circ} = 0 + j37.5 \text{ V}$$
$$E_2' = E_2 + I_2 R_2 + I_2 X_2$$
$$= 180.1 + j104 + 12.5 + j37.5 = 192.6 + j141.5$$
$$E_2' = 192.6 + j141.5 = 239 \underline{/36.3^\circ} \text{ V}$$

Thus the voltage induced in the secondary winding leads the secondary current by 36.3°.

EXAMPLE 22.4   The transformer of Example 22.3 has a 2:1 step-down ratio. Its primary resistance is 1 Ω and its leakage reactance is 3 Ω. What is the applied voltage?

Solution

$$\frac{E_1}{E_2'} = \frac{N_1'}{N_2'} \quad \text{[see Equation (22.1)]}$$

The voltage induced in the primary winding is then

$$E_1' = \frac{N_1'}{N_2'} E_2' = \frac{2}{1} \times 239 \underline{/36.3^\circ} = 478 \underline{/36.3^\circ} \text{ V}$$

The *counter EMF* that opposes the voltage applied to the input terminals is then $-E_1' = 478 \underline{/36.3^\circ} + 180^\circ = 478 \underline{/215.3^\circ} \text{ V} = -385.2 - j283.0$.

Adding to this the $I_1 R_1$ and $I_1 X_1$ drops, we get the applied voltage E_1. But we do not as yet know the magnitude and phase of I_1, the total primary current. We could measure the exciting current I_0 and the no-load power factor by applying about 500 V to the primary when the secondary is unloaded. Should the calculated terminal voltage turn out to be 5 or 10 percent larger or smaller than 500 V, the error in measuring I_0 would be negligible in its effect on the computation of E_1. So let us assume I_0 to be 1.25 A at 80° behind $(-E')$.

$$I_1' = \frac{N_2}{N_1} I_2 = \frac{1}{2} \times 50 = 25 \text{ A}$$

$$I_1' = 25 \underline{/180^\circ} = -25 + j0 \text{ A}$$

By assumption (or measurement), the phase of I_0 lags the phase of $-E'$ by 80°, so its phase is $\underline{/216.3^\circ} - 80^\circ = \underline{/136.3^\circ}$. The complex form of I_0 is then

$$I_0 = 1.25 (\cos 136.3^\circ + j \sin 136.3^\circ)$$
$$I_0 = -0.904 + j0.864 \text{ A}$$

The total primary current is

$$\mathbf{I}_1 = \mathbf{I}_1' + \mathbf{I}_0 = -25 + j0 - 0.904 + j0.864 = -25.9 + j0.864 \text{ A}$$

The primary resistance drop is in phase with $\mathbf{I}_1$ and is

$$\mathbf{I}_1 R_1 = 0.25(-25.9 + j0.864) = -6.475 + j0.216 \text{ V}$$

The primary reactance drop is $j\mathbf{I}_1 X_1$. Note that the reactance drop must be 90° ahead of $\mathbf{I}_1$. Therefore we must multiply by j.

$$\mathbf{I}_1 X_1 = j0.75(-25.9 + j0.864) = 0.648 - j19.43 \text{ V}$$

The voltage applied to the primary is

$$\begin{aligned}
\mathbf{E}_1 &= -\mathbf{E}_1' + \mathbf{I}_1 R_1 + \mathbf{I}_1 X_1 \\
&= -385.2 - j283 - 6.475 + j0.216 - 0.648 - j19.43 \\
\mathbf{E}_1 &= -392.3 - j302.2 = 495.3 \underline{/217.6°} \text{ V}
\end{aligned}$$

It must be remembered that all of these phase positions are specified on the assumption that $\mathbf{I}_2$ is at zero phase. That is, we said we would rotate the whole phasor diagram counterclockwise until the $\mathbf{I}_2$ phasor was horizontally to the right, which made $\mathbf{I}_2 = 50 + j0$. Since $\mathbf{I}_2$ is measurable, simply by means of an ac ammeter in series with the load, we may assume it has zero phase.

We may now prefer to redraw the phasor diagram with $\mathbf{E}_2$, the load voltage (at the secondary terminals) chosen to have zero phase. This would mean we must subtract the secondary power-factor angle (30°) from all of the phasors. None of these rotations affect the magnitudes of the quantities.

The positions of the phasors in Fig. 22.7 result from the logical development of the theory. After the student becomes familiar with the relationships among the quantities involved, he can readily start drawing the phasor diagram with $\mathbf{E}_2$, the secondary terminal voltage, at zero phase. He must know which voltages are 180° out of phase with each other, and which currents are 180° out of phase with each other also.

Problem

4. Redraw the phasor diagram for Example 22.3, placing the phasor $\mathbf{E}_2$ horizontally to the right so that $\mathbf{E}_2 = 208 + j0$. Label all phasors with angles with respect to the phase of $\mathbf{E}_2$, which is zero degrees.

22.7   Design Equation of a Power Transformer

In the design of power transformers it is recognized that Equation 22.2, which defines the EMF induced in a winding, also defines the terminal voltage of the winding at light loads, for all practical purposes. We reproduce the equation here, but use terminal voltage E instead of induced voltage $-E'_1$:

$$E = 4.44 f N \phi_m \text{ volts}$$

Now we know that total flux is equal to flux density times cross-section area of the iron path:

$$\phi = BA$$

Then

$$E = 4.44 f N B_m A \text{ volts}$$

From which

$$NA = \frac{E}{4.44 f B_m} \tag{22.5}$$

in which N = number of turns and A is the cross-section area of the flux path. this is the principal design equation for a power transformer.

EXAMPLE 22.5   Obtain some preliminary basic design data for a transformer which is to operate on 60 Hz frequency and deliver power at 120 V.

Solution. Magnetization curves for transformer steel which show flux density B in thousands of lines per square inch plotted against H, in ampere-turns per inch, are available. From these curves it is determined that a workable value at the knee of the curve might be $B = 90,000$ lines $(9 \times 10^{-4} \text{ Wb})/\text{in}^2$. Using these values in Equation (22.5),

$$N_2 A = \frac{120}{4.44 \times 60 \times 9 \times 10^{-4}} = 500$$

Because flux density is expressed in webers per square inch in this equation, area must be in square inches.

N_2, the number of turns on the secondary winding, may be chosen to give the proper proportion of copper to iron to provide minimum cost. A choice of 100 turns means an area of 5 sq in. will be required; 200 turns will cut the area in half, obviously. A desire to achieve minimum weight or volume may enter into the choice of N_2. The size of wire is determined by the amount of current to be carried.

Equation (22.5) shows, also, that doubling the frequency will cut the area requirement in half once the number of turns is fixed. This is one reason that 60-cycle transformers sell for less money than 25-cycle transformers.

22.8   Transformer Losses and Efficiency

We learned about the nature of hysteresis and eddy currents in magnetic materials in Chapter 8. Energy losses in transformer cores caused by these effects

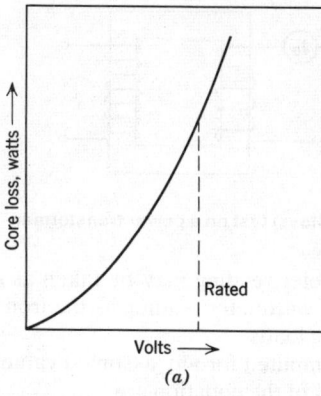

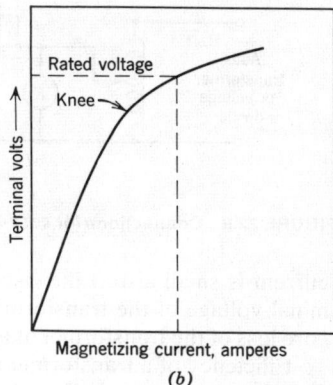

FIGURE 22.8 **(a)** Variation of core loss with operating voltage of a transformer. **(b)** A form of magnetization curve showing the relation between operating voltage of a transformer and magnetizing current required.

are significant. It has been found that the loss attributed to eddy currents varies as the square of the flux density and also as the square of the voltage. Loss owing to hysteresis varies as the 1.6 power of the flux density, and as the 1.6 power of the voltage. Therefore the core loss will increase nearly as the square of the voltage, as illustrated in Fig. 22.8a.

In Fig. 22.8b it is seen that transformers are operated at magnetizing current values which correspond to points well beyond the knee of the saturation curve. This is done to procure as high efficiency as possible compatible with maximum load-carrying capacity at allowable temperature. If a power transformer is operated at a voltage much above rated value, not only will the core loss increase and cause excessive temperature rise, but the magnetizing current increases very rapidly with a small increase in voltage. This contributes to more temperature rise and to a decrease in efficiency.

Iron loss in a transformer (loss attributed to hysteresis and eddy currents) is assumed to be constant at all loads if the applied voltage E_1 is kept constant (which is usually the case). The iron loss, usually termed *core loss*, is readily measured by an *open-circuit test*. The connections for this test are shown in Fig. 22.9. *Rated voltage E at rated frequency* is applied to the low-voltage winding with the high-voltage terminals left open. The power input is then equal to the constant iron loss plus a negligible amount of copper loss (I^2R) in the winding. Depending on the voltage ratings of the transformer, it may be more practical to take readings on the high voltage winding while the low-voltage winding is left open.

The reading of the wattmeter includes the power loss in the ammeter, which is ignored because the resistance of the ammeter coil is so very small and the

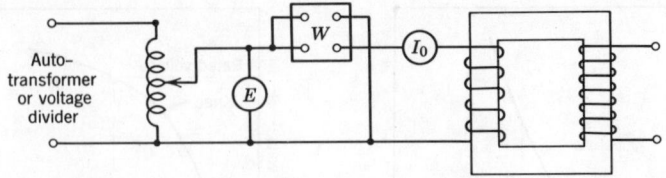

FIGURE 22.9 Connections for core-loss (or iron-loss) test on a power transformer.

current is small also. Likewise the voltmeter reading may be taken as the terminal voltage of the transformer, and the wattmeter reading as the iron loss or core loss of the transformer at all operating loads.

Efficiency of a transformer may be computed for any assumed value of current if the iron loss is known. Useful forms of the equation are

$$\text{Efficiency} = \frac{\text{output power}}{\text{input power}} = \frac{\text{output}}{\text{output} + \text{losses}} = \frac{\text{input} - \text{losses}}{\text{input}}$$

in which the losses are iron loss $+ I_1{}^2 R_1 + I_2{}^2 R_2$, the latter two being copper losses in the primary and secondary windings.

EXAMPLE 22.6 A distribution transformer is rated at 50 kVA, 2300 to 230 V. Its primary-winding resistance is 0.5 Ω, its secondary-winding resistance is 0.005 Ω, and its iron loss is 290 W. Determine the efficiency under each of the following load conditions: (a) 50 kW, unity power factor, (b) 5 kW, unity power factor, (c) 50 kVA, 80 percent power factor.

Find the *all-day* efficiency when the load is as specified in part c but connected for an 8-h day, although the primary is left on the 2300-V line 24 h of the day.

Solution

(a) $I_2 = \dfrac{50,000}{230} = 217.4 \text{ A}, I_1 = \dfrac{50,000}{2300} = 21.74 \text{ A (approximately)}$

To get I_1 in this way we must assume this is an ideal transformer. Otherwise we would have to work with the phasor diagram, and after we finished we would find that the true I_1 value is so near 21.74 A that the result is not worth the effort.

Copper loss $= 217.4^2 \times 0.005 + 21.74^2 \times 0.5 = 473$ W
Iron loss $= 290$ W
Total loss $= 763$ W
Output $= 50,000$ W
Input $= 50,000 + 763 = 50,763$ W
Efficiency $= 50,000 \div 50,763 = 0.985 = 98.5$ percent

(b) $I_2 = \dfrac{5000}{230} = 21.74$ A, $I_1 = \dfrac{5000}{2300} = 2.174$ A (approximately)

Copper loss $= 21.74^2 \times 0.005 + 2.174^2 \times 0.5 = 4.73$ W
Iron loss $= 290$ W
Total loss $= 294.73$ W
Output $= 5000$ W
Input $= 5294.73$ W
Efficiency $= 5000 \div 5294.73 = 0.944 = 94.4$ percent

(c) $I_2 = \dfrac{50,000}{230} = 217.4$ A, $I_1 = \dfrac{50,000}{2300} = 21.74$ A (approximately)

Copper loss $= 217.4^2 \times 0.005 + 21.74^2 \times 0.5 = 473$ W
Iron loss $= 290$ W
Total loss $= 763$ W
Output $= 50,000 \times 0.8 = 40,000$ W
Input $= 40,000 + 763 = 40,763$ W
Efficiency $= 40,000 \div 40,763 = 0.982 = 98.2$ percent

(d) All-day efficiency is defined as the ratio of the total *energy output* during 24 h to the total *energy input* in the same period. We assume, in this example, that the load is constant during the 8 h. For the load conditions in part c:

$$\text{All-day efficiency} = \frac{50,000 \times 0.8 \times 8 \times 100}{50,000 \times 0.8 \times 8 + 290 \times 24 + 473 \times 8} = 96.8 \text{ percent}$$

Certainly, an industrial load is not expected to be constant during two 4-h periods in a day. To compute all-day efficiency in a practical case, one may read current values at suitable time periods, every half hour for example, and square each value. The sum of these squares divided by the number of readings gives the average value of I^2. This value, multiplied by the resistance of the proper winding, will give the copper loss in that winding. Later we shall see how to express the total resistance of both windings as a single value by "transferring" a winding resistance to the "other side" of the transformer, as explained in the following section.

22.9 Equivalent Circuits of a Power Transformer

Much time and labor is saved, and sufficient accuracy achieved, through the use of equivalent circuits in solving problems dealing with transformer operation. This procedure is preferred to the use of phasor diagrams, like that of Fig. 22.7, which would lead to more precise answers.

Because the iron loss of a transformer is practically independent of the load current, the shunt path in the equivalent circuit of a transformer shown in

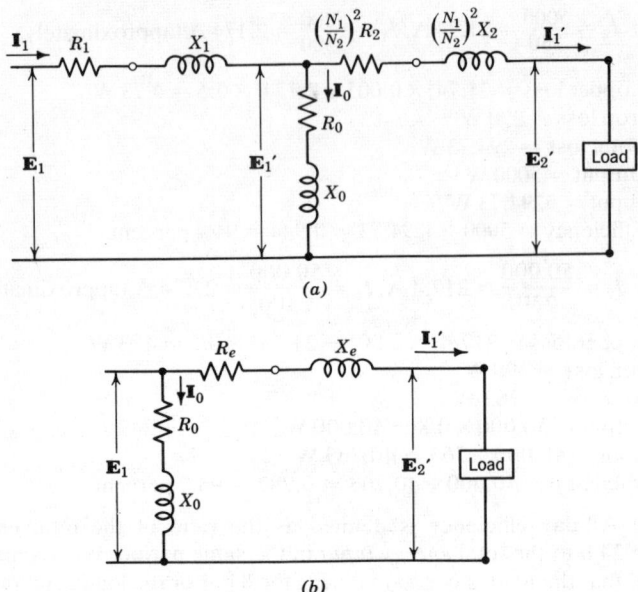

(a)

(b)

FIGURE 22.10 *(a)* Equivalent circuit of a power transformer. *(b)* Approximate equivalent circuit of a power transformer.

Fig. 22.10 represents well the path of the exciting current I_0 which is practically constant. We have just mentioned the transfer of secondary resistance and reactance into the primary circuit, so the remainder of Fig. 22.10 should cause no concern. The current I_1' to the *transferred load* and the voltage E_2' at this load represent equivalent primary-circuit values that are easily determined, as will be shown in the following example.

A word must be given about the transfer of load values to the primary. When the load is expressed in power units or in volt-amperes, the transfer is made using the given values. Power and volt-amperes are the same in either the high- or low-voltage side. If the load is given in terms of R_L and X_L, these must be transferred by using the $(N_1/N_2)^2$ multiplier to get equivalent load values for the primary equivalent circuit.

Computations with the equivalent circuit may be greatly simplified if an *approximate equivalent circuit*, like that in Fig. 22.10b, is used. The voltage applied to $\mathbf{Z}_0$, the shunt impedance made up of R_0 and X_0, is only a few percent more than it is in the more exact circuit in part a, but $\mathbf{I}_0$ is very small so the errors in computed values of currents, voltage drops, and power factor are negligible.

EXAMPLE 22.7   A 50-kVA, 2300-230 V transformer was tested on open circuit to obtain the following data: iron loss = 290 W, $E = 230$ V, $I_0 = 7.8$ A. On the short-circuit test the data were: applied voltage for full-load current = $E_1 = 53.6$ V, $I_1 = 21.74$ A, $P_{SC} = 473$ W.

(*a*) Determine the values of resistances and reactances in the equivalent circuit of Fig. 22.10*b*.

Solution. Since I_0 was measured in the secondary, we must use the equivalent primary value because the equivalent circuit has all primary constants.

This value is $7.8/10 = 0.78$ since the turns ratio is $10:1$.

$$R_0 = \frac{\text{iron loss}}{I_0^2} = \frac{290}{0.78^2} = 476 \ \Omega$$

$$Z_0 = \frac{E}{I_0} = \frac{2300}{0.78} = 2950 \ \Omega$$

$$X_0 = \sqrt{Z_0^2 - R_0^2} = \sqrt{2950^2 - 476^2} = 2912 \ \Omega$$

$$R_e = \frac{P_{SC}}{I_1^2} = \frac{573}{21.74^2} = 1.21 \ \Omega$$

$$Z_e = \frac{E_1}{I_1} = \frac{53.6}{21.74} = 2.46 \ \Omega$$

$$X_e = \sqrt{Z_e^2 - R_e^2} = \sqrt{2.46^2 - 1.21^2} = 2.14 \ \Omega$$

(*b*) Suppose this transformer operates at full load (50 kVA) and 80 percent power factor lagging. What value of input voltage E_1 is required to produce 230 V (V_2) at the load? What would be the power factor at the input of the transformer?

Solution. First draw the phasor diagram for the circuit (Fig. 22.10*b*). This is shown in Fig. 22.11. $\mathbf{I}_1'$ lags $\mathbf{E}_2'$ by α_2 which is evaluated from the power factor. In this case, $\alpha_2 = \cos^{-1} 0.8 = 36.9°$.

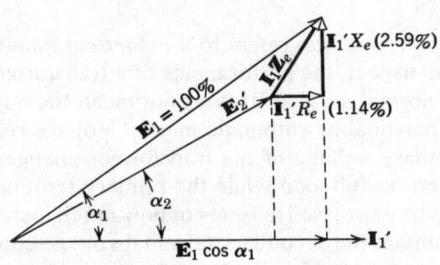

FIGURE 22.11   For Example 7.

After drawing the phasors $\mathbf{E}'_1$ and $\mathbf{I}'_1$, it is evident that $\mathbf{I}'_1 R_e$ and $\mathbf{I}'_1 X_e$ are added as shown.

The percentage quantities are evident when it is noted that 2300 V is considered 100 percent. The percent IR drop and IX drop are usually given on transformer name plates.

The turns ratio in power transformers is the same as the voltage ratio given on the nameplate. Therefore,

$$E'_2 = 230 \times 10 = 2300 \text{ V}$$

$$I'_1 = \frac{50,000}{2300} = 21.74 \text{ A}$$

$$I'_1 R_e = 21.74 \times 1.21 = 26.3 \text{ V} \ (1.14 \text{ percent of } 2300)$$

$$I'_1 X_e = 21.74 \times 2.74 = 59.6 \text{ V} \ (2.59 \text{ percent of } 2300)$$

The value of E_1 is readily obtained by adding the total resistance and reactance drops to E'_2:

$$\begin{aligned} E_1 &= [(E'_2 \cos \alpha_2 + I'_1 R_e)^2 + (E'_2 \sin \alpha_2 + I'_1 X_e)^2]^{1/2} \\ &= [(2300 \times 0.8 + 26.3)^2 + (2300 \times 0.6 + 59.6)^2]^{1/2} \\ &= (1866^2 + 1440^2)^{1/2} = 2356 \text{ V} \end{aligned}$$

The power factor at the input of the transformer is $\cos \alpha_1$:

$$1866 \div 2356 = 0.792 \text{ percent}$$

The power-factor angle is

$$\alpha_1 = \text{arc cos } 0.792 = 37.6°$$

These results were obtained without including the effect of the exciting current. The solution that takes this into account is much more involved and gives a result that is only about 1 percent lower in power factor.

22.10 Voltage Regulation of a Power Transformer

The term *voltage regulation* refers to a *numerical quantity* that characterizes, in one important aspect, the performance of a transformer when it is operating undisturbed in normal service. It does not mean the adjustment or control of the voltage by personal or automatic means. Voltage regulation refers to how much the secondary voltage of the transformer changes when the load is increased from zero to full load while the primary terminal voltage is kept constant. Or it may be expressed in terms of how much increase in primary voltage is needed to maintain the secondary terminal voltage constant while the load is increased from zero to full-load value. The power factor must be stated when regulation is specified.

Computation of voltage regulation is done with reference to the approximate equivalent circuit of Fig. 22.10*b* and the phasor diagram in Fig. 22.11. Voltage regulation is expressed in percent.

At no load, I_1' is zero and so is $I_1'Z_e$ so that $\mathbf{E}_1 = \mathbf{E}_2'$. At full load, $\mathbf{E}_1$ must be increased until it is the sum of phasors $\mathbf{E}_2'$ and $\mathbf{I}_1Z_e$ in order that $\mathbf{E}_2'$ may may remain constant. Note that the load voltage $\mathbf{E}_2'$ must remain constant from no load to full load. This is the alternative way, as indicated in the first paragraph in this section, to express the voltage regulation of a transformer. The actual voltage difference between $\mathbf{E}_1$ and $\mathbf{E}_2$ expressed as a percent of $\mathbf{E}_2$ is the percent voltage regulation. Therefore,

$$\text{Percent voltage regulation} = \frac{E_1 - E_2'}{E_2'} \times 100 \qquad (22.6)$$

EXAMPLE 22.8   A certain transformer has 2.6 percent reactance and 1.14 percent resistance. Compute the voltage regulation at 80 percent power factor.

Solution. Since the rated voltages of the transformer are not given, we may work with the symbols E_1 and E_2' and assume that E_2' is held at 100 percent of its no-load value while the load is increased from zero to full load. We are then entitled to use percentages for the voltage drops. Referring to Fig. 22.11 we see that

$$E_1 = [(E_2' \cos \alpha_2 + I_1'R_e)^2 + (E_2' \sin \alpha_2 + I_1'X_e)^2]^{1/2}$$

$$E_1 = [(100 \times 0.80 + 1.14)^2 + (100 \times 0.60 + 2.6)^2]^{1/2} = 102.4$$

Using Equation (22.6)

$$\text{Voltage regulation} = \frac{102.4 - 100}{100} \times 100 = 2.4 \text{ percent}$$

A qualitative explanation of what goes on inside a transformer *when the load impedance is decreased* is in order now. All that is necessary to decrease the load impedance is to turn on more lights or add to the motor load. This results in an increase in load current. Since the magnetic flux produced by the load current opposes the flux of the primary current, the *net flux* is decreased. This lowers the *back EMF* in the primary winding and more current is allowed to enter the primary. This additional current and the terminal voltage provide the additional power required to handle the increased power demand at the load. We can understand, from this, that the back EMF in the primary (the *induced EMF* in that winding produced by the *net flux*) acts as a "governor" to control the amount of input current the transformer receives.

When lights are turned off or the motor load on a transformer is decreased, the secondary current decreases and so does its own flux. The net core flux *in-*

creases and this increases the back EMF in the primary. The result is a *decrease* in primary current. With constant applied voltage to the primary, the current must decrease so that the decrease in input power will balance the decrease of power to the load.

Problem

4. A factory test on a power transformer revealed, among other things, that the secondary terminal voltage dropped from 240 V to 231 V while the load was increased from zero to 100 percent full-load value. What was the voltage regulation of the transformer?

22.11 Identification and Phasing of Transformer Windings

In the application of transformers to power generation, transmission, and distribution, those used in power-distribution circuits outnumber all of the rest put together. A distribution circuit serves a local area—a residential area, for example—where the voltages are of the order of 120 to 240 V. Distribution transformers also supply industrial areas that use electrical energy at 208 and 416 V, and even more in most large plants.

The primary and secondary windings of a power transformer usually consist of at least two separate coils each. This makes possible operation with more than one turns ratio and therefore more than one voltage ratio. Figure 22.12 illustrates coils, terminals, and connections that are standard for a distribution transformer.

The letter code for identifying terminals uses *H* on the high-voltage side and *X* on the low-voltage side. Note that pairs of successive odd numbers, beginning with 1, are assigned to individual coils. Suppose the identification letters

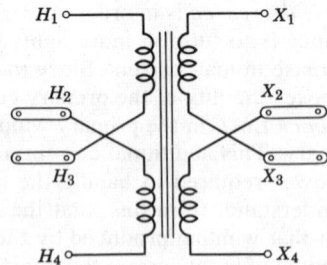

FIGURE 22.12 Coils and terminal symbols of a power-distribution transformer. Straps for interconnecting coils are shown on the inner terminals.

and numbers are not present. It is easy to tell which pair of terminals belongs to a single coil by means of an ohmmeter or other continuity checker. Or we could connect one lead to a 120-V supply line and, after connecting one wire of a socket containing a 115-V lamp to the other supply line, we could use the free wire from the socket to "explore" by touching it to other "unknown" terminals of the transformer. The lamp will light when the circuit is completed through the coil being identified. This is illustrated in Fig. 22.13. How can we tell whether this is a high-voltage coil or a low-voltage coil? If the bare leads of the coils can be seen, the wire size is an indication. The high-voltage coils are wound with smaller cross-section wire than that used for the low-voltage coils. Also, the smaller wire and larger number of turns cause the resistance of the high-voltage coil to be several times larger than the resistance of the low-voltage coil.

If it is known that the transformer has a voltage ratio greater than $1:1$, which is true in an overwhelming majority of cases, an ac voltmeter set on a safe high range may be used to check the voltages of the remaining coils. A voltmeter with insufficiently high range ordinarily would not be damaged if its connection across a high-voltage coil were made for only a fraction of a second—by a flick of the wire on the coil terminal, so to speak. This should be done only with due caution. Usually, a voltmeter with adequate range is available and this should be used.

After the leads from the coils are paired, that is, the coils identified, it is necessary to determine how they are polarized. Suppose the first coil identified is tagged H_1, H_3 on its leads as shown in Fig. 22.13. These tags could be applied by the operator if they are missing. The coil is then energized from the 120-volt line by connecting H_3 to the bottom line in Fig. 22.13. The lamp is disconnected. One lead of another coil on the same side of the transformer—H_2, for example— should then be connected to H_1. Because coil H_1, H_3 is energized, magnetic flux oscillates in the transformer core and induces voltage *in all the other wind-*

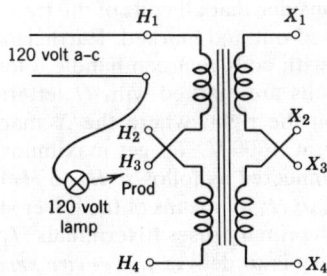

FIGURE 22.13 Use of lamp to "phase-out" windings. Obviously, the lamp will not light unless the prod is touched to the open end of the H_1–H_3 coil.

ings. The problem now is to determine which of the coil terminals go positive when H_1 goes positive and negative when H_1 goes negative. This is called *phasing-out.* Such information is necessary so that coils on the same side of the transformer may be connected in parallel or in series with an assurance of proper operation. Incorrect connection results in either zero voltage or a short circuit. The latter is sure to cause a fuse to blow, or a circuit breaker to open, or one or more coils to burn out.

With coil H_1, H_3 energized, there is no hazard in having terminal H_2 connected to H_1. It is recommended that this connection be made with the power shut off. If H_4 goes positive while H_3 is going positive, a voltmeter connected between H_3 and H_4 will read zero. If H_4 goes negative while H_3 is going positive, the voltmeter will read 240 volts. In this case, the power should be shut off and H_2 removed from H_1. Then H_4 should be connected to H_1. The voltmeter should now read zero between H_2 and H_3. This means H_2 and H_3 are going positive and then negative together, so it would be safe to connect them together which would put the two coils in parallel properly. A polarity mark should be attached to one lead of each coil to indicate that those leads may be connected together *for parallel* operation. Proper operation of transformer coils in parallel requires that the voltages in the coils *be in phase*, as well as instantaneously equal in magnitude. All coil voltages in a single-phase transformer are either in phase or 180° out of phase. If two coils connected in parallel have their voltages 180° out of phase, a *circulating current of dangerously high value* is set up in both coils.

Going back to the trial connections of terminals H_2 and H_4 of Fig. 22.13, when the meter reads zero between one of these terminals and H_3, it would be proper to tag that terminal and also H_3 with a plus (+) sign, or with some other designation to indicate that they are always of like polarity. This means that whenever one of them is positive the other is positive, and when either is negative the other is negative. This polarity determination is necessary for all windings on both the primary and secondary of the transformer.

Let us now consider that all coils of the transformer illustrated in Fig. 22.12 are properly phased-out and marked. Furthermore, assume it to be a distribution transformer with coils that can handle a maximum of 3600 V each at the left, where the coils are marked with H letters, and that is to deliver 120 V from each coil on the right, where the X markings are. The transformer is to receive power at 3600 V. To get maximum performance the high-voltage coils would be connected as follows: H_1 to H_2 by means of the upper convenient strap, and H_3 to H_4 by means of the lower strap. The 3600-V line would be connected through primary fuses to terminals H_1 and H_3.

If the secondary is to deliver *three-wire service*, which is standard in residential districts, one of the two straps on the right would be used to connect X_2 to X_3, while the other would serve as the terminal for the middle wire of the three-wire line which would serve several houses or other kinds of premises that use the energy supplied through the transformer. Three-wire service is discussed in Section 22.12.

Service to electric ranges and to some air conditioners requires 240 V. This is available at terminals X_1, X_4 of the three-wire system just mentioned. Lights are operated on 120 V, which is available at X_1, X_3 and at X_2, X_4. As stated earlier, X_2 and X_3 are strapped together.

It is more common to distribute power in a residential area at 2400 V on the high side instead of at 3600 V. In this case each primary winding would have to be rated at either 2400 V or 1200 V. Rural lines have primary voltages of 7200 V. If, in our example, the primary voltage had been 7200, the 3600-V coils would have to be connected in series instead of in parallel. In that case H_2 and H_3 would have been strapped together with nothing else connected to them. The secondary connections would not be changed because each primary coil would be getting its 3600 V, as before. If it is desirable to deliver power at only 120 V at maximum possible current, the secondary coils may be connected in parallel by strapping X_1 and X_2 together and X_3 and X_4 together. It is assumed that, by previous test, the corresponding terminals of each of these two pairs are always of like polarity, that is, X_1, X_2 have like polarity and X_3, X_4 have like polarity.

It is possible, although it is never done in ordinary distribution systems, to connect primary coils in series with secondary coils so that all voltages will add. In some kinds of testing, or in other special operations, such a procedure may be justified. Suppose a transformer with two 120-V primary coils and two 440-V secondary coils is used to deliver more than 880 V for a certain application. With only 120 V available from a supply, 1120 V could be produced by connecting all of the coils in series with additive polarities and energizing one of the 120-V coils from the line. Problem 15 at the end of the chapter relates to this kind of arrangement.

22.12 Three-Wire Distribution System

Distribution transformers are normally connected to deliver single-phase power at 120 V, 3 wire, to residential areas. This is illustrated in Fig. 22.14. The wiring in a residence is arranged so that the amount of load (lamps and appliances) that could be connected between one outside wire and the neutral is approximately the same as the amount of load that could be connected between the other outside line and the neutral.

Separate circuits in the house, from a few to many, are connected in a distribution box to the three main wires that extend from the transformer to the terminal straps in the distribution box. A fuse or a circuit breaker is connected in the "hot" line in the distribution box for protection. The hot line, covered with black insulation, is connected to one of the two service lines on either side

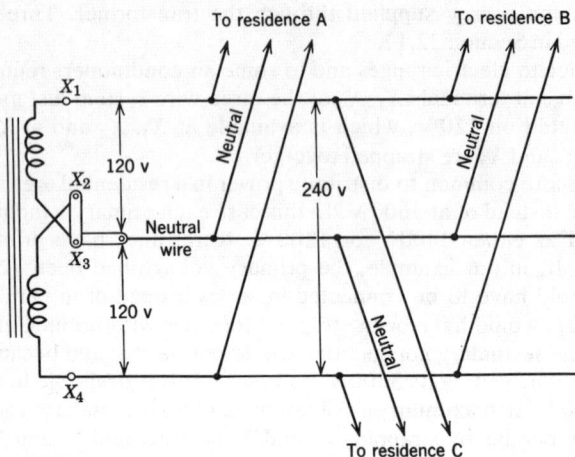

FIGURE 22.14   Secondary circuits of a distribution transformer.

of the neutral. No fuse or circuit breaker is allowed in the neutral wire, which is grounded both at the transformer and in the house. Modern service lines that run to the house from the power company's distribution lines consist of two insulated copper wires wrapped around a bare aluminum wire which serves as the neutral. This design provides the strength necessary to support the service lines when they extend from the pole to the house. Service lines in some housing areas are run underground.

If the electrical load in a house were perfectly balanced in the sense that the two "outside wires" of a three-wire system carried the same amount of current, the neutral wire would not be carrying any current. For example, if each side of the system demands I amperes, we can visualize the current flowing to the house from terminal X_1 and returning through the neutral wire to X_3, while the same amount of current goes to the house from X_2 through the neutral wire but in the opposite direction to that of the current which is returning to X_3. This means complete cancellation and, therefore, no current in the neutral wire.

One may conclude that some, or all, the current that goes to the house from X_1 returns to the transformer over the other outside wire to X_4. It is true that, at any instant, the amount of neutral-wire current is the difference between the amounts of current in the two outside wires of the three-wire system.

22.13   Metering of Power and Energy

Charges for the supply of electricity to households are made only on the basis of energy consumed over a fixed period of time, usually a calendar month. Some electric utility companies send out bills only every two months. Industrial customers are charged for total energy and also for the *maximum demand* in kilowatts that occurred in any time interval of predetermined length, which is usually 15 minutes. In addition, consumers of large amounts of power are penalized with extra charges for operating at low power factor. The reason for this is that the lower the power factor the more costly it is to supply power and the larger the transformer must be to deliver it. It is possible to improve the operating power factor of an industrial plant by connecting large capacitors to the wiring system so that their current, which leads the line voltage by 90°, will counteract the quadrature (90° lag) component of the line current drawn by the plant's low-power-factor equipment, and thus reduce the power-factor angle of the whole system. A quantitative treatment of power-factor improvement is given in Chapter 17.

The *watthour* meter is described in detail in Section 18.7. Figure 22.15 shows how a three-wire meter is connected in a residence installation. Large-capacity meters of this type for use in industrial plants are connected in the same way. Three-phase watthour meters have two sets of current and voltage coils. Both sets contribute driving torque to a single, vertical, rotating shaft which drives the hands on the dial that registers energy consumption in kilowatthours.

Some industrial plants frequently have need for extra large amounts of power during relatively brief periods of time. Much of the time their power

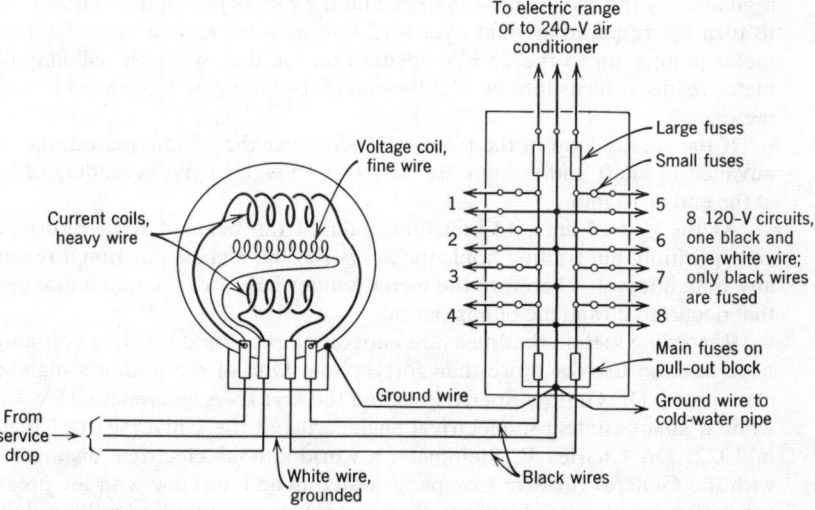

FIGURE 22.15 Connections in three-wire watthour meter and distribution box. Meter case and distribution-box case are grounded. A typical residence installation.

demand may be appreciably less than this maximum amount. Nevertheless, the electric power company is obliged to install, on the customer's premises, large transformer capacity and power-delivering capability that is ready for use whenever the customer wants it, regardless of how little total energy is used during the high-demand periods. Because of this, the power company is permitted to charge the customer an amount in addition to the charge for total kilowatt hours of energy delivered. This amount is based on the so-called *maximum demand* in kilowatts registered on a special dial on the watthour meter.

The metering of maximum demand is accomplished by means of a second drive shaft that moves a pointer upward around a circular dial. The shaft cannot turn more than about three-quarters of a revolution, and it turns this far only when the full capacity of the meter to register demand is in use. A disc segment, mounted on the demand shaft and riding very close to the rotating disc, receives torque corresponding to the instantaneous kilowatt load, and this makes the demand shaft try to turn. It is allowed to turn slowly, since an escapement pawl and ratchet wheel that inhibits its advance is actuated to serve as a timing system. The principle is best made clear by means of an illustration.

Suppose the demand meter has a 25-kW capacity, and that a steady load of 25 kW exists in an industrial plant. This power, in terms of current and voltage, passes through the meter and produces a constant torque on the disc segment that drives the demand shaft. The rate of turning of this shaft is regulated by the escapement system, and the rate is just right to allow the shaft to turn the required amount over a *15-min time interval* to push the demand-meter pointer up to the 25-kW position on the dial, where it will stay until a meter reader returns it to zero at the end of the billing period when he reads the meter.

If the steady load in the plant is 20 kW over the 15-min period, the rate of advance of shaft and pointer are only fast enough to give a reading of 20 kW at the end of 15 min.

At the end of each 15-min time interval the demand *shaft* returns to its zero position, but friction holds the *pointer* at the highest position it reaches in any time interval. Therefore the meter pointer indicates the maximum demand that occurred during the billing period.

Recording meters that measure kilovolt-amperes and reactive volt-amperes have been in use for more than forty years. One of the author's high-school classmates, Dr. George Sperti, invented the first kVA and reactive kVA meter while a junior student in electrical engineering at the University of Cincinnati in 1922. Dr. Charles P. Steinmetz, a world-famous electrical engineer, then with the General Electric Company, stated in an interview with the press that the meter could save electric-power companies one hundred million dollars a year.

22.14   Review Questions

1. Iron-core transformers have much less leakage flux than air-core transformers, but why are they not used in high-frequency circuits?†

2. How are the terminal voltages of a power transformer related to the numbers of turns? To the currents in the windings?

3. By what means is energy in the primary of a power transformer transferred to the secondary?

4. How could you identify or distinguish between the high-voltage and low-voltage windings of a transformer without energizing them?

5. What determines how much current flows in the secondary of a power transformer that is operating under load?

6. Upon what three quantities does the amount of EMF induced in a transformer winding depend?

7. A theoretical "ideal" power transformer has no losses. Sketch the phasor diagram for the case when it is connected to a source and has its secondary open. How must the diagram be altered to fit the case of an actual transformer operating at no load?

8. Define (a) leakage reactance, (b) exciting current, (c) magnetizing component of current, (d) core-loss component.

9. Write the equation for the voltage induced in a transformer winding. This involves the three quantities requested in Question 6. From this derive the basic design equation of the transformer.

10. Name the kinds of losses that are present in a power transformer. Which is considered to be independent of load and why?

11. How is each kind of transformer loss determined, with the help of laboratory measurements?

12. Sketch the approximate equivalent circuit for a power transformer. In what way does this circuit differ from the more accurate equivalent circuit?

13. (a) Define all-day efficiency of a transformer. (b) How do the constant losses of a transformer enter into the computation of all-day efficiency?

14. Define voltage regulation of a transformer in two different ways. Do you think the results of computations in these two ways would be identical or that they would differ? Would the differences be negligible, in a practical sense?

15. Explain why the ampere-turns of the secondary of a transformer must always be directed so that this MMF would buck the MMF of the primary.

16. Explain qualitatively what goes on in a transformer when the load impedance is suddenly increased.

17. Sketch two primary and two secondary coils of a transformer. Explain how you can be sure you are right the first time you connect the two primary coils in parallel. What would happen if you turned on the power when the connections were wrong?

18. Suppose the power transformer referred to in Question 17 has its primary properly energized and the secondary leads are left open. Could any harm be

†The iron-slug tuned IF transformers used in radio receivers are an exception. In this type, the iron serves as a very small part of the magnetic flux path.

done to the transformer by connecting one lead of one secondary winding (chosen at random) to a lead of the other winding? What two results of this procedure are possible? What should be done to assure that a parallel connection of the coils can be made properly?

19. Why are residences supplied with three-wire electrical service? What relation does the amount of current in the neutral service wire have to the amounts in the outside wires?

20. What is the function of a maximum-demand meter? Explain how it performs in indicating maximum demand.

22.15   Problems

1. A power transformer has 2500 turns in its primary and 250 turns in its secondary. *Neglecting losses*, what is its voltage ratio? Its current ratio? Its efficiency?

2. How much core flux would be needed in the transformer of Problem 1 to give it a rated primary voltage of 2500 V at 60 Hz? How much at 400 Hz?

3. A power transformer has rated voltage applied to its secondary, which is the low-voltage side, and the primary is open. The readings of meters in the secondary circuit are: volts = 240, amperes = 2.1, watts = 73.58. The voltage ratio of the transformer is 10 : 1. (a) What are the values of the magnetizing component and the core-loss component of the current? (b) What is the no-load power factor? (c) Draw the phasor diagram.

4. The transformer of Problem 3 supplies a load of 100 A at 240 V, 89 percent power factor. Determine the value of the primary current and the primary power factor if impedance drops are neglected.

5. A power transformer rated at 4800 V is operated at 5200 V, which is an $8\frac{1}{3}$ percent increase, in order to provide higher than rated voltage at its output terminals at full-load current. Compute the approximate percent increase in its core loss. What will be the effect on the transformer?

6. A power transformer is rated at 25 kVA, 7200 to 240 V. Its primary has 1.4-ohms resistance, its secondary has 0.05-Ω resistance, and its core loss is 200 W. Compute the efficiency when the load is 25 kW at unity power factor.

7. Solve Problem 6 with the load at 2.5 kW, unity power factor.

8. Solve Problem 6 with the load at 25 kVA 85 percent power factor.

9. Find the all-day efficiency of the transformer of Problem 6, which is used at a constant load of 25 kVA, 85 percent power factor for 8 h a day.

10. A transformer with a core loss of 435 W has an equivalent resistance of 3.2 Ω and an equivalent reactance of 12 Ω referred to the high-voltage winding. At full load this winding takes 5 A at 7200 V. Compute the full-load efficiency, assuming operation with unity power factor load.

11. A current of 4 A at 240 V was sent through the low-voltage winding of the transformer in Problem 6 to measure the core loss. With the low-voltage winding short-circuited, 180 V sent full load current through the primary at 285 W of power. Compute R_0, X_0, R_e, X_e, Z_e and draw the equivalent circuit.

12. A power transformer has a turns ratio $N_1/N_2 = 30$. Owing to the fact that the rated voltage of the primary is so high, measurements were taken on the

low-voltage secondary. The following values were obtained in terms of secondary quantities: $R_e = 0.1\,\Omega$, $X_e = 0.5\,\Omega$, $R_0 = 400\,\Omega$, $X_0 = 1800\,\Omega$. Convert these values to primary quantities and draw the equivalent circuit. Assume this to be a 100-kW, 13,200/440-V transformer. Compute the required input voltage to deliver 440 V to a 100-kW unity power factor load.

13. A factory has a connected load of 15 kW for lighting (100 percent power factor) and motors that have full-load ratings as follows: 10 motors, 5 hp each at 80 percent full-load efficiency; 2 motors, 25 hp each at 87 percent full-load efficiency. Only 6 of the 5-hp motors are ever operating at the same time, but both of the 25-hp motors run almost all day long. Compute the largest possible maximum demand that would be shown on the utility company's demand meter.

14. Under normal operating conditions, an industrial plant is furnished 50 kW of single-phase power at 70 percent power factor. The voltage is 240 V, 60 Hz. How can the management improve the power factor? What equipment might be chosen and what would its ratings be if the power factor after improvement is 87 percent?

15. A distribution transformer has two 480-volt windings on its high-voltage side and two 120-V windings on its low-voltage side. It is to be used on a special project where voltages as high as 1200 V are to be used for tests. The supply is 120 V rms.
    (a) Sketch the connections of the windings, labeling the terminals so that correct polarities will exist.
    (b) What values of voltage will be available at the output?
    (c) Show where a control resistance (or reactance) may be added to provide for some variation of output voltage. If the applied voltage may be reduced as much as 25 percent by this means, what ranges of output voltage (on open circuit) will then be available?

16. How much would the installation of only 500 μF improve the power factor of the electrical system of Problem 14?

CHAPTER TWENTY-THREE

Polyphase Systems

All of our studies of alternating-current sources, circuits, and loads up to this point have been about sinusoidal voltages and currents supplied by generators with two terminals. This is termed *single-phase* operation in contrast to what we shall encounter in this chapter known as *polyphase* operation. By far the most popular of polyphase systems, and for good reason, are *three-phase* systems. We say systems, rather than system, because two different kinds of connections of ac generator (*alternator*) windings provide so-called *delta* (Δ) and *wye* (Y) subsystems.

First we shall justify industry's preference for three-phase operation to single-phase operation.

23.1   Advantages and Disadvantages

GENERATION AND TRANSMISSION

Three-phase alternators are used, instead of single-phase alternators, for the generation of large amounts of power because they (a) are more efficient, (b) are smaller in physical size, (c) require smaller amounts of copper for their windings, and (d) are subject to less mechanical vibration because the instantaneous value of output power is practically constant instead of pulsating.

519

The amount of copper in a three-phase transmission line is substantially less, for a given amount of power transmitted, than the amount required for single-phase operation, even though three conductors are required instead of two. This means a saving in the number and strength of transmission line towers.

The step-up and step-down transformers required at the ends of the transmission line cost less with three-phase transmission than with single phase.

Three-phase operation is not as practical for residence use where motors are usually smaller than 1 hp and where lighting circuits supply most of the load.

POWER UTILIZATION

We shall find that three-phase motors are much smaller and less expensive than single-phase motors because less material (copper, iron, insulation) is required. Another advantage is that three-phase motors are self-starting, that is, they do not require special provisions to get them started.

Very small single-phase motors like those used to drive small fans, food mixers, vacuum cleaners, hand drills, and clocks do not require internal starting devices. Single-phase motors in sizes from about $\frac{1}{4}$ hp upward have special starting provisions inside their frames. Single-phase motors larger than 2 hp are not generally available. Fractional-horsepower three-phase motors are not manufactured, except for special applications such as automatic control.

Three-phase power has an important advantage in applications where large amounts of direct-current power are required, such as radio and television transmitters, and the production of aluminum from ore. The dc voltage for transmitter operation should be as constant as possible, that is, it should have a minimum of ripple. Rectified three-phase voltage is smoother than rectified single-phase voltage when the smoothing filters are comparable in cost and size. Another way to look at it is to conclude that for a given ripple tolerance in the dc voltage, the filter for three-phase operation costs much less than a filter for single-phase.

23.2 Two-Phase Systems

The demand for two-phase power has almost disappeared. It is quite safe to say there are no two-phase transmission systems in this country, because they do not have any advantages that are not equaled or surpassed by three-phase systems in generation, transmission, or utilization.

We shall devote very little time to a discussion of two-phase operation in this book.

23.3   Generation of Three-Phase Voltages

A two-pole alternator is represented in Fig. 23.1a. It has three separate coils held in slots in the inside surface of the stationary part, called the *stator*, which is fastened to the frame of the machine. We shall give our attention to the coil of phase A, whose sides are represented by groups AA'. Coil ends A_1 and A_2 are the beginning and end of the continuous path in this coil.

The *rotor* of the alternator has permanent north (N) and south (S) poles, produced by a direct-current winding on the rotor. The current, supplied by an outside dc source called the *exciter*, enters and leaves the direct-current winding through slip rings.

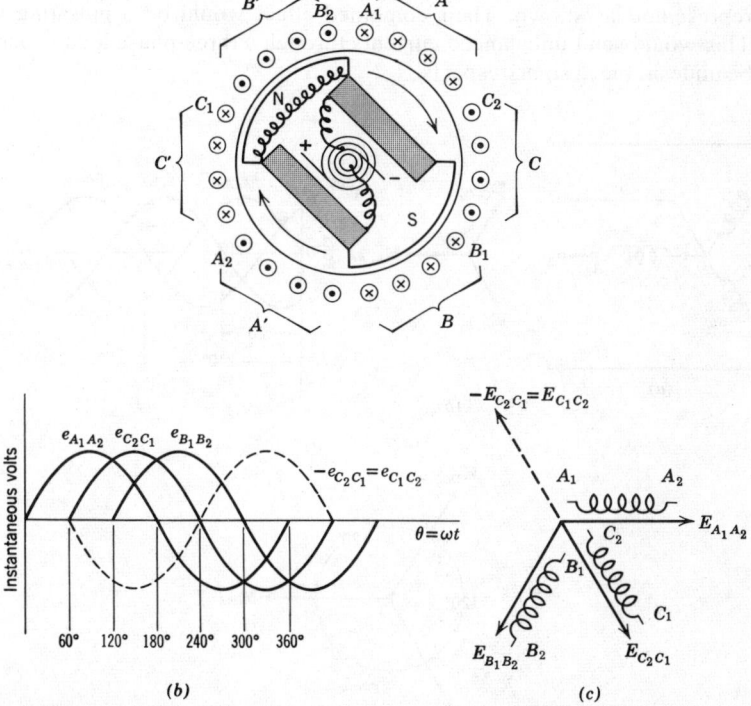

(b)   (c)

FIGURE 23.1   Generation of three-phase voltages. Note that $E_{C_2C_1}$, when reversed in connection, is 120 behind $E_{B_1B_2}$. This is actually done in practice, giving a perfectly balanced set of voltages as shown in Fig. 23.2. (a) Constant-strength magnetic field rotating inside an ac stator winding. Direct current is supplied to field winding through +, − brushes on slip rings. (b) One cycle of each generated voltage. (c) Rms values and relative phases of generated voltages in sequence.

When the rotor is turned at constant speed, the instantaneous EMF generated in coil AA' is sinusoidal. It is represented by $e_{A_1A_2}$ in Fig. 23.1b. At the instant $e_{A_1A_2}$ reaches 86.6 percent of its peak value, the rotating field begins generating EMF in coil C. The reversal of these subscripts will be justified in a moment. The sine wave for $e_{C_2C_1}$ is shown starting 60°, after $e_{A_1A_2}$ started because the coil C is placed a 60° turn from the position of coil A_1A_2. Next the rotating field reaches coil B and begins to generate voltages in it. This is represented by the sine wave $e_{B_1B_2}$ which lags $e_{A_1A_2}$ by 120°. *And this is important.*

We show only one cycle of each generated voltage for simplicity of explanation. As the voltages are being generated they come up in the positive polarity in the order $e_{A_1A_2}$. $e_{C_2C_1}$, $e_{B_1B_2}$, as shown in Fig. 23.1c. If coil ends A_1, C_2, and B_1 were connected to a common terminal and ends A_2, C_1, and B_2 were brought out to three separate terminals, rms values of the EMFs would be properly represented as shown. Their combined effect would be a pulsating voltage. This would send unbalanced currents through a three-phase load which would be undesirable in some respects.

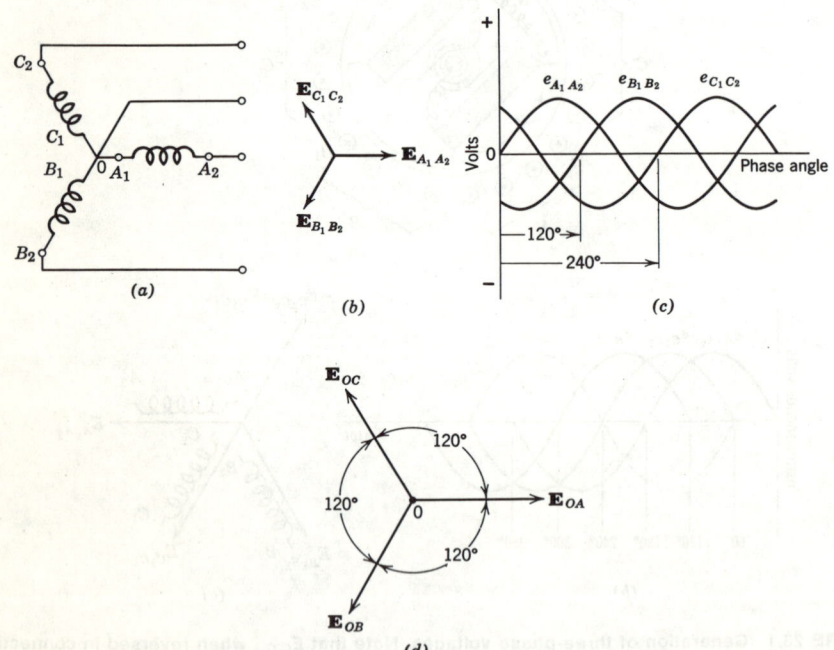

(a)

(b)

(c)

(d)

FIGURE 23.2 Coils and voltages of three-phase alternator, Wye-connection. (a) Connections of alternator stator (stationary armature) windings. (b) Rms values of coil voltages. (c) Instantaneous voltages generated in coils. (d) Neutral-to-terminal voltages.

When A_1, C_1, and B_1 are connected to one terminal on the alternator and the other ends of the coils are connected to separate terminals, the voltage of coil C is reversed and the resulting set have 120° phase displacements as shown in Fig. 23.2. The sum of the voltages is then zero (surprisingly enough) as may be seen by adding the three phasors (tail to head) in (b) or adding vertical distances *at any chosen phase angle* in (c).

Now how can we expect to get power from the terminals of an alternator if the sum of the voltages it delivers is zero at every instant? This will be explained soon.

In Fig. 23.2a we show the correct coil connections and the letter O used for the common terminal called the "neutral." In (b) we use the standard labeling method which we shall use henceforth. The stator windings are connected for a four-wire Y (wye) power system.

23.4 Coil Connections for Delta Operation

Some alternators have their windings connected so that they form a closed path inside the machine. This is practical since *there is no circulating current in the coils* because the sum of the generated voltages at any instant is zero. The connections and resulting voltages are shown in Fig. 23.3. When power is sent to a load from terminals A, B, and C, the *line-to-line* **voltages** *are the same as the coil* **voltages**. But, line **currents** are not the same as coil **currents** in a delta connection.

In contrast, *with the wye connection line* **currents** are the same as alternator coil **currents**, but line-to-line voltages are larger than single coil voltages. We shall see that the constant $\sqrt{3}$ enters into the calculations of line currents in the delta connection and line-to-line voltages in the wye connection.

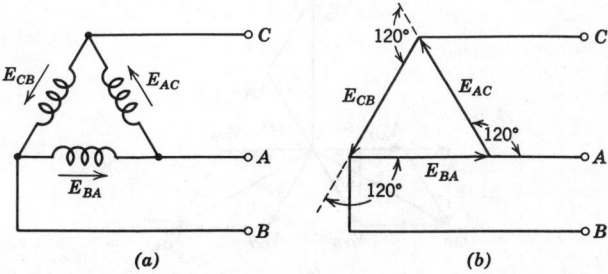

FIGURE 23.3 Coils and voltages of three-phase alternator, Delta connection. (a) Connections of alternator stator windings. (b) Rms values of terminal voltages.

23.5 Voltages and Currents in a Balanced-Y System

A three-phase load consisting of three equal impedances, Z_1, Z_2, and Z_3, is supplied by a Y-connected generator as shown in Fig. 23.4a. The voltage drops *across the load terminals* are V_{AB}, V_{BC}, and V_{CA}. The voltage drops through the separate impedances are shown in (b). To verify that the sum of the three impedance voltage drops is zero, all we need to do is add them graphically by maintaining their correct angles and draw the arrows tail-to-point. This gives a closed triangle similar to that of Fig. 23.3b. Another way is to note that the vertical components of $V_{CO'}$ and $V_{BO'}$ cancel each other and their horizontal components, each being half the full voltage value, combine to cancel $V_{AO'}$.

In Fig. 23.4c we have shown graphically the value and phase angle of V_{AB}. Note that to get from terminal A to terminal B we must go through Z_1 from A to O' which gives us $V_{AO'}$ and from O' to B which gives us $V_{O'B} = -V_{BO'}$. We shall now show that the magnitude of $V_{AB} = \sqrt{3}V_{AO'}$, and express V_{AB} in polar and complex form.

The horizontal component of $V_{BO'}$ has a magnitude $V_{BO'}\cos 60° = 0.5\,V_{BO'}$. Its vertical component has a magnitude $V_{BO'}\sin 60° = (\sqrt{3}/2)V_{BO'}$. The magni-

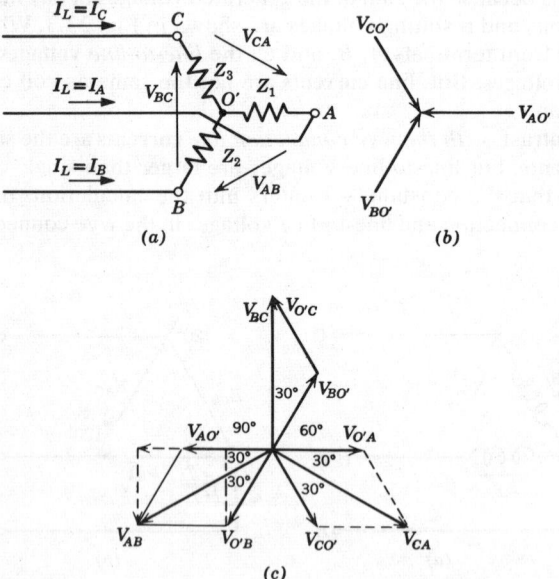

(a)   (b)

(c)

FIGURE 23.4 Three-phase Y-connected load. (a) Three equal impedances. (b) Phase voltages to neutral. (c) Line-to-line voltages = $\sqrt{3}$ × phase voltages. Line-to-line voltages are 120° apart.

tude of V_{AB} is then

$$V_{AB} = \sqrt{(V_{AO'} + 0.5V_{BO'})^2 + \left(\frac{\sqrt{3}}{2}V_{BO'}\right)^2}$$

Let's call the voltage drops across the impedances the phase voltages, each magnitude represented by V_ϕ and the voltages V_{AB}, V_{BC}, V_{CA} the line voltages, each magnitude being V_L. In this case

$$V_L = \sqrt{(V_\phi + 0.5V_\phi)^2 + \left(\frac{\sqrt{3}}{2}V_\phi\right)^2}$$

$$= V_\phi\sqrt{1.5^2 + \tfrac{3}{4}} = V_\phi\sqrt{2.25 + 0.75}$$

$$V_L = \sqrt{3}V_\phi \tag{23.1}$$

The line-to-line voltage of a Y-connected 3-phase system $= \sqrt{3}$ times the phase voltage.

By simple geometry we see that V_{AB} is 30° below the horizontal. Therefore

$$\mathbf{V}_{AB} = \sqrt{3}V_\phi\underline{/-150°}$$
$$= V_\phi\sqrt{3}\,[\cos(-150°) + j\sin(-150°)]$$
$$\mathbf{V}_{AB} = V_\phi(-1.5 - j0.866)$$

EXAMPLE 23.1   Let the rms value of the voltage induced in each phase, as shown in Fig. 23.1, be 100 V. Express $\mathbf{E}_{A_1A_2}$, $\mathbf{E}_{C_2C_1}$, and $\mathbf{E}_{B_1B_2}$ in complex form.

Solution

$$\mathbf{E}_{A_1A_2} = 100 + j0 \text{ V}$$
$$\mathbf{E}_{C_2C_1} = 100[\cos(-60°) + j\sin(-60°)]$$
$$\mathbf{E}_{C_2C_1} = 50 - j86.6 \text{ V}$$
$$\mathbf{E}_{B_1B_2} = 100[\cos(-120°) + j\sin(-120°)]$$
$$\mathbf{E}_{B_1B_2} = -50 - j86.6 \text{ V}$$

EXAMPLE 23.2   After reversing the polarity of the voltage induced in phase coil C_2C_1, determine the sum of the three voltages of Example 23.1.

Solution

$$\mathbf{E}_{C_1C_2} = -\mathbf{E}_{C_2C_1} = -50 + j86.6 \text{ V}$$

Sum of $\mathbf{E}_{A_1A_2} + \mathbf{E}_{B_1B_2} + \mathbf{E}_{C_1C_2}$

$$= 100 + j0 - 50 - j86.6 - 50 + j86.6 = 0$$

EXAMPLE 23.3 In the balanced Y-connected load of Fig. 23.4, assume the line-to-line voltages are 100 V rms each, and each of the phase impedances Z_1, Z_2, and Z_3 is 10 Ω pure resistance. Determine the magnitude of the voltage drop across each phase and the amount of current in it.

Solution

Line-to-line voltage = $\sqrt{3}$ × phase voltage
$$100 = \sqrt{3}\, V_\phi$$
$$V_\phi = 57.7 \text{ V}$$

Phase current = line current in Y connection
$$I_\phi = I_L = 57.7 \div 10 = 5.77 \text{ A}$$

EXAMPLE 23.4 Express each phase voltage in Fig. 23, and each phase current of Example 23.3 in complex form.

Solution

$$V_{AO'} = -57.7 + j0 = 57.7 \underline{/180°} \text{ V}$$
$$V_{BO'} = 57.7\,(\cos 60° + j \sin 60°)$$
$$V_{BO'} = 28.85 + j49.97 = 57.7 \underline{/60°} \text{ V}$$
$$V_{CO'} = 57.7\,(\cos 60° - j \sin 60°)$$
$$V_{CO'} = 28.85 - j49.97 = 57.7 \underline{/-60°} \text{ V}$$
$$I_{AO'} = 57.7 \underline{/180°}/10 \underline{/0°} = 5.77 \underline{/180°} \text{ A}$$
$$I_{BO'} = 57.7 \underline{/60°}/10 \underline{/0°} = 5.77/60° \text{ A}$$
$$I_{CO'} = 57.7 \underline{/-60°}/10 \underline{/0°} = 5.77 \underline{/-60°} \text{ A}$$

The phase currents (which are identical to the line currents) are in phase with the phase (line-to-neutral) voltages.

EXAMPLE 23.5 Express the line-to-line voltages of Examples 23.3 and 23.4 in complex form.

Solution. In Fig. 23.4c we can see that each line-to-line voltage is the phasor sum of two phase voltages. The phase voltages involved are ready for us in Example 23.4.

$$\mathbf{V}_{AB} = \mathbf{V}_{AO'} + \mathbf{V}_{O'B}$$
$$= -57.7 + j0 - (28.85 + j49.97)$$
$$\mathbf{V}_{AB} = -86.55 - j49.97 = 100 \underline{/-150°} \text{ V}$$
$$\mathbf{V}_{BC} = \mathbf{V}_{BO'} + \mathbf{V}_{O'C}$$
$$= 28.85 + j49.97 - (28.85 - j49.97)$$
$$\mathbf{V}_{BC} = j100 = 100 \underline{/90°} \text{ V}$$
$$\mathbf{V}_{CA} = \mathbf{V}_{CO'} + \mathbf{V}_{O'A}$$
$$= 28.85 - j49.97 - (57.7 + j0)$$
$$\mathbf{V}_{CA} = 86.55 - j49.97 = 100 \underline{/-30°} \text{ V}$$

Problems

1. Assume the phasor diagram in Fig. 23.4c is rotated counterclockwise until V_{AB} is defined by $V_{AB} = 100 + j0$ V. Express V_{BC} and V_{CA} in both polar and complex form.
2. Now that we have, in Problem 1, new expressions for the line-to-line voltages (which we now shall call simply line voltages), what will be the new expression in both polar and complex form for each of the line and phase currents if every one of the phase impedances is 10 Ω pure resistance?
3. A three-phase supply that has line voltages of 240 V feeds a balanced Y-connected load of three equal impedances of $24 + j0$ Ω. Calculate the phase currents and give them in polar and vector forms. Draw the phasor diagrams, letting $V_{AB} = 240 \underline{/0°}$ V.

23.6   Voltages and Currents in a Balanced Delta System

After having learned to express the voltages and currents, and their phase relationships, of a Y-connected balanced load, it will be very easy to understand these characteristics of a Δ-connected load. In view of this situation we

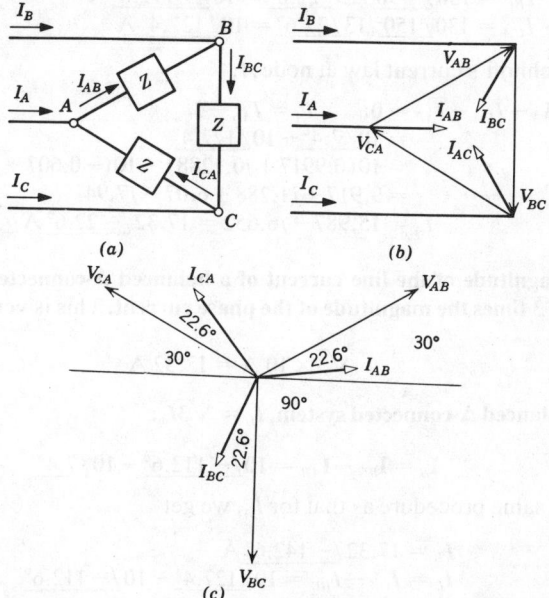

FIGURE 23.5   Balanced three-phase inductive load, Δ-connected. (a) Schematic. (b) Phase voltages and currents, phase voltage = line voltage. Line current = $\sqrt{3}$ × phase current. (c) Phasor diagram to make analysis easier.

shall illustrate, in our next example, a balanced Δ-connected load that has lagging power factor.

Figure 23.5a shows a balanced impedance $\mathbf{Z} = 12 + j5 = 13\underline{/22.6°}\ \Omega$ supplied by a balanced three-phase supply of which the line voltage is $\mathbf{V}_L = 130$ V rms. The voltages and currents of the three impedances are shown in (b). Because a shifting of the positions of phasors in a graphical representation does not change their relationships to one another so long as their respective directions (angles) are not changed, we show them in a much more favorable arrangement in (c).

EXAMPLE 23.6   Determine the magnitudes and phase angles of the phase currents and line currents in the balanced Δ-connected load of Fig. 23.5. Each phase impedance is $\mathbf{Z} = 12 + j5 = 13\underline{/22.6°}\ \Omega$ and each line voltage is 130 V rms.

Solution

$$V_{AB} = 130\underline{/30°}\ V;\ V_{BC} = 130\underline{/-90°}\ \text{V};\ V_{CA} = 130\underline{/150°}\ \text{V}$$
$$I_{AB} = 130\underline{/30°}/13\underline{/22.6°} = 10\underline{/7.4°}\ \text{A}$$
$$I_{BC} = 130\underline{/-90°}/13\underline{/22.6°} = 10\underline{/-112.6°}\ \text{A}$$
$$I_{CA} = 130\underline{/150°}/13\underline{/22.6°} = 10\underline{/127.4°}\ \text{A}$$

Using Kirchhoff's current law at node A,

$$\mathbf{I}_A + I_{CA} - I_{AB} = 0;\qquad I_A = I_{AB} - I_{CA}$$
$$= 10\underline{/7.4°} - 10\underline{/127.4°}$$
$$= 10(0.9917 + j0.1288) - 10(-0.607 + j0.794)$$
$$= 9.917 + j1.288 + 6.07 - j7.94$$
$$I_A = 15.987 - j6.652 = 17.32\underline{/-22.6°}\ \text{A}$$

The magnitude of the line current of a balanced Δ-connected 3-phase load is equal to $\sqrt{3}$ times the magnitude of the phase current. This is verified here:

$$\sqrt{3} \times 10\ \text{A} = 17.32\ \text{A}$$

In a balanced Δ-connected system, $I_L = \sqrt{3}I_\phi$:   (23.2)

$$\mathbf{I}_B = \mathbf{I}_{BC} - \mathbf{I}_{AB} = 10\underline{/-112.6°} - 10\underline{/7.4°}$$

Using the same procedure as that for I_A, we get

$$I_B = 17.32\underline{/-142.6°}\ \text{A}$$
$$I_C = I_{CA} - I_{BC} = 10\underline{/127.4°} - 10\underline{/-112.6°}$$
$$I_C = 17.32\underline{/97.4°}$$

If you will sketch I_A, I_B, and I_C phasors to scale at their correct phase angles, you will see that their phasor sum is zero. They are, mutually, 120° apart.

Problems

4. A balanced three-phase Δ-connected load has a line current of 25 A. How much current flows in each phase impedance? What is the line voltage if each phase impedance is 13.85 Ω?

5. If each phase impedance in Problem 4 is inductive and has 10 Ω resistance, what is its inductive reactance? What is the phase angle between each phase current and the voltage drop across that phase?

6. Construct the phasor diagram for the conditions in Problems 4 and 5. Label the junctions *A*, *B*, and *C*. Determine the phase angle between the line current I_A and the line voltage V_{AB}.

23.7   Power and Power Factor

When we know the amounts and phase angles of the *phase voltages and currents* of a three-phase load (or generator, for that matter) we can readily calculate the total power. It is, of course, the sum of the powers in the three separate phases. These can be computed using I^2R or $VI \cos \theta$.

We shall continue our discussion of balanced loads only. In this case the three separate identical phase impedances will carry the same amount of current at the same power factor angle. Therefore, we need to calculate the power in only one phase and multiply it by three. When the load is unbalanced we must calculate all three "phase powers" separately and then add them together. We should remind ourselves that I^2R still applies to each phase regardless of the phase angle.

EXAMPLE 23.7   In Example 23.3 we had the easy case of the three identical impedances, each of 10 Ω resistance. The voltage drops were all 57.7 V and the currents were 5.77 A. What is the total power?

Solution. In this unity power factor case, the total power is

$$P = 3 \times 57.7 \times 57.7 \cos 0° = 1000 \text{ W}$$
$$Z = 10 \, \Omega = R, \quad I^2R = 5.77^2 \times 10 \times 3 = 1000 \text{ W}$$

as expected.

EXAMPLE 23.8   Calculate the total power in the balanced load in Example 23.6.

Solution

$$P_T = 3V_\phi I_\phi \cos \theta = 3 \times 130 \times 10 \times \tfrac{12}{13} = 3600 \text{ W}$$
$$P_T = 3I_\phi^2 R_\phi = 3 \times 10^2 \times 12 = 3600 \text{ W}$$

We shall soon prove that the total power in a balanced three-phase load is given by $\sqrt{3}E_L I_L \cos\theta$ in which the *line voltage* and *line current* are used. This holds for both delta and wye connections. Let's try it out here.

$$P_T = \sqrt{3} \times 130 \times 17.32 \cos 22.6°$$
$$P_T = 3900 \times 0.923 = 3600 \text{ W}$$

The power factor of a *balanced* three-phase load may be calculated as R/Z from a phase impedance. It is also given by true power $\div$ apparent power, that is,

$$\text{three-phase power factor} = \frac{\text{total watts}}{\text{total voltamperes}} \text{ or } \frac{\text{kW}}{\text{kVA}} \quad (23.3)$$

This quotient will give power factor of an unbalanced load also. It is an overall power factor and it has no specific relation to the power factors of the separate phases.

REACTIVE VOLTAMPERES

Let us first recall that voltamperes, power, and reactive voltamperes are related to one another in the same sense that the lengths of sides of a right triangle are related. With a power factor represented by $\cos\theta$, we have

$$\text{Real power} = P_R = \text{voltamperes} \times \cos\theta \quad (23.4)$$

$$\text{Reactive voltamperes} = P_X = \text{voltamperes} \times \sin\theta \quad (22.5)$$

With this in mind, we may express reactive voltamperes as follows:

$$\text{Y connection: } \mathbf{VAR} = 3V_\phi I_\phi \sin\theta = 3\frac{V_L}{\sqrt{3}} I_L \sin\theta \quad (23.6)$$

$$\mathbf{VAR} = \sqrt{3}V_L I_L \sin\theta$$

$$\Delta \text{ connection: } \mathbf{VAR} = 3V_\phi I_\phi \sin\theta = 3V_L \frac{I_L}{\sqrt{3}} \sin\theta \quad (23.7)$$

$$\mathbf{VAR} = \sqrt{3}V_L I_L \sin\theta$$

Because $\sin\theta$ is the factor by which $\sqrt{3}V_L I_L$ is multiplied to get reactive voltamperes, $\sin\theta$ is called the *reactive factor*. This means that the reactive factor may be computed thus

$$\text{Reactive factor} = \sin\theta = \frac{\mathbf{VAR}}{\sqrt{3}V_L I_L} \quad (23.8)$$

Reactive voltamperes may be measured, or it may be calculated from $3I^2X$ in which X is the reactance per phase of a balanced system.

Problems

7. A three-phase Δ-connected load has three identical impedances $Z = 15 + j5\,\Omega$ and 200 V between terminals. How much power does the load receive? Calculate it two ways.

8. How many voltamperes are delivered to the load of Problem 7? Calculate the power factor in two ways.

9. What is the reactive factor of the load of Problem 7? Construct the triangle that represents the VA, VAR, and true power.

23.8   Measurement of Three-Phase Power

We have seen that the total power delivered to a three-phase load is, quite logically, the sum of the powers in the separate load phases. Inasmuch as the wattmeter connections for measuring single-phase power (and that is what we have in each of the load phases) were described and illustrated in Chapter 18, we shall not devote space and time to show how to do this in each of the phases of a three-phase system. We shall go on to the more practical and less expensive method of measuring power with two wattmeters. The two-wattmeter method holds whether the three-phase load is balanced or not.[1] The proof of this is too complicated for reproduction in this book.

In a balanced, three-phase system the power-factor angle θ is the same in the three phases (load phases or generator-armature coils). We have noticed this before and we see it again in Fig. 23.6b. The connections of the two-wattmeter method of measuring three-phase power are shown in Fig. 23.6a. Load impedances may be Δ connected instead of Y connected. Of course, if *these particular impedances* were Δ connected and supplied from the same lines, the total power would be greater than in this case because $\sqrt{3}$ times as much voltage would be applied *to each phase of the load*, that is, to each impedance branch.

Wattmeter W_1 in Fig. 23.6a carries line current $\mathbf{I}_A$ and uses line voltage $\mathbf{V}_{AC}$. In setting up the expression for the computation of each wattmeter reading, it is necessary to take the direction of the voltage from line to line through the circuit the same as that taken for current. We see, then, that the readings of W_1 and W_2 are given by

$$W_1 = V_{AC} I_A \cos (30° - \theta) \tag{23.9}$$

$$W_2 = V_{BC} I_B \cos (30° + \theta) \tag{23.10}$$

---

[1]R. M. Kerchner and G. F. Corcoran, *Alternating-Current Circuits*, 4th ed.. New York: Wiley. 1960. pp. 338–339.

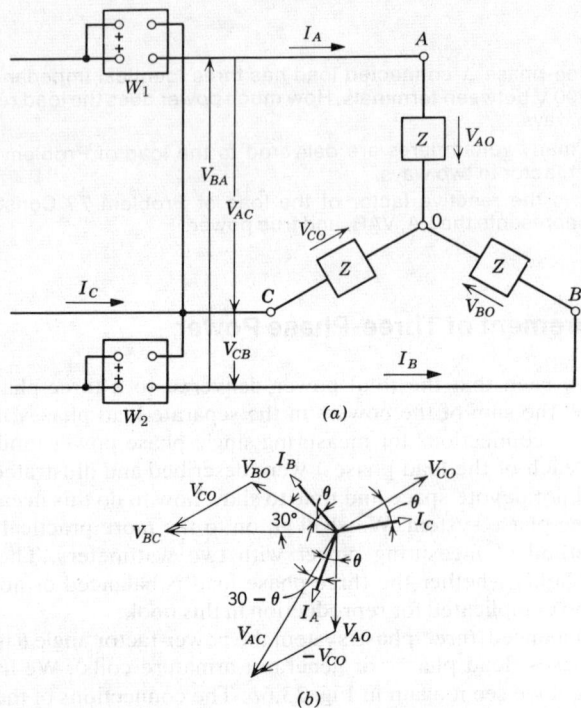

FIGURE 23.6 Two-wattmeter method of measuring power to a three-phase load. This method applies to delta-connected three-phase loads also. (a) Connections. (b) Phasor diagram.

We may now drop the subscripts of V and I because all the line-to-line voltages are equal and all the phase currents are equal. Only magnitudes are involved.

$$W_1 = VI \cos (30° - \theta)$$
$$W_2 = VI \cos (30° + \theta) \tag{23.11}$$

$$W_1 + W_2 = VI [\cos (30° - \theta) + \cos (30° + \theta)] \tag{23.12}$$

$$= VI [\cos 30° \cos \theta + \sin 30° \sin \theta + \cos 30° \cos \theta - \sin 30° \sin \theta]$$

$$= VI \left[\frac{\sqrt{3}}{2} \cos \theta + \frac{\sqrt{3}}{2} \cos \theta \right]$$

$$W_1 + W_2 = \sqrt{3} \, VI \cos \theta \tag{23.13}$$

Thus it is seen that the algebraic sum of the readings of the two wattmeters gives the total power in a balanced, three-phase system at any power factor. In more advanced books proof is given that this method of measuring total three-phase power yields the correct value under any conditions of load balance or current (or voltage) wave form.

Let us examine Equations (23.10) and (23.11). At unity power factor, θ is zero and the two meter readings are equal. At 50 percent power factor, $\cos \theta = 0.5$ and $\theta = 60°$. Then wattmeter W_2 reads zero. For all lagging power factors below 50 percent, the reading of W_2 is negative and its reading *must be subtracted* from the reading of W_1 to get the actual total power.

The question now arises: How can one determine if the power factor is less than 50 percent. If the meters are properly connected in the circuit, with due regard to the plus-minus ($\pm$) polarity marks as shown in Fig. 23.6a, the pointer of one of the meters (in this example, W_2) will be off-scale to the left of the zero point if the power factor is below 50 percent. It can then be made to read up-scale by reversing the connections to its potential coil. The operator must not forget to subtract this reading from the reading of the other meter to get the value of true power.

When the wattmeter coils are connected to the power line through *instrument transformers*, as is the case of high voltage or large values of line current, correct connections are possible if polarity markings are observed in the procedure. A good method for checking will be described with reference to Fig. 23.6a. Open the top line A. This will cause all power to reach the load over lines B and C. If meter W_2 reads up-scale, it must be properly connected to read power going to the load. Now reconnect line A and disconnect line B. If W_1 is not reading up-scale, reverse its voltage leads so that it does. Then reconnect line B. Now the meters are properly connected so that up-scale readings indicate that power is flowing to the load. If, during actual measurements, the pointer of either meter is below the zero of the scale, power through that meter is being transferred back to the source through that part of the circuit, and this power must be considered negative with respect to the power measured by the other meter. In order to read this negative power value, the pointer must be made to move up-scale. A polarity-reverse switch is usually mounted on the meter panel.

23.9   Two or More Balanced Loads on a Three-Phase System

We shall consider a balanced, three-phase load of pure resistance, which might be an electric oven, and motor loads at lagging power factor on one three-phase system. The analysis of this kind of situation perhaps is best explained by means of an example.

EXAMPLE 23.9   Figure 23.7a shows a three-phase system with a balanced unity-power-factor load consisting of three heating elements in an electric furnace connected in Δ and taking 3.74 kW each, and a balanced load consisting of induction motors which, as a group, total 25 hp at an average efficiency of 83 percent and an average power factor of 80 percent. It is desired to compute the total power, the power factor, and the line current.

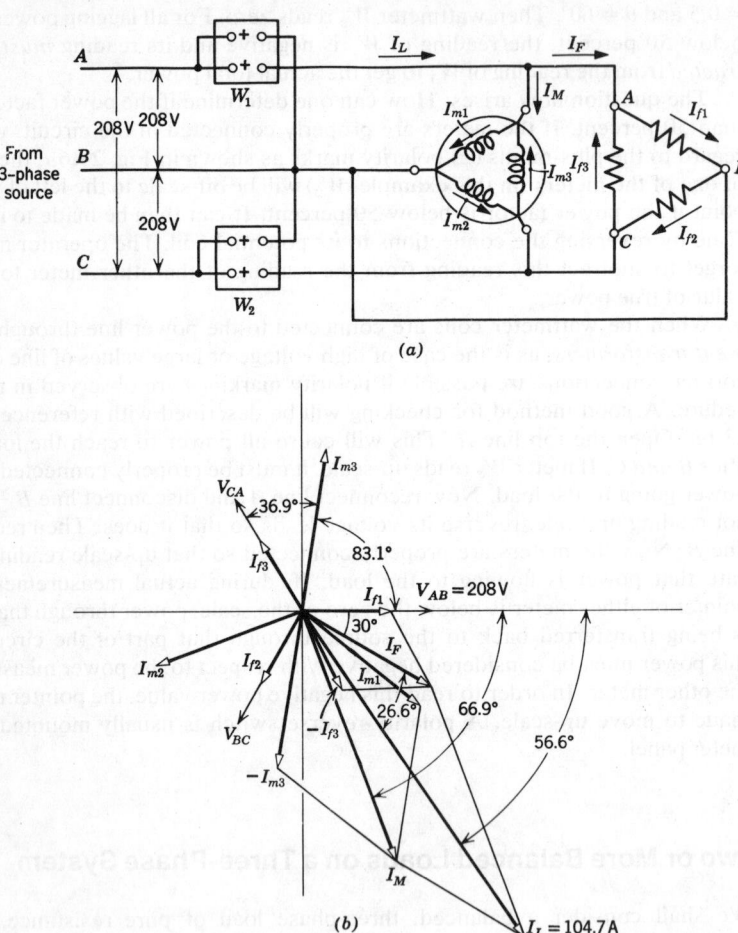

FIGURE 23.7   Metering of power to two balanced three-phase loads. (a) Schematic. (b) Phasor diagram.

Solution. The current in each element of the furnace is

$$I_f = \frac{3740}{208} = 18 \underline{/0°} \text{ A}$$

The line current I_F to the furnace is

$$I_F = \sqrt{3} \times 18 = 31.17 \underline{/30°} \text{ A}$$
$$= 26.9 - j15.58 \text{ A}$$

Note that

$$I_F + I_{f3} - I_{f1} = 0$$

$$I_F = I_{f1} - I_{f3}$$

The power input, P_M, to the motor is

$$P_M = 25 \times 746 / 0.83 = 22,500 \text{ W}$$

$$\text{Total power} = 22,500 + 3 \times 3740 = 33,720 \text{ W}$$

The part of the line current taken by the motor is found from $P_M = \sqrt{3} \, V_M I_M$ cos θ_M, where cos $\theta_M = 0.80$, the power factor of the balanced load the motor puts on the line. $\theta_M = $ arc-cos $0.8 = 36.9°$

$$P_M = \sqrt{3} \times 208 \times I_M \times 0.8 = 22,500$$

from which

$$I_M = \frac{22,500}{288.2} = 78 \text{ A}$$

Figure 23.7a shows that $I_M + I_{m3} = I_{m1}$. From this, $I_M = I_{m1} - I_{m3}$.

$$I_{m1} = 78/\sqrt{3} \text{ A and lags } V_{AB} \text{ by } \theta_M = 36.9°.$$

$$I_{m1} = 45 \underline{/-36.9°} \text{ A} = 36 - j27 \text{ A}$$

I_{m3} lags V_{CA} by 36.9°, and $V_{CA} = 208 \underline{/120°} \text{ V}$

$$I_{m3} = 45 \underline{/120° - 36.9°} = 45 \underline{/83.1°} = 5.4 + j44.7 \text{ A}$$

$$I_M = 36 - j27 - 5.4 - j44.7 = 30.6 - j71.7 = 78 \underline{/-66.9°} \text{ A}$$

The line current $I_L = I_M + I_F$:

$$I_L = 30.6 - j71.7 + 26.9 - j15.58$$

$$I_L = 57.5 - j87.28 = 104.7 \underline{/-56.6°}$$

The wattmeter readings may be calculated by means of Equations (23.11):

$$W_1 = 208 \times 104.7 \cos -56.6° = 11,980 \text{ W}$$

Note that $-56.6° = -(30° + 26.6°)$. Because we have a balanced load, the angle between I_C, which flows in W_2, and V_{CB} is $30° - 26.6° = 3.4°$. Furthermore, the line currents are all equal in magnitude.

$$W_2 = 208 \times 104.7 \cos 3.4° = 21,700 \text{ W}$$

The total power is

$$P_T = 11,980 + 21,740 = 33,720 \text{ W}$$

It is not difficult to show that I_C lags V_{CB} by 3.4°, but the addition of these phasors to Fig. 23.7b would make it more complicated and it is already complicated enough. The reasoning is as follows:

The current to the furnace at junction C lags V_{CB} by 30°, just as I_F lags V_{AB} by 30°. Of course, $V_{CB} = -V_{BC}$. The current entering the motor from line C lags V_{CA} by 56.6°. This means the phasor for I_C *leads* V_{CB} by $60° - 56.6° = 3.4°$. This is precisely like the situation in Fig. 23.6b where I_A leads V_{AC} by $30° - \theta$. The angle between V_{CA} and V_{CB} is 60°. Wattmeter W_2 uses V_{CB} and I_C, as is evident in its connections in Fig. 23.7a.

The overall power factor is given by watts ÷ voltamperes

$$\text{P.F.} = 33,720/\sqrt{3} \times 208 \times 104.7 = 0.894$$

23.10   Measurement of Energy in Three-Phase Systems

Watthour meters for use on three-phase systems are of the polyphase type. These contain two sets of coils which provide magnetic fields that cause rotation of a common shaft.

A meter installation for a system like that shown in Fig. 23.7 is illustrated in Fig. 23.8. Current transformers are used in the three main lines. They deliver current to the current coils of the meter. This current amounts to 5 A in each transformer secondary when its primary is carrying 100 percent rated current. The transformer capacity must be compatible with the meter capacity and with the full-load current value in the main lines.

Potential transformers are not used unless the voltage exceeds 600 V. None is needed on this 208-V system. When used, each primary is connected between a line wire and the neutral just as the potential coils are directly connected to these wires in Fig. 23.8.

The current coils of the watthour meter have 2.5 A current capacity even though the current transformers deliver 5 A when the system is operating at full load. This takes advantage of two important practical situations. One is that at 100 percent overload, which the current coils are able to carry owing to safety factors in design, the meter accuracy does not decrease. The other is that at light loads the accuracy is better with the 2.5 A coil than it would be with a 5-A coil.

A four-wire, Δ-connected supply for 240-V, three-phase motors and 120-V

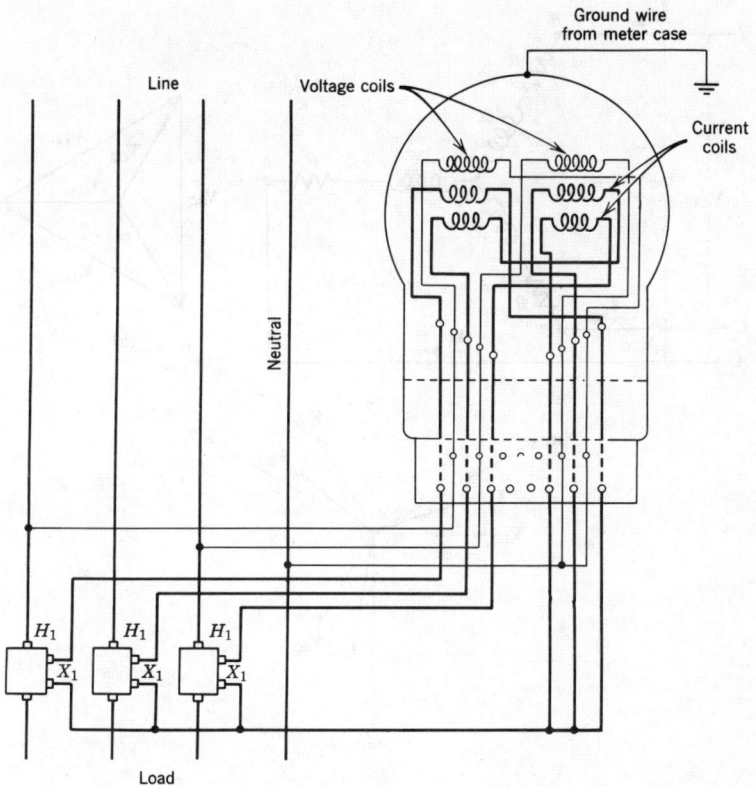

FIGURE 23.8 Polyphase watthour-meter connections on a three-phase, four-wire, Y-connected system. Current transformers supply 5 A to each current coil in the meter when the load is 100 percent of full value at unity power factor. Potential transformers are used when the line voltage exceeds 600 V.

lighting circuits is used in common practice also. A center tap on one of the Δ-connected secondaries provides two "channels" to serve branch circuits for 120-V lighting

23.11   Application of Complex Algebra to Analysis of Three-Phase Systems

EXAMPLE 23.10   (a) Assume a balanced, three-phase load, as shown in Fig. 23.9, to which 208 V are applied between each pair of line wires in the phase sequence V_{AB}, V_{CA}, V_{BC}. We shall use the complex forms for

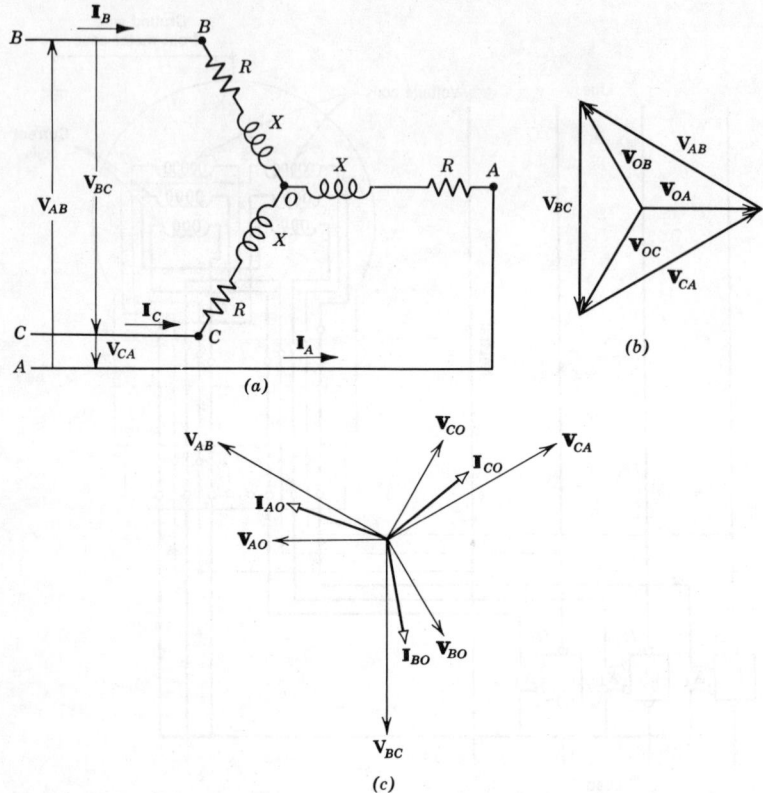

FIGURE 23.9 A balanced, three-phase inductive load, Y-connected. (a) Schematic. (b) Voltage phasors. (c) Phasor diagram of phase and line voltages and currents. $R = 12\,\Omega$, $X = 5\,\Omega$.

voltage, current, and impedance, and compute the line current, power per phase, and total power.

Solution. We start by drawing the phasor diagram of the voltages (Fig. 23.9*b*) and note that $\mathbf{V}_{AB}$, followed by $\mathbf{V}_{CA}$, then by $\mathbf{V}_{BC}$ make 120° phase angles with one another. These phasors are drawn from a common point in Fig. 23.9*c*. Because the phase currents lag the voltages applied between the *phase terminals*, and because we show each current flowing from the outer terminal to the neutral, we represent the phase voltages, $\mathbf{V}_{AO}$, $\mathbf{V}_{BO}$, $\mathbf{V}_{CO}$, as they actually combine to equal the line-to-line voltages. We can now write the phase voltages in complex form:

Power per phase,

$$\mathbf{V}_{AO} = 120\underline{/-180°} = -120 + j0 \text{ V}$$

$$\mathbf{V}_{BO} = 120\underline{/-60°} = 120 \cos 60 - j120 \sin 60$$
$$= 60 - j104 \text{ V}$$

$$\mathbf{V}_{CO} = 120\underline{/+60°} = 120 \cos 60 + j120 \sin 60$$
$$= 60 + j104 \text{ V}$$

$$\mathbf{V}_{AB} = \mathbf{V}_{AO} + \mathbf{V}_{OB} = \mathbf{V}_{AO} - \mathbf{V}_{BO}$$
$$= -120 + j0 - 60 + j104 = -180 + j104 \text{ V}$$

$$\mathbf{V}_{BC} = \mathbf{V}_{BO} + \mathbf{V}_{OC} = \mathbf{V}_{BO} - \mathbf{V}_{CO}$$
$$= 60 - j104 - 60 - j104 = -j208 \text{ V}$$

$$\mathbf{V}_{CA} = \mathbf{V}_{CO} + \mathbf{V}_{OA} = \mathbf{V}_{CO} - \mathbf{V}_{AO}$$
$$= 60 + j104 + 120 - j0 = 180 + j104 \text{ V}$$

$$\mathbf{I}_A = \mathbf{I}_{AO} = \frac{\mathbf{V}_{AO}}{\mathbf{Z}_{AO}} = \frac{-120}{12+j5} = \frac{-120(12-j5)}{169}$$
$$= -8.53 + j3.55 = 9.23\underline{/157.4°} \text{ A}$$

$$\mathbf{I}_B = \mathbf{I}_{BO} = \frac{\mathbf{E}_{BO}}{\mathbf{Z}_{BO}} = \frac{60-j104}{12+j5} = \frac{200-j1548}{169}$$
$$= 1.18 - j9.17 = 9.23\underline{/-82.6°} \text{ A}$$

$$\mathbf{I}_C = \mathbf{I}_{CO} = \frac{\mathbf{V}_{CO}}{\mathbf{Z}_{CO}} = \frac{60+j104}{12+j5} = \frac{1240+948}{169}$$
$$= 7.33 + j5.61 = 9.23\underline{/37.4°} \text{ A}$$

$$P_{AO} = VI + V'I' = 120 \times 8.53 + 0 \times 3.55 = 1023 \text{ W}$$

or

$$P_{AO} = V_{AO}I_{AO} \cos \theta = 120 \times 9.23 \cos (180° - 157.4°)$$
$$= 1107.6 \cos 22.6° = 1023 \text{ W}$$

or

$$P_{AO} = I_{AO}^2 R_{AO} = 9.23^2 \times 12 = 1022 \text{ W}$$

Total power $= 3 \times 1022 = 3066$ watts $= 3.066$ kW

(b) Assume that the three load impedances of the above circuit are connected to the lines in Δ instead of in Y. The circuit and phasor diagrams are shown in Fig. 23.10. To illustrate the choice of subscripts, we note that at the instant of time when the voltages and currents are represented, terminal B is higher in potential than terminal A. This causes $\mathbf{I}_{BA}$ to have the direction from B to A. The arrow V_{AB} indicates a potential *rise* from the line on point A to the line on point B.

The voltages causing flow of phase currents are represented in the phasor

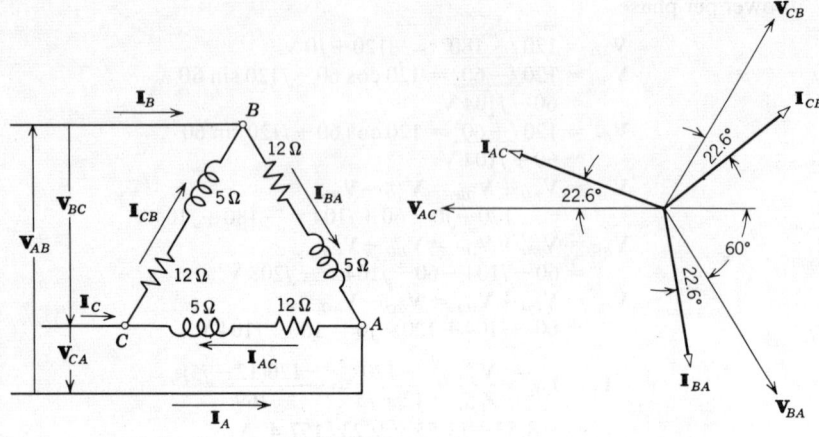

FIGURE 23.10 Same load elements shown in Fig. 23.9, but connected in Δ.

diagram, and their complex expressions are

$$\mathbf{V}_{BA} = 208 \cos 60° - j208 \sin 60° = 104 - j180 \text{ V}$$
$$\mathbf{V}_{CB} = 208 \cos 60° + j208 \sin 60° = 104 + j180 \text{ V}$$
$$\mathbf{V}_{AC} = -208 + j0 \text{ V}$$

The currents are

$$\mathbf{I}_{BA} = \frac{\mathbf{V}_{BA}}{\mathbf{Z}_{BA}} = \frac{104 - j180}{12 + j5} = \frac{348 - j2680}{169} = 2.06 - j15.86 = 16\underline{/-82.6°} \text{ A}$$

$$\mathbf{I}_{CB} = \frac{\mathbf{V}_{CB}}{\mathbf{Z}_{CB}} = \frac{104 + j180}{12 + j5} = \frac{2148 + j1640}{169} = 12.7 + j9.7 = 16\underline{/37.4°} \text{ A}$$

$$\mathbf{I}_{AC} = \frac{\mathbf{V}_{AC}}{\mathbf{Z}_{AC}} = \frac{-208}{12 + j5} = \frac{-2496 + j1040}{169} = -14.76 + j6.15 = 16\underline{/157.4°} \text{ A}$$

Power per phase,

$$P_{BA} = VI + V'I' = 104 \times 2.06 + (-180)(-15.86) = 3070 \text{ W}$$

or

$$P_{BA} = V_{BA}I_{BA} \cos \theta = 208 \times 16 \cos (82.6° - 60°) = 3070 \text{ W}$$

or

$$P_{BA} = I_{BA}^2 R_{BA} = 16^2 \times 12 = 3072 \text{ watts}$$

Total power $= 3 \times 3072 = 9216$ watts $= 9.21$ kW

This should be (and is) three times the power to the same load that was Y connected on the same supply lines, because in Δ the phase voltage is $\sqrt{3}$ times larger and the phase current is $\sqrt{3}$ times larger. This causes the quantity $3EI \cos \theta$ to be three times larger.

23.12   The Polyphase Wattmeter

The polyphase wattmeter has two sets of single-phase wattmeter coils contained in one case with the moving coils mounted on a single shaft. The deflection of the single pointer is produced by the sum of the torques produced by force on the two moving coils. It is, therefore, proportional to the total power.

One advantage of the polyphase wattmeter is that its single reading is taken easier than the readings of two single-phase meters and it will be more accurate if the load fluctuates. Inaccuracies caused by changes in load after one meter is read and the second is about to be read are eliminated.

The polyphase wattmeter is lighter in weight and smaller in physical size than two single-phase meters, but mutual induction between the coils causes some inaccuracy in the readings. This mutual induction is not present when two single-phase wattmerers are used.

If, when a polyphase wattmeter is used, we want to know the power factor, we can obtain the necessary readings that two single-phase wattmeters would give. It is only necessary to disconnect the voltage-coil lead of one element to get the watts indicated by the other element, then replace it and disconnect the voltage lead of the other element. The watt-ratio curve* would then be used to get the power factor.

23.13   Scott Connection

Before presenting the Scott system we must state that a *two-phase* system has only two voltages that are 90° *apart*.

Sometimes it is necessary to operate one or more *two-phase* motors when only *three-phase* power is available. Or, we may wish to operate three-phase equipment on a two-phase supply. Both of these operations are made possible by means of the *Scott connection*, shown in Figure 23.11a.

Two transformers are used which are identical except that one has a tap on its secondary winding at the turn which is 86.6 percent ($\sqrt{3}/2$) of the distance (turn-wise) from the end that is connected to the center tap of the other trans-

*H. Alex Romanowitz, *Electrical Fundamentals and Circuit Analysis*, New York, Wiley, 1966, pp. 614–615.

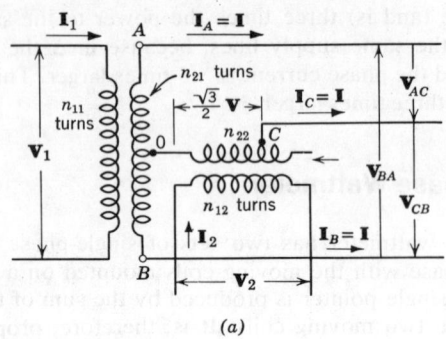

(a)

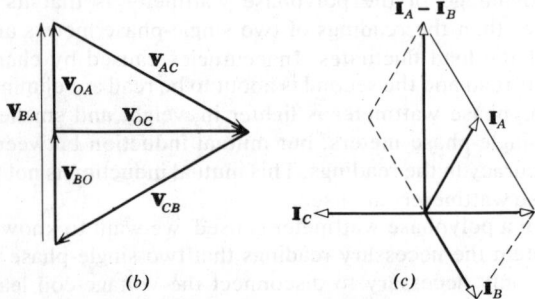

(b)   (c)

FIGURE 23.11 Illustrating Scott connection. (a) Scott connection, two phase to three phase. All windings have same number of turns, giving 1:1 ratio. Other ratios are possible. (b) Voltage phasor diagram. (c) Current phasor diagram. $I_C = -(I_A + I_B)$

former. Because $\mathbf{V}_2$, one of the two-phase voltages, supplies the primary, the secondary voltage $\mathbf{V}_{OC}$ is phased at 90° with respect to $\mathbf{V}_{AB}$, the other secondary voltage. The relative magnitudes of $\mathbf{V}_{AO}$ and $\mathbf{V}_{OC}$ cause the magnitude of $\mathbf{V}_{AC}$ to be equal to that of $\mathbf{V}_{AB}$

$$\sqrt{\left(\frac{\sqrt{3}}{2}V\right)^2 + \left(\frac{V}{2}\right)^2} = \sqrt{\frac{3V^2}{4} + \frac{V^2}{4}} = V$$

in which V represents the magnitude of the line-to-line voltages (rms) on the three-phase side.

Let us now consider the amount of current $I_C = I$ in one three-phase line. Remembering that $n_1 I_1 = n_2 I_2$ in a transformer,

$$n_{12} I_2 = n_{22} I \tag{23.14}$$

Also

$$\frac{n_{12}}{n_{22}} = \frac{V_2}{\sqrt{3}V} = \frac{2V_2}{\sqrt{3}V} \qquad (23.15)$$

In a 1:1 transformer, which would be used if no change in magnitude is desired, $V_2 = V$. Using this and the relations in Equations (23.14) and (23.15), we get

$$I = \frac{n_{12}}{n_{22}} I_2 = \frac{2}{\sqrt{3}} I_2 \qquad (23.16)$$

That is, the line current and coil current on the three-phase side is larger than the coil current on the two-phase side by the factor $2/\sqrt{3} = 1.154$.

The currents $\mathbf{I}_A$ and $\mathbf{I}_B$ (each has the same magnitude, namely that of I) are *not* 180° out of phase, as one might imagine. Current from the other secondary is forced through the halves of the winding. The three-phase line currents must be equal and 120° apart in phase for a balanced load, because the line voltages are equal and balanced.

Because $\mathbf{I}_A$ and $\mathbf{I}_B$ come out of opposite ends of the n_{21} secondary, their magnetizing effect is produced by their *difference*, which is $\sqrt{3}I_A = \sqrt{3}I$. We then express the magnetizing effect in ampereturns as before:

$$n_{11}I_1 = n_{21}I \frac{\sqrt{3}}{2} \qquad (23.17)$$

Recalling that the voltage transformation ratio is 1 : 1, which means

$$\frac{n_{11}}{n_{21}} = \frac{V_1}{V_{AB}} = 1$$

Using this in Equation (23.17),

$$I_1 = I \frac{\sqrt{3}}{2} = I_2 \qquad \text{[see Equation (23.16)]}$$

The total two-phase volt amperes is $2VI_1$. Using $I_1 = I(\sqrt{3}/2)$, as just derived, this becomes $\sqrt{3}VI$, which is the same voltampere capacity as that on the three-phase side.

EXAMPLE 23.11   A 50-hp, 460-V, *three-phase* motor is supplied from a 2300-V, *two-phase* line. The motor efficiency is 92 percent and the power factor is 87 percent. Determine the line current to the motor and current in the primary windings of the Scott-connected transformers.

Solution

$$\text{Motor input} = \frac{50 \times 746}{0.92 \times 0.87} = 46{,}625 \text{ VA}$$

This three-phase input is expressed as

$$\sqrt{3}VI = \sqrt{3} \times 460I = 46,625 \text{ VA}$$
$$I = 58.52 \text{ A} \qquad \text{(in the three-phase line)}$$

Neglecting losses in the transformer, which has a voltage ratio of 5 : 1,

$$2 \times 2300 \times I_{2\phi} = 46,625$$
$$I_{2\phi} = 10.13 \text{ A}$$

23.14   Open-Δ Connection (V Connection)

Suppose trouble develops in one of three single-phase, Δ-connected transformers which are supplying a three-phase load. That transformer may be taken out of service and the other two will continue to supply the load but at reduced capacity. They will deliver somewhat less than two-thirds the capacity of three transformers. This will now be proved.

The phasor diagram will show that the secondary voltages **V** in the open-Δ case are balanced and 120° apart.

Voltage induced in the secondary of a transformer is not dependent on the size of wire in its winding, but the current capacity is limited by this. Therefore the kVA rating of the transformer is determined by the full-load current value.

Let *I* be the rated current of each transformer (Fig. 23.12*b*) and *V* be the

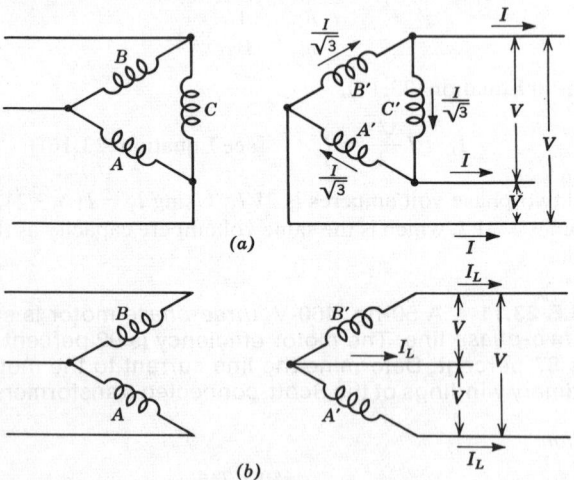

(a)

(b)

FIGURE 23.12   Open-Δ (or V) operation. (a) Normal Δ. (b) Open Δ, or V.

line voltage. The total kVA capacity of two transformers connected open Δ is $\sqrt{3}VI$. The total kVA capacity of three transformers connected in Δ is $3VI$. The ratio of open-Δ capacity to closed-Δ capacity is

$$\frac{\sqrt{3}VI}{3VI} = \frac{1}{\sqrt{3}} = 0.578 \text{ or } 57.8 \text{ percent}$$

This means that if one transformer of a closed-Δ set of three 25-kVA transformers operating in Δ were taken out of service, the remaining two would furnish $0.578 \times 75 = 43.35$ kVA when operating at full load, and not $\frac{2}{3} \times 75$ which is 50 kVA. With a balanced load, each transformer rated at 25 kVA single phase would be fully loaded when it supplies $\frac{1}{2} \times 43.35 = 21.68$ kVA. This is a 13.28 percent reduction in capacity. The reasons for this reduction, when operation is open Δ, is that the current in one of the transformers is shifted 30° in one direction from its phase position when in closed-Δ operation, and the current in the other is shifted 30° in the opposite direction. This causes the transformers to operate at 0.866 power factor (cos 30°).

Electric utility companies install distribution transformers in open Δ where there is likely to be an increase in the demand for power in a few years. The addition of one transformer would give a 73 percent increase in capacity with only a 50 percent increase in investment. This capacity increase is arrived at by using the 25-kVA transformer installations in an example.

Operation in closed Δ: 75 kVA

Operation in open Δ: 43.35 kVA

Increase $75 - 43.35 = 31.65$

Percent increase $\frac{31.65}{43.35} \times 100 = 73$ percent

23.15 Review Questions

1. Name some features of three-phase power operation that make it more practical than single-phase operation.
2. Explain how a rotating field of constant strength is maintained in a large alternating-current generator.
3. Why are the connections to one of the coil groups in a three-phase alternator reversed? See Fig. 23.1 (*b*) and (*c*).
4. How do the coil connections in a three-phase delta system differ from those of a wye system?
5. Express the magnitude of the line-to-line voltage of a balanced wye-connected load in terms of the voltage across one of the phases.
6. How does the magnitude of the line current of a balanced delta-connected load compare with the current in one of the phases?

7. What is the formula for the total voltamperes of a balanced three-phase load? Does this hold for both delta and wye connections?

8. What is the formula for total power to a balanced three-phase load? Does this hold for both delta and wye connections?

9. Express the power factor of a balanced three-phase load in terms of voltamperes and watts. Do this both in words and in equation form.

10. Write the formula for reactive voltamperes of a three-phase system.

11. Sketch connections of two wattmeters in a three-phase line that will measure the total power. Does this method apply to unbalanced as well as to balanced loads?

12. How would you measure the power factor of an unbalanced three-phase load?

13. Suppose you are measuring three-phase power by the two wattmeter method and you found that even with correct meter connections, one of the meters is reading backward. What does this signify and what would you do to get a reading on that meter? After you got its reading what would you do with it?

14. Assume it is necessary to measure the total power supplied to a load that has 1200 V between lines and 200 V in each line. What additional equipment, besides the wattmeter, would be needed?

15. How does a polyphase wattmeter differ from a single-phase wattmeter?

16. What is the Scott connection of transformers used for? Will it transform power both ways?

17. What are the advantages of the open-delta (or vee) connection of transformers. When a third transformer (identical in capacity) is added, how much is the power capacity increased? How much is the transformer cost increased?

23.16   Problems

10. A Y-connected generator has phase (line-to-neutral) voltages as follows: $V_{OA} = 120 + j0$, $V_{OB} = -60 - j103.9$, $V_{OC} = -60 + j103.9$. Express the following line voltages in complex form: $V_{AB} = V_{AO} + V_{OB}$: V_{BC}; V_{CA}.

11. Express the line voltages requested in Problem 10 in polar form.

12. The terminal voltages of the alternator of Problem 10 are applied to a delta-connected load, each element of which is a 25-Ω resistance. What are the magnitudes of the line currents?

13. Calculate the power supplied to the load of Problem 12 by two methods.

14. A Y-connected three-phase load has each phase impedance $Z = 20 + j10 \, \Omega$. The three-phase supply lines have 208 V between them. (a) What is the amount of current in each impedance? (b) Calculate the total power received by two methods.

15. What are the power factor and reactive factor of the load in problem 14? Can you calculate each of these by two methods?

16. A three-phase motor with Y-connected windings takes 76.2 A line current on 208 line volts. The measured power input is 21.95 kW. At what power factor does it operate? What is the horsepower output if the efficiency is 85 percent?

17. A three-phase 10-hp motor operates at full load on 416 V at 0.8 power factor and 85 percent efficiency. Calculate the line current and the input power.

18. Two 50 kVA single-phase transformers are operating in open Δ. What is their combined kVA capacity? How many kVA would be added to this capacity by the addition of a third 50 kVA single-phase transformer and operation in closed delta effected?

19. Three inductive impedances, all alike, are connected in Y across a 240-V, three-phase, 60-Hz system. Their total power is 4800 W at 6000 VA. What are the (a) power factor, (b) line current (c) current in each impedance, (d) resistance and reactance of each impedance?

20. Here's a long one: A three-phase, 115-V line supplies 150/50-W lamps connected in groups of 50 to form a balanced lighting load. Also a 10-hp three-phase induction motor is connected to the line and is operated fully loaded at 0.8 power factor and 85 percent efficiency. Calculate the (a) current in each group of lamps, (b) line current to the lamps, (c) line current to the motor, (d) total line current, (e) total power. Show connections to two wattmeters that measure the power and calculate the readings of the meters.

CHAPTER TWENTY-FOUR

Nonsinusoidal waves

Electrical and electronics technicians, especially those working on communications equipment, often encounter voltage and current waveforms that are complex rather than sinusoidal. Sound amplification invariably must involve reproduction of input electrical signals that have very complex wave forms.

A tuning fork that produces the "C" note on the musical scale vibrates at the rate of 256 vibrations per second. This is a single-frequency tone. However, if the key "middle C" is struck on a piano the vibrating string will send out not only the tone of the 256 Hz frequency (called the fundamental) but also a double-frequency tone of 512 Hz, called the second harmonic. In all likelihood a triple-frequency tone of $3 \times 256 = 768$ Hz would exist also. The string is capable of vibrating as if it were fixed at its midpoint and also at other fractions of its length. The higher frequency notes, called *harmonics*, give richness and improved quality to the effect. A microphone actuated by air pressures that vary in frequency and strength generates a complex electric wave that must be amplified.

The flying spot that produces the image on a cathode ray oscilloscope tube is caused to move from left to right at a constant speed by a deflection voltage that has the shape of a saw tooth. Such a nonsinusoidal wave has many harmonics in its makeup.

24.1   Composition of a Complex Wave

Figure 24.1 shows a complex wave that contains a fundamental, a second harmonic, and a third harmonic. This can represent a voltage wave whose second harmonic has half the amplitude of its fundamental and whose third harmonic has one fifth the amplitude of its fundamental. If we let the fundamental component be represented by $e_1 = E_m \sin \omega t$, the second harmonic is then $e_2 = 0.5\ E_m \sin 2\ \omega t$ and the third harmonic is $e_3 = 0.2\ E_m \sin \omega t$. The complex wave is then represented by

$$e = E_m(\sin \omega t + 0.5 \sin 2\omega t + 0.2 \sin 3\omega t) \qquad (24.1)$$

Note that all three components (the fundamental and the two harmonics) come up to positive values at the start of the positive half cycle. This requires + signs on all of the terms of Equation (24.1). A different complex wave form would result if one or more of the components did not behave as we just indicated. Suppose the second harmonic had gone negative at $\omega t = 0$ instead of positive and had reached *negative maximum* at $\omega t = 45°$. Its expression would then have been $e_2 = -0.5 \sin 2\ \omega t$, and that term in Equation (24.1) would have had the minus sign.

In order to draw the complex wave when its components are given, as in Equation (24.1), all we need do is plot the components to scale on the ωt axis and then add together the vertical distances to the component waves at

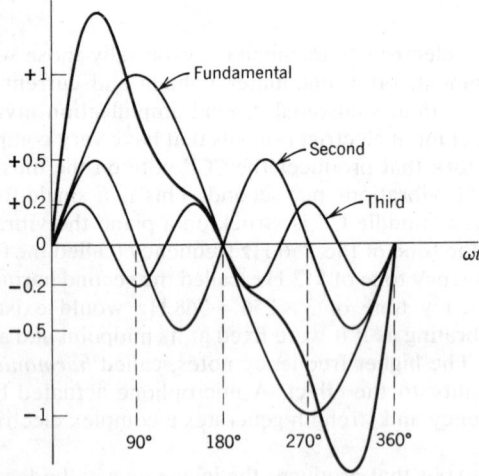

FIGURE 24.1   A complex wave that contains a fundamental frequency component, a second harmonic, and a third harmonic.

each of many points. We should select points that let us use the maximum instantaneous values of the components and the zero values as well as some in between. In cases where there are many components of a complex wave (the half-sine wave is an example), it is usually necessary to plot only the fundamental and the three lowest harmonics (2nd, 3rd, and 4th) to get a good graphical representation of the whole wave. The half-sine wave happens to have a dc (constant value) component which is included in the graphical summation. A constant term in the equation of the total wave represents this component.

24.2 Cosine Wave

The equations of some complex wave forms have cosine terms as well as sine terms. In fact, the terms in the equations of some other complex waves have only cosine terms. A cosine wave has the same form as a sine wave except that it *leads* its related sine wave by 90°. The consine wave in Figure 24.2 has its positive maxim at $\omega t = 0°$ and 360°. Because it reaches its maxima 90° before the sine wave does, we say it leads the sine wave by 90°. The equation of the cosine wave is

$$e = E_m \cos \omega t \tag{24.2}$$

Note that when $\omega t = 0$, $e = E_m$ the maximum value. When $\omega t = 90°$, $e = E_m$, $\cos 90° = 0$; when $\omega t = 180°$, $e = E_m \cos 180° = -E_m$ and when $\omega t = 360°$, $e = E_m \cos 360° = E_m$.

EXAMPLE 24.1   What is the equation of the complex wave whose components are shown in Fig. 24.3?

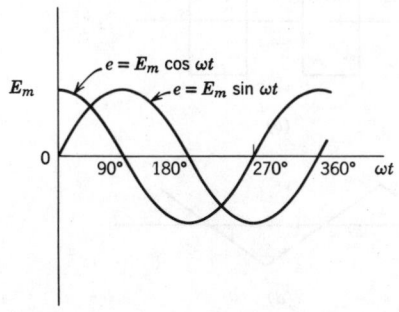

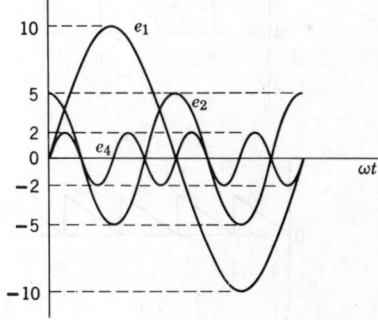

FIGURE 24.2   Cosine wave and sine wave.
Cosine wave leads sine wave by 90°.

FIGURE 24.3   Components of a complex wave.
For Example 24.1.

Solution

$$e_1 = 10 \sin \omega t; \; e_2 = 5 \cos 2\omega t; \; e_4 = 2 \sin 4\omega t$$
$$e = 10 \sin \omega t + 5 \cos 2\omega t + 2 \sin 4\omega t$$
$$= 10 (\sin \omega t + 0.5 \cos 2\omega t + 0.2 \sin 4\omega t)$$

24.3   Examples of Nonsinusoidal Waves

Two complex wave forms that are frequently encountered are those delivered to a load by rectifiers used in radio and television practice. When a voltage with sine-wave form is applied to a full-wave rectifier, the output of the rectifier has two loops in the positive direction and none in the negative direction, as shown in Fig. 24.4 *a*. This means the current in the load is a direct current and it flows in one direction only. If the sine-wave voltage is applied to a half-wave rectifier, the negative loops will be cut off because the rectifier is not capable of reversing them. It simply prevents them from reaching the output, and a "half-sine wave" results, as shown in Fig. 23.4*b*.

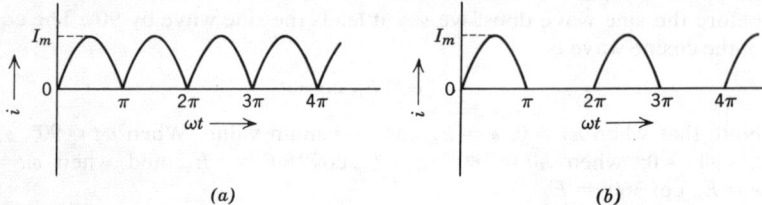

FIGURE 24.4   (a) A rectified sine-wave form. (b) A half-rectified sine-wave form.

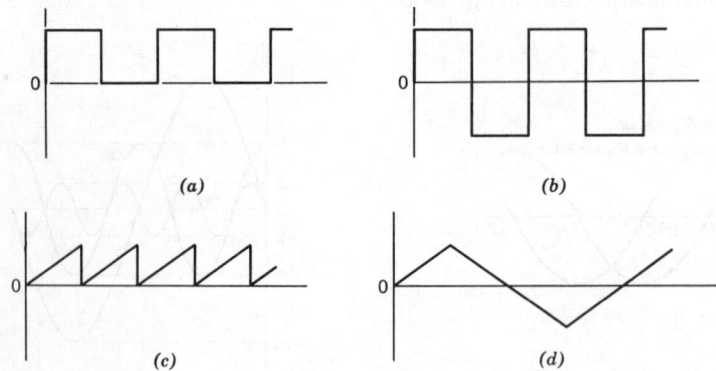

FIGURE 24.5   (a) Square wave with only positive polarity. (b) Square wave with positive and negative polarity. (c) Saw-tooth wave. (d) Triangular wave.

Some other types of nonsinusoidal wave forms encountered in practice are the *square wave*, the *saw-tooth* wave, and the *triangular* wave shown in Fig. 24.5.

Complete mathematical analysis of a recurring complex wave of any form is made by applying principles which lead to the equation for the wave in a form called *Fourier series*. In order to represent the maximum value of the components of the wave so that they may represent not just volts but also amperes or pressures or any other quantity, the letters A_1, A_2, A_3, etc. are used to represent the coefficients of the sine terms. A_0 is the symbol for the constant term. B_1, B_2, B_3, etc., are used to represent the coefficients of the cosine terms.

Analysis of a half-sine wave in Fig. 24.6 leads to the following amplitudes:

$$A_0 = 0.318, A_1 = 0.5, A_2, A_3, A_4, \text{etc. are all zero}$$

$B_1 = 0, B_2 = -0.212, B_3 = 0, B_4 = -0.0424$, and the equation for a half-sine wave *current* is

$$i = I_m(0.318 + 0.500 \sin \omega t - 0.212 \cos 2 \omega t - 0.0424 \cos 4 \omega t \ldots) \qquad (24.3)$$

Figure 24.6 shows the actual half-sine wave and Fig. 24.7 shows how close to the actual wave we can get by plotting the dc component term and only three of the many harmonics that are in the series. The curve designated in the graph by Equation (24.3) is quite close, in shape to the original (dash) curve which is the same as the half-sine curve in Fig. 24.6.

As a matter of interest we should look at the equation for the square wave in Fig. 24.5a [Equation (24.4)]. With a maximum value of 10 V as the ordinate and ωt as the abscissa, the equation is

$$e = 5 + 6.37 \sin \omega t + 2.12 \sin 3 \omega t + 1.27 \sin 5 \omega t + 0.909 \sin 7 \omega t + \cdots \qquad (24.4)$$

The dots that follow these equations indicate that other terms exist in the series. The general term giving any coefficient in Equation (24.4) is $10/n\pi$ $(1 - \cos n\pi)$ in which n must be an odd number such as 1, 3, 5, 7, etc. Note that when $n = 1$, representing the fundamental frequency, the coefficient of

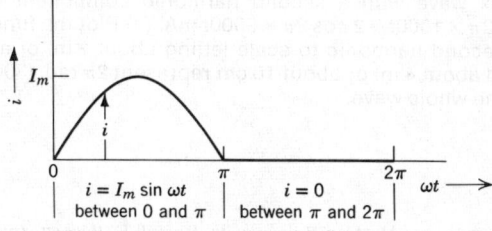

FIGURE 24.6   Wave form of instantaneous current of the output of a half-wave rectifier.

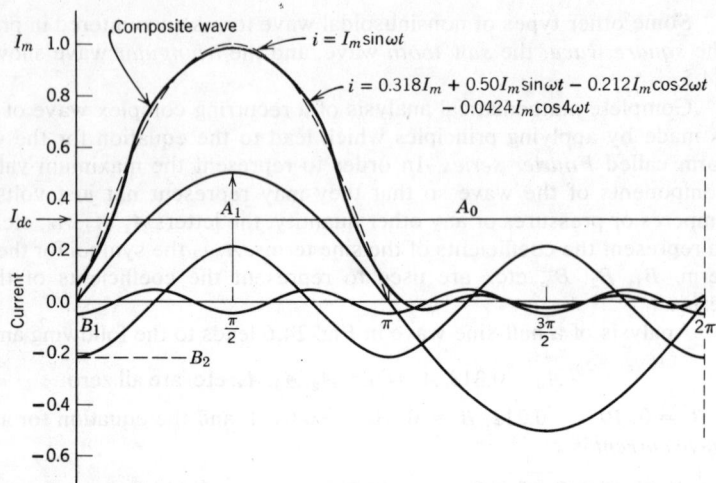

FIGURE 24.7 DC component, fundamental, second, and fourth harmonics in half-sine wave combine to produce half-sine form to a good approximation. The dashed half-sine wave is a plot of the total wave shown in Fig. 24.6.

$\sin \omega t$ is $(10/\pi)2 = 6.37$. When $n = 5$ the coefficient of $\sin 5\omega t$ is $(10/5\pi)$ $\times (1 - \cos 5\pi) = (2/\pi)(2) = 1.27$.

In an elementary study of electronics we shall learn that a single-tube amplifier generates a second harmonic† in its output current. A third harmonic component of current exists in the exciting current component of an iron-core transformer. Also, a third harmonic component of current circulates in the closed-delta path of a three-phase power transformer.

Problem

1. A complex wave with a second harmonic component has for its equation $i = 10 \sin 2\pi \times 1000t - 2 \cos 2\pi \times 2000t$ mA. (a) Plot the fundamental conponent and the second harmonic to scale letting about 2 in. or about 5 cm represent 10 mA and about 4 in. or about 10 cm represent 2π rad. (360°). (b) Add ordinates and plot the whole wave.

---

†For an explanation, see H. Alex Romanowitz, Russell E. Puckett, *Introduction to Electronics*, John Wiley & Sons, Inc., New York, 1968, pp. 381–383.

24.4 Effective Value of a Nonsinusoidal Wave

The mathematical analysis that gives $I_m/\sqrt{2}$ or $\sqrt{I_m{}^2/2}$ for the effective value of a sine wave with I_m as its maximum instantaneous value is applicable to each of the separate harmonic components of a nonsinusoidal wave. That is, the effective value of the fundamental component is given by $\sqrt{I_m{}^2/2}$, and of the second harmonic it is $\sqrt{I_{mz}^2/2}$. Remembering that the direct current component is constant, we represent its value by I_0. The zero subscript indicates zero frequency.

Rigorous mathematical analysis gives the following expression for the effective value of a complex wave of current:

$$I_{\text{eff}} = \sqrt{I_0{}^2 + \frac{I_{m1}^2 + I_{m2}^2 + I_{m3}^2 + \cdots + I_{mn}^2}{2}} \tag{24.5}$$

in which the subscripts $m1$, $m2$, $m3 \ldots mn$ indicate that the instantaneous maximum values of the fundamental and corresponding harmonic components are used.

An alternate way of deriving the equivalent of Equation (24.5) is begun by using the expression for the power in a resistance in terms of the rms current value.

$$P = I_{rms}^2 R$$

When dc and ac components are present,

$$P = I_{dc}^2 R + I_{1\,rms}^2 R + I_{2\,rms}^2 R + \cdots$$

$$P = (I_{dc}^2 + I_{1\,rms}^2 + I_{2\,rms}^2 + \cdots)R$$

Therefore the *rms value* of the nonsinusoidal current is

$$I_{rms}^2 = \sqrt{I_{dc}^2 + I_1{}^2 + I_2{}^2 + \cdots} \tag{24.6}$$

This is the same as Equation (24.5) when I, *the rms value of each ac component*, is replaced by $I_m/\sqrt{2}$.

We get the same expression as Equation (24.6) for the effective value of a nonsinusoidal voltage wave.

Using V^2/R for each power contribution, we have

$$P = \frac{V_{dc}^2}{R} + \frac{V_1{}^2}{R} + \frac{V_2{}^2}{R} + \cdots + \frac{V_n{}^2}{R}$$

But

$$P = \frac{V_{\text{eff}}^2}{R}$$

so

$$\frac{V_{\text{eff}}^2}{R} = \frac{V_{dc}^2 + V_1{}^2 + V_2{}^2 + \cdots + V_n{}^2}{R}$$

Therefore

$$V_{\text{eff}} = \sqrt{V_{\text{dc}}^2 + V_1^2 + V_2^2 + \cdots + V_n^2} \qquad (24.7)$$

in which $V_1, V_2, \ldots V_n$ are rms values.

EXAMPLE 24.2 Determine the effective value of the complex voltage wave defined by Equation (24.4)

Solution

$$V_{\text{eff}} = \sqrt{5^2 + \frac{6.37^2 + 2.12^2 + 1.27^2 + 0.455^2}{2}}$$

Remember that $6.37/\sqrt{2}$ is $V_{1\text{eff}}$, $2.12/\sqrt{2}$ is $V_{3\text{eff}}$, etc. The dc voltage is its own "effective" value.

$$V_{\text{eff}} = \sqrt{25 + \frac{46.88}{2}} = 6.96 \text{ V}$$

These equations hold whether the expression for the complex wave has only sine terms, only cosine terms, or a mixture of sine and cosine terms. Remember that $I_{\text{eff}}^2 R$ is power, regardless of the phase angle of the current or voltage.

Now that we know how to calculate the power in an impedance that carries a nonsinusoidal current, let's look at the voltampere situation. We know that *voltamperes* are calculated by multiplying together effective values of voltage and current.

$$VA = VI = \sqrt{V_0^2 + \frac{V_1^2 + V_2^2 + V_3^2 + \cdots + V_n^2}{2}}$$
$$\times \sqrt{I_0^2 + \frac{I_1^2 + I_2^2 + I_3^2 + \cdots + I_n^2}{2}} \qquad (24.8)$$

Problems

2. The current of Problem 1, defined by $i = 10 \sin 2000\,\pi t - 2 \cos 4000\,\pi t$ mA, flows in a load resistor. Calculate its effective value.
3. The current of Problem 2 flows in a 1000-Ω load resistor. How much is the rms voltage drop across the resistor?
4. Calculate the voltamperes in the resistor of Problem 3 and verify that the power power factor is unity.
5. A voltage $e = 10 + 5 \sin \omega t - 3 \cos 2\,\omega t + 2 \sin 3\,\omega t$ exists across a resistance of 100 Ω. Determine the amount of the current and the power dissipated.

24.5   Nonsinusoidal Waves with Shifted Harmonics

The subject of this section means nonsinusoidal waves which have harmonics that are not in phase with the fundamental. But that was too long a title to use.
So far we have studied waves that are simpler in form than one defined by

$$e = E_m \,[\sin \omega t + 0.5 \sin (2\,\omega t + 30°) + 0.1 \sin (3\,\omega t - 60°)]$$

The 30° and $-60°$ angles do not bother us when we calculate effective values of voltage, or when we use E^2/R to get power contributions of each component. The same is true when we deal with current waves of this form. Furthermore, VI is still voltamperes and total power divided by VI is still power factor. We are making this comment because we are now ready to discuss the problem of a nonsinusoidal voltage applied to a series circuit containing R, L, and C. We shall find out how to set up the equation for the current in the circuit, and this is where these phase shift angles must be taken into account.

24.6   Nonsinusoidal Currents in Series Circuits

When a voltage that has a nonsinusoidal wave form is impressed on a series circuit, the current that flows will contain, in general, a harmonic component for each component in the voltage wave. That is, if the voltage wave has a second, third, fourth, and fifth harmonic in addition to its fundamental component, the current wave will have a fundamental with the same frequency as that of the voltage wave and it will have a second, third, fourth, and fifth harmonic.

When inductance and capacitance values of circuit elements are given, the *reactances* at all frequencies must be computed separately for use in obtaining the magnitude of the component current waves.

We may work with either maximum instantaneous or effective values. Because we are usually interested in computing power and the voltage or current that a meter would read, it is advantageous to work with effective values. This will be done in the examples that follow.

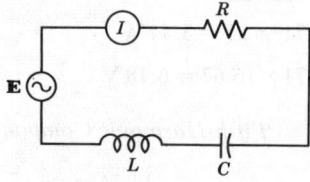

FIGURE 24.8   For Example 24.3. $R = 25\ \Omega$, $L = 0.07$ H, $C = 53\ \mu$F.

EXAMPLE 24.3   A nonsinusoidal voltage $e = 141.4 \sin \omega t + 35.35 \sin (3\omega t + 30°) - 14.14 \sin (5\omega t - 30°)$, in which $\omega = 377$ rad/s, is applied to the series circuit of Fig. 24.8. Compute I, the ammeter reading the total power dissipated, and the voltage drop across the capacitor. Write the equation of the current wave.

Solution. The effective value of the fundamental component of the voltage is $V_1 = 141.4/\sqrt{2} = 100$ V.

$$X_{L1} = \omega L = 377 \times 0.07 = 26.39 \ \Omega$$

$$X_{C1} = \frac{10^6}{377 \times 53} = 50 \ \Omega$$

$$Z_1 = 25 + j26.39 - j50 = 25 - j23.61 = 34.4 \underline{/- 43.4°} \ \Omega$$

The effective value of the fundamental component of current is

$$I_1 = \frac{100}{34.4 \underline{/- 43.4°}} = 2.91 \underline{/43.4°} \text{ A, leading } V_1 \text{ by } 43.4°$$

$$P_1 = 2.91^2 \times 25 = 211.7 \text{ W}$$

$$V_{C1} = IX_{C1} = 2.91 \times 50 = 145.5 \text{ V}$$

Third-Harmonic Component

$$V_3 = \frac{35.35}{\sqrt{2}} = 25 \text{ V}$$

$$X_{L3} = 3 \times 377 \times 0.07 = 79.17 \ \Omega$$

$$X_{C3} = \frac{10^6}{3 \times 377 \times 53} = 16.67 \ \Omega$$

$$Z_3 = 25 + j79.17 - j16.67 = 25 + j62.5 = 67.4 \underline{/68.2°} \ \Omega$$

The effective value of the third-harmonic component of current is

$$I_3 = \frac{25}{67.4 \underline{/68.2°}} = 0.371 \underline{/- 68.2°} \text{ A lagging } V_3 \text{ by } 68.2°$$

$$P_3 = 0.371^2 \times 25 = 3.44 \text{ W}$$

$$V_{C3} = 0.371 \times 16.67 = 6.18 \text{ V}$$

Fifth-Harmonic Component

$$V_5 = \frac{14.14}{\sqrt{2}} = 10 \text{ V}$$

$$X_{L5} = 5 \times 377 \times 0.07 = 132 \ \Omega$$

$$X_{C5} = \frac{10^6}{5 \times 377 \times 53} = 10 \ \Omega$$

$$\mathbf{Z}_5 = 25 + j132 - j10 = 25 + j122 = 124.5 \underline{/78.4°}$$

The effective value of the fifth-harmonic component of current is

$$\mathbf{I}_5 = \frac{10}{124.5 \underline{/78.4°}} = 0.08 \quad \underline{/-78.4°} \ \text{V, lagging } V_5 \text{ by } 78.4°$$

$$P_5 = 0.08^2 \times 25 = 0.16 \ \text{W}$$

$$V_{C5} = 0.08 \times 10 = 0.8 \ \text{V}$$

Total current: $I = \sqrt{2.91^2 + 0.371^2 + 0.08^2} = 2.94$ A, effective value.

Total power: $P = P_1 + P_2 + P_5 = 211.7 + 3.44 + 0.16 = 215.3$ W. We also get total power by I^2R:

$$P = 2.94^2 \times 25 = 216 \ \text{W}$$

The fundamental component of *current* leads the fundamental component of *voltage* by 43.4° degrees. The expression for its instantaneous value is $i_1 = 2.91\sqrt{2} \sin(\omega t + 43.4°)$.

For the other harmonics we have

$$i_3 = 0.371\sqrt{2} \sin(3\omega t + 30° - 68.2°) = 0.371\sqrt{2} \sin(3\omega t - 38.2°)$$

$$i_5 = -0.08\sqrt{2} \sin(5\omega t - 30° - 78.4°) = -0.08\sqrt{2} \sin(5\omega t - 108.4°)$$

The complete equation representing instantaneous total current is then,

$$i = 4.12 \sin(\omega t + 43.4°) + 0.524 \sin(3\omega t - 38.2°)$$
$$- 0.113 \sin(5\omega t - 108.4°) \ \text{A}$$

24.7 Nonsinusoidal Currents in Parallel Circuits

Suppose we now connect the series circuit we have just analyzed in parallel with another series circuit and then apply the same voltage to both circuits. We could determine I_1', I_3', and I_5', the effective values of current components, in the newly added branch in the same manner as we obtained I_1, I_3, and I_5.

The *fundamental component* of the *line current* would be the *phasor sum* of the fundamental components of the branch currents. In like manner the other components of the line current would be obtained. Power values and voltage across elements in a branch would be computed in the same manner as we obtained them in the series circuit. An example, using two-element branches, will now be given

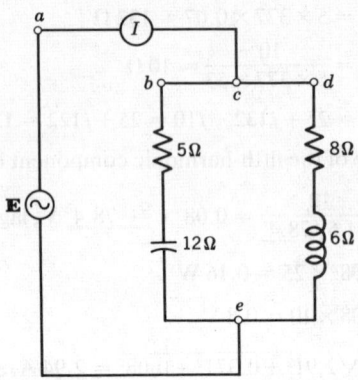

FIGURE 24.9   A parallel circuit with reactance evaluated at 60 Hz.

EXAMPLE 24.4   A parallel circuit of resistances and reactances is shown in Fig. 24.9. Assume that the complex voltage wave $e = 141.4 \sin \omega t + 35.35 \sin (3\omega t + 30°) - 14.14 \sin (5\omega t - 30°)$ is impressed. The frequency of the fundamental component is 60 Hz. Determine the values of the total current in each branch, the total power dissipated, and the equation for the instantaneous value of the total current.

Solution

Fundamental

$$V_1 = \frac{141.4}{\sqrt{2}} = 100 \text{ V rms} \qquad \mathbf{V}_1 = 100 + j0 \text{ V}$$

$$\mathbf{I}_{be1} = \frac{100(5 + j12)}{(5 - j12)(5 + j12)} = 2.96 + j7.1, = 7.69\underline{/67.4°} \text{ A}$$

$$\mathbf{I}_{de1} = \frac{100(8 - j6)}{(8 + j6)(8 - j6)} = 8 - j6, = 10\underline{/-36.9°} \text{ A}$$

The total fundamental component of current is

$$\mathbf{I}_{ac1} = 2.96 + j7.1 + 8 - j6 = 10.96 + j1.1 = 11.0\underline{/5.83°} \text{ A rms}$$

The total power of fundamental in 5 Ω is

$$P_{be1} = 7.69^2 \times 5 = 296 \text{ W}$$

The total power of fundamental in 8 Ω is

$$P_{de1} = 10^2 \times 8 = 800 \text{ W}$$

Third Harmonic

The capacitor offers one-third as much reactance at 180 Hz as it does at 60 Hz. Therefore $X_{C3} = 4$ ohms. The inductance offers three times as much reactance, which is 18 ohms.

The effective value of the third-harmonic component voltage $V_3 = 35.35/\sqrt{2} = 25$ V. We shall choose the conventional axis (horizontal to the right) as the reference axis, so that $\mathbf{V}_3 = 25 + j0$ V. Remember $X_{C3} = 4\,\Omega$ and $X_{L3} = 18\,\Omega$ at third-harmonic frequency.

$$\mathbf{I}_{be3} = \frac{25}{5 - j4} = \frac{125 + j100}{41} = 3.05 + j2.44 = 3.9\underline{/38.70°}\ \text{A}$$

$$\mathbf{I}_{de3} = \frac{25}{8 + j18} = \frac{200 - j450}{388} = 0.515 - j1.16 = 1.269\underline{/-66.1°}\ \text{A}$$

$$\mathbf{I}_{ac3} = 3.05 + j2.44 + 0.515 - j1.16 = 3.565 + j1.28$$
$$= 3.79\underline{/19.75°}\ \text{A}$$
$$P_{be3} = 3.9^2 \times 5 = 76.05\ \text{W}$$
$$P_{de3} = 1.269^2 \times 8 = 12.88\ \text{W}$$

Fifth Harmonic

The capacitive reactance at five times the fundamental frequency is $\frac{1}{5}$ that at fundamental frequency or 2.4 Ω. The reactance is five times as large, or 30 Ω.

$$\mathbf{V}_5 = \frac{14.14}{\sqrt{2}} = 10\ \text{V} \qquad \mathbf{V}_5 = 10 + j0$$

$$\mathbf{I}_{be5} = \frac{10}{5 - j2.4} = \frac{50 + j24}{30.76} = 1.627 + j0.781 = 1.80\underline{/25.6°}\ \text{A}$$

$$\mathbf{I}_{de5} = \frac{10}{8 + j30} = \frac{80 - j300}{964} = 0.083 - j0.311 = 0.322\underline{/-75.1°}\ \text{A}$$

$$\mathbf{I}_{ac5} = 1.627 + j0.781 + 0.083 - j0.311 = 1.71 + j0.470$$
$$= 1.77\underline{/15.4°}\ \text{A}$$
$$P_{be5} = 1.80^2 \times 5 = 16.2\ \text{W}$$
$$P_{de5} = 0.322^2 \times 8 = 0.83\ \text{W}$$

Effective value of current in $be = \sqrt{7.69^2 + 3.9^2 + 1.80^2} = 8.81$ A
Effective value of current in $de = \sqrt{10^2 + 1.269^2 + 0.322^2} = 10.08$ A
Effective value of total current $= \sqrt{11^2 + 3.79^2 + 1.78^2} = 11.75$ A

$$P_{be} = 296 + 76.05 + 16.2 = 388.25\ \text{W}$$
$$P_{de} = 800 + 12.88 + 0.83 = 813.71\ \text{W}$$
$$\text{Total power dissipated} = 1201.96\ \text{W}$$

The fundamental component of the total current is $11.0\underline{/5.73°}$ A effective value. The instantaneous value of the fundamental component is then defined by $i_1 = 11\sqrt{2} \sin (\omega t + 5.73°)$ A.

The expressions for the instantaneous values of the other components are:

$$i_3 = 3.79\sqrt{2} \sin (3\omega t + 30° + 19.75°) = 3.79\sqrt{2} \sin (3\omega t + 49.75°) \text{ A}$$
$$i_5 = 1.78\sqrt{2} \sin (5\omega t - 30° + 15.4°) = 1.78\sqrt{2} \sin (5\omega t - 14.6°) \text{ A}$$

The expression for instantaneous total current is then

$$i = i_1 + i_3 + i_5$$
$$= 15.55 \sin (\omega t + 5.73°) + 5.36 \sin (3\omega t + 49.75°)$$
$$+ 2.52 \sin (5\omega t - 14.6°) \text{ A}$$

24.8   Nonsinusoidal Currents Meeting at a Junction

Consider two current paths that join at a junction, as indicated in Fig. 24.10a. Currents i_1 and i_2 unite to become i_3 in the single outgoing path.

These currents may be assumed and defined in terms of instantaneous values:

$$i_1 = 5 \sin (\omega t + 60°) - 10 \sin (3\omega t - 20°) \text{ A}$$
$$i_2 = 10 \sin (\omega t - 30°) + 5 \sin (3\omega t + 40°) \text{ A}$$

Let us determine the expression for i_3.

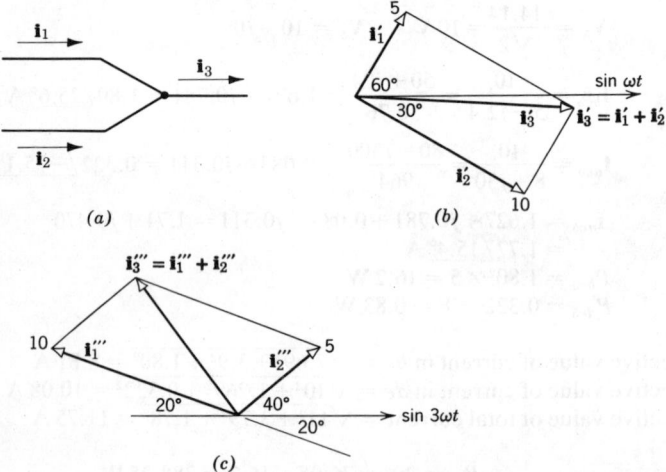

FIGURE 24.10   Nonsinusoidal currents at a junction. (a) $i_1 + i_2 = i_3$. (b) Phasors of fundamental-frequency components. (c) Phasors of third harmonic components.

FUNDAMENTAL COMPONENTS

The fundamental components are shown in phasor form in Fig. 24.10b where the phase of sin ωt is the reference. The phasor labeled $i'_3 = i'_1 + i'_2$ represents the *fundamental component* of the "outgoing current" i_3 in part a.

We now express the steady-state maximum values of i'_1 and i'_2 as complex quantities. The symbols I'_{m1} and I'_{m2} will be used.

$$I'_{m1} = 5(\cos 60° + j \sin 60°) = 2.5 + j4.33 \text{ A}$$
$$I'_{m2} = 10(\cos 30° - j \sin 30°) = 8.66 - j5 \text{ A}$$
$$I'_{m3} = I'_{m1} + I'_{m2} = 11.16 - j0.67 = 11.18\underline{/-3.45°} \text{ A}$$

We may now express the instantaneous value of the fundamental of the total current:

$$i'_3 = 11.18 \sin (\omega t - 3.45°) \text{ A}$$

THIRD HARMONIC COMPONENTS

The steady-state maximum values of i'''_1 and i'''_2, which are the third-harmonic components of i_1 and i_2 are:

$$I'''_{m1} = -10 \cos 20° + j10 \sin 20° = -9.40 + j3.42 \text{ A}$$

$$I'''_{m2} = 5 \cos 40° + j5 \sin 40° = 3.83 + j3.22$$

$$I'''_{m3} = I'''_{m1} + I'''_{m2} = -5.57 + j6.64 = 8.67 \underline{/130°} \text{ A}$$

TOTAL CURRENT

The expression for i'''_3 is, therefore,

$$I'''_3 = 8.67 \sin (3\omega t + 130°) = 8.67 \cos (3\omega t + 40°) \text{ A†}$$

The complete expression for i_3, the instantaneous value of the sum of i_1 and i_2 is

$$i_3 = 11.18 \sin (\omega t - 3.45°) + 8.67 \cos (3\omega t + 40°) \text{ A}$$

24.9 Review Questions

1. What is a harmonic with reference to a nonsinusoidal wave?
2. When is a harmonic component in phase with the fundamental of a complex wave?
3. How does a cosine wave differ from a sine wave?

†You will see that this is true if you let $\omega t = 0$: Sin 130° = sin 50° = cos 40°.

4. Any complex wave may be represented by a mathematical expression composed, in general, of a constant term and some sine and/or cosine terms. What is this mathematical expression called?

5. A complex current wave has a constant (dc) component, a fundamental component, and two harmonics with maximum instantaneous values I_0, I_{m1}, I_{m2}, I_{m3}. Write the expression for the effective value.

6. A complex voltage wave has two components. One is $e_1 = E_{1m} \sin \omega t$ and the other, e_2, lags e_1 by 30°. Write the expression for e_2.

7. A complex voltage wave is impressed upon a coil that has $R = 5\,\Omega$ and $X_L = 2\,\Omega$ at the frequency of the fundamental component. The wave has a second harmonic and a third harmonic. What are the values of R, the coil resistance at these two harmonic frequencies? What are the corresponding values of X_L at the two harmonic frequencies?

8. Suppose a capacitor with $12\,\Omega$ capacitive reactance were in series with the coil of Question 7. What would be its reactance at each harmonic frequency?

9. A series circuit containing the coil and capacitor of Questions 7 and 8 carries a current of complex wave form whose component effective values are $I_1 = 4$ A (fundamental), $I_2 = 2$ A (second harmonic), $I_3 = 1$ A (third harmonic). What is the total power dissipated in the series circuit?

10. Tell, *in words*, how you would proceed to calculate the power factor of the series circuit whose constants and currents are those given in Questions 7, 8, and 9.

24.10 Problems

6. A $5000\,\Omega$ resistor has a voltage drop $V = 20 \sin 377t + 5 \sin 1131t$ across it. Write the equation for the instantaneous current.

7. Plot the voltage and current waves for Problem 6.

8. The voltage wave of Problem 6 is applied to a series circuit in which $R = 5000\,\Omega$ and $L = 6$ H. What are the maximum values of the two current components?

9. Write the equation for the current in the circuit of Problem 8.

10. A series circuit with $R = 600\,\Omega$, $X_L = 300\,\Omega$ at 60 Hz has $e = 100 \sin 377t + 30 \cos 754t - 10 \sin 1508t$. (a) Calculate the impedances at the three frequencies. (b) Determine the expression for the instantaneous value of the current wave.

11. How much power is being supplied to the circuit of Problem 10?

12. Add an 8.84 μF capacitor in series with the resistor and the inductance of Problem 10. (a) Calculate the maximum value of the capacitive reactance at each frequency. (b) Determine the equation of the current wave.

13. How much power is dissipated in the circuit of Problem 12?

14. Suppose the voltage wave of Problem 10 were shifted on the ωt axis so that its maximum instantaneous value occurred 20° later than it does in that problem. Write the new equation for e which would indicate this.

15. Two currents i_1 and i_2 meet at a junction and combine to form a single current i_3. What is the equation for i_3 if $i_1 = 5 \sin \omega t + 2 \sin 3 \omega t$ and $i_2 = 3 \cos \omega t + 4 \sin (3 \omega t + 30°)$?

CHAPTER TWENTY-FIVE

Motors and Generators

We shall devote practically all of this chapter to a discussion of principles of motor operation. Technicians seldom come in contact with generator problems, but they should be acquainted with elementary facts about their construction and performance.

25.1   Direct-Current Generators

In Chapter 12 we learned about the generation of EMF in a loop of wire rotating in a magnetic field. A rotating armature on which many loops of wire are wound, and properly connected, will produce a direct voltage if the wire loops are connected to insulated segments of a commutator and stationary carbon brushes are made to press against it. A steady voltage of constant magnitude and unchanging polarity is thus made available between the positive and negative brushes.

The generator is provided with a magnetic field by sending a direct current through *field coils* mounted on laminated iron poles that extend from inside the frame of the machine toward the armature. A short air gap separates the surface of the rotating armature from the stationary pole surfaces. The magnetic flux coming out of one or more *north poles* crosses the air gap, passes through the armature iron near the gap, and then goes back through another gap into one or more adjacent *south poles*. There are very few generators with only one pair of poles (a north and a south). When there are two pairs, each north-pole flux separates into two halves after crossing the air gap and entering the armature iron. One half goes to the right (let us say) and enters the adjacent south pole on the right, after passing through the air gap under it, of course. The other half of the flux goes to the left, passes through the air gap under the adjacent south pole on that side, and enters that pole.

Direct current leaves the generator at the positive brush (or brushes), passes through the external circuit, and returns to the negative brush (or brushes).

The terminal voltage of a dc generator may be increased by increasing the current in the field coils, and reduced by decreasing the current. Generators are generally run at practically constant speed by their "prime movers." The prime mover may be some kind of motor, engine, or turbine. In the early days reciprocating steam engines were almost universally used. Gasoline engines are still used in isolated places, such as farms, that are not served with power from a central station.

Until recent years, every automobile was built with a dc generator which kept the battery charged and helped it supply current to the car lights. Many new cars now have alternators, which generate an ac voltage. The alternator output current is changed to dc by means of solid-state rectifiers before it reaches the battery.

25.2 Alternating-Current Generators

Large central power stations generate and deliver three-phase alternating current. The principle is as discussed at the beginning of Chapter 23 (Fig. 23.1a).

The rotating member of a large alternator in a central power station consists of an assembly of circular disks mounted on a heavy shaft. The disks, pressed firmly together, are electrically insulated from one another by a coat of insulating varnish. Slots in the cylindrical surface thus formed contain coils that carry direct current. The coils are arranged so that they produce two pairs of strong magnetic fields that are equal in strength. They alternate in polarity, that is, each north pole has two south poles beside it.

The rotating magnetic fields sweep through windings mounted in slots inside the frame of the machine and generate alternating voltages in them, as described in reference to Fig. 23.1a. Three-phase ac power is taken from the three stationary terminals on the machine.

As indicated on Fig. 23.1a, direct current for the rotating field coils is fed to them through two brushes that ride on two slip rings. The big alternator has a small dc generator (called an *exciter*) mounted on its shaft. This provides the direct current to the rotating coils. The exciter must have its field coils supplied with direct current, also. DC generators of this size, and those described in Section 24.1, build up their own dc field current in an interesting way. *Residual magnetism* that remains in iron cores after the direct current of their exciting coils is shut off is large enough to provide a small amount of dc voltage. This sends a small amount of current through the field coils in the proper direction to increase the magnetic field strength. The larger induced voltage this produces causes an increase in the current. The current thus builds up through this cumulative process until magnetic saturation in the iron prevents any further increase. Conditions are such that the correct value of current is achieved. This process is called *self excitation*. A variable resistance, called a *field rheostat*, in series with the field coils is adjusted to control the field strength so that the output current is just right for the field coils of the alternator. This means, of course, that the small dc generator has the desired voltage at its terminals also.

25.3 DC Motor

The production of torque (twisting effort) on a dc meter coil that carries current in a magnetic field is by the same mechanism that causes a dc motor armature to turn. The coils on the motor armature are caused to carry current in the proper direction (whether the conductors are under a north or a south pole) so that the forces on the conductors are always in the direction of forward rotation. The commutator bars passing under stationary brushes take care of sending the current in the proper direction through the coils. The current must be reversed each time a coil side leaves the field of one of the magnetic poles and enters the reverse-direction field of the adjacent one. The commutator does this.

25.4 Torque, Armature Current, and Speed of a DC Motor

We shall not devote much space to a discussion of the principles of operation of a dc motor, but they deserve some attention because ac motor operation (which will receive most of our attention) is based on essentially the same kinds of phenomena.

TORQUE

In Chapter 9 we found that the torque on a coil of wire carrying a current I in a magnetic field of constant flux ϕ is equal to the produce of the flux and the current.

$$T = \Phi I \tag{9.3}$$

In a motor there are many coils on the armature and there are at least two paths the current can take as it passes through the coils from the positive brush, where it enters them, to the negative brush where it leaves them. Also, there may be 2, 4, 6, or more poles in the machine. The number of paths and the number of coils are constants, so that letting K be a multiplying factor that represents the effects of all of the constants involved, we get the torque equation for a dc motor:

$$T = K\phi I_a \qquad (25.1)$$

in which ϕ is the flux per pole and I_a is the armature current.

ARMATURE CURRENT

One of the first things we must realize when we turn our attention to the current that flows through the armature of a dc motor delivering mechanical power is that most of the product *armature current times armature terminal voltage* is available for *output power* and that only a small fraction of it is lost. Part of this loss is $I_a^2 R_a$, the so-called *armature copper loss*. Beginning students are sometimes inclined to equate $V_t I_a$ to $I_a^2 R_a$. *If this were true there would not be any of the armature input power available for mechanical output.*

Figure 25.1 shows the circuit of a *shunt motor*. The field coils, in series with one another, are connected across the armature terminals. This is commonly referred to as the *field circuit*, and its current symbol is I_f. Obviously, $I_m = I_a + I_f$.

When the motor is running the armature coils surround magnetic fields which alternate in direction through the coils. First the magnetic flux passes through a coil from the north pole downward toward the shaft, and then, a fraction of a second later, flux is passing upward through the same coil into an adjacent south pole. This means an EMF E_g is generated (induced) in the conductors of every coil on the armature. The *direction of this generated EMF is such that it opposes the inflow of armature current*. The way to take care of this opposition quantitatively is to subtract E_g from the voltage V_t which is applied to the brushes (and, therefore, to the armature). The difference between

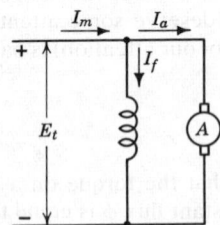

FIGURE 25.1   Circuit of a shunt motor.

these two voltages is the voltage drop due to armature resistance.

$$V_t - E_g = I_a R_a \tag{25.2}$$

From this we get the expression for armature current.

$$I_a = \frac{V_t - E_g}{R_a} \tag{25.3}$$

So we see that I_a is not equal to V_t divided by R_a.

ARMATURE INPUT POWER

Let's find out how to get the power loss due to the resistance of the armature. It must be given by $I_a{}^2 R_a$. Multiply Equation (25.2) by I_a.

$$V_t I_a - E_g I_a = I_a{}^2 R_a \tag{24.4}$$

This is a very important equation, not only because it relates the armature copper loss, $I_a{}^2 R_a$, to the other two expressions containing I_a, but because it says, in effect, that the input power $V_t I_a$ is the sum of the armature copper loss and $E_g I_a$ which must be the rest of the power put into the armature. $E_g I_a$ is the sum of *the other losses chargeable to the armature* and the output power (mechanical) at the armature shaft. This output power is delivered either by direct connection to the machine, or other device, it drives, or by means of a belt or gears. Note, especially, that the generated EMF, E_g, often called *back EMF*, is an important factor in the determination of power put into the armature.

POWER TO THE SHUNT FIELD

It is easy to see, in Fig. 25.1, that the power supplied to the field windings of the shunt motor is given by

$$P_f = V_t I_f \tag{25.5}$$

This is also equal to the power loss due to the resistance in the field circuit.

$$P_f = I_f^2 R_f \tag{25.6}$$

No power is required to maintain the magnetic field in the motor.

The other losses in a dc motor are losses caused by brush friction on the commutator, bearing friction, windage friction, and a small amount of iron loss like that in a transformer. These losses are small and considered to be constant.

From what we have just learned we can calculate quite a bit of important quantitative information about a shunt motor under load.

SPEED EQUATION

We have learned that the back EMF of a motor is an induced voltage produced by changes in flux linkages with the revolving armature coils. This EMF, E_g, is proportional to the flux per pole and the speed. If we let the constant k represent, numerically, the combined effect of the number of field poles, armature-coil turns, number of current paths from positive brush to negative brush, and other constant factors, we can express the back EMF as follows:

$$E_g = k\phi S \text{ volts} \tag{25.7}$$

S represents the motor speed in revolutions per minute and ϕ is the flux per pole.

Substituting this equation into Equation (25.1), we get the speed equation of a shunt motor

$$k\phi S = V_t - I_a R_a$$
$$S = \frac{V_t - I_a R_a}{k\phi} \tag{25.8}$$

Now, what does this equation tell us? Perhaps the most important thing is that we can cause the speed to *increase* if we can *decrease* the amount of magnetic flux per pole. This is easily done by putting a variable resistance unit in series with the field coils and increasing the amount of the resistance. This causes the field current, I_f, to decrease, which reduces the amount of magnetic flux. If we want to slow up the motor we can cut out some of the variable resistance.

Reduction of the applied (terminal) voltage will decrease the numerator of Equation (25.8) and decrease the speed. It is not very practical, or economical, to change the amount of the applied voltage.

As the load on a shunt motor is increased from zero to full-load value, the motor speed drops gradually. The total decrease is of the order of 4 percent of full-load speed. Motors are rated in full-load values. For example, a 5 hp 1500 rpm motor runs at that speed when delivering 5 hp output. Such a motor would run at about 1560 rpm at no load.

EXAMPLE 25.1 A dc shunt motor receives 15 A of armature current at 240 V. The resistance through the armature from one brush terminal to the other is 1 Ω. The shunt field circuit resistance is 155 Ω. Assume the friction and iron losses (called *stray losses*) are 3 percent of the armature input power under load. Calculate the (a) copper losses, (b) back EMF, (c) armature input power, (d) output mechanical power, (e) motor efficiency. How much of the armature input power is converted into output mechanical power?

Solution

Shunt field current: $I_f = 240/155 = 1.55$ A
Shunt field copper loss $= I_f^2 R_f = 1.55^2 \times 155 = 372$ W
Armature copper loss $= I_a^2 R_a = 15^2 \times 1 = 225$ W
Back EMF $= V_t - I_a R_a = 240 - 15 \times 1 = 225$ V
Armature input power $= V_t I_a = 240 \times 15 = 3600$ W
Iron, friction, and windage losses $= 0.03 \times 3600 = 108$ W
Output mechanical power $= 3600 - (225 + 108) = 3267$ W
In horsepower units, this is $3267/746 = 4.38$ hp
Total motor input power $= 240(15 + 1.55) = 3972$ W
Motor efficiency $= 3267/3972 = 0.825 = 82.2$ percent
The amount of the armature input power converted into mechanical power is

$$V_t I_a - I_a^2 R_a = 3600 - 225 = 3375 \text{ W}$$

Of this, 108 W are lost in supplying the stray losses.

This question may, quite logically, arise: "Why are there iron losses when we are dealing with direct current? Transformers carry only alternating current." The answer is that the magnetic flux in the iron laminations on the armature *reverses direction* every time the armature coils pass from their position under one field pole to their position under the next field pole which, of course, has opposite magnetic polarity. This means the armature iron carries an *alternating flux*.

Problems

1. A shunt motor is delivering 10 hp to a machine. The terminal voltage is 240 V. The field circuit has 200 Ω resistance and the armature circuit resistance is 0.4 Ω. The armature current is 34 A. Calculate the total current taken by the motor.

2. How much back EMF is generated by the motor armature of Problem 1? What is the armature input power?

3. Calculate the total input power to the motor of Problem 1. What is the efficiency of the motor?

4. Calculate the copper losses and the stray losses in the motor of Problem 1.

25.5   Performance of a DC Motor under Load

We shall now explain, briefly, how a motor automatically takes more power from its supply lines when it is required to deliver an increase in horsepower output and less power when some of the mechanical load it is carrying is removed. How do such load changes come about?

Suppose the motor is driving a lathe, a milling machine, or a boring machine. As the cutting tool moves from air (cutters turning in air) to the metal, the motor output power suddenly changes from a very small amount to what may be near its full-load value. At the instant cutting begins the motor speed is sharply reduced. *At that instant the back EMF is reduced.* More current is allowed to flow into the armature, and this increases the torque (twisting effort on the motor shaft). This follows as indicated by Equation (25.1) ($T = K\phi I_a$). The motor needs this larger torque to handle the load.

At the end of the cut the load on the motor abruptly decreases. At that instant the motor speed increases and so does the back EMF. This reduces the armature current [Equation (25.3)] and the input power is thus decreased because much less is needed when no metal is being cut.

Compare this automatic action with what an automobile engine will do when you idle along on a level road and let the car start up a hill without feeding it any more gasoline!

Figure 25.2 shows performance curves of a shunt motor. The efficiency reaches maximum value near, or at, full load. It is very low at light loads

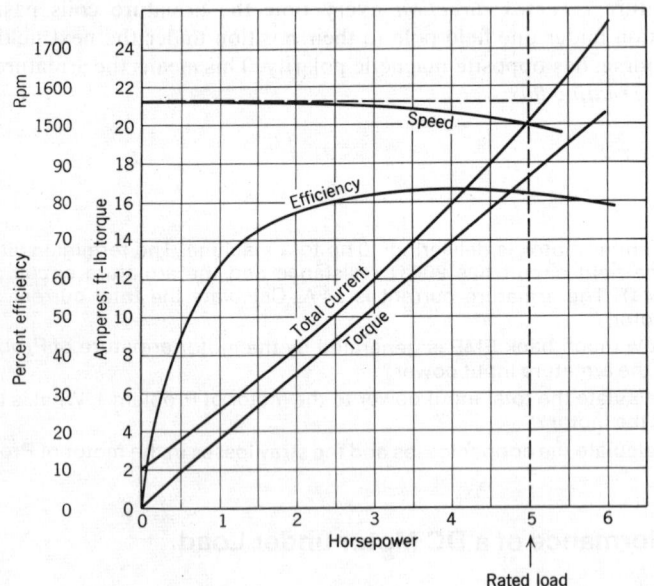

FIGURE 25.2   **Typical characteristic curves of a 5-hp, 230 V, 1500-rpm shunt motor.**

because the stray losses are practically the same as at heavy loads and thus are responsible for most of the input power requirements.

25.6   Series and Compound Motors

Direct-current motors which have windings that carry the armature current, and which have no shunt-field windings, are called *series motors*. They develop much higher starting torque than do shunt motors because the flux increases almost directly as the armature current increases. The torque is thus proportional to the *square of the armature current* (practically) rather than to the first power of the armature current as in the shunt motor.

Resistance must be inserted in series with a series motor at starting to prevent excessive "in-rush" current. The back EMF in the armature is very small immediately after the motor is energized because the flux is so small in value. Contrast this with the case of the shunt motor that has full value of flux almost instantly after the motor is energized.

A series motor will run at a dangerously high speed if its load is reduced to zero. This result is caused by the absence of sufficient back EMF, owing to exceedingly small values of flux.

A compound motor has both a series field and a shunt field. Its operating characteristics are, as expected, a compromise between those of a shunt motor and those of a series motor. It has much more starting torque than that of a shunt motor of the same horsepower rating and voltage, but not as much as that of a series motor. It will not "run away" at light loads. See the curves in Fig. 25.3.

DC motors may be constructed with *interpoles*, which are small auxiliary poles mounted in the frame between the main field poles, to improve operation. The action of an interpole is to so reshape the magnetic field (in which coils lie as the commutator segments to which they are connected pass under the brush) that no sparking will occur at any value of load from light load to somewhat above full load.

25.7   Starting a Shunt Motor

Direct-current motors of about one horsepower and larger must be supplied with reduced armature voltage at starting. This is done by means of a "starting box" which is either hand operated or automatic. The physical appearance and precise connections of components of starting devices are presented in many books on electric-power engineering. Only the circuit diagram will be presented here.

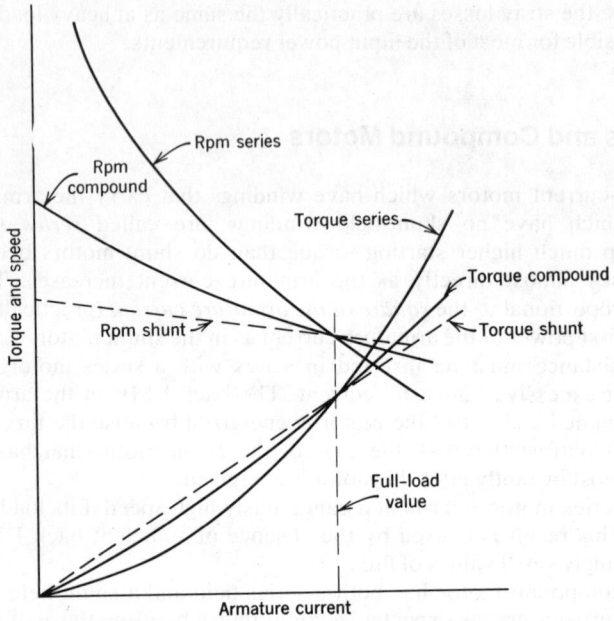

FIGURE 25.3 **Curves of torque and speed of dc motors which have identical full-load values.**

It is desirable to supply somewhat more than normal field current to the motor at starting in order to produce maximum torque. During the starting process the current is reduced automatically to normal as the armature series resistance is added in series with the field windings.

Figure 25.4 shows the basic circuit of a starting device for a dc shunt motor. After the switch is closed the sliding contact is moved to the first button on the resistance in series with the armature. This puts all the resistance in series with the armature and limits the armature current to a safe

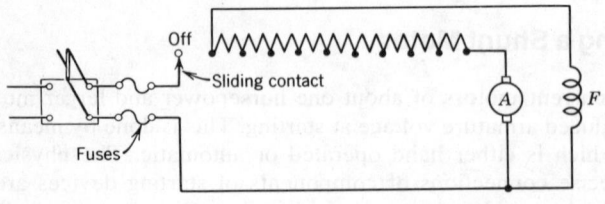

FIGURE 25.4 **Circuit for starting a dc motor.**

value. At the same time full voltage is applied to the shunt field. As the armature begins to rotate and gain speed, the sliding contact is advanced slowly to
successive buttons until all the resistance is cut out. The motor is then operating at full speed for the load it is driving.

Suppose the supply voltage is interrupted while the motor is operating. It
is important, in that case, to *provide for the automatic return of the sliding
contact to the "off" position*. Otherwise, if, and when, the voltage comes
back on, the motor will draw excessive current because full voltage is applied
to the armature at standstill. This will blow a fuse, open a circuit breaker, or
burn out the armature winding of the motor. A mechanical spring is provided
to throw the arm, on which the sliding contact is mounted, back to the "off"
position at the instant the supply voltage is interrupted.

Automatic starters have magnetic devices that close auxiliary circuits and
cut out sections of the starting resistance in a properly timed sequence. Their
delayed action is usually controlled by the amount of armature current flowing
during the starting process.

Starters are designed to allow the motor to take approximately 1.5 to 2
times full-load current in starting. This is needed when the motor must start
while connected to a machine that is loading the motor appreciably. Starters
for motors that must start under heavy loads are usually called *controllers*.
They are built to carry overload currents continuously.

REVERSING DIRECTION OF ROTATION

The easiest way to reverse the direction of rotation of a dc motor is to reverse
the polarities of its field poles. This means reversing the connections of the
field winding of a shunt motor, reversing the connections of the field winding
of a series motor, or reversing the connections of both series field and shunt
field of a compound motor. Reversing the connections of the two supply wires
will not do it. This will reverse the polarities of the field poles and the direction
of current in the armature conductors, with the result that the forces on the
conductors will remain in the same direction as before.

SAFETY DEVICES

The arm carrying the sliding contact of the motor starter must be moved against
an opposing force created by a spiral spring attached to the arm. When the arm
is in the full "on" position, it makes direct contact with the armature terminal
and cuts out all of the series resistance. It is held in this position by an electromagnet that is energized by current that flows in the field coils. This is a very
important safety feature because there is a critical minimum value below which
the field current should not fall. If it should, either the motor speed would
become dangerously high or the armature current would exceed safe limits.
Severe sparking at the brushes would also occur.

When the field current is too low for safe operation, the electromagnet becomes too weak to hold the starter arm in the "on" position and it flies back and shuts off the power. This feature is used, also, when the motor becomes overloaded.

An overload electromagnet, carrying armature current, will pull up a plunger that carries a copper strip fastened to its top. The strip makes contact across the terminals to which the field-current electromagnet is connected and thus short circuits it. This, of course, completely deenergizes the electromagnet and lets the arm fly back to "off."

25.8 Motor Testing by a Prony Brake

A method of measuring the output of a motor employs a special pulley and a brake arm as shown in Fig. 25.5. The output torque is computed from data of speed, net force at the point of measurement, and the distance L, which is constant.

The net force is obtained by subtracting the *brake tare* from the force measured during operation. The *tare* is the average of two readings taken with the brake riding loosely on the pulley. A reading of force exerted when the motor is running in one direction (clockwise in Fig. 25.5) will be larger than the reading taken when the rotation is reversed. The average of these two readings is the tare value, which becomes part of every reading taken during the test. Instead of driving the motor electrically, it may be turned slowly by hand first in one direction and then in another in measuring the break tare.

The net force at the existing load (varied by tightening the band on the pulley) will be denoted by F. The distance the force must act to absorb the

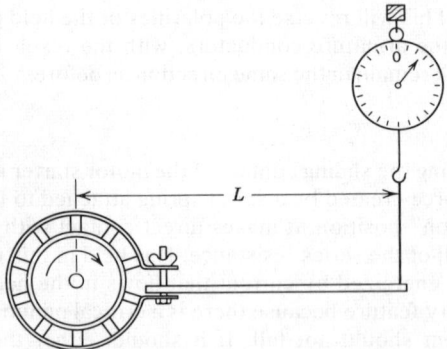

FIGURE 25.5   Prony brake clamped on special pulley on motor shaft. Used for load tests.

output energy of each revolution is the circumference of a circle L feet in radius. The work in foot-pounds per revolution is $2\pi FL$ when L is in feet and F is in pounds. The work done per minute is this value times S if the speed is in revolutions per minute. Because there are 33,000 ft-lb per min in 1 hp, the horsepower output is given by

$$\text{Hp} = \frac{2\pi LFS}{33,000} \qquad (25.9)$$

EXAMPLE 25.2   In a test with a Prony brake, the following data were taken: for brake tare: $f_1 = 1.2$ lb, $f_2 = 0.9$ lb, total force at 1560 rpm = 7.63 lb, $L = 2.5$ ft, applied voltage = 228, input current = 19.0 A. Compute (a) output power, (b) input power, (c) efficiency.

Solution

$$\text{Brake tare} = \frac{1.2 + 0.9}{2} = 1.05 \text{ lb}$$

$$\text{Output horsepower} = \frac{2\pi \times 2.5(7.63 - 1.05)1560}{33,000} = 4.88$$

$$\text{Input horsepower} = \frac{228 \times 19}{746} = 5.81$$

$$\text{Efficiency} = \frac{4.88}{5.81} \times 100 = 84 \text{ percent}$$

Problem

5. A 3-hp 1750-rpm motor is to be tested with a Prony brake that has an arm 18 in. long. The brake tare is 12 oz (a) How much force will be exerted on a spring balance supporting the arm when the motor runs at full load? (b) What is the output power when the balance reads 5 lb at 1760 rpm? (c) What is the *full-load* efficiency if the motor takes 11.88 A at 230 V while delivering 3 hp?

25.9   Three-Phase Motors

The rotating part of an ac motor is called the *rotor*. The stationary part, called the *stator*, is a set of stacked laminations that are solidly mounted inside the frame of the machine. They have coils of the three-phase windings mounted in slots in the inside surface. These windings are arranged in the same way as those of a three-phase alternator. They are connected either in wye or in delta and conductors are brought out from the terminals of the windings to main terminals on the frame of the machine. Power is delivered by the ac supply lines to the motor at these external terminals.

The rotor of a three-phase motor may have three groups of single-phase windings fastened securely in slots near the surface, or it may have solid copper bars in the slots instead of the sides of wound coils. In that case, the bars are parallel to the shaft and extend from an *end ring* on one end of the stack of laminated disks to an identical *end ring* on the other end. This bar and end ring construction is like the rotating cage of a squirrel cage, and the rotor called a *squirrel-cage rotor* gets its name from this similarity. The ends of the bars are brazed or welded to the end rings. The rotor current flows from one ring to the other through a group of bars and returns to the first ring through an adjacent set of bars. The current is produced by induced EMFs, in the bars, that are set up by a *revolving* magnetic field.

25.10   Revolving Magnetic Field

We could go into a somewhat lengthy explanation of how a three-phase current in the windings (coil groups — one for each phase) of a stator produces a *revolving* magnetic field. Instead, let's take a look at Fig. 23.1*a* which shows how a permanent two-pole magnetic field in an ac generator is made to revolve by the turning of a shaft on which the iron magnetic path and the flux-producing coils are mounted. We learned that this revolving field induces voltages in the coils and a three-phase current flows which is supplied to a load on the generator. *Here is an important point*: as the currents build up toward their maximum instantaneous values and then decrease and reverse in direction, they produce magnetic fields of their own that oppose the changes in the amount of flux enclosed by their coils, and they must follow the rotating field poles as they turn. This, we say, is what Lenz's law is about. Thus, we see that *the alternating currents in the stator produce rotating magnetic fields*.

Now, let's turn our thoughts back to a three-phase ac motor which carries alternating current that the supply lines send into its stator windings. This current, made up of three single-phase parts (one in each of the three coil groups) will produce a rotating magnetic field that will go into the rotor, and perpendicular to its surface, along a small part of its circumference and leave it along a part of equal area that is diametrically opposite. There will be a two-pole revolving field if the stator winding has only one group of coils for each phase. This is the case in Fig. 23.1*a* where group *AA'* is for one of the three phases, *BB'* for the second, and *CC'* for the third. A four-pole revolving field is common in motors. These have two groups of coils for each phase. One group occupies half of the stator slots alloted to it and the other group occupies the other half.

The rotating field caused by the stator current induces EMFs in the rotor coils, or bars, and the resulting currents have forces on them and their conductors which combine to produce turning effort.

25.11   Speed of an AC Motor

A *two-pole revolving field* produced by stator-winding current *requires one complete current cycle to produce one revolution of the field*. A four-pole revolving field will make only one-half revolution in one current cycle. The field of a two-pole motor operating on a 60 Hz supply will turn at a speed of 60 rps which is 3600 rpm. A four-pole field will turn at 1800 rpm. The speed of the revolving field is called *synchronous speed*. The formula for synchronous speed is then

$$S = \frac{120\,f}{P}\,rpm \qquad\qquad (25.10)$$

in which **P** is the number of poles and **f** is in Hz. This equation applies for any number of poles.

The synchronous speeds and normal full-load speeds of three-phase motors operating on a 60-Hz line are given in Table 25.1.

TABLE 25.1   **Speeds of Three-Phase, 60-Cps Induction Motors**

Poles	Synchronous Speed	Normal Full-Load Speed
2	3,600	3,500
4	1,800	1,750
6	1,200	1,160
8	900	875
10	720	700
12	600	585

SLIP

The rotor of an induction motor can never attain synchronous speed. The torque that keeps it running depends on induced currents in the rotor. These will not exist unless there is *relative motion* between the rotating field (produced by stator currents) and the rotor bars or windings.

The difference between the speed of the rotating field (the synchronous speed) and the speed of the rotor is called the *revolutions slip* of the motor. In Table 25.1 we see that the normal full-load speed of a four-pole induction motor on 60 Hz is 1750 rpm. The revolution slip of the motor is then 1800 − 1750 = 50 rpm.

Slip is usually expressed in percent. It is denoted by *s* (lower-case *s*) in this book, and the percentage refers to synchronous speed. Let *S* represent the actual speed and S_s the synchronous speed, then

$$s = \frac{S_s - S}{S_s} \times 100 = \text{percent slip} \qquad (25.11)$$

In the example of the four-pole motor, we have

$$s = \frac{1800 - 1750}{1800} \times 100 = 2.78 \text{ percent}$$

Evidently, *the motor speed at any loading is given by*

$$\mathbf{S = S_s[1 - (s/100)] \text{ rpm}} \qquad (25.12)$$

25.12 Rotor Current and Torque

The equations for rotor current and torque and their derivations are somewhat involved and will not be presented in this book. Graphs showing how these two quantities vary with motor speed are shown in Fig. 25.6.

The parts of the curves below about 90 to 92 percent of synchronous speed apply to the starting period. Note that at full load the slip is only about 3 percent. At slips around 10 percent the motor may stall. If it does not, it will draw very large current and will probably overheat.

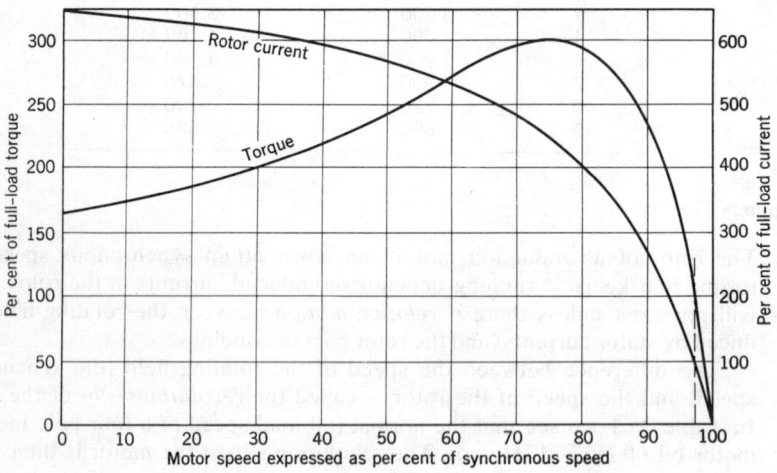

FIGURE 25.6 Torque and rotor-current variations with motor speed. Full-load conditions exist at about 97 percent of synchronous speed, as indicated.

Analysis and experience show that the torque increases when the rotor resistance is increased. If the rotor is the squirrel cage type its resistance can be increased by putting it in a lathe and taking a cut off an end ring. If too much is taken off it is not practical to try to put some of it back on. A wound-rotor motor may be provided with taps at its winding junctions from which leads can be brought out to the under sides of three slip rings. Brushes riding on the rings are then connected to three sliding contacts on a wye-connected bank of resistors in a starting box. The motor is energized from the power line with all of the resistances cut in. As it comes up to speed the three resistances are gradually cut out by a three-armed control device. When full-load speed is reached all of the starter resistance is out.

25.13   Starting a Squirrel-Cage Motor

Small squirrel-cage motors may be started with full line voltage. Large ones have such small resistance in their rotors that they must be started with reduced line voltage. Autotransformers are used for this purpose. A triple-pole double-throw switch, with its three middle posts (which carry the blades) connected to the motor terminals, is used to connect to the taps during the starting period. When the motor reaches full operating speed the switch is quickly thrown to the other side whose terminals are connected to the supply lines. This applies full line voltage to the motor.

25.14   Reversing Rotation of Three-Phase Motors

Alternating-current motors that operate on the induction principle, that is, those whose rotor current is obtained by electromagnetic induction, are often called induction motors. Three-phase motors of the squirrel-cage and wound rotor types operate on this principle, as we have learned.

Soon we shall learn about single-phase motors, most types of which are induction motors, but there is a type that gets its rotor current through brushes and a commutator. This is a *series type* ac motor which is different from an induction motor.

From what we have learned we should be able to conclude that if the *direction of rotation* of the magnetic field can be reversed the motor would reverse its direction of rotation. This is correct, but how do we manage to do it? The answer is simple: *interchange the connections of any two of the three wires* of the three-phase power line where they are connected to the machine. This causes the currents to change their relative timing in going positive and negative, with the result that the magnetic field reverses its direction of rotation.

25.15   Single-Phase Series Motor

Perhaps the most popular type of single-phase motor is the series motor. It is self-starting and develops relatively high torque. It has a commutator on its rotor (armature) shaft, the insulated segments of which are connected to the coils wound on the armature. Field windings are mounted on magnetic pole pieces similar to the field poles of a dc motor. Alternating current produces north and then south poles as the coil current continues its normal reversal of direction of flow. This current flows through the armature windings, entering through one brush during the positive half cycle and then through the other during the negative half cycle. The current in a conductor reverses at the same time the magnetic field, in which it lies, reverses, and so the force on the conductor does not change its direction. This means the armature of a single-phase series motor continues turning in one direction.

Single-phase series motors are used to drive food mixers, household fans, television antenna rotators, and hand tools such as drills, hedge trimmers, sanders, and buffers. Larger appliances such as refrigerators, washing machines, and central air-conditioning compressors employ different kinds of single-phase motors which will be described later.

The vast majority of single-phase motors are small and are usually classified as **fractional-housepower motors**.

25.16   Split-Phase Motor

The magnetic field in a single-phase motor does not rotate, it merely oscillates back and forth (or up and down) through the rotor. Only the series type of single-phase motor is self-starting without any special starting mechanism. To be inherently self-starting the motor must have a rotating magnetic field.

The circuit components of a *split-phase* motor are represented in Fig. 25.7. The single-phase currents in windings A and B are out of phase with each other because winding A has a separate resistance R connected in series with it or its winding resistance is R. The two windings are displaced 90 electrical degrees from each other. A is called the auxiliary winding, or the starting winding. B is the main winding. The rotor is of the conventional squirrel-cage type.

The switch S is closed when the rotor is at rest. When it attains about 75 to 80 percent of synchronous speed, a centrifugal device opens the switch. The motor continues to run as a single-phase induction motor and accelerates to normal speed.

The flux produced by the current $\mathbf{I}_A$ may be resolved into two components, at least for the purposes of analysis, such that one is in phase with the flux of $\mathbf{I}_B$ and the other 90° ahead of that flux. These component fluxes would be produced by the current components $I_A \cos \alpha$ and $I_A \sin \alpha$, where α is the

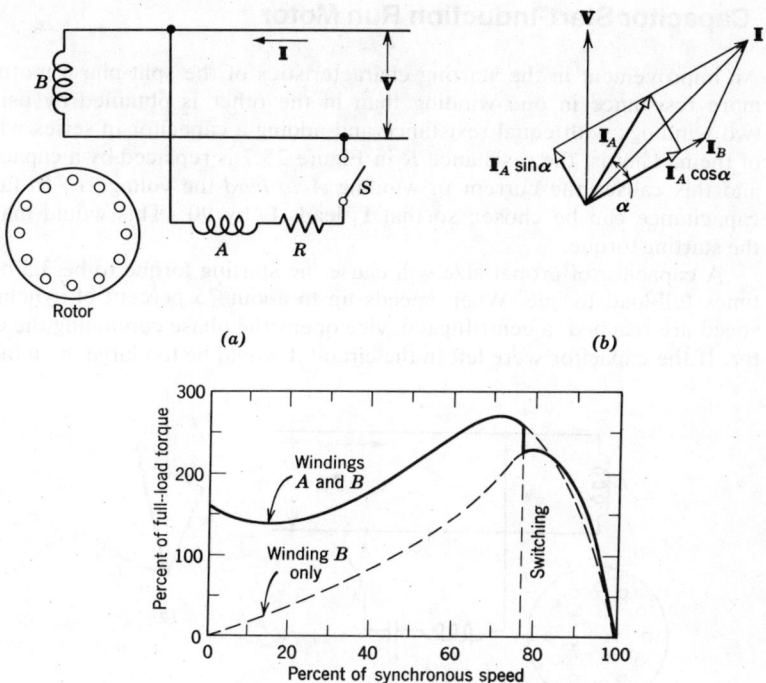

FIGURE 25.7   Split-phase motor. (a) Circuit and (b) phasor diagram at starting. (c) Variations of torque with speed.

phase angle between the two phase currents. Recall that the fluxes produced by two currents that are in phase rise and fall together and thus do not produce a rotating resultant. This means the flux of $I_A \cos \alpha$ does not contribute to torque. But that produced by $I_A \sin \alpha$ is at right angles to the flux of $\mathbf{I}_B$, so the starting torque produced is a function of the product of I_B and $I_A \sin \alpha$:

$$T = kI_AI_B \sin \alpha \qquad (25.13)$$

The starting torque of general-purpose, split-phase motors is between 75 and 200 percent of full-load torque, and the starting current is about 6 times full-load value. A high-torque motor suitable only for intermittent service produces up to about 2.5 times full-load torque at starting. The use of smaller wire for coil A is a practical means of providing more resistance than that of coil B, and this makes the separate resistance unit R unnecessary. Figure 25.7c shows variations of torque with speed.

25.17 Capacitor Start-Induction Run Motor

An improvement in the starting characteristics of the split-phase motor with more resistance in one winding than in the other is obtained by using the two windings with equal resistance and adding a capacitor in series with one of them. That is, the resistance R in Figure 25.7 is replaced by a capacitor C, and this causes the current in winding A *to lead* the voltage V. In fact, the capacitance can be chosen so that $\mathbf{I}_A$ leads $\mathbf{I}_B$ by 90°. This would maximize the starting torque.

A capacitor of proper size will cause the starting torque to be 3.5 or more times full-load torque. When speeds up to about 75 percent of synchronous speed are reached, a centrifugal device opens the phase containing the capacitor. If the capacitor were left in the circuit it would be too large, by a factor of

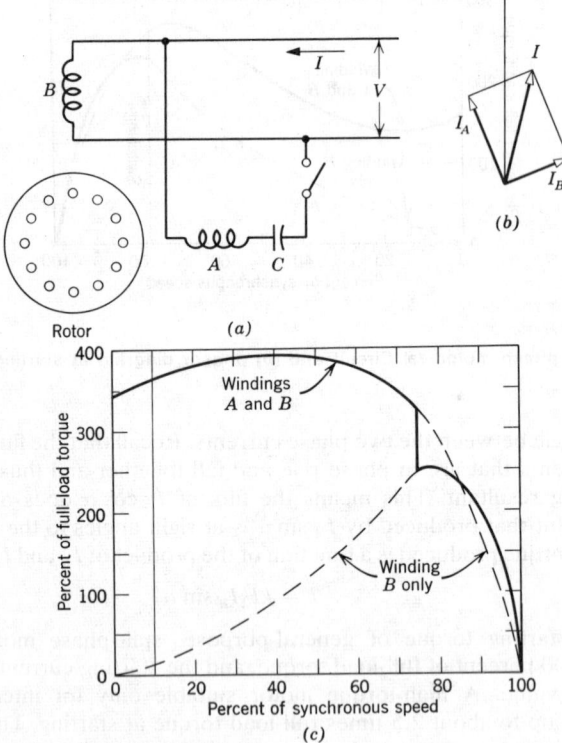

FIGURE 25.8 Capacitor-start motor. (a) Circuit. (b) Phasor diagram at start. (c) Variations of torque with speed.

about four, for best operation. Also, a capacitor for starting operation only can be of the electrolytic type which is smaller and cheaper than one that can stand continuous duty.

A capacitor-start, capacitor-run induction motor has, as one may surmise, two capacitors in parallel during the starting process. At the proper speed, a centrifugal device cuts one of them out, but leaves the other and its winding connected. The leading current in this winding, combined with the lagging current in the other, results in operation at nearly 100 percent power factor.

In applications where the motor starts with practically zero load, the expense of a centrifugal switch and an extra capacitor can be avoided by leaving the one capacitor in the circuit. The starting torque is low because the current in the auxiliary winding is limited to a small value by the size of capacitor which is dictated by the requirements of good running characteristics.

Figure 25.8 shows the circuit and phasor diagrams, and the performance curves for the capacitor start-induction run motor. The torque is about 3.25 times full load torque when the auxilairy winding is cut out.

Some capacitor motors are built with two capacitors in parallel. One is left in after the switch removes the other. Although it has less starting torque than the single-capacitor-start motor, it operates under load at better power factor and efficiency.

25.18   Shaded-Pole Motor

An inexpensive method of producing, in a motor, a magnetic field that moves in a rotational direction is to place a *single turn of copper* around part of each pole face. Figure 25.9*a* illustrates this construction in a *shaded-pole* motor and Fig. 25.9*b* is its torque-speed characteristic.

Current induced in the single-turn coil produces an opposing flux which causes the flux density near the pole tips to build up slower than the flux density in the main parts of the poles builds up. This produces the effect of a rotating flux, with the result that currents are induced in the rotor, and starting torque is developed. The moving field varies substantially in magnitude and has varying angular velocity.

The starting torque is relatively small – only about 40 percent of full-load value. The slip is about 15 or 20 percent at full load, but it is classified as a constant-speed motor. The speed does remain almost constant when the load variations are small.

The shaded-pole motor, in very small sizes of $\frac{1}{20}$ hp and less, is widely used for small fans, some electric clocks, and other light-load applications. Its efficiency is low, but this is of little consequence at such small power levels. It is usually available with synchronous speed of only 1800 rpm and 3600 rpm, 60 cycles.

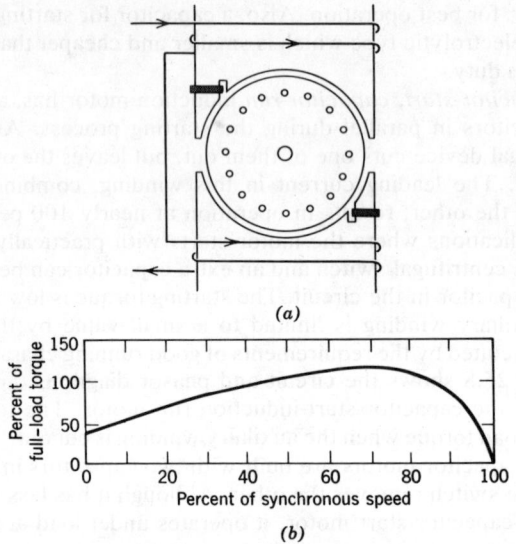

FIGURE 25.9 Shaded-pole motor.

25.19 Single-Phase Synchronous Motors

Motors used to drive electric clocks must run at constant speed. Fortunately, the load on them is very light and the efficiency is not a factor because the total power required is only two or three watts. Motors designed for such light service fall in the single-phase, synchronous-motor classification. The reluctance motor is a popular type.

Strictly speaking, a synchronous motor runs at synchronous speed at all loads. This is made possible in the larger sizes by building a dc field structure with salient poles into a motor, and supplying the windings on this structure with *direct current*. Large synchronous motors that operate on three-phase systems are used where truly constant speed is required. Such a motor can be made to draw leading currents from the supply lines, which will have the effect of neutralizing some of the effect of lagging currents taken by induction motors, and as a result the power factor of the whole system will be improved.

25.20 The Reluctance Motor

A metal disk with extended sectors and free to turn on a shaft will rotate when placed inside the stator of a shaded-pole motor as shown in Fig. 25.10. Starting

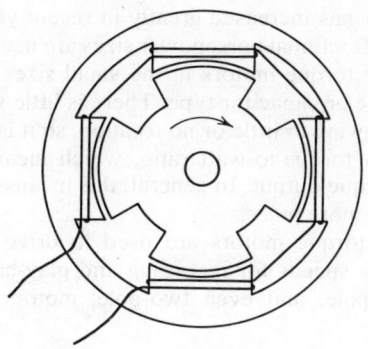

FIGURE 25.10   A four-pole reluctance motor.

torque is provided by the rotating field caused by shaded-pole action. The protrusions of the rotor, which is said to have salient-pole construction, follow the revolving field at exactly synchronous speed.

It is characteristic of magnetic objects, when free to move in a magnetic field, to align themselves into the position that will result in their containing the maximum magnetic flux.

The four-pole motor illustrated here runs at 1800 rpm on 60 Hz. Electric-clock motors run slower to reduce the amount of gearing necessary and to prevent noticeable noise and vibration. Their rotors have many projecting poles. One with 30 poles would run at only 240 rpm.

25.21   Torque Motors

Some tire-service mechanics use small portable torque motors to remove and and replace bolts when removing and replacing automobile wheels. Torque motors are also used in assembly plants to speed up the installation of nuts and bolts.

A torque motor is generally used for intermittent duty, although some are designed so that they are capable of remaining stalled on the supply line at rated voltage. In actual practice, a torque motor may be called on to extert its torque with almost no rotation, or to rotate very slowly. There are applications in which it may be required to allow itself to be driven in reverse by the mechanism to which it is connected.

Torque motors are built with wound armatures and brushes for use on either alternating or direct current, and also with induction-type construction for ac use only. The maximum torque available from the squirrel-cage type is less than that from the brush type. Nevertheless, the use of the induction-type torque motor has increased greatly in recent years in applications where fractional and subfractional horsepower sizes are needed.

Induction-type torque motors in the small sizes are usually of the single-phase, shaded-pole or capacitor type. There is little ventilation of the windings of such a motor, owing to little or no rotation, so it is important to design them for the maximum torque-to-watt ratio, which means the lowest watts input for the highest torque output. In general, this means low-speed motors, that is, motors with six or more poles.

Subfractional torque motors are used to drive magnetic-tape recorders. These require low speeds for recording and play-back, but higher speeds for rewinding. Four-pole, and even two-pole, motors are used in this kind of application.

25.22   Hysteresis Motor

We learned about hysteresis when we studied iron-core transformers. Hysteresis can be employed to produce mechanical torque. A smooth solid-steel cylinder without slots or windings serves as the rotor of the hysteresis motor. The stator has two coils in slots inside its surface, so distributed that a flux distribution that is as nearly sinusoidal as possible is passed through the rotor. The stator windings are the capacitor-start type but the capacitor is left permanently in series with one of them. The stator then produces a rotating field that revolves at synchronous speed.

In Fig. 25.11, the center line of the rotating flux produced by the stator windings is the line $\phi_s\phi_{s'}$. This flux rotates at constant angular velocity ω_s. The line $\phi_R\phi_{R'}$ is the center line of the rotor flux, which lags behind the stator flux by the angle α, also rotates at the same angular velocity in the same direction. We can readily understand that the rotating north and south poles of the stator flux will drag the corresponding south and north poles of the rotor flux around with them.

When the rotor is stationary the torque is proportional to the product of the fundamental components of the stator magnetomotive force and the rotor flux times the sine of α, the angle between the two, called the *torque angle*. This angle does not change as the speed changes because its value is determined by the form of the hysteresis loop of the rotor. Therefore, the motor develops constant torque throughout its speed range.

The constant-speed characteristic of the hysteresis motor is one of its outstanding advantages. When the load is suddenly — or gradually — changed the speed will vary only during a minute adjustment period and its torque angle will almost instantly change to the amount required to produce synchronous speed.

Applications of hysteresis motors are in electric clocks because of their

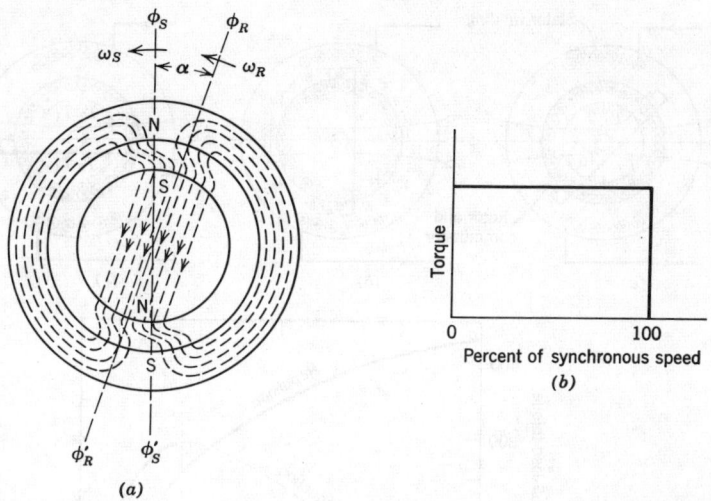

FIGURE 25.11   Hysteresis motor. (*a*) Field pattern in a two-pole motor. (*b*) Torque-speed curve.

low power requirements as well as the constant-speed characteristic, and in record-player turntables where quietness and constant speed are of paramount importance.

25.23   Repulsion-Start Induction Motor

A single-phase motor of the repulsion-start type has a dc type armature winding with commutator bars and brushes. The stator has a conventional single-phase winding that is usually wound for four or more poles. For simplicity and convenience, a two-pole motor will be used as an example.

Figure 25.12*a* illustrates the stator and rotor windings and actual position of the short-circuited brushes during starting. Torque is produced, with a net amount in the clockwise direction. The curved arrows marked T_o indicate that there is a small amount of opposing torque in the upper-left and lower-right sections of the rotor cross section. The manner in which these torques are produced will now be described.

Assume the brushes to be in the plane perpendicular to the flux lines produced by the stator field, as shown in part *b*. The rotor conductors are paired by numbers so that each loop may be identified. That is, individual coils bear numbers 1, 1′, 2, 2′, 3, 3′, and so on. The dots and arrows indicate the directions of induced voltages produced in the loops by the oscillating flux of the

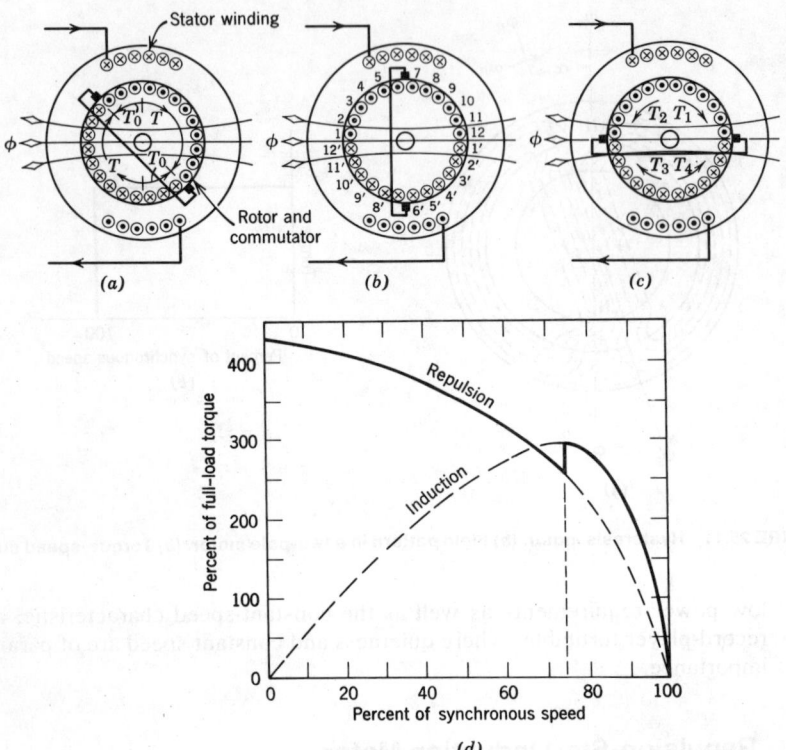

FIGURE 25.12 Repulsion motor currents, flux, and torque. (*a*) Brush position for normal operation. (*b*) Brush axis perpendicular to flux; zero current, zero torque. (*c*) Brush axis parallel to flux. Equal and opposite torques. (*d*) Performance curves.

stator winding. These directions are dictated by Lenz's law. In following through the rotor loops from one brush to the other, we find that the voltage induced in half of the path is in one direction and in the other direction in the remaining half. The result is zero current in the rotor conductors, and zero torque.

In part *c* the brushes are parallel to the field, and here we expect the induced voltage between brushes to be at maximum value. The current will then be at maximum. Careful analysis will reveal that the loops whose sides are in quadrants 1 and 3 have torques that are trying to produce clockwise rotation while coils with sides in quadrants 2 and 4 have torques that are trying to produce rotation in the counterclockwise direction. The result is, again, no net torque.

If, now, we have the brushes in an intermediate position, as in part *a*, one directional torque is decreased and the other is increased. The net torque will be in the direction in which the brushes are shifted *from the position that is parallel to the field produced by stator-winding current*. The axis parallel to the field flux is called the neutral axis.

The repulsion motor has load and speed characteristics similar to those of a dc series motor. It has the advantages of high starting torque and small starting current when compared with other single-phase motors. These characteristics may be combined with the better speed regulation of the induction motor when provision is made to convert the repulsion-start motor into an induction motor after it has come up to full operating speed.

The *repulsion-start, induction-run motor* has a centrifugal mechanism that short-circuits all of the commutator bars when full speed is attained. At the same time the brushes are lifted from the commutator surface to prevent wear. The motor then becomes a squirrel-cage induction motor.

The *repulsion-induction motor* has an armature winding with commutator and brushes of the standard repulsion motor and also a squirrel-cage structure of the induction motor on the same rotor. It has high starting torque and it takes only a little more starting current (i.e., line current at start) than the repulsion motor.

25.24 Review Questions

1. How does the current that enters a positive brush of a dc motor reach the armature coils?

2. One side of single coil in an armature winding lies in a slot in the armature and the other side lies in another slot located one-fourth the way around the armature from the first slot. The motor has four poles. Why is it necessary that the sides of the coils be separated a distance along the armature surface which is approximately equal to the distance between centers of adjacent poles?

3. How does doubling the armature current of a motor affect the torque, all other quantities remaining fixed?

4. Assuming that armature current remains constant, what is the effect on torque when the flux per pole is halved?

5. A shunt motor is running steadily and delivering full-load torque. What would be the effect on the back EMF of suddenly reducing the flux per pole? What would happen to the armature current? Would the motor attain a higher or a lower speed? Why?

6. A 5-hp, 230-V, dc motor, running under normal operating conditions, has a total input of 20 A. What happens, inside the machine, to most of the input power? About what per cent of the input power is lost? Where does this energy go?

7. A shunt motor has an armature with $\frac{1}{2}$-Ω resistance. What is the theoretical value of the armature current that would flow if 120 V were impressed across

the motor while it is at rest? Assume there is no overload protection such as a fuse or a circuit breaker.

8. Explain the sequence of events that happen in a motor that is operating at a constant load, say about half-full value, and, suddenly the load is nearly doubled.

9. Suppose a motor is operating at full load and suddenly the load is halved. Describe the sequence of events that takes place in the motor.

10. Write the speed equation for a shunt motor.

11. What part does back EMF play in determining the speed of a dc motor? How can it be changed?

12. A hand starter for use with a shunt motor has an arm, pivoted on one end, which is turned to the right so that a contact on the other end will press against brass buttons arranged along the arc of a circle. It is held in the full "on" position by an electromagnet that carries field current. What two situations that may possibly happen while the motor is running would cause the electromagnet to be deenergized? A mechanical spring on the arm returns the handle to the "off" position (see Fig. 25.4) when the electromagnet is deenergized. Justify the presence of this safety feature.

13. Compare a series motor with a shunt motor in the following respects: (*a*) windings, (*b*) torque, (*c*) variation of speed with load.

14. How do the windings of a compound motor differ from those of a shunt motor and from those of a series motor?

15. In what respect is a compound motor a more desirable choice than a series motor for a job on which the load may become as small as one-tenth of full-load value?

16. What would happen if a break should occur in the field circuit of a shunt motor while it is driving a load? Answer for two cases: (*a*) motor load heavy, (*b*) motor load very light.

17. What is a Prony brake? What is meant by brake tare? Write the equation for horsepower output as measured by means of a Prony brake.

18. How is current produced in the rotor of an induction motor at standstill? Explain in some detail.

19. What determines the speed of rotation (rpm) of the magnetic field produced by alternating current in a stator winding? What is the special name for this speed?

20. Write the equation for the synchronous speed of a 60-Hz induction motor. What is the synchronous speed of a six-pole, 25-cycle motor?

21. What is meant by the revolutions slip of an induction motor? How is the slip value obtained when it is expressed as a fraction rather than as a number of revolutions per unit time?

22. Can an induction motor ever run under its own power at synchronous speed? Why?

23. Name an important reason for the popularity of the series motor used in small household appliances. Is it an *induction* motor?

24. Explain how the windings of a split-phase motor are arranged so that starting torque is provided.

25. Explain the principle of starting a single-phase induction motor using a capacitor. In what two important respects is a capacitor-start, capacitor-run motor superior to a split-phase motor?

26. Describe the shaded-pole motor and its means of producing starting torque. Is the starting torque as large as that of the split-phase or capacitor motor?

27. How do large synchronous motors differ from induction motors in construction? What are two important operating characteristic of a synchronous motor which an induction motor does not have?

28. What are the physical forms of the stator and rotor of a reluctance motor? Comment on its starting torque, its speed characteristic, its efficiency, and its principal applications in miniature sizes.

29. Describe the construction and principle of operation of the hysteresis motor. name its most important performance characteristics.

30. Describe the construction of the repulsion motor. What advantages does it have over other single-phase motors?

31. How is the repulsion motor converted into a repulsion-start induction-run motor? What are the advantages gained when this conversion is made?

25.25 Problems

6. A shunt motor running at no load takes 1.6 A armature current at 230 V. The armature resistance is $1.0\,\Omega$. (a) What is the back EMF? (b) How much power is lost due to armature resistance? (c) How much power is absorbed in iron losses and friction? (d) Theoretically, how much current would enter the armature upon starting, if no series resistance were used?

7. A shunt motor takes 12 A while delivering full load. Armature and field resistances are $R_a = 0.75\,\Omega$, $R_f = 150\,\Omega$. The line supplies 125 V. (a) Assuming the stray losses to be 280 W, how much mechanical power, expressed in watts, is developed in the armature? (b) What are the total copper losses? (c) What is the horsepower output?

8. Many electric power stations in Europe operate at a frequency of 50 Hz. How fast would a four-pole alternator have to run to deliver power at that frequency?

9. What would be the full load speed of a 6-pole induction motor if its slip is 5 percent when operating on a 50 Hz power line?

10. An old ac generator driven by a single-cylinder reciprocating steam engine was found to have 16 poles. What was its synchronous speed when delivering power at 50 Hz?

11. A machine is to be driven at about 700 rpm by a three-phase induction motor. How many poles would the motor have if it operated on (a) 60 Hz? (b) 25 Hz? (c) Assuming the slip at full load to be 4 percent, compute the rated full-load speed of each.

12. A large three-phase, 60-Hz induction motor drives a machine at 390 rpm when fully loaded. (a) How many poles does it have? Determine (b) the slip, (c) the time required for the rotor to slip 1 revolution, (d) the cycles in 1 revolution of the rotating field.

13. A 15-hp, 440-V, three-phase, four-pole, squirrel-cage induction motor has a full-load speed of 1725 rpm. A Prony brake test was made at approximately full load, and the following readings taken: 440 V, 18.94 A per terminal, 23.5 lb at the end of a 2-ft brake, brake tare + 1.25 lb, slip 75 rpm, wattmeter readings: $P_1 = 8400\,W$, $P_2 = 4300\,W$ by the two-meter method. Compute (a) torque, (b) percent slip, (c) horsepower output, (d) horsepower input, (e) efficiency, (f) power factor.

14. A small four-pole, 60-cycle, 1725-rpm, single-phase capacitor motor was tested by means of a Prony brake and the data in the following table were obtained. The voltage and frequency were constant at 120 V and 60 Hz. Compute (1) horsepower output, efficiency, power factor, and (2) plot efficiency, torque, power factor, percent slip, and current versus horsepower output. The name plate on the machine specifies it to be $\frac{1}{15}$ hp. Indicate this value by a dashed line on the horsepower scale.

Amperes	0.71	0.77	0.82	0.90	1.00	1.2	1.3	1.5
Watts input	65.0	72.0	80	92.0	103	120	130	150
Torque, oz-in.	0	8.9	17.9	29.5	40	52	60	70
Rpm slip	15	30	50	69	110	152	232	390

Appendix

A.1 Phasor Representation; Slide Rule Application

Let us consider specific cases involving voltage and current phasors instead of dealing with letter symbols only.

Figure A.1 shows a current phasor lagging an associated voltage phasor by a 30° phase angle as they rotate counterclockwise around the origin O. It may be replaced, so far as its effect and use in calculations are concerned, by its *vertical component* and its *horizontal component*. An arrow head could be drawn at B indicating that the effect of the vertical component OB is downward, and an arrow head could be drawn at A indicating that the effect of the horizontal component OA (also called the *in-phase* component) is to the right.

Because $OB/8 = \sin 30°$

$$OB = 8 \sin 30° = 8 \times 0.5 = 4 \text{ A}$$

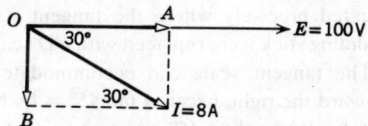

FIGURE A.1

595

Because $OA/8 = \cos 30°$

$$OA = 8 \cos 30° = 8 \times 0.866 = 6.928 \text{ A}$$

In our problems we find that the phase angle is almost always different from 30°, so in this example we could have called it θ and our expressions would have been $OB = 8 \sin \theta$, $OA = 8 \cos \theta$. The sine and cosine of any angle we encounter could be looked up in trigonometric tables. Let's see how we can read the answers from slide-rule settings.

USE OF SLIDE RULE

Most slide rules that are useful in electrical engineering technology have scales whose divisions are marked in degrees and fractions of degrees (some with minutes instead of tenths of degrees). One scale called the *S* (or *Sine*) scale has markings on the sliding stick precisely placed so that when the scale is lined up with the *D* scale on the stationary part, a reading on that scale is the sine of the angle on the *S* scale at the point of the reading. Saying it another way, the sine of any angle chosen on the *S* scale can be read from the *D* scale at the point where the chosen angle value is located. Now let's use the slide rule to obtain 8 sin 30° and 8 cos 30°, which we have already determined. Each *end point* of the scales on the sliding stick is called the *index*.

To evaluate 8 sin 30° we first set the right index on the *8 on the D scale*. To do this we pull the sliding stick toward the left. Either index is usable in many operations of this kind, but if we try to use the left index in this case (which would require pulling the stick toward the right) most of the *S* scale would not be available for use with the *D* scale. Now move the slider so that the hair line is at 30° on the *S* scale. The answer is 4, as read on the *D* scale under the hair line.

The *S* scale has a second set of numbers which increase in value from right to left and which are printed in red on some rules. These are angles whose *cosines* can be read on the *D* scale. Note that sine 30° = cos 60°. This means that *the setting that gave us 4 for 8 sin 30° also gives us the product 8 cos 60° = 4.*

Now let's find 8 cos 77°. Leave the stick index on 8, move the slider over to 77° *on the cosine scale* (13° on the sine scale) and read 1.80.

Try 25 sin 20°: (Ans. 8.55), 25 cos 67°: (Ans. 9.77), 85 sin 10°: (Ans. 14.75), 15 cos 50°: (Ans. 9.65).

Now for the tangent scale. Remember that the degree marks on the tangent scale are located precisely where the tangent values would be if the tangent scale on the sliding stick were replaced with a *D* scale.

Important. The tangent scale can accommodate angles only from 0° to 45° increasing toward the right, because tan 45° = 1 which is *the right index* of the *D* scale. For angles larger than 45° we must use the second set of numbers (often found printed in red) on the tangent scale. These increase from 45° at the right end, to 90° at the left end.

When working with angles larger than 45° we must use the *inverted D scale.*

which is called the *DI* scale and is usually printed in red. Remember that the tangents of angles larger than 45° are greater than unity.

For all angles less than about 5.6° the tangent is practically equal to the sine. In so far as the accuracy of slide-rule calculations are concerned, they are equal. The sine, tangent scale (*S*, *T*) is used directly with the *D* scale but the values on the *D* scale should be divided by 10.

As an example, let's obtain 12 sin 4° = 0.837. Place the left index at 12, then place the hair line on 4 on the *ST* scale. Read 8.37 on the *D* scale. This, divided by 10, gives us 12 sin 4° = 0.837. We get the same answer if we evaluate 12 tan 4°. Computations by means of the *ST* scale give results that differ by 0.1 percent, or less, from those obtained with five-place log tables.

Let's get 24 tan 20°. Set the left index on 24, move the slider hair line to 20° on the *T* scale, and read 8.74 on the *D* scale. Next, 75 tan 30°. Set the right index of the *T* scale on 75 and move the hair line to 30°. Read 4.33 on the *D* scale. Try 13.5 tan 60°. Set the right index on 13.5 on the *DI* scale (we have an angle larger than 45°) and move the hair line to 60° on the red *T* scale. Read 23.4 on the *DI* scale. Get 65 tan 83° (Ans. 529). Try 19.5 tan 26° (Ans. 9.5).

Perhaps the easiest way to remember how to divide by a sine, a cosine, or a tangent is to think of it as the reverse of a multiplying operation. Set the numerator on the *D* scale and use the *trig* scale as if it were a *C* scale.

To divide 4 by sin 30°, simply set the hair line on 4 on the *D* scale, draw the sine scale under it until the 30° mark is under the hair line and read 8 at the right index. (4/0.5 = 8). To divide by cos 30°, set the red 30° mark under the hair line and read 4.62. Note that this setting is identical with the setting for 4.62 cos 30° = 4.

To divide by the tangent of an angle use a similar procedure. We had 24 tan 20° = 8.74. To divide 8.74 by tan 20° we set the hair line on 8.74 on the *D* scale and pull the tangent scale under it until the 20° mark coincides with the hair line. The left index gives us 24 as the answer.

Now divide 24 by tan 55°. Set the hair line on 24 on the *DI* (red) scale. Why? Draw the red tan scale until 55° is under the hair line and read 16.8.

Should you encounter the cotangent function (cot θ) you can handle it by using the tangent of $(90° - \theta)$, thus working with tangents. For example, let's obtain 25 cot 70°. This is the same as 25 tan (90 − 70°) = 25 tan 20° = 9.1.

A.2   Solution of Differential Equation in the Text

Equation (11.4)

$$E = Ri + L\frac{di}{dt}$$

$$L\frac{di}{dt} = E - Ri$$

$$\frac{di}{E - Ri} = \frac{dt}{L}$$

Multiply both sides by $-R$. This makes the numerator on the left a differential of the denominator

$$\int \frac{-R\,di}{E-Ri} = \int -\frac{R\,dt}{L} \tag{1}$$

Integrating,

$$\ln\,(E-Ri) = -\frac{Rt}{L} + C \tag{2}$$

The constant C may be evaluated by inserting the values of i and t at $t=0$, At $t=0, i=0$

$$\ln E = C$$

Substituting into (2)

$$\ln\,(E-Ri) = -\frac{Rt}{L} + \ln E$$

$$\ln\left(\frac{E-Ri}{E}\right) = -\frac{Rt}{L}$$

$$\frac{E-Ri}{E} = \epsilon^{-Rt/L}$$

$$i = \frac{E}{R}\,(1-\epsilon^{-Rt/L})$$

A.3   Solution of Differential Equation in Text

Equation 11.18

$$Ri + L\frac{di}{dt} = 0, \qquad L\frac{di}{dt} = -Ri$$

$$\int \frac{di}{i} = \int -\frac{R}{L}\,dt, \qquad \ln i = -\frac{R}{L}t + C$$

$$\ln i = -\frac{R}{L}t + \ln\frac{E}{R}$$

When $t=0, i=\ln\dfrac{E}{R}$, so $\ln\dfrac{iR}{E} = -\dfrac{R}{L}t$

$$\frac{iR}{E} = \epsilon^{-(R/L)t}$$

$$i = \frac{E}{R}\epsilon^{-(R/L)t}$$

A.4   Solution of Differential Equation in Text

Equation (12.40)

$$\frac{di_c}{i_c} = -\frac{dt}{RC}$$

$$ln\ i_c = -\frac{t}{RC} + k_1$$

$$i_c = \epsilon^{-(t/RC - k_1)} = \epsilon^{-t/RC}\epsilon^{k_1}$$

$$i_c = k_2\epsilon^{-t/RC}$$

because ϵ^{k_1} is a constant $= k_2$
When $t = 0$, $i_c = E/R$, so $k_2 = E/R$.
This, substituted for k_2, gives

$$i_c = \frac{E}{R}\epsilon^{-t/RC} \qquad\qquad (12.41)$$

A.5 Capacitor Energy Dissipated in a Resistance

When a charged capacitor of C farads discharges through a resistor of R ohms, the energy dissipated in the resistor is given by

$$W = \int_0^\infty i^2 R\ dt$$

in which i is the instantaneous value of the current. The instantaneous value of the current is given by Equation (12.48) as $(E_0/R)\epsilon^{-t/RC}$ and derived in Appendix A.6

$$W = \int_0^\infty \frac{E_0^2}{R^2}\epsilon^{-2t/RC}(R)\ dt$$

$$= \frac{E_0^2}{R}\int_0^\infty \left(-\frac{RC}{2}\right)\epsilon^{-2t/RC}\left(-\frac{2}{RC}\right)dt$$

$$= -\frac{E_0^2 C}{2}\int_0^\infty \epsilon^{-2t/RC}\left(-\frac{2}{RC}\right)dt$$

$$= -\frac{E_0^2 C}{2}[\epsilon^{-2t/RC}]_0^\infty$$

$$= -\frac{E_0^2 C}{2}(0 - 1)$$

$$W = \tfrac{1}{2}CE_0^2$$

This is, as expected, the total energy stored in the capacitor when its terminal voltage is E_0.

A.6 Solution of Differential Equation in Text

Equation (12.47)

$$R\frac{di_c}{dt} + \frac{1}{C}i_c = 0$$

$$R\frac{di_c}{dt} = -\frac{1}{C}i_c$$

$$\frac{di_c}{i_c} = -\frac{1}{RC}dt$$

Integrating,

$$\ln i_c = -t/RC + k_1$$

$$i_c = \epsilon^{-t/RC+k_1} = \epsilon^{-t/RC}\epsilon^{k_1}$$

Let $\epsilon^{k_1} = k_2$,

$$i_c = k_2\epsilon^{-t/RC}$$

at $t = 0$, $\quad i = \frac{E_0}{R}, \quad \frac{E_0}{R} = k_2$

Substituting for k_2,

$$i_c = \frac{E_0}{R}\epsilon^{-t/RC} \tag{12.48}$$

A.7 Average Value (Average Ordinate) of a Half-Cycle Sine Wave

Equation of sine wave of current:

$$i = I_m \sin \omega t$$

Average over only one-half cycle:

$$I_{avg} = \frac{1}{\pi}\int_0^\pi I_m \sin \omega t \, d(\omega t)$$

$$= \frac{I_m}{\pi}(-\cos \omega t)_0^\pi$$

$$I_{avg} = \frac{2}{\pi}I_m = 0.637 \, I_m$$

This is the average value of the rectified sine wave shown in Fig. 24.4a. A half-wave rectifier delivers a half-sine wave represented by Fig. 24.6. Its average value is half of the value of the full wave rectifier output (Fig. 24.4a) and is $0.5 \times 0.637 I_m = 0.318 I_m$.

A.8 Effective Value of an Alternating Current

Derivation of $I = I_m/\sqrt{2}$, $I =$ effective value, $I_m =$ maximum instantaneous value of current when the wave form is sinusoidal.
Let

$$i = I_m \sin \omega t \text{ (instantaneous value of current)}$$

$$i^2 = I_m^2 \sin^2 \omega t$$

This is the ordinate at any point of the i^2 curve of Fig. 13.7. A thin vertical strip $d(\omega t)$ wide is imagined erected under this curve at any value of ωt. The area of this strip is $i^2 \, d(\omega t) = I_m^2 \sin^2 \omega t \, d(\omega t)$. The total area under this curve over a half period, from $\omega t = 0$ to $\omega t = \pi$, is obtained by integration. The average value of i^2 is then obtained by dividing the expression for total area by π, thus:

$$\text{average } i^2 = \frac{1}{\pi} \int_0^\pi I_m^2 \sin^2 \omega t \, d(\omega t)$$

$$= \frac{I_m^2}{2\pi} \int_0^\pi (1 - \cos 2\omega t) \, d(\omega t)$$

$$= \frac{I_m^2}{2\pi} \int_0^\pi d(\omega t) + \frac{I_m^2}{2\pi} \int_0^\pi -\cos 2\omega t \, d(\omega t)$$

$$= \frac{I_m^2}{2\pi} [\omega t - \tfrac{1}{2} \sin 2\omega t]_0^\pi$$

$$= \frac{I_m^2}{2\pi} (\pi - 0) = \frac{I_m^2}{2}$$

The rms value of a current with sinusoidal wave form is then

$$I = \sqrt{\frac{I_m^2}{2}} = \frac{I_m}{\sqrt{2}} = 0.707 I_m$$

The rms value of a voltage with sinusoidal wave form is determined in the same kind of procedure to be

$$E = \frac{E_m}{\sqrt{2}} = 0.707 E_m$$

A.9   Showing mathematically that current in a pure inductance lags 90° in phase behind the applied voltage

The voltage applied to a pure inductance, L, is equal to the induced EMF. This comes from the application of Kirchhoff's voltage law. Summing the voltage aroused a closed path:

$$V_L - L\frac{di}{dt} = 0; \qquad V_L = L\frac{di}{dt}$$

Let V_L be $V_m \sin \omega t$,

$$L\frac{di}{dt} = V_m \sin \omega t$$

$$di = \frac{V_m}{L} \sin \omega t \, dt$$

$$di = \frac{V_m}{\omega L} \sin \omega t \, d(\omega t)$$

$$i = \frac{V_m}{\omega L} \int \sin \omega t \, d(\omega t)$$

$$i = \frac{V_m}{\omega L} (-\cos \omega t)$$

Note that $\sin(\omega t - \pi/2) = \sin \omega t \cos \pi/2 - \cos \omega t \sin \pi/2 = -\cos \omega t$.

$$i = \frac{V_m}{\omega L} \sin\left(\omega t - \frac{\pi}{2}\right)$$

which shows that the current lags the applied voltage ($\pi/2$) or 90°.
This result also shows that the maximum value of the current is

$$I_m = \frac{V_m}{\omega L}$$

Therefore, the effective value is

$$I = \frac{V}{\omega L} = \frac{V}{X_L}$$

A.10 Showing mathematically that current in a pure capacitance leads 90° in phase the applied voltage

By definition,

$$v_c = \frac{q}{C}$$

$$\frac{dv_c}{dt} = \frac{1}{C} \frac{dq}{dt} = \frac{i}{C}$$

Let $v_c = V_m \sin \omega t$,

$$i = C \frac{dv_c}{dt} = C \frac{d}{dt} (V_m \sin \omega t)$$

$$i = \omega C V_m \cos \omega t$$

Note that $\sin\left(\omega t + \frac{\pi}{2}\right) = \sin \omega t \cos \frac{\pi}{2} + \cos \omega t \sin \frac{\pi}{2} = \cos \omega t$

Therefore we can write

$$i = \omega C V_m \sin\left(\omega t + \frac{\pi}{2}\right)$$

Therefore i_c is ($\pi/2$) radians (90°) ahead of V_c. From the last equation is is evident that the maximum instantaneous value of the capacitor current is

$$I_m = \omega C V_m = \frac{V_m}{1/\omega C}$$

The effective value is

$$I = \frac{V}{1/\omega C} = \frac{V}{X_c}$$

A.11   Q of an *RLC* Circuit

Figure 19.4*e* has a series *RLC* network connected to the terminals of a generator with internal impedance R_g. The derivations of the expressions for Q are as follows;

Consider the series network alone. The maximum energy stored per cycle in the capacitance is

$$W_{cm} = \frac{V_{cm}^2 C}{2} = V_c^2 C = \frac{I^2}{\omega C}$$

The maximum energy stored per cycle in the inductance is

$$W_{Lm} = I^2 L \qquad \text{(Equation 20.7)}$$

At frequencies below resonance, $f < f_r$ and $\omega L < 1/10\, G$, therefore

$$W_{cm} > W_{Lm}$$

Thus, from Equation (20.14), *below resonance*

$$Q_s = \frac{1}{\omega C R} \quad (f < f_r)$$

And above resonance, $W_{Lm} > W_{Cm}$ and the quality factor of the series circuit becomes

$$Q_s = \frac{\omega L}{R} \quad (f > f_r)$$

At resonant frequency the maximum energy stored is the same for both the inductance and the capacitance. The quality factor at resonance may then be expressed in terms of either *L* or *C*:

$$Q_s (\text{at resonance}) = \frac{\omega L}{R} = \frac{1}{\omega C R} \quad (f = f_r)$$

If Q of the *complete circuit* is required, that is, including the generator resistance, the total series resistance must be used:

$$Q_r = \frac{\omega_r L}{R_g + R} \text{ at resonance}$$

A.12 Common Logarithms

	0	1	2	3	4	5	6	7	8	9	proportional parts 1 2 3	4 5 6	7 8 9
10	0000	0043	0086	0128	0170	0212	0253	0294	0334	0374	*4 8 12	17 21 25	29 33 37
11	0414	0453	0492	0531	0569	0607	0645	0682	0719	0755	4 8 11	15 19 23	26 30 34
12	0792	0828	0864	0899	0934	0969	1004	1038	1072	1106	3 7 10	14 17 21	24 28 31
13	1139	1173	1206	1239	1271	1303	1335	1367	1399	1430	3 6 10	13 16 19	23 26 29
14	1461	1492	1523	1553	1584	1614	1644	1673	1703	1732	3 6 9	12 15 18	21 24 27
15	1761	1790	1818	1847	1875	1903	1931	1959	1987	2014	*3 6 8	11 14 17	20 22 25
16	2041	2068	2095	2122	2148	2175	2201	2227	2253	2279	3 5 8	11 13 16	18 21 24
17	2304	2330	2355	2380	2405	2430	2455	2480	2504	2529	2 5 7	10 12 15	17 20 22
18	2553	2577	2601	2625	2648	2672	2695	2718	2742	2765	2 5 7	9 12 14	16 19 21
19	2788	2810	2833	2856	2878	2900	2923	2945	2967	2989	2 4 7	9 11 13	16 18 20
20	3010	3032	3054	3075	3096	3118	3139	3160	3181	3201	2 4 6	8 11 13	15 17 19
21	3222	3243	3263	3284	3304	3324	3345	3365	3385	3404	2 4 6	8 10 12	14 16 18
22	3424	3444	3464	3483	3502	3522	3541	3560	3579	3598	2 4 6	8 10 12	14 15 17
23	3617	3636	3655	3674	3692	3711	3729	3747	3766	3784	2 4 6	7 9 11	13 15 17
24	3802	3820	3838	3856	3874	3892	3909	3927	3945	3962	2 4 5	7 9 11	12 14 16
25	3979	3997	4014	4031	4048	4065	4082	4099	4116	4133	2 3 5	7 9 10	12 14 15
26	4150	4166	4183	4200	4216	4232	4249	4265	4281	4298	2 3 5	7 8 10	11 13 15
27	4314	4330	4346	4362	4378	4393	4409	4425	4440	4456	2 3 5	6 8 9	11 13 14
28	4472	4487	4502	4518	4533	4548	4564	4579	4594	4609	2 3 5	6 8 9	11 12 14
29	4624	4639	4654	4669	4683	4698	4713	4728	4742	4757	1 3 4	6 7 9	10 12 13
30	4771	4786	4800	4814	4829	4843	4857	4871	4886	4900	1 3 4	6 7 9	10 11 13
31	4914	4928	4942	4955	4969	4983	4997	5011	5024	5038	1 3 4	6 7 8	10 11 12
32	5051	5065	5079	5092	5105	5119	5132	5145	5159	5172	1 3 4	5 7 8	9 11 12
33	5185	5198	5211	5224	5237	5250	5263	5276	5289	5302	1 3 4	5 6 8	9 10 12
34	5315	5328	5340	5353	5366	5378	5391	5403	5416	5428	1 3 4	5 6 8	9 10 11
35	5441	5453	5465	5478	5490	5502	5514	5527	5539	5551	1 2 4	5 6 7	9 10 11
36	5563	5575	5587	5599	5611	5623	5635	5647	5658	5670	1 2 4	5 6 7	8 10 11
37	5682	5694	5705	5717	5729	5740	5752	5763	5775	5786	1 2 3	5 6 7	8 9 10
38	5798	5809	5821	5832	5843	5855	5866	5877	5888	5899	1 2 3	5 6 7	8 9 10
39	5911	5922	5933	5944	5955	5966	5977	5988	5999	6010	1 2 3	4 5 7	8 9 10
40	6021	6031	6042	6053	6064	6075	6085	6096	6107	6117	1 2 3	4 5 6	8 9 10
41	6128	6138	6149	6160	6170	6180	6191	6201	6212	6222	1 2 3	4 5 6	7 8 9
42	6232	6243	6253	6263	6274	6284	6294	6304	6314	6325	1 2 3	4 5 6	7 8 9
43	6335	6345	6355	6365	6375	6385	6395	6405	6415	6425	1 2 3	4 5 6	7 8 9
44	6435	6444	6454	6464	6474	6484	6493	6503	6513	6522	1 2 3	4 5 6	7 8 9
45	6532	6542	6551	6561	6571	6580	6590	6599	6609	6618	1 2 3	4 5 6	7 8 9
46	6628	6637	6646	6656	6665	6675	6684	6693	6702	6712	1 2 3	4 5 6	7 7 8
47	6721	6730	6739	6749	6758	6767	6776	6785	6794	6803	1 2 3	4 5 5	6 7 8
48	6812	6821	6830	6839	6848	6857	6866	6875	6884	6893	1 2 3	4 4 5	6 7 8
49	6902	6911	6920	6928	6937	6946	6955	6964	6972	6981	1 2 3	4 4 5	6 7 8
50	6990	6998	7007	7016	7024	7033	7042	7050	7059	7067	1 2 3	3 4 5	6 7 8
51	7076	7084	7093	7101	7110	7118	7126	7135	7143	7152	1 2 3	3 4 5	6 7 8
52	7160	7168	7177	7185	7193	7202	7210	7218	7226	7235	1 2 2	3 4 5	6 7 7
53	7243	7251	7259	7267	7275	7284	7292	7300	7308	7316	1 2 2	3 4 5	6 6 7
54	7324	7332	7340	7348	7356	7364	7372	7380	7388	7396	1 2 2	3 4 5	6 6 7

	0	1	2	3	4	5	6	7	8	9	proportional parts								
											1	2	3	4	5	6	7	8	9
55	7404	7412	7419	7427	7435	7443	7451	7459	7466	7474	1	2	2	3	4	5	5	6	7
56	7482	7490	7497	7505	7513	7520	7528	7536	7543	7551	1	2	2	3	4	5	5	6	7
57	7559	7566	7574	7582	7589	7597	7604	7612	7619	7627	1	2	2	3	4	5	5	6	7
58	7634	7642	7649	7657	7664	7672	7679	7686	7694	7701	1	1	2	3	4	4	5	6	7
59	7709	7716	7723	7731	7738	7745	7752	7760	7767	7774	1	1	2	3	4	4	5	6	7
60	7782	7789	7796	7803	7810	7818	7825	7832	7839	7846	1	1	2	3	4	4	5	6	6
61	7853	7860	7868	7875	7882	7889	7896	7903	7910	7917	1	1	2	3	4	4	5	6	6
62	7924	7931	7938	7945	7952	7959	7966	7973	7980	7987	1	1	2	3	3	4	5	6	6
63	7993	8000	8007	8014	8021	8028	8035	8041	8048	8055	1	1	2	3	3	4	5	5	6
64	8062	8069	8075	8082	8089	8096	8102	8109	8116	8122	1	1	2	3	3	4	5	5	6
65	8129	8136	8142	8149	8156	8162	8169	8176	8182	8189	1	1	2	3	3	4	5	5	6
66	8195	8202	8209	8215	8222	8228	8235	8241	8248	8254	1	1	2	3	3	4	5	5	6
67	8261	8267	8274	8280	8287	8293	8299	8306	8312	8319	1	1	2	3	3	4	5	5	6
68	8325	8331	8338	8344	8351	8357	8363	8370	8376	8382	1	1	2	3	3	4	4	5	6
69	8388	8395	8401	8407	8414	8420	8426	8432	8439	8445	1	1	2	2	3	4	4	5	6
70	8451	8457	8463	8470	8476	8482	8488	8494	8500	8506	1	1	2	2	3	4	4	5	6
71	8513	8519	8525	8531	8537	8543	8549	8555	8561	8567	1	1	2	2	3	4	4	5	5
72	8573	8579	8585	8591	8597	8603	8609	8615	8621	8627	1	1	2	2	3	4	4	5	5
73	8633	8639	8645	8651	8657	8663	8669	8675	8681	8686	1	1	2	2	3	4	4	5	5
74	8692	8698	8704	8710	8716	8722	8727	8733	8739	8745	1	1	2	2	3	4	4	5	5
75	8751	8756	8762	8768	8774	8779	8785	8791	8797	8802	1	1	2	2	3	3	4	5	5
76	8808	8814	8820	8825	8831	8837	8842	8848	8854	8859	1	1	2	2	3	3	4	5	5
77	8865	8871	8876	8882	8887	8893	8899	8904	8910	8915	1	1	2	2	3	3	4	4	5
78	8921	8927	8932	8938	8943	8949	8954	8960	8965	8971	1	1	2	2	3	3	4	4	5
79	8976	8982	8987	8993	8998	9004	9009	9015	9020	9025	1	1	2	2	3	3	4	4	5
80	9031	9036	9042	9047	9053	9058	9063	9069	9074	9079	1	1	2	2	3	3	4	4	5
81	9085	9090	9096	9101	9106	9112	9117	9122	9128	9133	1	1	2	2	3	3	4	4	5
82	9138	9143	9149	9154	9159	9165	9170	9175	9180	9186	1	1	2	2	3	3	4	4	5
83	9191	9196	9201	9206	9212	9217	9222	9227	9232	9238	1	1	2	2	3	3	4	4	5
84	9243	9248	9253	9258	9263	9269	9274	9279	9284	9289	1	1	2	2	3	3	4	4	5
85	9294	9299	9304	9309	9315	9320	9325	9330	9335	9340	1	1	2	2	3	3	4	4	5
86	9345	9350	9355	9360	9365	9370	9375	9380	9385	9390	1	1	2	2	3	3	4	4	5
87	9395	9400	9405	9410	9415	9420	9425	9430	9435	9440	0	1	1	2	2	3	3	4	4
88	9445	9450	9455	9460	9465	9469	9474	9479	9484	9489	0	1	1	2	2	3	3	4	4
89	9494	9499	9504	9509	9513	9518	9523	9528	9533	9538	0	1	1	2	2	3	3	4	4
90	9542	9547	9552	9557	9562	9566	9571	9576	9581	9586	0	1	1	2	2	3	3	4	4
91	9590	9595	9600	9605	9609	9614	9619	9624	9628	9633	0	1	1	2	2	3	3	4	4
92	9638	9643	9647	9652	9657	9661	9666	9671	9675	9680	0	1	1	2	2	3	3	4	4
93	9685	9689	9694	9699	9703	9708	9713	9717	9722	9727	0	1	1	2	2	3	3	4	4
94	9731	9736	9741	9745	9750	9754	9759	9763	9768	9773	0	1	1	2	2	3	3	4	4
95	9777	9782	9786	9791	9795	9800	9805	9809	9814	9818	0	1	1	2	2	3	3	4	4
96	9823	9827	9832	9836	9841	9845	9850	9854	9859	9863	0	1	1	2	2	3	3	4	4
97	9868	9872	9877	9881	9886	9890	9894	9899	9903	9908	0	1	1	2	2	3	3	4	4
98	9912	9917	9921	9926	9930	9934	9939	9943	9948	9952	0	1	1	2	2	3	3	4	4
99	9956	9961	9965	9969	9974	9978	9983	9987	9991	9996	0	1	1	2	2	3	3	3	4

A.13 Natural Trigonometric Functions For Decimal Fractions of a Degree

deg	sin	cos	tan	cot	
0.0	.00000	1.0000	.00000	∞	90.0
.1	.00175	1.0000	.00175	573.0	.9
.2	.00349	1.0000	.00349	286.5	.8
.3	.00524	1.0000	.00524	191.0	.7
.4	.00698	1.0000	.00698	143.24	.6
.5	.00873	1.0000	.00873	114.59	.5
.6	.01047	0.9999	.01047	95.49	.4
.7	.01222	.9999	.01222	81.85	.3
.8	.01396	.9999	.01396	71.62	.2
.9	.01571	.9999	.01571	63.66	.1
1.0	.01745	0.9998	.01746	57.29	89.0
.1	.01920	.9998	.01920	52.08	.9
.2	.02094	.9998	.02095	47.74	.8
.3	.02269	.9997	.02269	44.07	.7
.4	.02443	.9997	.02444	40.92	.6
.5	.02618	.9997	.02619	38.19	.5
.6	.02792	.9996	.02793	35.80	.4
.7	.02967	.9996	.02968	33.69	.3
.8	.03141	.9995	.03143	31.82	.2
.9	.03316	.9995	.03317	30.14	.1
2.0	.03490	0.9994	.03492	28.64	88.0
.1	.03664	.9993	.03667	27.27	.9
.2	.03839	.9993	.03842	26.03	.8
.3	.04013	.9992	.04016	24.90	.7
.4	.04188	.9991	.04191	23.86	.6
.5	.04362	.9990	.04366	22.90	.5
.6	.04536	.9990	.04541	22.02	.4
.7	.04711	.9989	.04716	21.20	.3
.8	.04885	.9988	.04891	20.45	.2
.9	.05059	.9987	.05066	19.74	.1
3.0	.05234	0.9986	.05241	19.081	87.0
.1	.05408	.9985	.05416	18.464	.9
.2	.05582	.9984	.05591	17.886	.8
.3	.05756	.9983	.05766	17.343	.7
.4	.05931	.9982	.05941	16.832	.6
.5	.06105	.9981	.06116	16.350	.5
.6	.06279	.9980	.06291	15.895	.4
.7	.06453	.9979	.06467	15.464	.3
.8	.06627	.9978	.06642	15.056	.2
.9	.06802	.9977	.06817	14.669	.1
4.0	.06976	0.9976	.06993	14.301	86.0
.1	.07150	.9974	.07168	13.951	.9
.2	.07324	.9973	.07344	13.617	.8
.3	.07498	.9972	.07519	13.300	.7
.4	.07672	.9971	.07695	12.996	.6
.5	.07846	.9969	.07870	12.706	.5
.6	.08020	.9968	.08046	12.429	.4
.7	.08194	.9966	.08221	12.163	.3
.8	.08368	.9965	.08397	11.909	.2
.9	.08542	.9963	.08573	11.664	.1
5.0	.08716	0.9962	.08749	11.430	85.0
.1	'08889	.9960	.08925	11.205	.9
.2	.09063	.9959	.09101	10.988	.8
.3	.09237	.9957	.09277	10.780	.7
.4	.09411	.9956	.09453	10.579	.6
.5	.09585	.9954	.09629	10.385	.5
.6	.09758	.9952	.09805	10.199	.4
.7	.09932	.9951	.09981	10.019	.3
.8	.10106	.9949	.10158	9.845	.2
.9	.10279	.9947	.10334	9.677	.1
6.0	.10453	0.9945	.10510	9.514	84.0

| | cos | sin | cot | tan | deg |

deg	sin	cos	tan	cot	
6.0	.10453	0.9945	.10510	9.514	84.0
.1	.10626	.9943	.10687	9.357	.9
.2	.10800	.9942	.10863	9.205	.8
.3	.10973	.9940	.11040	9.058	.7
.4	.11147	.9938	.11217	8.915	.6
.5	.11320	9936	.11394	8.777	.5
.6	.11494	.9934	.11570	8.643	.4
.7	.11667	.9932	.11747	8.513	.3
.8	.11840	.9930	.11924	8.386	.2
.9	.12014	.9928	.12101	8.264	.1
7.0	.12187	0.9925	.12278	8.144	83.0
.1	.12360	.9923	.12456	8.028	.9
.2	.12533	.9921	.12633	7.916	.8
.3	.12706	.9919	.12810	7.806	.7
.4	.12880	.9917	.12988	7.700	.6
.5	.13053	.9914	.13165	7.596	.5
.6	.13226	.9912	.13343	7.495	.4
.7	.13399	.9910	.13521	7.396	.3
.8	.13572	.9907	.13698	7.300	.2
.9	.13744	.9905	.13876	7.207	.1
8.0	.13917	0.9903	.14054	7.115	82.0
.1	.14090	.9900	.14232	7.026	.9
.2	.14263	.9898	.14410	6.940	.8
.3	.14436	.9895	.14588	6.855	.7
.4	.14608	.9893	.14767	6.772	.6
.5	.14781	.9890	.14945	6.691	.5
.6	.14954	.9888	.15124	6.612	.4
.7	.15126	.9885	.15302	6.535	.3
.8	.15299	.9882	.15481	6.460	.2
.9	.15471	.9880	.15660	6.386	.1
9.0	.15643	0.9877	.15838	6.314	81.0
.1	.15816	.9874	.16017	6.243	.9
.2	.15988	.9871	.16196	6.174	.8
.3	.16160	.9869	.16376	6.107	.7
.4	.16333	.9866	.16555	6.041	.6
.5	.16505	.9863	.16734	5.976	.5
.6	.16677	.9860	.16914	5.912	.4
.7	.16849	.9857	.17093	5.850	.3
.8	.17021	.9854	.17273	5.789	.2
.9	.17193	.9851	.17453	5.730	.1
10.0	.1736	0.9848	.1763	5.671	80.0
.1	.1754	.9845	.1781	5.614	.9
.2	.1771	.9842	.1799	5.558	.8
.3	.1788	.9839	.1817	5.503	.7
.4	.1805	.9836	.1835	5.449	.6
.5	.1822	.9833	.1853	5.396	.5
.6	.1840	.9829	.1871	5.343	.4
.7	.1857	.9826	.1890	5.292	.3
.8	.1874	.9823	.1908	5.242	.2
.9	.1891	.9820	.1926	5.193	.1
11.0	.1908	0.9816	.1944	5.145	79.0
.1	.1925	.9813	.1962	5.097	.9
.2	.1942	.9810	.1980	5.050	.8
.3	.1959	.9806	.1998	5.005	.7
.4	.1977	.9803	.2016	4.959	.6
.5	.1994	.9799	.2035	4.915	.5
.6	.2011	.9796	.2053	4.872	.4
.7	.2028	.9792	.2071	4.829	.3
.8	.2045	.9789	.2089	4.787	.2
.9	.2062	.9785	.2107	4.745	.1
12.0	.2079	0.9781	.2126	4.705	78.0

| | cos | sin | cot | tan | deg |

deg	sin	cos	tan	cot		deg	sin	cos	tan	cot	
12.0	0.2079	0.9781	0.2126	4.705	78.0	18.0	0.3090	0.9511	0.3249	3.078	72.0
.1	.2096	.9778	.2144	4.665	.9	.1	.3107	.9505	.3269	3.060	.9
.2	.2113	.9774	.2162	4.625	.8	.2	.3123	.9500	.3288	3.042	.8
.3	.2130	.9770	.2180	4.586	.7	.3	.3140	.9494	.3307	3.024	.7
.4	.2147	.9767	.2199	4.548	.6	.4	.3156	.9489	.3327	3.006	.6
.5	.2164	.9763	.2217	4.511	.5	.5	.3173	.9483	.3346	2.989	.5
.6	.2181	.9759	.2235	4.474	.4	.6	.3190	.9478	.3365	2.971	.4
.7	.2198	.9755	.2254	4.437	.3	.7	.3206	.9472	.3385	2.954	.3
.8	.2215	.9751	.2272	4.402	.2	.8	.3223	.9466	.3404	2.937	.2
.9	.2233	.9748	.2290	4.366	.1	.9	.3239	.9461	.3424	2.921	.1
13.0	0.2250	0.9744	0.2309	4.331	77.0	19.0	0.3256	0.9455	0.3443	2.904	71.0
.1	.2267	.9740	.2327	4.297	.9	.1	.3272	.9449	.3463	2.888	.9
.2	.2284	.9736	.2345	4.264	.8	.2	.3289	.9444	.3482	2.872	.8
.3	.2300	.9732	.2364	4.230	.7	.3	.3305	.9438	.3502	2.856	.7
.4	.2317	.9728	.2382	4.198	.6	.4	.3322	.9432	.3522	2.840	.6
.5	.2334	.9724	.2401	4.165	.5	.5	.3338	.9426	.3541	2.824	.5
.6	.2351	.9720	.2419	4.134	.4	.6	.3355	.9421	.3561	2.808	.4
.7	.2368	.9715	.2438	4.102	.3	.7	.3371	.9415	.3581	2.793	.3
.8	.2385	.9711	.2456	4.071	.2	.8	.3387	.9409	.3600	2.778	.2
.9	.2402	.9707	.2475	4.041	.1	.9	.3404	.9403	.3620	2.762	.1
14.0	0.2419	0.9703	0.2493	4.011	76.0	20.0	0.3420	0.9397	0.3640	2.747	70.0
.1	.2436	.9699	.2512	3.981	.9	.1	.3437	.9391	.3659	2.733	.9
.2	.2453	.9694	.2530	3.952	.8	.2	.3453	.9385	.3679	2.718	.8
.3	.2470	.9690	.2549	3.923	.7	.3	.3469	.9379	.3699	2.703	.7
.4	.2487	.9686	.2568	3.895	.6	.4	.3486	.9373	.3719	2.689	.6
.5	.2504	.9681	.2586	3.867	.5	.5	.3502	.9367	.3739	2.675	.5
.6	.2521	.9677	.2605	3.839	.4	.6	.3518	.9361	.3759	2.660	.4
.7	.2538	.9673	.2623	3.812	.3	.7	.3535	.9354	.3779	2.646	.3
.8	.2554	.9668	.2642	3.785	.2	.8	.3551	.9348	.3799	2.633	.2
.9	.2571	.9664	.2661	3.758	.1	.9	.3567	.9342	.3819	2.619	.1
15.0	0.2588	0.9659	0.2679	3.732	75.0	21.0	0.3584	0.9336	0.3839	2.605	69.0
.1	.2605	.9655	.2698	3.706	.9	.1	.3600	.9330	.3859	2.592	.9
.2	.2622	.9650	.2717	3.681	.8	.2	.3616	.9323	.3879	2.578	.8
.3	.2639	.9646	.2736	3.655	.7	.3	.3633	.9317	.3899	2.565	.7
.4	.2656	.9641	.2754	3.630	.6	.4	.3649	.9311	.3919	2.552	.6
.5	.2672	.9636	.2773	3.606	.5	.5	.3665	.9304	.3939	2.539	.5
.6	.2689	.9632	.2792	3.582	.4	.6	.3681	.9298	.3959	2.526	.4
.7	.2706	.9627	.2811	3.558	.3	.7	.3697	.9291	.3979	2.513	.3
.8	.2723	.9622	.2830	3.534	.2	.8	.3714	.9285	.4000	2.500	.2
.9	.2740	.9617	.2849	3.511	.1	.9	.3730	.9278	.4020	2.488	.1
16.0	0.2756	0.9613	0.2867	3.487	74.0	22.0	0.3746	0.9272	0.4040	2.475	68.0
.1	.2773	.9608	.2886	3.465	.9	.1	.3762	.9265	.4061	2.463	.9
.2	.2790	.9603	.2905	3.442	.8	.2	.3778	.9259	.4081	2.450	.8
.3	.2807	.9598	.2924	3.420	.7	.3	.3795	.9252	.4101	2.438	.7
.4	.2823	.9593	.2943	3.398	.6	.4	.3811	.9245	.4122	2.426	.6
.5	.2840	.9588	.2962	3.376	.5	.5	.3827	.9239	.4142	2.414	.5
.6	.2857	.9583	.2981	3.354	.4	.6	.3843	.9232	.4163	2.402	.4
.7	.2874	.9578	.3000	3.333	.3	.7	.3859	.9225	.4183	2.391	.3
.8	.2890	.9573	.3019	3.312	.2	.8	.3875	.9219	.4204	2.379	.2
.9	.2907	.9568	.3038	3.291	.1	.9	.3891	.9212	.4224	2.367	.1
17.0	0.2924	0.9563	0.3057	3.271	73.0	23.0	0.3907	0.9205	0.4245	2.356	67.0
.1	.2940	.9558	.3076	3.251	.9	.1	.3923	.9198	.4265	2.344	.9
.2	.2957	.9553	.3096	3.230	.8	.2	.3939	.9191	.4286	2.333	.8
.3	.2974	.9548	.3115	3.211	.7	.3	.3955	.9184	.4307	2.322	.7
.4	.2990	.9542	.3134	3.191	.6	.4	.3971	.9178	.4327	2.311	.6
.5	.3007	.9537	.3153	3.172	.5	.5	.3987	.9171	.4348	2.300	.5
.6	.3024	.9532	.3172	3.152	.4	.6	.4003	.9164	.4369	2.289	.4
.7	.3040	.9527	.3191	3.133	.3	.7	.4019	.9157	.4390	2.278	.3
.8	.3057	.9521	.3211	3.115	.2	.8	.4035	.9150	.4411	2.267	.2
.9	.3074	.9516	.3230	3.096	.1	.9	.4051	.9143	.4431	2.257	.1
18.0	0.3090	0.9511	0.3249	3.078	72.0	24.0	0.4067	0.9135	0.4452	2.246	66.0
	cos	sin	cot	tan	deg		cos	sin	cot	tan	deg

A.13 Natural Trigonometric Functions For Decimal Fractions of a Degree

deg	sin	cos	tan	cot		deg	sin	cos	tan	cot	
24.0	0.4067	0.9135	0.4452	2.246	**66.0**	**30.0**	0.5000	0.8660	0.5774	1.7321	**60.0**
.1	.4083	.9128	.4473	2.236	.9	.1	.5015	.8652	.5797	1.7251	.9
.2	.4099	.9121	.4494	2.225	.8	.2	.5030	.8643	.5820	1.7182	.8
.3	.4115	.9114	.4515	2.215	.7	.3	.5045	.8634	.5844	1.7113	.7
.4	.4131	.9107	.4536	2.204	.6	.4	.5060	.8625	.5867	1.7045	.6
.5	.4147	.9100	.4557	2.194	.5	.5	.5075	.8616	.5890	1.6977	.5
.6	.4163	.9092	.4578	2.184	.4	.6	.5090	.8607	.5914	1.6909	.4
.7	.4179	.9085	.4599	2.174	.3	.7	.5105	.8599	.5938	1.6842	.3
.8	.4195	.9078	.4621	2.164	.2	.8	.5120	.8590	.5961	1.6775	.2
.9	.4210	.9070	.4642	2.154	.1	.9	.5135	.8581	.5985	1.6709	.1
25.0	0.4226	0.9063	0.4663	2.145	**65.0**	**31.0**	0.5150	0.8572	0.6009	1.6643	**59.0**
.1	.4242	.9056	.4684	2.135	.9	.1	.5165	.8563	.6032	1.6577	.9
.2	.4258	.9048	.4706	2.125	.8	.2	.5180	.8554	.6056	1.6512	.8
.3	.4274	.9041	.4727	2.116	.7	.3	.5195	.8545	.6080	1.6447	.7
.4	.4289	.9033	.4748	2.106	.6	.4	.5210	.8536	.6104	1.6383	.6
.5	.4305	.9026	.4770	2.097	.5	.5	.5225	.8526	.6128	1.6319	.5
.6	.4321	.9018	.4791	2.087	.4	.6	.5240	.8517	.6152	1.6255	.4
.7	.4337	.9011	.4813	2.078	.3	.7	.5255	.8508	.6176	1.6191	.3
.8	.4352	.9003	.4834	2.069	.2	.8	.5270	.8499	.6200	1.6128	.2
.9	.4368	.8996	.4856	2.059	.1	.9	.5284	.8490	.6224	1.6066	.1
26.0	0.4384	0.8988	0.4877	2.050	**64.0**	**32.0**	0.5299	0.8480	0.6249	1.6003	**58.0**
.1	.4399	.8980	.4899	2.041	.9	.1	.5314	.8471	.6273	1.5941	.9
.2	.4415	.8973	.4921	2.032	.8	.2	.5329	.8462	.6297	1.5880	.8
.3	.4431	.8965	.4942	2.023	.7	.3	.5344	.8453	.6322	1.5818	.7
.4	.4446	.8957	.4964	2.014	.6	.4	.5358	.8443	.6346	1.5757	.6
.5	.4462	.8949	.4986	2.006	.5	.5	.5373	.8434	.6371	1.5697	.5
.6	.4478	.8942	.5008	1.997	.4	.6	.5388	.8425	.6395	1.5637	.4
.7	.4493	.8934	.5029	1.988	.3	.7	.5402	.8415	.6420	1.5577	.3
.8	.4509	.8926	.5051	1.980	.2	.8	.5417	.8406	.6445	1.5517	.2
.9	.4524	.8918	.5073	1.971	.1	.9	.5432	.8396	.6469	1.5458	.1
27.0	0.4540	0.8910	0.5095	1.963	**63.0**	**33.0**	0.5446	0.8387	0.6494	1.5399	**57.0**
.1	.4555	.8902	.5117	1.954	.9	.1	.5461	.8377	.6519	1.5340	.9
.2	.4571	.8894	.5139	1.946	.8	.2	.5476	.8368	.6544	1.5282	.8
.3	.4586	.8886	.5161	1.937	.7	.3	.5490	.8358	.6569	1.5224	.7
.4	.4602	.8878	.5184	1.929	.6	.4	.5505	.8348	.6594	1.5166	.6
.5	.4617	.8870	.5206	1.921	.5	.5	.5519	.8339	.6619	1.5108	.5
.6	.4633	.8862	.5228	1.913	.4	.6	.5534	.8329	.6644	1.5051	.4
.7	.4648	.8854	.5250	1.905	.3	.7	.5548	.8320	.6669	1.4994	.3
.8	.4664	.8846	.5272	1.897	.2	.8	.5563	.8310	.6694	1.4938	.2
.9	.4679	.8838	.5295	1.889	.1	.9	.5577	.8300	.6720	1.4882	.1
28.0	0.4695	0.8829	0.5317	1.881	**62.0**	**34.0**	0.5592	0.8290	0.6745	1.4826	**56.0**
.1	.4710	.8821	.5340	1.873	.9	.1	.5606	.8281	.6771	1.4770	.9
.2	.4726	.8813	.5362	1.865	.8	.2	.5621	.8271	.6796	1.4715	.8
.3	.4741	.8805	.5384	1.857	.7	.3	.5635	.8261	.6822	1.4659	.7
.4	.4756	.8796	.5407	1.849	.6	.4	.5650	.8251	.6847	1.4605	.6
.5	.4772	.8788	.5430	1.842	.5	.5	.5664	.8241	.6873	1.4550	.5
.6	.4787	.8780	.5452	1.834	.4	.6	.5678	.8231	.6899	1.4496	.4
.7	.4802	.8771	.5475	1.827	.3	.7	.5693	.8221	.6924	1.4442	.3
.8	.4818	.8763	.5498	1.819	.2	.8	.5707	.8211	.6950	1.4388	.2
.9	.4833	.8755	.5520	1.811	.1	.9	.5721	.8202	.6976	1.4335	.1
29.0	0.4848	0.8746	0.5543	1.804	**61.0**	**35.0**	0.5736	0.8192	0.7002	1.4281	**55.0**
.1	.4863	.8738	.5566	1.797	.9	.1	.5750	.8181	.7028	1.4229	.9
.2	.4879	.8729	.5589	1.789	.8	.2	.5764	.8171	.7054	1.4176	.8
.3	.4894	.8721	.5612	1.782	.7	.3	.5779	.8161	.7080	1.4124	.7
.4	.4909	.8712	.5635	1.775	.6	.4	.5793	.8151	.7107	1.4071	.6
.5	.4924	.8704	.5658	1.767	.5	.5	.5807	.8141	.7133	1.4019	.5
.6	.4939	.8695	.5681	1.760	.4	.6	.5821	.8131	.7159	1.3968	.4
.7	.4955	.8686	.5704	1.753	.3	.7	.5835	.8121	.7186	1.3916	.3
.8	.4970	.8678	.5727	1.746	.2	.8	.5850	.8111	.7212	1.3865	.2
.9	.4985	.8669	.5750	1.739	.1	.9	.5864	.8100	.7239	1.3814	.1
30.0	0.5000	0.8660	0.5774	1.732	**60.0**	**36.0**	0.5878	0.8090	0.7265	1.3764	**54.0**
	cos	sin	cot	tan	deg		cos	sin	cot	tan	deg

deg	sin	cos	tan	cot	
36.0	0.5878	0.8090	0.7265	1.3764	54.0
.1	.5892	.8080	.7292	1.3713	.9
.2	.5906	.8070	.7319	1.3663	.8
.3	.5920	.8059	.7346	1.3613	.7
.4	.5934	.8049	.7373	1.3564	.6
.5	.5948	.8039	.7400	1.3514	.5
.6	.5962	.8028	.7427	1.3465	.4
.7	.5976	.8018	.7454	1.3416	.3
.8	.5990	.8007	.7481	1.3367	.2
.9	.6004	.7997	.7508	1.3319	.1
37.0	0.6018	0.7986	0.7536	1.3270	53.0
.1	.6032	.7976	.7563	1.3222	.9
.2	.6046	.7965	.7590	1.3175	.8
.3	.6060	.7955	.7618	1.3127	.7
.4	.6074	.7944	.7646	1.3079	.6
.5	.6088	.7934	.7673	1.3032	.5
.6	.6101	.7923	.7701	1.2985	.4
.7	.6115	.7912	.7729	1.2938	.3
.8	.6129	.7902	.7757	1.2892	.2
.9	.6143	.7891	.7785	1.2846	.1
38.0	0.6157	0.7880	0.7813	1.2799	52.0
.1	.6170	.7869	.7841	1.2753	.9
.2	.6184	.7859	.7869	1.2708	.8
.3	.6198	.7848	.7898	1.2662	.7
.4	.6211	.7837	.7926	1.2617	.6
.5	.6225	.7826	.7954	1.2572	.5
.6	.6239	.7815	.7983	1.2527	.4
.7	.6252	.7804	.8012	1.2482	.3
.8	.6266	.7793	.8040	1.2437	.2
.9	.6280	.7782	.8069	1.2393	.1
39.0	0.6293	0.7771	0.8098	1.2349	51.0
.1	.6307	.7760	.8127	1.2305	.9
.2	.6320	.7749	.8156	1.2261	.8
.3	.6334	.7738	.8185	1.2218	.7
.4	.6347	.7727	.8214	1.2174	.6
.5	.6361	.7716	.8243	1.2131	.5
.6	.6374	.7705	.8273	1.2088	.4
.7	.6388	.7694	.8302	1.2045	.3
.8	.6401	.7683	.8332	1.2002	.2
.9	.6414	.7672	.8361	1.1960	.1
40.0	0.6428	0.7660	0.8391	1.1918	50.0
.1	.6441	.7649	.8421	1.1875	.9
.2	.6455	.7638	.8451	1.1833	.8
.3	.6468	.7627	.8481	1.1792	.7
.4	.6481	.7615	.8511	1.1750	.6
40.5	0.6494	0.7604	0.8541	1.1708	49.5
	cos	sin	cot	tan	deg

deg	sin	cos	tan	cot	
40.5	0.6494	0.7604	0.8541	1.1708	49.5
.6	.6508	.7593	.8571	1.1667	.4
.7	.6521	.7581	.8601	1.1626	.3
.8	.6534	.7570	.8632	1.1585	.2
.9	.6547	.7559	.8662	1.1544	.1
41.0	0.6561	0.7547	0.8693	1.1504	49.0
.1	.6574	.7536	.8724	1.1463	.9
.2	.6587	.7524	.8754	1.1423	.8
.3	.6600	.7513	.8785	1.1383	.7
.4	.6613	.7501	.8816	1.1343	.6
.5	.6626	.7490	.8847	1.1303	.5
.6	.6639	.7478	.8878	1.1263	.4
.7	.6652	.7466	.8910	1.1224	.3
.8	.6665	.7455	.8941	1.1184	.2
.9	.6678	.7443	.8972	1.1145	.1
42.0	0.6691	0.7431	0.9004	1.1106	48.0
.1	.6704	.7420	.9036	1.1067	.9
.2	.6717	.7408	.9067	1.1028	.8
.3	.6730	.7396	.9099	1.0990	.7
.4	.6743	.7385	.9131	1.0951	.6
.5	.6756	.7373	.9163	1.0913	.5
.6	.6769	.7361	.9195	1.0875	.4
.7	.6782	.7349	.9228	1.0837	.3
.8	.6794	.7337	.9260	1.0799	.2
.9	.6807	.7325	.9293	1.0761	.1
43.0	0.6820	0.7314	0.9325	1.0724	47.0
.1	.6833	.7302	.9358	1.0686	.9
.2	.6845	.7290	.9391	1.0649	.8
.3	.6858	.7278	.9424	1.0612	.7
.4	.6871	.7266	.9457	1.0575	.6
.5	.6884	.7254	.9490	1.0538	.5
.6	.6896	.7242	.9523	1.0501	.4
.7	.6909	.7230	.9556	1.0464	.3
.8	.6921	.7218	.9590	1.0428	.2
.9	.6934	.7206	.9623	1.0392	.1
44.0	0.6947	0.7193	0.9657	1.0355	46.0
.1	.6959	.7181	.9691	1.0319	.9
.2	.6972	.7169	.9725	1.0283	.8
.3	.6984	.7157	.9759	1.0247	.7
.4	.6997	.7145	.9793	1.0212	.6
.5	.7009	.7133	.9827	1.0176	.5
.6	.7022	.7120	.9861	1.0141	.4
.7	.7034	.7108	.9896	1.0105	.3
.8	.7046	.7096	.9930	1.0070	.2
.9	.7059	.7083	.9965	1.0035	.1
45.0	0.7071	0.7071	1.0000	1.0000	45.0
	cos	sin	cot	tan	deg

609

Answers to Selected Problems

Chapter 1

   1. 6×10^8 **2.** 9.109×10^{-28} g   **3.** 3.6   **5.** 10^3 mm   **7.** 16.387 cm³
   9. 88 ft/s, 26.83 m/s   **13.** 49.8×10^{28} electrons   **15.** 5.2×10^{18} e/s   **17.** 746 J/s
   19. 107.6 kgm, 107.6×10^5 g cm   **20.** 1.57 in³, 12.5%   **21.** 25% less skin
   23. 31% more volume, 12.3% more cost   **24.** \$21,478,000   **25.** 2796.5 ft, 0.529 mi

Chapter 2

   1. 10 C/s   **2.** 800 s   **4.** 7200 C, 8.05 g   **6.** (a) 0.25 V, (b) 11.8 V, (c) 12.3 V
   8. 2 h, 23 min, 6 s   **10.** 0.00831 in/s   **12.** 416.7 mA   **15.** 0.6 J/s = 0.6 W
   17. 25 C   **19.** 74.88×10^{14} e/s   **20.** 24.03×10^{-18} J

Chapter 3

   1. 2.9 Ω   **3.** 4.56 mi   **4.** 173,200 cmil   **6.** 5.39 Ω, 5.16 Ω   **14.** 1500 Ω
   16. 0.09 V   **19.** 10 MΩ   **21.** 187,500 mils², 2,387,000 cmil   **23.** 188.4 ft
   25. 3.67 ft   **28.** 1.724×10^{-4} Ω   **31.** 220°C   **33.** 7.5 A surge, 0.5 A continuous
   36. 1.77 Ω   **39.** 4.125 ft   **41.** 46 Ω   **43.** A = 476 (brass)

Chapter 4

   1. 40 J   **3.** 8 J/s, 4 J/s   **5.** 86.9%   **7.** 100 W, 50 V, 100 W   **10.** 20 V
   12. 1 hp   **14.** 10 ft-lb, net change = zero   **17.** 7.34 hp   **21.** 418,600 J
   23. 17.54 kWh   **26.** 0.25 W   **28.** 26.99 db   **31.** 8.7 nep   **33.** \$23.67
   34. 375 Btu   **38.** 25.9 A   **40.** 2 ¢   **42.** 5 W   **44,** 2 W   **45.** 3 db

Chapter 5

2. (a) 2 Ω (b) 200 W **4.** 12.65 V **6.** (a) 750 Ω (b) 0.08 A **8.** 0.0035 mho
10. (a)$G_1 = 0.333 \times 10^{-3}$, $G_2 = 0.167 \times 10^{-3}$, $G_T = 0.5 \times 10^{-3}$,(b) 0.1 A, (c) 2000 Ω
12. (a) $\frac{1}{9}$ A, (b) $\frac{2}{9}$ A **15.** 388 Ω **17.** $I_T = 0.2$ A, $V_{BC} = 50$ V **19.** 181.8 Ω,
1.375 A
22. 61.5 mA, 38.5 mA **24.** 105.6 Ω **28.** 18,750 V **31.**$\frac{2}{5}$ A **33.** 256 times
35. 6000 Ω **38.** 7.5 Ω **41.** (a) 240 V 255 V (b) V_2 is 5 V greater: Yes
44. $I = 55.2$ mA, $V = 1.38$ V, $P = 76.2$ mW
46. (b) 178.5 V (c) 150 V (d) 16.7% **47.** 25 V

Chapter 6

1. 1.182 A **2.** $I_A = 7.9$ A, $I_B = 0$, $I_C = 7.9$ A **4.** $\frac{1}{9}$ A **9.** $I_{R3} = 3.67$ A,
$I_{R4} = 6.33$ A, $I_G = 0.33$ A **13.** $I_1 = 0.476$ A, $I_2 = -1.62$ A, $I_1 - I_2 = 2.096$ A
15. $I_{1,2} = 2.096$ A **17.** 1.387 A **21.** (a)(b) 2 A **24.** $V_2 = 3.16$ V **26.** $1\frac{1}{19}$ A
28. (a)$\frac{1}{34}$ A upward, (b) $I_G = 2.5$ A **30.** $R_T = 4.687$ Ω, $I = 38.4$ mA
upward, (b) $I_G = 2.5$ A **30.** $R_T = 4.687$ Ω, $I = 38.4$ mA
32. $V = 116.70$ V, $I_1 = 8.35$ A, $I_2 = 3.25$ A

Chapter 7

5. 5 Ω, 1.25 W, 9.58 W **7.** 200 Ω **9.** 900, 900, 1440 Ω **14.** 1.102 A
downward **17.** 8 Ω **19.** 1.2 A **24.** 3.16 mA **30.** 24.8 Ω

Chapter 8

4. 0.05 wbT, 6.25 V **6.** 200 At **8.** 28 μ Wb/m² **10.** 68.2 Wb
12. (a) 245,000 At/Wb (b) 102,300 At/Wb **15.** (a) 52.4 μ Wb/m² (b) 41.7 At/m
17. 1250 At **19.** (a) 390, (b) 1400, (c) 960,000 At/m **22.** (a) 8189 At (b) 8090 At
23. 0.2 A **25.** 8.7 At/m **27.** 627 At/m **29.** 8.44 W **32.** 1.626 A
33. 5570 μ Wb **35.** (a) 268 J/m³/cycle, (b) 322 W **38.** 1.47 Wb/m²
39. 26.18 V **40.** 55.5 in.²

Chapter 9

2. (a) 3.75 N (NW) (b) 3.75 N (30°E of N) **4.** 0.25 Nm (N to S), East side moves
downward. **6.** 0.050 Ω **8.** 2980 Ω **11.** 148.18 V, 1.2% **13.** 100 kΩ
15. 18.98 μA **17.** 4.5 kΩ **19.** 1.575 MΩ **21.** $\frac{1}{15}$ mA, $\frac{7}{15}$ mA, 7.875 MΩ
23. 100 Ω **25.** 26.47 cm **27.** $R_s = 59,960$ Ω **29.** 3960 Ω (0–10 V),
40 kΩ (0–150 V), 180 kΩ (0–750 V)
31. Use 0.25 W for R_1, R_2, R_3, 0.5 W for R_4, 1.5 W for R_5 **33.** 2.7 MΩ
34. 2.4 MΩ **36.** Many possibilities. Example: Let $R_1 = 1000 = R_2$, then $R_3 =$
1234 Ω (b) $R_x = 1230$ or 1240 — exact balance not possible. Error = 0.323%
38. $R = 149,950$ Ω. Must add 2000 Ω in series with the original resistor.

Chapter 10

1. −5 V **3.** −80 A/s **6.** (a) 0.049 V reverse direction, (b) −0.0018 V forward
8. 1500 μH, 1.5 mH, 0.0015 H **10.** 400 mH **12.** 110 mH **14.** (a) 75 J
(b) 2.21 ft **16.** 18.06 H **18.** 39 H **19.** 4×10^{-9} Wb **20.** 56 μH

Chapter 11

2. 6 A/s, zero   **4.** 3 V, zero   **5.** 2.633 A   **6.** $V_R = 2.633$ V, $V_L = 0.367$ V
8. (a) 12.5 A/s, (b) 2.5 A/s   **10.** $t = 0.554$ s, $V_L = 50$ V   **13.** $t = 0.416$ s
15. (a) 293.6 mA, $i = 0$ at 100 s, (b) 0.862 W, (c) 2.936 V   **17.** (a) 14.35 ms,
(b) 69 ms   **19.** (a) 0.59J, (b) 1.5J   **20.** 7.29 V   **22.** 1.386 s   **24.** (a) -20 A/s,
(b) 240 V   **27.** 149.7 ms   **28.** 600,000 V (theoretically)   **29.** 80 V
30. 100 Ω

Chapter 12

1. (a) 3.375 N, (b) -1.35 N, (c) Repulsion in (a); Attraction in (b).   **3.** (a) 3.37 $\times$
10^5 N/C, (b) 0.538×10^5 N/C, Both radially outward from 150 μC
5. 44.4 μC   **7.** (a) 10^5 V/m, (b) 8.854×10^{-7} C/m²   **9.** $\bar{D} = -0.01272\bar{a}_r\mu$C/m²
11. 31.4×10^3 N/C at 153.5° with the X axis.   **14.** 62.5 V on 0.06, 37.5 V on 0.1
16. (a) $V_C = 0$, $V_R = 100$ V. At $t =$ infinity, $V_C = 100$ V, $V_R = 0$ **18.** -5 mA/s
20. (a) 44.04 V, 22.02 μA (b) 16.36 V, 8.18 μA   **22.** 300 ms   **25.** (a) 2 μC,
(b) -0.08 A/s, (c) 6.93 ms   **26.** 1.063 ns.   **28.** (a) 0.96×10^{-11} N, (b) 1.053 $\times$
10^{19} m/s toward center of path   **30.** 180 μC/m²   **32.** (a) $4\pi \times 10^{-5}$ C, (b) $\bar{D} =$
$(-0.59\bar{a}_x + 2.59\bar{a}_y)10^{-6}$ C/m²   **34.** 60.6 V, 24.2 V, 15.2 V   **35** Nos 1 and 3
up, Nos 2 and 7 down   **36.** (a) $Q_1 = 70$, $Q_2 = 140$ μC, (b) 70 V, (c) 0.00735 J
37. (a) 125 kV/m, (b) 5.53×10^{-6} C/m², (c) 6.94 nC, (d) 13.88 pF, (e) 1.735 μJ
39. (a) 0.255 J, (b) zero   **40.** (a) 6.075×10^{-3} W, (b) 0.823 mW   **42.** 435,500Ω
44. 560 V   **46.** 40 pF   **47.** 32.4 nF, 800 V   **48.** 272.7 V   **50.** 87 Ω
52. 4870 V, 18.25 s, 1187.5 J

Chapter 13

2. 7.36 A   **4.** 1800   **5.** 11.18 rad/s, 1.78 Hz   **7.** 0.45 cm   **10.** 5.77 Ω,
20.83 A   **12.** $V_m I_m \sin^2 \omega t$ **13.** (a) $i = 5 \sin 2513$ t, (b) 35.35 V, 3.535 A,
(c) 125 W, (d) 250 W   **15.** (a) $e = 10 \cos (377t + \pi/2)$ **17.** (b) $v = 0$, $i = 2$ A,
(c) 70.7 V   **18.** (a) 0.081J, (b) zero   **20.** 600 W   **22.** (a) Phasor directly upward, 5 V max; (b) same as (a); (c) 5 cos ωt; (d) They are identical.

Chapter 14

1. (a) $70.7 + j70.7$, (c) $-25 + j43.3$, (d) $-103.92 - j60$
2. (b) $8\underline{/120°}$, (e) $90.1\underline{/-123.7°}$ **4.** $30.7 + j140$, $143\underline{/77.6°}$ **7.** $-225 + j475$
9. (a) $600\underline{/105°}$, (b) $320\underline{/28°}$ **10.** (b) $5\underline{/-15°}$ **11.** $100/30°$, $86.6 + j50$
12. $100\underline{/-60°}$, $50 - j86.6$ **14.** $1.715\underline{/-29.35°}$ **16.** ϵ^{-j20} **18.** 32.3 m
20. (a) $20\underline{/-45°}$ V, (c) $120\underline{/-80°}$V   **21.** (a) $16\underline{/55°}$, (b) $4\underline{/30°}$, (c) $169\underline{/134.8°}$

Chapter 15

1. 60.3, 65.3 Ω **3.** 38 V, 91.65 V   **5.** 357 Ω **7.** $V_R = 3.71$ mV,
$V_L = 9.28$ mV, V_L leads I by 90°, V_T by 21.8°   **9.** 10.61 nF   **10.** $4 - j7.95 =$
$8.9\underline{/-63.3°}$, $I = 2.81$ leading V by 63.3°   **12.** 2 μF   **14.** z $= 15.81\underline{/18.4°} =$
$15 + j5$ Ω, $I = 6.32\underline{/-18.4°} = 6 - j2$ A   **17.** $V_R = 16 + j8$, $V_L = -4 + j8$, $V_C = 8 - j/6$
22. $0.12 + j0.16$ **24.** (c) $100\underline{/240°} = -50 - j86.6$
26. $0.159\underline{/-88.1°}$ A   **28** (a) $z = 4.47\underline{/26.6°} = 4 + j2$, (b) $I = 2 - j = 2.23\underline{/-26.6°}$,
(c) 8.92 V, 22.3 V, 17.84 V, (d) 10^{-5} H, 0.125 μF   **29.** 132 μF, $j200$ V, $-j200$ V,
$150 + j0$ V

Chapter 16

1. $I_R = 2\underline{/0°}, I_C = 5\underline{/90°}, I_T = 5.38\underline{/68.2°}, z_T = 9.3\underline{/-68.2°}$
3. $I = 3.2\underline{/51.3°}, 15.62\underline{/-51.3°}\,\Omega$ **6.** $G = 0.02$ mho, $B = 0.05$ mho, $Y = 0.02 + j0.05$ mho **7.** $6.89 - j17.24\,\Omega$ **10.** $Y = 0.02 + j0.025$ mho, $I = 1 + j1.25 = 1.60\underline{/51.3°}$ A **12.** $Y_{BCD} = 0.0784 + j0.1804, Z_T = 11.5\underline{/64.1°}$
13. $I_{BD} = 0.886\underline{/-40.6°}, I_{BC} = 0.356/-144.7°, I_T = 0.87\underline{/-64.1°}$ **15.** $5140\underline{/49.5}\,\Omega$
17. $0.251 + j0$ A, $j0.251$ A **19.** $17.24 - j6.89\,\Omega$ **21.** $19.9\underline{/5.7°}\,\Omega$
23. $Y_T = 0.0333 - j0.0666, I_T = 1.49\underline{/-63.4°}$ A **25.** $I_L = 7.14\underline{/-46.9°}$ A, $I_C = 5.49\underline{/28.8°}$ A, $V_{RL} = 42.84\underline{/-46.9°}, V_{XL} = 57.12\underline{/43.1°}, V_{DE} = 23.79$ V
26. $R_3 = 5, X_3 = -j1$ **28.** $I = 1.41\underline{/-45°}$ **30.** $19.1\underline{/16.7°}$ mA, $10.54\underline{/76.8°}$ V
31. $I_1 = 4.5\underline{/45°}$ A, $I_2 = 9.6\underline{/36.9°}$ A, $I_3 = 11.27\underline{/-63.4°}$ A, (b) $PF_1 = 0.707, PF_2 = 0.8, PF_3 = 0.448$, (c) $P_1 = 753$ W, $P_2 = 1843$ W, $P_3 = 1197$ W, (d) $P_T = 3793$ W,
(e) $I_T = 17.6\underline{/-25.3°}$ A, (f) $\theta_T = -25.3°$, (g) $VI \cos \theta = 3820$ (0.71% difference)

Chapter 17

2. 4.87 W, $PF = 0.625$, **4.** $\frac{1}{6}$ Wh **6.** 340 W **8.** $PF = 0.75, VAR = 88,100$
at $0.9\,PF, VAR = 48,300$, Reduction $= 39,800\,VAR$ **10.** $610\,\mu$F **11.** $169\underline{/0°}$ V,
$PF = 0.949$ **13.** $I_T = 10.8 - j1.21 = 10.9\underline{/-6.4°}, PF = 0.994$ **15.** (a) 24 V,
(b) 105.2 W, (c) $0.976\,PF$ **17.** $P = 1180$ W $0.97\,PF$ **18.** $KVAR = 10.5, C = 6440\,\mu$F **20.** $C = 663\,\mu$F **22.** (a) $KVA = 57$, (b) 274 A **24.** 25 W, 25 Ω
25. $I = 93.4\underline{/26.9°}, PF = 0.892$

Chapter 18

1. (b) $P_V = 0.96$ W, (c) $I_L = 12.492$ A, $P_L = 1199.04$ W, (d) $0.8\,PF$ **2.** 1201.44 W
3. (a) $z = 5 + j94.24\,\Omega, L = 0.015$ H **4.** $R_4 = 800, X_4 = 11,000, L_4 = 0.175$ H
5. $z = 1,000 - j31,830$ **6.** $4.1\,\mu$F **7.** $z = 2.4$ mH, $z = 7.54 + j15.08\,\Omega$
8. $L_4 = 100\,\mu$H, $R_4 = 24\,\Omega$

Chapter 19

1. $z' = 3.846 + j0.769, E_{OC} = 76.9 + j15.4, Z_L = 20 + j0$ **2.** $I_L = 3.27\underline{/9.55°}$
4. $I_L = 3.78\underline{/-6.4°}$ **6.** $I_A = 20$ A **7.** $I_L = +3.78\underline{/-6.4°}$ A **9.** $z_L = 2 - j5\Omega$,
$P = 12.5$ W **11.** $Z_L = 89.4\underline{/+26.6°}\,\Omega = 80 - j40$ **13.** $5.31\underline{/-11.3°}$ A
15. $4.88/81.7°$

Chapter 20

1. $f_r = 454.57$ kHz, $Q_r = 14.28, V_C = 14.28$ V **3.** $C = 0.9$ nF, $Q = 23.5, V_c = 23.5$ V
6. $BW = 13,210$ Hz **9.** $R_s = 0.016\,\Omega, C_s = 0.5\,\mu$F **11.** (a) 1 mHz, (b) $Z_r = 2,530\,\Omega$, (c) $Q = 15.9$, (d) $I_c = 6.28$ mA, (e) 6.28 mA, (f) $395 \times 10^{-12}J$ **13.** $120\,\mu$H
14. $R = 12.56\,\Omega, z = 3185\,\Omega$ **16.** 1.73 mHz, $Q = 14.5$ **17.** (a) 33.5 kHz,
(b) 335 kHz **19.** 25.4 mHz (approx), 26.7 mHz (exact), 4.87% diff. **20.** 127.2
mHz **21.** 3.76 V, 2.46 V, 98.9% **22.** $f_0 = 1452$ Hz, $R_k^2 = 3 \times 10^5$
23. $f_0 = 0.89$ mHz, $R_k = 559\,\Omega$ **24.** (a) $BW = 500$ Hz, (b) 5050 Hz **25.** $R = 1$ MΩ

Chapter 21

1. $I_1 = 1$ A, $I_L = \frac{2}{3}$ A, $k = 0.739, Z_{tr} = 150\,\Omega, I_2/I_1 = \frac{2}{3}$ **3.** $I_1 = 7.19$ mA, $Z_{tr} = 27,000\underline{/42.6°}\,\Omega, I_2/I_1 = 0.515\underline{/31.1°}, I_2 = 3.70\underline{/-42.6°}$ mA, $k = 0.62\underline{/15.5°}$

5. $Z_1 = 1.846, Z_2 = 0.932, Z_3 = Z_m = 1.384\ \Omega, I_1 = 1.9\ \text{A}, I_2/I_1 = 0.166, I_2 = 0.315\ \text{A},$
$Z_{tr} = 3.17\ \Omega, k = 0.192$ **9.** $Z_{12} = 5\ \Omega = Z_{21}, Z_{11} = 8.67\ \Omega, Z_{22} = 12.5\ \Omega, I_G = 0.75\ \text{A},$
$I_L = 0.3\ \text{A}$ **12.** $I_1 = 1.645\ \underline{/-26.9°}, I_2 = 0.797\ \underline{/-37.2°}, I_3 = 2.43\ \underline{/-30.5°}$
14. $I_G = 2.94\ \underline{/-43.8°}, I_L = 0.667\ \underline{/-115.4°}, P_G = 34.53\ \text{W}, P_L = 2.22\ \text{W}$
15. (a) $k = 0.822$, (b) $Z = 401\ \underline{/86.3°}$, (c) $I = 0.25\ \underline{/-86.3°}$ **17.** Referred $Z =$
$6.56\ \underline{/-66.8°}, Z_1' = 4.94 - j2.03 = 5.34\ \underline{/-22.4°}\ \Omega$ **19.** (a) 3.46 mW, (b) $N_1/N_2 = 4$
(c) 15.6 mW

Chapter 22

2. (a) 3.75×10^{-3} Wb, (b) 0.564×10^{-3} Wb **4.** $I_1 = 10\ \text{A}, 0.89\ \text{PF}$ **6.** 0.97
8. 0.964 **10.** 0.985 **11.** (a) $R_0 = 11{,}250\ \Omega$, (b) $52{,}700\ \Omega = X_0, Z_0 = 54{,}000\ \Omega$
(c) $R_e = 2.36\ \Omega$, (d) $X_e = 45.2, Z_e = 51.8\ \Omega$. **14.** Connect 1050 μF across the power
line. **16.** PF raised to 77.9%

Chapter 23

1. $V_{BC} = 100\ \underline{/-120°} = -50 - j866\ \text{V}, V_{CA} = 100\ \underline{/120°} = -50 + j86.6\ \text{V}$
2. $I_A = 5.77\ \underline{/-30°} = 5 - j2.9\ \text{A}, I_B = 5.77\ \underline{/-150°} = -5 - j2.9\ \text{A}, I_C = 5.77\ \underline{/90°} =$
$j5.77\ \text{A}$ **4.** $I = 14.42\ \text{A}, V = 200\ \text{V}$ **5.** $X = 9.58\ \Omega, \theta = -43.8°$ **7.** 7200 W
8. 7580 VA, 0.949 PF **12.** $I = 14.39\ \text{A}$ **14.** $I = 5.37\ \underline{/-26.6°}\ \text{A}, P = 1728\ \text{W}$
16. 0.8 PF, 25 hp **17.** 8.77 kW input, $I_L = 15.2\ \text{A}$ **18.** (a) 86.7 kVA (2 transformers), (b) 63.3 kVA added **19.** (a) 0.8 PF, (b) 14.425 A (c) 14.425 A (d) 9.6 Ω
(e) 7.56 Ω reactance **20.** (a) 62.5 A, (b) $112.92\ \underline{/-30°}\ \text{A}$ if $V_{AB} = 115\ \underline{/0°}\ \text{V}$, (c) $I_{LM} =$
$42.5\ \underline{/96.9°}\ \text{A}$, (d) $I_{LT} = 93.8\ \underline{/-8.7°}\ \text{A}$, (e) $P_T = 31{,}280\ \text{W}$ calculated. Metered power:
$P_1 = 11{,}700\ \text{W}, P_2 = 18{,}500\ \text{W}, P_T = 30{,}200\ \text{W}$. There is a 3.5% difference (slide rule
calculations).

Chapter 24

2. 7.21 mA **4.** 7.21 V, 7.21 mA, $VA = 51.98 \times 10^{-3}, P = I^2R = 51.98 \times 10^{-3}$ W
5. $I = 0.109\ \text{A}, P = 1.19\ \text{W}$ **6.** $i = 4\sin 377\ t + \sin 1131\ t$ mA **8.** $I_{m1} = 3.53$
$\underline{/-28°}$ mA, $I_{m3} = 0.119\ \underline{/-53.6°}$ mA **10.** (a) $z_1 = 671\ \underline{/26.6°}, Z_2 = 848\ \underline{/45°}$,
$z_3 = 1342\ \underline{/63.4°}$, (b) $i = 0.149\sin(377t - 26.6°) + 0.0352\cos(754t - 45°) - 746 \times 10^{-6}$
$\sin(1508t - 63.4°)$ **11.** 7 W **13.** $P = 13.98\ \text{mW}$ **15.** $i_3 = 5.83\sin(\omega t + 31°)$
$+ 5.78\sin(3\omega t + 21.5°)$

Chapter 25

1. $I_T = 35.2\ \text{A}$ **2.** $E_g = 226.4\ \text{V}$ **3.** 8448 W, 0.883 PF **4.** 750.4 W copper
loss, 237.6 stray loss **5.** $F = 6.75\ \text{lb}, 2.135\ \text{hp}$ output, 82% eff.
6. (a) $E_g = 228.4\ \text{V}$, (b) 2.56 W, (c) 365.44 W, (d) 230 A **7.** (a) 1021.5 W, (b) 197.6 W,
(c) 1.37 hp **8.** 1500 rpm **9.** 950 rpm **10.** 350 rpm **11.** (a) 10 poles.
(b) 4 poles, (c) 10 poles, 691.2 rpm, 4 poles, 720 rpm **12.** (a) 18 poles, 400 rpm sync.
speed (b) 2.5% slip (c) Rotor slips 1 rev. in 6 s, (d) 9 cycles required for 1 rev. of field
13. 44.5 ft lb torque, 4.16% slip, 14.6 hp, (d) 17.02 hp, (e) 85.8% eff, (f) 0.88 PF
14. Sample: At 1.3 A input: 0.0933 hp output, 53.5% efficiency, 0.833 power factor.

INDEX

MOLDED MICA TYPE CAPACITORS

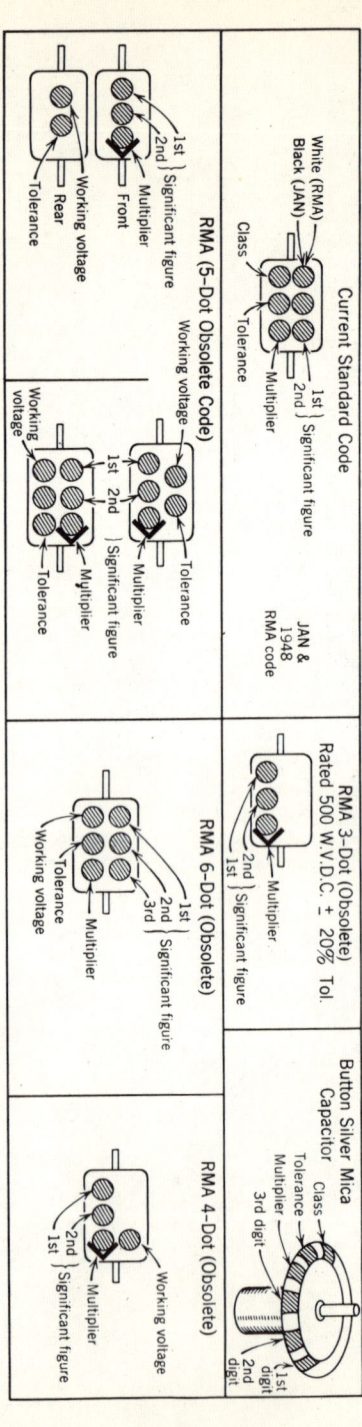

Current Standard Code

White (RMA)
Black (JAN)

1st } Significant figure
2nd
Class
Multiplier
Tolerance

JAN & 1948 RMA code

1st } Significant figure
2nd

1st } Significant figure
2nd
Rear
Front
Working voltage
Tolerance

RMA (5-Dot Obsolete Code)

Working voltage

Tolerance
1st } Significant figure
2nd
Multiplier
Tolerance

1st } Significant figure
2nd
Working voltage
Multiplier
Tolerance

RMA 3-Dot (Obsolete) Rated 500 W.V.D.C. + 20% Tol.

1st } Significant figure
2nd
Multiplier

RMA 6-Dot (Obsolete)

1st
2nd } Significant figure
3rd
Tolerance
Multiplier
Working voltage

Button Silver Mica Capacitor

Class
Tolerance
Multiplier
3rd digit
1st digit
2nd digit

RMA 4-Dot (Obsolete)

1st } Significant figure
2nd
Multiplier
Working voltage

MOLDED PAPER TYPE CAPACITORS

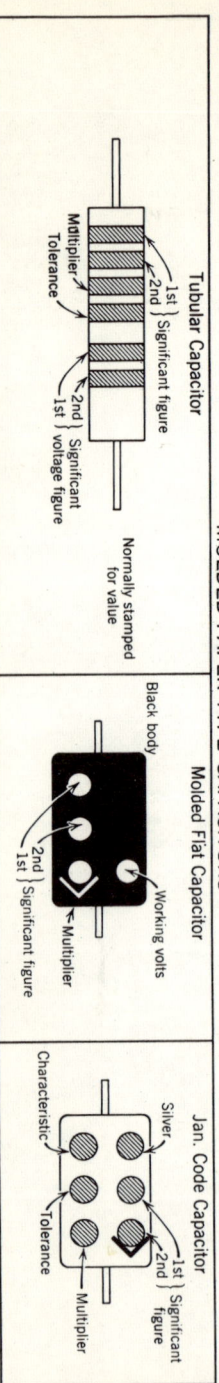

Tubular Capacitor

Multiplier
Tolerance
1st } Significant figure
2nd

2nd } Significant
1st } voltage figure

Normally stamped for value

Molded Flat Capacitor

Black body

2nd } Significant figure
1st
Multiplier
Working volts

Jan. Code Capacitor

Silver
1st } Significant figure
2nd
Characteristic
Tolerance
Multiplier

The tolerance rating of capacitors is determined by the color code. For example: red = 2 per cent, green = 5 per cent, etc. The voltage rating of capacitors is obtained by multiplying the color value by 100.

For example: orange = 3 × 100 or 300 volts. Blue = 6 × 100 or 600 volts.